ESTATES GAZETTE LAW REPORTS

Estates Gazette Law Reports

2012

Volume 3

Edited by
HH Judge Hazel Marshall QC

Routledge
Taylor & Francis Group

EG Books

First published 2013
by Routledge
2 Park Square, Milton Park, Abingdon, Oxon OX14 4RN

Simultaneously published in the USA and Canada by Routledge
711 Third Avenue, New York, NY 10017

EG Books is an imprint of the Taylor & Francis Group, an informa business

British Library Cataloguing in Publication Data
A catalogue record for this book is available from the British Library

Library of Congress Cataloging in Publication Data
A catalog record for this book has been requested

ISBN 978-0-415-64266-8
ISSN 0951-9289

Typeset in Times New Roman by Estates Gazette

CONTENTS

Cases reported in "Estates Gazette" [2012] issues 35-50

SUPPLEMENT

Cases not reported in "Estates Gazette"

TABLE OF CASES

Titles of cases shown in bold type
Names of parties reversed shown in ordinary type

INDEX OF SUBJECT MATTER

right to remain at end of lease term by virtue of Sched 10 to Local Government and Housing Act 1989 – Whether hope value properly awarded in respect of non-participating flats with reversions of less than 80 years – Held, decision of UT not vitiated by failure to consider 1989 Act where parties accepting UT's approach at the time – UT entitled to give guidance on deferment rate for reversions of less than five years – Guidance open to qualification in case where party raising issue of possible application of 1989 Act at end of lease term – Hope value of non-participating flats properly taken into account so far as attributable to possibility of tenants seeking new leases by negotiation rather than as of right under 1993 Act

Carey-Morgan* v *Trustees of Sloane Stanley Estate (CA), 38

LEASEHOLD REFORM ACT 1967 – See **Leasehold enfranchisement**

LEASEHOLD REFORM, HOUSING AND URBAN DEVELOPMENT ACT 1993 – See **Leasehold enfranchisement**

LEASEHOLD VALUATION TRIBUNAL

Decision

Reasons

Leaseholder challenging reasonableness of service charge levied by landlord for provision of television and radio signals – LVT determining amount to be unreasonable – LVT giving further reasons when refusing permission to appeal – Whether LVT's reasons adequate – Whether entitled to expand on reasons in decision on permission to appeal – Held, LVT entitled to amplify reasons in decision on permission to appeal in order to remedy defect in original reasons – Amplified reasons to be confined to matters on which original decision based without second or additional thoughts or alteration to ultimate decision – LVT entitled to rely on own expertise provided parties given opportunity to deal with issues raised – Amplified reasons inadequate where relying on LVT's general knowledge and expertise on matters not fully aired before parties

Havering London Borough Council* v *Macdonald (UT:LC), 49

Jurisdiction

Service charge

Landlord and Tenant Act 1985 – Application to leasehold valuation tribunal (LVT) under s27A of Landlord and Tenant Act 1985 – Tenants disputing amount of service charge demanded by landlords for window replacement and alleging poor workmanship – LVT determining charge unreasonable since not reasonable to replace windows – Whether LVT entitled to reach decision on point not advanced by parties – Whether exceeding jurisdiction – Held, jurisdiction of LVT under s27A confined to determining matters in dispute as pleaded or otherwise specifically identified in application form and parties' statements of case – No jurisdiction to conduct own inquisitorial process and determine matters that parties not asking it to resolve – No jurisdiction to determine reasonableness of replacing windows where that matter not in dispute between parties

Birmingham City Council* v *Keddie (UT:LC), 53

LICENCE

Termination

Notice

Agricultural land – Administrators of deceased's estate granting licence to one of them to occupy certain land – Licensee failing to pay licence fee – Claim for possession of land – Whether licence effectively terminated by notice in accordance with its terms – Whether notice ineffective where served on behalf of only one of two surviving licensors – Whether requirement for all licensors to give notice excluding licensor who was also licensee – Held, licensee expressly included within phrase "the Licensor" – Not possible to construe that phrase differently in clause dealing with giving of notice – Potential difficulties in obtaining licensee's concurrence in giving of notice not rendering notice machinery unworkable – No scope for implying term – Notice ineffective

Fitzhugh* v *Fitzhugh (CA), 57

LOCAL GOVERNMENT AND HOUSING ACT 1989 – See **Leasehold enfranchisement** (Leasehold Reform, Housing and Urban Development Act 1993)

NOTICES – See also **Landlord and tenant** (Assignment); **Landlord and tenant** (Service charge); **Licence**

Service

Time limit

Share sale agreement specifying that advance written notice be given of any claim for breach of warranty with any proceedings to be issued and served within 12 months thereafter – Agreement specifying certain methods of service including personal delivery to recipient's address – Whether notice served at time when delivered to recipient's home or later on deemed receipt by recorded delivery – Whether subsequent claim served in time – Held, service by personal delivery effected only where documents handed over to recipient personally – No service by personal delivery where documents simply left at recipient's address – Methods of service specified in agreement permissive not exclusive – Service of notice therefore occurring when actually seen by recipient not later date of service by recorded delivery – Claim documents served outside time limit

Ener-G Holdings plc* v *Hormell (CA), 61

NUISANCE

Rule in *Rylands* v *Fletcher*

Fire

Fire starting on business premises fuelled by tyres stored there for purposes of business of motor vehicle tyre supplier and fitter – Fire destroying adjoining property – Whether strict liability arising under *Rylands* v *Fletcher* for damage to adjoining property – Held, liability under *Rylands* v *Fletcher* arising only where dangerous thing brought onto defendant's property and subsequently escaping – No special rule for cases involving damage caused by spread of fire – No liability where escape of fire started or increased by dangerous thing rather than of thing itself – Escape of fire fuelled by tyres not giving rise to strict liability since tyres themselves not dangerous and not escaping

Stannard (t/a Wyvern Tyres)* v *Gore (CA), 129

OCCUPIERS' LIABILITY – See **Defective premises** (Duty of Landlord)

POSSESSION – See **Licence**

PRACTICE AND PROCEDURE

Disclosure

Expert witnesses

Claim in nuisance and/or negligence for contamination of land by escape of oil – Damages claimed for lost sale of land or diminution in value – Claimants alleging that land now unsaleable and of no value – Each party obtaining expert report on valuation – Whether claimants obliged to disclose second report obtained from same expert – Held, expert evidence not required for claim for loss of actual sale once sale price agreed to be reasonable – Remaining issue of whether unsaleable property having no value an issue of law on which no expert evidence needed – Expert evidence needed on alternative claim for diminution in value – Inappropriate to order disclosure of second expert report where possibly based on misunderstanding of issues and further report on correct basis needed in any event – No general principle that everything disclosable regardless of privilege

Odedra* v *Ball (TCC), 69

RATING

Non-domestic rates

Hereditament

Office premises – Ratepayer occupying second and sixth floors in modern office building – Whether each floor comprising separate hereditament – Whether two floors to be listed together as single hereditament – Whether appropriate to make end allowance as part of valuation process to reflect inconvenience of separation within building – Held, *Gilbert (VO)* v *S Hickinbottom & Sons Ltd* [1956] 2 QB 40 not establishing incontrovertible formula for identification of hereditament – Concepts of curtilage and essential functional link not relevant when identifying hereditaments within modern office building – Separation of floors not relevant where communication between all floors gained through common parts of building – Separate floors within building occupied by same occupier therefore

A

Dutta and another v Hayes

CHANCERY DIVISION: BIRMINGHAM

31 May 2012

HH JUDGE SIMON BARKER QC, SITTING AS A HIGH COURT JUDGE

[2012] EWHC 1727 (Ch)

Estates Gazette 22 September 2012

[2012] 38 EG 93

B

Easements – Excessive use – Right of way – Right of way over track limited by reference to agricultural use – Way used for equestrian purposes – Owners of track objecting to use of way in connection with stud farm business – Whether this constituting unlawful use of way – Whether meaning of agricultural use extended by convention of parties to grant

In 2003, the defendant purchased several acres of farmland for the purpose of breeding horses. Vehicular access to the defendant's land from the highway could be gained only by a track across the claimants' adjoining land. A right of way over the track, with or without vehicles, existed "for the purpose only of the pursuance of the agricultural use C of" the defendant's land. The relevant right was contained in a 1991 transfer effected by a common predecessor in title of the parties and was subject to a requirement to pay a proportion of the cost of maintenance and repair of the track.

The claimants did not object to the defendant using the right of way in connection with the grazing or exercise of horses; such use had been made of the track in the past. They also permitted the defendant to graze horses on their land pursuant to a grazing licence, and leased some stables to him, and they were content for him to use the track for those purposes. However, when the defendant began to run a stud farm business on his land in 2010, the claimants contended that use of the D track for such purposes was unlawful. They objected to the passage and repassage of horse lorries, diggers, dumper trucks, cars and 4x4 vehicles along the track and to any use of the track by vets, farriers, feed suppliers, commercial visitors, oil suppliers and manure contractors.

The defendant accepted that the concept of "agricultural use" did not ordinarily extend to the running of a stud farm. However, he contended that the right of way should be construed in the context of the past use of the land for equestrian purposes; he contended that, by granting the stable lease and grazing licence, the claimants had acknowledged that equestrian use was consistent with the right of way.

Held: **The claim was allowed. (1). When deciding the meaning E of an easement or grant, it is permissible to take account of the use to which the dominant tenement was put at the time of the grant, or the use that might reasonably have been expected to be made of it. Although the established meaning of "agricultural use" does not include the stabling and keeping of horses, as opposed merely to grazing them, the concept of agricultural use can be given an extended meaning by convention of the parties. It was relevant that there had been recreational equestrian use of the defendant's land at the time of the grant and the track had been used for the purpose of accessing the defendant's land. The agricultural activity of grazing would not cease to be agricultural if there was a stable on the land and the horses were kept there overnight or in inclement weather, or if the land were insufficiently fertile or productive to F yield sufficient fodder and hay or other food had to be brought in, provided that the land was suitable for grazing. The use of the track to bring horses to and from the defendant's land for grazing, for the carriage of fodder and for access by vets, farriers, other suppliers or contractors and staff were all permitted uses of the track. That included transporting the horses by vehicle to and from the land. However, the defendant was not entitled to use the track as a means of access to and from his land for the purposes of the stud farm. The operation of a stud farm fell outside the ordinary and statutory meanings of agriculture, the meanings established by case law and the scope of any convention between the parties or G their predecessors in title in 1991. There was no evidence that such use had been considered as a likely future use of the land at that time. (2) There were exceptional circumstances that justified an award of damages in lieu of an injunction. It would be impractical to grant an injunction since it was not possible simply to turn off the operation of a stud farm. Damages of £100 per calendar month were fair and just compensation for the intrusion caused to the claimants. No higher figure was likely to result from a hypothetical negotiation for the grant of the necessary rights. The defendant should also contribute 60% of the repair costs in respect of the track, pursuant to his obligation under the right of way.** H

The following cases are referred to in this report.

Gainsborough-Field v *Hyde* [2005] EWHC 2229 (QB)
Moncrieff v *Jamieson* [2007] UKHL 42; [2007] 1 WLR 2620
Partridge v *Lawrence* [2003] EWCA Civ 1121
Rainy Sky SA v *Kookmin Bank* [2011] UKSC 50
Stadium Capital Holdings (No 2) Ltd v *St Marylebone Property Co* [2011] EWHC 2856 (Ch); [2012] 1 EGLR 103
Sykes v *Secretary of State for the Environment* [1980] 42 P&CR 19; [1981] 1 EGLR 137
Toddrick v *Western Omnibus Co* [1934] Ch 190
Vitol BB v *Compagnie European des Petroles* [1988] 1 Lloyd's Rep 574
Whitsbury Farm and Stud Ltd v *Hemens* [1988] 1 AC 601; [1988] 1 EGLR 159; J [1988] 13 EG 43-48

This was the hearing of a claim by the claimants, Mohit Dutta and Amanda Queiroz, against the defendant, Thomas Hayes, alleging unlawful use of a right of way.

Jamal Demachkie (instructed by Actons Solicitors, of Nottingham) appeared for the claimants; Frances Pigott (instructed by Whatley West & Fox Solicitors, of Worcester) represented the defendant.

Giving judgment, **HH Judge Simon Barker QC** said:

1. This action concerns a right of way along a track ("the track") K running across the claimants' land from a public highway known as Stocks Lane to the defendant's land.

2. Although the defendant's land has a small section of frontage to the same public highway the only route for vehicular access to the defendant's land is said to be along the track.

3. The defendant's land comprises approximately 37 acres of field or farmland and approximately one acre of woodland. It is irregular in shape. Apart from a frontage of approximately 100m or thereabouts to Stocks Lane, the defendant's land is surrounded to the north by Leigh Brook, to the east by the continuation of Leigh Brook and by woodland,

A to the south by the claimants' land and to the west by the claimants' land.

4. For some years prior to 1991 the claimants' land and the defendant's land were in the common ownership of a Mr Timothy Siviter, who has given evidence in this trial. The right of way was created on 1 November 1991 when he sold the land over which the track runs in two parcels, one to a Mr and Mrs Stokes and the other to a Panamanian company as nominee for the Swiss trustee of a settlement created by and/or for Mr and Mrs Stokes.

5. In the recent past, and in order to access his land other than by the track, the defendant has attempted to build a hardcore-based track that B would support the weight of vehicles along the line of a public footpath running across his land from the point where it meets Stocks Lane. However, the Malvern Hills District Council had not given permission for the construction of this route and they required the defendant to remove the hardcore he was laying down and to reinstate the land, which he has done.

6. The right of way, as reserved by Mr Siviter for the benefit of what is now the defendant's land, is, so far as material, in the following terms: "a right of way at all times with or without vehicles over and along the track" and then there is an identification plan reference "for the purpose only of pursuance of the agricultural use of" and then there C is a reference to what is now the defendant's land.

7. This right of way was subject to a condition to pay a proportion of the cost of maintenance and repair of the track, to which I shall refer later.

8. At the time, the land retained by Mr Siviter included in addition to what is now the defendant's land other land to the south of the track and to the south of the defendant's land, some of which is now owned by the claimants.

9. The defendant purchased his land in 2003 with a view to breeding horses on the land. Although horses had been grazed and exercised on D the land, there is no evidence that prior to that time it had been used for stabling or breeding horses.

10. The best evidence of use of the land prior to and at the time of the November 1991 conveyances creating the right of way was given by Mr William Lambert, a witness for the defendant, who was born at Lower Tundridge Farm, which is now the claimants' home and on the claimants' land, in July 1943. By that time, Lower Tundridge Farm had been occupied by Mr Lambert's forebears, initially as tenants and then as owners, for upwards of a century. After marrying in 1973 Mr Lambert moved to a nearby cottage, where he still lives. Thus he is a valuable source of some 70 years' worth of direct knowledge of the E land in the locality.

11. Mr Lambert notes that the right of way is also a public footpath; although Mr Demachkie submitted that it is not, he did not challenge Mr Lambert's evidence that it is.

12. Mr Lambert makes clear that, until 1978, the fields now owned by the defendant were hop yards. In 1978, the hop yards were removed and the fields were mainly used for growing corn, with some fields being left for pasture for grazing horses. Following the purchase of the claimants' and defendant's land by Mr Siviter in 1987, some of the fields were let to a local farmer, who grazed sheep and cattle until 1991.

F 13. Throughout all this period there was a variety of equestrian use of and activity on what is now the claimants' and the defendant's land. Mr Lambert also spoke of farm horses being stabled on the claimants' land and working the fields, including what is now the defendant's land, during his family's ownership of Lower Tundridge Farm.

14. Adults and children are said to have exercised and ridden horses and ponies across the claimants' and the defendant's land and along the track, and the North Ledbury Hunt did (and still does, according to Mr Lambert) ride across the claimants' and the defendant's land.

15. After 1991, and in the years leading up to the defendant's purchase of the fields accessed by the track, the fields (or some of them) were used for growing oilseed rape.

16. Mr Lambert's evidence is clear. Having observed him during G cross-examination, I have no hesitation in accepting his evidence as reliable.

17. The claimants do not object to the defendant using the right of way in connection with or for the furtherance of the grazing or the exercising of horses.

18. Indeed, over the period 2008-2010 the claimants permitted the defendant to graze horses on their land to the south of the track and to the south of the defendant's land pursuant to a grazing licence; and, over the same two-year period, the claimants also granted the defendant a lease of stables originally built, according to Mr Lambert, as hop workers' barracks, situated adjacent to and, in my judgment H and notwithstanding evidence to the contrary from the first claimant, necessarily and inevitably accessed via the track in which stables the defendant kept thoroughbred horses.

19. In March 2009, the defendant was granted planning permission to use the barn which he had built on his land as stables, and in 2010 a retrospective application for an all-weather manège was also granted.

20. At about this time the defendant also brought a mobile home on to his land for use by an employee during the foaling season. Planning permission was refused for the mobile home and it has been removed.

21. These events in and around 2010 mark the establishment of "Tundridge Stud" as a business. The claimants object to the use of the J track in connection with this business. Examples of use objected to by the claimants include the passing and re-passing of horse lorries, JCBs, dumper trucks, cars and 4x4 vehicles. The claimants contend that every journey along the path by vets, farriers, feed suppliers, commercial visitors, oil suppliers and manure contractors also constitute unlawful use of the track.

22. The defendant's land is for the most part registered with DEFRA as grade 3 or grade 4 agricultural land. Grade 4 is poor quality land mainly suited to grass or pasture. Grade 3 is land in a somewhat better condition suitable for crops but subject to moderate limitations.

23. Against that summary of the factual background I return to the K relevant words of the right of way by which the defendant is entitled to use the track:

- at all times
- with or without vehicles
- for the purpose only of pursuance of the agricultural use of the land.

24. Mr Demachkie, the claimants' counsel, submits correctly that the starting point for the court is to consider whether or not the parties have used unambiguous language and that if they have the court must simply apply the language used to the facts of the case, (see *Rainy Sky SA* v *Kookmin Bank* [2011] UKSC 50 and in particular the speech of Lord Clarke at [23]). L

25. The issue here is whether the phrase "pursuance of the agricultural use of the land" is unambiguous.

26. Mr Demachkie submits that there is no ambiguity about the phrase because (1) the key to the permitted use of the track is the meaning of the word "agricultural" and (2) whether reference is made to definitions in statute, standard dictionaries or case law, the word 'agricultural' does not encompass the keeping (ie the stabling, rearing or breeding) of horses unless that use is ancillary to the main agricultural use of the land. As a fallback submission, Mr Demachkie submits that even if that is wrong and reference is properly to be made to the actual use of the defendant's land at the time of and prior to 1 November 1991 M conveyances, there is no evidence that the defendant's land had been used for the breeding or stabling or training of horses.

27. Mrs Pigott, the defendant's counsel, acknowledges that agricultural use does not ordinarily extend to running a stud farm. She submits however that: (1) the defendant's land is used as agricultural land, being grazing land; (2) since 9 March 2009 the defendant has had permission to keep horses in the building originally built as a barn, ie to stable horses on the land; (3) in and before 1991 the track was habitually used by Mr Siviter and others as well and before him for equestrian purposes; (4) that the claimants acknowledged that equestrian use was consistent with the right of way granted by leasing the stables on or adjacent to the track to the defendant and by granting

A a licence in respect of the fields to the south of the track as a gallop and as a paddock; and, (5) that in the context of Part III of the Highways Act 1980, which concerns the creation or adoption of new roads and paths, agriculture is defined as including the breeding or keeping of horses. That is a reference to section 29 of that Act.

28. In addition, Mrs Pigott submits that the words surrounding the phrase "agricultural use" in the right of way are not to be ignored or regarded as unnecessary verbiage. In particular she observes that (1) "pursuance" is a noun which is predicated upon there being an existing state of affairs; (2) the use of the definite article before "agricultural use" transforms an abstract concept into a particular situation; and, (3) B the reference to "the vendor's land" provides the context by reference to which the agricultural use is to be objectively understood.

29. Mrs Pigott refers to a passage in the judgment of Farwell J in *Toddrick* v *Western Omnibus Co* [1934] Ch 190 at 206 and also to the Scottish appeal case in the House of Lords of *Moncrieff* v *Jamieson* [2007] UKHL 42; [2007] 1 WLR 2620 and the speech of Lord Hope at [7], and takes them together as authority for the proposition that when deciding upon the meaning of an easement or a grant account may be taken of the use to which the dominant tenement was being put or might then reasonably have been expected to be put in the future.

30. Mr Demachkie did not argue to the contrary and I have no reason C to do other than to accept that proposition.

31. However, this must not be taken too far; and, the court must always bear in mind that, when faced with a dispute as to the meaning of a document, the task is to conduct an objective inquiry in order to ascertain what a reasonable person having all the background knowledge which would be available to the parties would understand the document to mean. The court may not attempt to improve the document but it must identify and state its objective meaning.

32. Mr Demachkie submits that in this case the words surrounding the phrase "agricultural use", including the preceding definite article "the" are merely stylistic and are of no significance. To support this D proposition he relies upon the use of the definite article later in the same right of way grant in the charging proviso.

33. In my judgment, the words used in the grant do, as Mrs Pigott submits, admit of more than one possible meaning. On the one hand, they may be, as Mr Demachkie submits, entirely neutral in their impact upon the phrase "agricultural use". On the other hand, there may have been something about the use of the land retained by Mr Siviter which was in context objectively to be understood to be agricultural and capable of continuing or being likely to be followed in the future. In reaching this conclusion, I have borne in mind the reliance placed by Mr Demachkie on *Partridge* v *Lawrence* [2003] EWCA Civ 1121, *Vitol* E *BB* v *Compagnie European des Petroles* [1988] 1 Lloyd's Rep 574, and the cited extracts from Sir Kim Lewison's book, *The Interpretation of Contracts*, 5th ed. However, I consider it necessary in this case to have regard to the factual matrix in order to ensure that the meaning of the right of way granted may be properly determined.

34. Mr Demachkie and Mrs Pigott have referred with different emphasis to three further authorities.

35. First, in *Sykes* v *Secretary of State for the Environment* [1980] 42 P&CR 19*, the Divisional Court had to consider whether planning permission was required for the keeping of horses on land. This question was determined by adopting a purposive approach to the use of F the land. Grazing horses did not require planning permission. However, keeping horses did. Both Donaldson LJ (as he then was) and Kilner-Brown J acknowledged that the use might be mixed, in which case their view was that the purpose was to be determined by the predominant or substantial use.

36. Next, in *Whitsbury Farm and Stud Ltd* v *Hemens* [1988] 1 AC 601**, the House of Lords was concerned with whether stud buildings were agricultural buildings for rating purposes and, if so, were thus exempt from rating. In that case, the land was used for pasturing the ratepayer's thoroughbred horses. The buildings were used for breeding, rearing and keeping the horses. The House of Lords unanimously held G that such was an independent purpose which was not agricultural.

37. Lord Keith, with whom Lords Brandon, Griffiths and Oliver agreed, held that when considering the buildings on agricultural land their use is to be accessory to the land which is to be regarded as principal. Lord Goff arrived at the same conclusion by a different route. In particular, he did not follow the principal/accessory analysis and cited as a reason for not doing so an example in modern day farming in which the buildings and land may be regarded as equally important. For Lord Goff, the litmus test was whether the operations being performed on the land, including the buildings, viewed as a whole were agricultural or not. Lord Goff expressly held that the operation of H breeding and rearing of thoroughbred horses for the purposes broadly described as recreational does not properly fall within the definition of an agricultural operation.

38. Finally, Mr Demachkie and Mrs Pigott referred to *Gainsborough-Field* v *Hyde* [2005] EWHC 2229 (QB), which concerned a right of way granted "for all purposes in connection with the present and any future agricultural use of the property both with or without vehicles, animals or equipment". The main issue was whether the right of way permitted access for the purposes of keeping and grazing ponies for recreational riding. The judge, HH Judge Seymour QC, was referred to both *Sykes* and *Whitsbury Farm and Stud*. J

39. The learned judge held that (1) whether or not it was logical, the authorities drew a well-established distinction between grazing and agricultural activity and keeping, stabling and looking after horses, so that keeping would have to be ancillary to grazing for it to be an agricultural use or purpose; and, (2) livestock, as a matter of ordinary English, does not include horses kept for recreation as distinct from working horses used in agriculture. The judge held that when construing the grant he could also consider whether the convention of the parties at the date of the grant was that keeping horses was an agricultural use. In fact both counsel before him agreed that this was a proper approach for the court to take. It was by that route that the judge K came to the conclusion that stabling and grazing for horses and ponies was by convention between the parties an agricultural use. It is the approach by reference to a convention which Mrs Pigott urges upon me as the correct approach.

40. In the case before me there is clear evidence of recreational equestrian use of the defendant's land at the time of the grant and of the use of the track for the purpose of accessing the defendant's land. In that context, I reject as unrealistic the claimants' evidence that the horses stabled in the converted hop barracks made or could have made their way to and from the stable without setting hoof on the track. However, that is quite different from operating a stud farm from the defendant's L land and using the track to access the stud farm.

41. Referring again to *Whitsbury Farm and Stud*, I note that Lord Keith identified as the use of building the breeding, rearing and keeping of thoroughbreds and held that that purpose did not serve the agricultural use of the land and that thoroughbred horses were not "livestock" in the context of a statutory reference to "buildings used for the keeping or breeding of livestock". However, Lord Goff excluded from agricultural operations the breeding and rearing of thoroughbreds but not – at least not expressly – keeping them in buildings.

42. Just as dairy farming had moved on in the century between M the publication of *Tess of the D'Urbervilles* and the House of Lords' decision in *Whitsbury Farm and Stud*, so too has the general approach of farmers to animal husbandry. Grazing horses, whether they be shires for agricultural use or point to point ponies or brewery shires or racing thoroughbreds, is an agricultural activity – in plain English and under the statutory definition to which I have been referred; and, this proposition is not contradicted by case law.

43. In my judgment, the activity of grazing or use of the land for grazing would not cease to be agricultural if there was a stable on the land and the horses were kept there overnight and/or in inclement weather. That would be a matter of good husbandry and not a matter which would dominate the grazing use of the land.

---

* Editor's note: Also reported at [1981] 1 EGLR 137

** Editor's note: Also reported at [1988] 1 EGLR 159; [1988] 13 EG 43-48

44. In my judgment, keeping horses on the land for grazing would not cease to be an agricultural activity if the land was insufficiently fertile or productive to yield sufficient fodder or if hay or other food had to be brought in provided the land was suitable itself for grazing, which in the case of the defendant's land appears to be the case according to the DEFRA classification.

45. Thus, the use of the track to bring horses to and from the defendant's land for grazing, for the carriage of fodder, for access by vets, farriers, other suppliers or contractors, and staff to bring in and turn out the horses grazing on the land are in my judgment all permitted uses of the track. For the avoidance of doubt, this includes transporting the horses by vehicle to and from the defendant's land for grazing.

46. Where the line is crossed is, my judgment, in the operation of a stud farm. That falls outside the ordinary meaning of agriculture, outside the statutory definition of agriculture to which I have been referred, and outside the case law definitions of agriculture and agricultural use or operations, and outside the scope of any convention between the parties or their predecessors in title in operation on or prior to 1 November 1991. Nor is there any evidence of such use being considered to be a likely future use of the land at that time in 1991.

47. Thus, in my judgment, it is not open to the defendant to use the track as a means of access to and from his land for the purpose of the Tundridge Stud Farm.

48. The remedies sought are declaratory, injunctive and pecuniary. As to the declaration sought, it seems to me that there is little point in formulating an alternative version of the right of way; and so, on the formula that has been proposed I am not inclined to make a declaration.

49. As to an injunction, that is the usual remedy and damages in lieu should only be awarded in exceptional circumstances. Exceptional circumstances however include whether it would be oppressive to the defendant to grant an injunction. In my judgment, it would be impractical – at least at present – as it is not possible simply to turn off the operation of a stud farm. Other arrangements will have to be made in the meantime (which is not, of course, a reference to an endless piece of string) to have the stud horses fed pending their removal elsewhere and existing contracts in relation to breeding or exercising may have to be honoured.

50. As to a pecuniary remedy, the first claimant has given evidence as to the nuisance effect of the passage of vehicles, including horse transporters, along the track. The second claimant has given written evidence but as she has decided not to attend for cross-examination I attach no weight to her statement. In terms of the disturbance, however, the rural idyll described by the first claimant is also blighted by what he has described in a planning objection referred to in evidence as constant overflying by test flights and RAF training flights. By comparison, traffic noise likened to that on a countryside public highway is not likely to be particularly intrusive.

51. The trespass began in the late spring of 2010 and has continued unabated since. It is not trifling and it does cause some intrusion. Viewed from this perspective, it seems to me that an award measured at the rate of about £100 per calendar month would be fair and just compensation. On the basis that the nuisance started in May 2010 at the latest we are now 25 months further on.

52. I must also have regard to the benefit which the defendant has derived from the abuse of the right of way. Use of the track was and is essential to the operation of the stud farm. In a hypothetical negotiation between persons in the claimants' position and a person in the defendant's position before the trespass, and taking into account the claimants' trump card of having the only viable means of access to and from the stud farm for vehicles, and also having regard to the defendant's obligation to contribute to the maintenance and the repair of the track, and on the basis that the defendant would expect to make some profit from the stud farm, I ask rhetorically would the resultant negotiation yield a tariff in excess of £100 per calendar month? If, as should be, it is a hypothetical prospective negotiation, the actual profit would be unknown and there would be a realistic risk that no profit at all would result from the operation of a stud business.

53. I was referred in the context of the hypothetical negotiations to both the Court of Appeal and the further decision of Vos J in the case of *Stadium Capital Holdings (No 2) Ltd* v *St Marylebone Property Co* [2011] EWHC 2856 (Ch)†, which is the reference back to Vos J following the Court of Appeal decision. There are, however, two points of distinction between that case and the position of the claimants and the defendant: (1) the claimants and defendant are neighbours; and (2) the trespass is not solely in furtherance of an operation geared to the maximisation of profit. I therefore think and hold that a higher tariff than £100 per calendar month would not be achieved as the result of a hypothetical negotiation.

54. I now come to the final issue, which is the apportionment of the agreed cost of the works to be undertaken to the track in accordance with the defendant's obligation under the right of way to pay on demand a fair proportion of the costs and expenses of maintenance and repair of the track. Such contribution is ordinarily to be calculated by the surveyor for the time being of (on the facts of this case) the claimants whose decision is to be final except in the case of manifest error.

55. As I understand the first claimant's evidence, his surveyor's view is that because only the defendant made use of the track it is obvious that the entire costs should fall on the defendant. Were that to be his decision it would in my judgment constitute a manifest error. Reasons would include that such a view (1) would entirely disregard the claimants' obligations as occupiers and servient tenement holders, and (2) would entirely disregard all factors other than actual use, which is of course relevant but by no means determinative.

56. Whatever view the claimants' surveyor may have expressed to the first claimant, upon advice the claimants' case is that the defendant should contribute 75% of the costs as his proportionate share. That proportion takes into account the fact that user is not the only relevant factor and makes allowance for the fact the track is an asset of the claimants in which they have an interest and which they have a further interest in preserving. It also takes account of the fact that over time weather will cause the surface of both the tarmacadam length of the track and the compressed stone length of the track to deteriorate and that is not something that is solely attributable to the defendant.

57. The first claimant's evidence is that the track was in good condition when he and his partner acquired Lower Tundridge Farm; that there was no drain alongside any length of the track; that he and his partner have made minimal use of the track and no vehicular use of the track; that all damage is attributable to the defendant's use; and, that the claimants have deferred attending to the track pending the outcome of this case. Mr Lambert gave evidence that there used to be a ditch alongside the track into which rainwater would run off, which evidence I accept.

58. Mrs Pigott submits that (1) the claimant is under no obligation to maintain the track but that he neglected to do so at his own risk and (2) a further consequence of such neglect given the right to repair and demand a proportion of payment from the defendant is that the claimant should bear the consequential cost of delay.

59. It is also relevant that when the defendant sought a self-help remedy by hiring labour to dig a ditch alongside the track he was politely told to desist by the first claimant.

60. The allocation of a proportion of the cost is not an exact science and there is little evidence about the effect of weather – rain and ice – coupled with delay in effecting repair. In my judgment though the effect of weather is not immaterial, neither is the delay, and as a burden this should fall significantly on the claimants as they have chosen to retain the track and exercise control over its maintenance and repair. I must also have regard to the defendant's use, which is unauthorised and extensive. Bearing in mind all the circumstances to which I have been referred, I consider that the defendant should contribute 60% of the repair costs pursuant to his obligation under the right of way. That is my judgment.

Claim allowed.

---

† Editor's note: Also reported at [2012] 1 EGLR 103

Giles v Tarry and another

COURT OF APPEAL

21 June 2012

WARD AND LEWISON LJJ AND NORRIS J

[2012] EWCA Civ 837

Estates Gazette 6 October 2012

[2012] 40 EG 114

Easements – Excessive use – Right of way – Right of way existing over driveway for purpose of gaining access to certain fields including paddock – Paddock also having gate directly onto highway – Additional land adjoining paddock not having benefit of right of way but accessible directly from paddock – Right of way used to drive sheep into paddock – Sheep then driven out of gate onto highway and back in before being allowed to graze on the additional land – Whether this a permissible use of the right of way

The respondent owned several fields, one of which contained an open barn or hovel. Access to the fields was primarily gained by a driveway that lay in the appellant's ownership but which was subject to a right of way, granted by a predecessor in title of the parties, for the benefit of the respondent's fields. One of the fields, a paddock, could additionally be accessed by a gate opening directly onto the public highway.

A dispute arose over the respondent's use of the driveway to take animals across the paddock to an adjoining field that he rented. There was no physical barrier between the paddock and the rented field; the respondent had covenanted to erect a fence along the boundary but had not done so. The rented field did not benefit from the right of way and had no gate onto the highway. By a compromise reached in county court proceedings, the respondent undertook not to use the right of way over the driveway for the purpose of gaining access to the rented field.

Thereafter, the respondent adopted a practice whereby, once he had driven sheep into the paddock using the right of way, he would then drive them out of the gate onto the highway and back in again, then leaving them free to cross from the paddock onto the rented field. The appellant disputed the lawfulness of that practice. In the court below, it was held that the practice did not involve any impermissible use of the right of way since the respondent was not putting the sheep onto the rented field via the right of way.

The judge further held that the respondent's use of the hovel for lambing, treatments, winter feeding or emergency purposes in respect of a maximum of 45 sheep was not a breach of a restrictive covenant not to use the hovel "otherwise than for normal agricultural purposes apart from the keeping of animals or poultry except on a temporary basis".

Held: **The appeal was allowed in part. (1) Interpreting the right of way in the light of the physical features of the land at the date of grant, the respondent was entitled to drive sheep along the driveway to the paddock and out through the gateway onto the highway, and to do the same in reverse. However, when determining whether a right of way is being used lawfully, it is necessary to consider what, in substance and intention, is the user claimed by the person having the right. It would be an error to focus on the actual movements rather than the ultimate intention of the user. The objective of the respondent was to graze his sheep both on the paddock and on the adjoining rented field. His manoeuvres with the sheep were an artificial device or expedient to achieve that objective. The movements of sheep should not be broken down into their constituent elements since, in reality, there was one continuous operation. It followed that the respondent was, in substance and intention, using the driveway for the purpose of gaining access to the paddock and the rented field as a single agricultural unit. He was not entitled to use the right of way in that manner. His practice of driving sheep along the right of way into the paddock, then out of the gate onto the highway and back in again, and thereafter grazing them on the paddock and the rented field, was unlawful. (2) The word "temporary" does not have a single meaning in all legal contexts. Its meaning in the restrictive covenant fell to be construed in the context of the relevant conveyance. Whether a particular use was "temporary" was a question of fact and degree and it was inappropriate to disturb the conclusion of the county court judge in applying the document to the facts that he found.**

The following cases are referred to in this report.

Alvis v *Harrison* (1990) 62 P&CR 10
Das v *Linden Mews Ltd* [2002] EWCA Civ 590; [2001] 1 EGLR 87; [2001] 13 EG 152
Finch v *Great Western Railway* (1879) LR 5 Ex D 254
Fitzroy House Epworth Street (No 1) Ltd v *Financial Times Ltd* [2006] EWCA Civ 329; [2006] 1 WLR 2207; [2006] 2 EGLR 13; [2006] 19 EG 174
Harris v *Flower* (1904) 91 LT 816
Peacock v *Custins* [2002] 1 WLR 1815; [2001] 1 EGLR 87; [2001] 13 EG 152
Skull v *Glenister* (1864) 16 CB (NS) 81
Williams v *James* (1867) LR 2 CP 577 *Williams* v *James* (1867) LR 2 CP 577

This was an appeal by the appellant, Edward Giles, from a decision of HH Judge Harris QC, sitting in Northampton County Court, dismissing a claim against the respondent, Geoffrey Tarry, for unlawful use of a right of way to certain agricultural land.

Alexander Hill–Smith (instructed by BrookStreet des Roches, of Abingdon) appeared for the appellant; Lindsay Johnson (instructed by JC Lawrence & Co, of Daventry) represented the respondent.

Giving judgment, **Norris J** said:

1. The trustees of Sir Thomas White's Charity owned a farm at Preston Capes in Northamptonshire. In February 1975 they split it up.

2. By a conveyance dated 24 February 1975 they sold the farmhouse and outbuildings to the claimant's ("Mr Giles'") predecessor in title. Access to and from the farmhouse and buildings was principally over a driveway that passed between fields on either side. The driveway was included in the conveyance, but the fields to the west and the east were not. In the western field immediately adjacent to the driveway and close to the farmhouse there was an open barn or hovel. On the plan to the conveyance the driveway was coloured brown and the fields between which it passed were edged in red. In the conveyance the trustees reserved a right of way,

> "For the benefit only of that part of the vendors' neighbouring land edged red on the said plan… a right of way at all times and for all purposes over the land coloured brown on the said plan…"

3. In the conveyance, the trustees also entered into a restrictive covenant for the benefit of the land thereby conveyed,

> "Not to use the hovel… otherwise than for normal agricultural purposes apart from the keeping of animals or poultry except on a temporary basis."

4. On 25 February 1975, the fields edged red (on one of which the hovel stood) were conveyed to the defendant ("Mr Tarry"). The conveyance was expressed to be "subject to but together with the benefit of… the exceptions and reservations and covenants…contained in [the conveyance dated 24 February 1975]". Mr Tarry took the benefit of the right of way reserved over the driveway and was subject to the restrictive covenant as to the use of the hovel. The field edged red to the east of the driveway ("the Paddock") adjoins another much larger field ("the Green Land"). This was also owned by the trustees at that time and is now rented by Mr Tarry. There was no physical division between the Paddock and the Green Land. In the conveyance of 25 February 1975 Mr Tarry covenanted to erect a fence along this boundary: but he has not done so, and the Paddock and the Green Land still form a single agricultural unit.

5. There are two ways of getting in and out of the Paddock. One is from a gate onto the driveway: so to use this Mr Tarry has to exercise his right of way. The other is at the northern end of the Paddock, where a gate opens to an Old Forge Lane, which is a public highway. These two gateways can be used differently. If Mr Tarry uses the gate onto the driveway he can only do so by exercising a private right of way and he must abide by the limits of the grant. Although the right of way may be exercised at all times and for all purposes the right exists "for the

A benefit only of [the land edged red]"; and the use of the driveway is so limited. But if Mr Tarry uses the gate which opens onto Old Forge Lane, that is a public highway upon which there are no restrictions: he can use that gate to gain access to any land which is physically served by that gateway. There is no gateway from the Green Land itself onto Old Forge Lane: so the most convenient way of getting to the Green Land is through the northern gateway on the Paddock.

6. In 1991, Mr Giles bought the farmhouse and outbuildings and driveway comprised in the conveyance dated 24 February 1975. Since that purchase a state of virtual war has existed between Mr Giles and Mr Tarry. I make no suggestion as to who is responsible for that state B of affairs. I simply record the fact. Court proceedings were started by Mr Giles in 1993. In those proceedings Mr Tarry brought his own counterclaim. Those disputes were compromised in 1995. As part of the compromise Mr Tarry undertook:

> "Not to exercise the right of way over the land coloured brown for the purpose of gaining access to [the Green Land]."

That compromise itself was back before the court in 1996 – twice. There were further differences which did not reach court in 2000, 2001, 2003, 2004 and 2008. But the proceedings were eventually revived again in a somewhat unorthodox way in 2010 by Mr Giles acting in C person. Mr Tarry (who was then an aged man) also acted in person with the assistance of his family. All procedural deficiencies were overcome and Judge Harris case-managed the dispute so as to identify what issues fell for decision. Amongst those issues were disputes over the use of the right of way and over the use of the hovel. But I would record that the trial judge was called upon also to determine a range of other disputes.

7. Mr Giles and Mr Tarry's family argued the case before the Judge without the benefit of legal representation. On 2 December 2010, Judge Harris delivered an oral judgment to which, in the circumstances, I would pay particular tribute. It made admirably clear findings of fact, and contained a considered analysis of the relevant legal principles D (which must have entailed much conscientious application on the part of the judge). On this appeal we are called upon to consider those same principles in very different circumstances, having the benefit of written and the oral advocacy and the opportunity to test in argument the suggested application of the relevant principles.

8. So far as the right of way is concerned, the judge recorded that Mr Tarry:

> "Wants to be able to graze his sheep on the green [land] from time to time as well as in the Paddock. It can only be got to from the Paddock."

E The judge noted that for Mr Tarry to drive sheep from the right of way into the Paddock for the purposes of going thence onto the Green Land to graze was not permissible (and he commented on the undertaking which Mr Tarry had given). He then continued:

> "Appreciating this, what is done by [Mr Tarry's] daughter and granddaughter is to drive the sheep into the Paddock and without going onto the Green Land, up to and through the gate onto the public highway at Forge Lane. There they pause more or less briefly, before driving them back into the Paddock where they are left free to pass, if they want to, onto the Green Land. Thus the entry into the Paddock immediately preceding the entry into the Green Land is not via the right of way. Is this practice sufficient to avoid the restriction on the right of F way to the benefit of the Paddock?...It is clearly a somewhat artificial device or expedient, but if it is efficacious in law [Mr Tarry] is entitled to make use of it. … [I] approach the question as one of principle and logic. The objective of the defendant has been to graze his sheep both on the Paddock and on the adjacent green land. To get to the Paddock with his sheep he is perfectly entitled to utilise the right of way. It is quite legitimate then for him to take his sheep out of the Paddock via the Forge Lane gate for whatever purpose… When the sheep are in Forge Lane they are not in the Paddock and not on the Green Land. He is then entitled to put them back into the Paddock and to allow or drive them onto the adjacent green land… because he is not then putting the sheep onto the Green Land via the right of way over Mr Giles' land… Usage of a right of way must be analysed in terms of actual movement, not the ultimate intention of the user".

9. The judge therefore made an order declaring:

> "That the defendant may make use of the right of way over the brown land G to drive sheep into the Paddock and that if the same sheep leave the Paddock via the northern end gate and then re-enter and thereafter graze in the green field there is no impermissible use of the right of way".

10. Against that order Mr Giles appeals. He submits that on the facts found by the judge the conclusion must embody an error of law. The error is that the Judge has assumed for the purposes of his argument that the use of the driveway to get to the Paddock was lawful: but if (as the judge found) Mr Tarry wants to be able to graze his sheep on the Green Land as well as in the Paddock, and if what he does is clearly a somewhat artificial device or expedient to achieve that end, then he is not using the right of way for the benefit only of the Paddock. H

11. In my judgment, this submission is sound. In the course of argument the individual steps in the judge's reasoning were examined, and the limits of the various propositions advanced were tested against differing hypothetical facts. I mean no disrespect to this approach by not doing the same in this judgment. But this is a difficult area of the law with sometimes fine distinctions being drawn. I think it better to ground my decision upon this appeal solely upon the facts of this case, and to decide only those questions that are necessary for the disposal of this appeal.

12. The classic statement of the applicable principle is that of Romer J LJ in *Harris* v *Flower* (1904) 91 LT 816 at p819 col 1:

> "The law really is not in dispute. If a right of way be granted for the enjoyment of close A, the grantee, because he owns or acquires close B, cannot use the way in substance for passing over close A to close B."

13. It is significant that the statement of the principle has regard to what "in substance" the person having the benefit of the grant is doing. This requirement reflects earlier statements of the law.

14. In *Skull* v *Glenister* (1864) 16 CB (NS) 81, a plot of land ("Wheeler's") had the benefit of the right of way to the highway. It was rented to Glenister. Glenister bought the adjoining plot (which did not have the benefit of the right of way). Glenister used the right of way K to carry building materials to Wheeler's where he deposited them, and then used them to build a house on his own plot. The question which the judge left to the jury was "Whether the defendants used the way as a way to Wheeler's land, or was it a mere colourable use of it for the purpose of getting at their own land?" Erle CJ said, at p103, "that was the correct way to leave the question". Williams J added, at p106, having reviewed the cases:

> "These authorities appear to establish the principle that, if the defendants here had directly used the road in question as a way over the grantor's land through Wheeler's close to Glenister's, that would have been an excess of the L right. The question was whether they had not substantially done so".

The other judges agreed.

15. In *Williams* v *James* (1867) LR 2 CP 577, the farmer, having the right of way to a field, used the right of way to carry hay that had been stacked on the field: but the hay stack had been composed in part of hay grown on the field and in part of hay grown on another field. In the Court of Common Pleas, Bovill CJ held, at p580:

> "It is… clear… that where a person has a right of way over one piece of land to another piece of land, he can only use such right in order to reach the latter place. He cannot use it for the purpose of going elsewhere. In most cases of this M sort the question has been whether there was a bona fide or mere colourable use of the right of way".

In his judgment, Willes J held, at p582, that:

> "To be a legitimate user of the right of way, it must be used for the enjoyment of the Nine Acre Field, and not colourably for other closes."

16. Those two cases were relied upon by Vaughan Williams LJ in his judgment in *Harris*, at p818 col 1, to formulate as the question for decision in that case:

> "We have to consider here what was in substance and intention the user claimed by the defendants in the present case."

He immediately went on to deal with how an appellate court should deal with such a question, saying:

> "The question of the user is a question of fact; but if we come to a conclusion different from that at which [the trial judge] arrived, it will not be on a question of fact such as those questions of fact on which the Court of Appeal very unwillingly and reluctantly comes to a different conclusion from that taken in the court below, because here the question of fact does not depend on any conflict of evidence, but is a question of the proper inference to be drawn from the facts which are not in dispute. Under these circumstances we are bound to deal with the matter according to our own view of the proper inference to be drawn from the facts."

17. *Harris* was considered in detail by this court in *Peacock* v *Custins* [2002] 1 WLR 1815*. A field ("the red land") had the benefit of a right of way ("the yellow strip"). The farmer farmed that field as one unit with an adjacent field ("the blue land"), using the yellow strip to reach both fields. After reviewing the cases the judgment of the court said (at [22] on p1823):

> "The law is clear at the extremes. To use the track for the sole purpose of accessing the blue land is outside the scope of the grant. However, in some circumstances a person who uses the way to access the dominant land but then goes off the dominant land, for instance to picnic on the neighbouring land, is not going outside the scope of the grant. The crucial question in the present case is whether those circumstances include a case where one of the essential purposes of the use of the way is to cultivate land other than the dominant land for whose benefit the grant was made".

The court held on the facts found that such user was not within the scope of the grant, holding (at [27] on p1824):

> "It is in our judgment clear that the grantor did not authorise the use of the way for the purpose of cultivating the blue land. This cannot sensibly be described as ancillary to the cultivation of the red land. We therefore allow the appeal and declare that the claimants are not entitled to use the yellow strip for the purpose of obtaining access to the blue land in order to cultivate it".

18. *Harris* was again considered by this court in *Das* v *Linden Mews Ltd* [2002] EWCA Civ 590**. The owner of a house had a right to pass over a carriageway from the highway to his property, and the right to halt a vehicle immediately adjacent to his property for the purpose of loading and unloading that vehicle. Instead of returning down the carriageway to the highway he drove off the end of the carriageway and on to a parking space, separate from his house and not part of that dominant land. At the trial HH Judge Cook had approached the matter in this way (see [26] of the judgment of Buxton LJ):

> "[Counsel for the owner says] that he merely intends to pass through the adjacent land and eventually to arrive at number 5… But that is not the reality of this case. What they want to do is to park on the gardens. This is the primary purpose and, that being so, it is an attempt to use the way to accommodate [numbers] 4 and 5. On the authorities, that cannot be done".

19. When reviewing that passage of the judgment Buxton LJ held, at [27]:

> "the overall finding of the judge is entirely clear, that the asserted use of the way is to access and to serve a separate use on a tenement that is not part of the dominant tenement. He found the parking on the garden ground was the primary purpose of this use of the way. That of course poses a formidable difficulty for the owner's case. But it was not necessary for the judge to go that far. Once he found that the use was to access an activity on land that was not the dominant tenement, and which could not be regarded simply as a minor, or indeed any, extension to the access already obtained to number 4, then that use, whether a primary purpose or not, fatally extends the dominant tenement."

20. The language in which the relevant question to be asked is expressed varies from case to case: use "in substance", "colourable use", "bona fide use", the "essential purpose" of the use, the "reality of the case" amongst others. It would not assist this area of the law to analyse the words as if they were in statute or put a gloss upon them. I regard the formulation in the classic statement of the rule in *Harris* to be entirely sufficient: what was in substance and intention the user claimed by the defendant?

21. Judge Harris did not have the advantage of being taken to the authorities cited to us, or the benefit of reading and hearing the arguments of counsel and of testing the propositions they advanced. The relevant question was not identified for him. In consequence, he did not ask himself that question, but instead focused on "actual movement, not the ultimate intention of the user". In my judgment, that was an error of law. Once he had identified that the objective of Mr Tarry was to graze his sheep both on the Paddock and on the adjacent Green Land; and once he had identified what Mr Tarry's family actually did (by which conduct that objective was manifest); and once he had correctly characterised those pointless and self-cancelling manoeuvres as a "somewhat artificial device or expedient" then he was bound to conclude that Mr Tarry was in substance and intention using the driveway for the purpose of gaining access to the Paddock and the Green Land as a single agricultural unit. I would therefore allow the appeal on this point.

22. The argument what might or might not be legitimate use of the way was tested against various sets of hypothetical facts. I have deliberately avoided expressing any view upon those. Mr Giles and Mr Tarry are neighbours: the law contemplates that the rights they have over each other's lands will be reasonably exercised and reasonably allowed. It is not for this court to create a detailed regulatory regime by reference to what either of them might or might not do.

23. For the same reason, I have addressed only the argument which, for me, disposes of the appeal. We heard other interesting arguments (in particular whether the use of a gate onto the highway might itself constitute a use ancillary to the use of the Paddock; and how the principle in *Harris* might apply if the right of way to close A is exercised for the purpose of passing over close A in order to gain access to a highway). Ruling upon these arguments it is unnecessary for the decision in this case.

24. This brings me to the second point of the appeal, namely the use of the hovel. The judge noted that the words of the relevant covenant were far from clear, but that the parties agreed that some animal usage was permissible, but only if it was "temporary". As to the meaning of that word the judge held:

> "The *Shorter Oxford English Dictionary* defines the word "temporary" as: "lasting for a limited time… transient, or made to supply a passing need". A strong element or constituent of the meaning of "temporary" is that it is not permanent. Something which is not permanent can, in many instances, properly be described as temporary. A state of affairs can be regular, but is still temporary."

25. As to the factual situation to which the word "temporary" (properly understood) fell to be applied, the judge held:

> "What the defendant, by Moira and Rosie Goodwin, does with the hovel is to use it for the lambing of some 43 ewes out of their total flock of some 120. They lamb late… between the end of April and July, for some three months or perhaps a little more. This seems a surprisingly long span of time for 43 ewes, some of whose births would doubtless coincide. I find that on not more than 43 days spread out over about three months ewes can give birth in the hovel and no doubt remain in it with their lambs for a few days depending on weather conditions. Between July and October its only ovine use has been for treatments, such as injections and worming or attention to feet or tails. From October to the end of March it is used daily, though only for an hour or two for feeding."

26. The conclusion of the judge was:

> "Ultimately, it seems to me the question is not one of law, but one of fact and degree. While the matter is not wholly straightforward, I have concluded that the use made of this hovel does not infringe the terms of the conveyance. For most of the hours in a year the hovel was not occupied by sheep and when it was occupied it was for limited periods for temporary or short-term purposes, such as giving birth, sporadic physical treatment, or the provision of winter feed. It follows that the claimant is not entitled to prevent the existing pattern of usage".

The judge accordingly declared that:

---

* Editor's note: Also reported at [2001] 1 EGLR 87; [2001] 13 EG 152

** Editor's note: Also reported at [2002] 2 EGLR 76; [2002] 28 EG 130

A "[Mr Tarry] is entitled to make use of the hovel for not more than 45 sheep to the extent and in the way that he presently does, that is for lambing, treatments, winter feeding or emergency purposes."

27. Mr Giles appeals against that order. On his behalf counsel argued that the judge's self-direction as to the meaning of the word "temporary" placed insufficient emphasis on the nuance that "temporary" use is *casual* (as opposed to a part of a regular pattern of events), in part because the word forms part of a qualification to an exception to a general ban on keeping animals in the hovel. He further submitted that whilst the judge did recognize the nuance of "transience" or "the fulfilling of a passing need", he failed to appreciate that the actual use B made by Mr Tarry could not properly be described in either way, since it was rather part of a planned cycle of husbandry.

28. I would dismiss the appeal against this part of the order. The meaning to be given to the word "temporary" is the meaning that the word has in the context of the conveyance. It plainly does not have a single legal meaning in all legal contexts: each party on the appeal recognised that the other's reading of the word was one which the word could bear. I consider that the judge correctly recognised that the word was an ordinary English word containing within it a variety of nuances (some of which would be more prominent than others when the word came to be applied to different sets of facts). I consider that he C correctly held that whether a particular use (or combination of uses) was "temporary" was a question of fact and degree: and I would not disturb the application by an experienced County Court judge of the working document he construed to the facts he found.

29. But I would allow the appeal in relation to the right of way. I would rescind the declaration made in para 3 of the judge's order and substitute (using the language of the original order) a declaration that it is unlawful for the defendant to make use of the right of way over the brown land by driving sheep into the Paddock and then out of the Paddock via the northern end gate and then to re-enter through that gate and thereafter graze the Paddock and the green field.

D

Lewison LJ said:

30. In *Elements of Land Law* (5th ed, para 5.1.73) Gray and Gray write:

"In an era of increasingly intense land use and housing development, it is predictable that awkward questions should arise over the extent to which an easement created or acquired for the benefit of one tenement, Greenacre, may also be used for the benefit of an adjoining tenement, Redacre."

31. We have to answer that awkward question, albeit in a rural context. Norris J has set out the facts, and I gratefully borrow his E description. I particularly wish to associate myself with his tribute to the careful and conscientious way in which Judge Harris dealt with the factual and legal questions that arose. The fact that the parties were self-represented before him cannot have made his task easier.

32. The judge described the use made of the right of way as follows:

"... what is done by [Mr Tarry's] daughter and granddaughter is to drive the sheep into the Paddock and without going onto the Green Land, up to and through the gate onto the public highway at Forge Lane. There they pause more or less briefly, before driving them back into the Paddock where they are left free to pass, if they want to, onto the Green Land. Thus the entry into the F Paddock immediately preceding the entry into the Green Land is not via the right of way. Is this practice sufficient to avoid the restriction on the right of way to the benefit of the Paddock?"

33. It is particularly to be noted that the sheep re-enter the Paddock through precisely the same gate as that through which they left. Although the judge described this practice as "a somewhat artificial device or expedient" he decided that it came within the terms of the granted easement. He reasoned as follows:

"The objective of [Mr Tarry] has been to graze his sheep both on the Paddock and on the adjacent green land. To get to the Paddock with his sheep he is perfectly entitled to utilise the right of way. It is quite legitimate then for him to take his sheep out of the Paddock via the Forge Lane gate for whatever purpose... When the sheep are in Forge Lane they are not in the Paddock and G not on the Green Land. He is then entitled to put them back in the Paddock and to allow or drive them onto the ... green land... because he is not then putting the sheep onto the Green Land via the right of way over Mr Giles' land."

34. He noted that if sheep were loaded onto a lorry in the yard and then driven round to Forge Lane it could not be (and was not) suggested that that was a misuse of the right of way. He said that usage of a right of way "must be analysed in terms of actual movement, not the ultimate intention of the user". He added that:

"In any case... the Paddock is not simply a bridge, it is itself grazing land which the sheep are expected to make use of. They are put there for that purpose as for the purpose of allowing them to move and graze into the Green Land. A H journey over the right of way is just as necessary to utilise the Paddock as it is to utilise the Green Land and the sheep would make the same journey, even if confined simply to the Paddock, though possibly... less often."

35. By way of preliminary point, Mr Johnson, appearing for Mr Tarry, said in his written argument (although he did not develop the point orally) that the judge should not have entertained the action at all. The previous litigation had raised the question of Mr Tarry's use of the right of way; and that action had been compromised on terms including the undertaking Norris J has quoted. In those circumstances Mr Giles should have applied to enforce the undertaking rather than starting J another action. I do not consider that we should summarily dismiss the appeal on this ground. First, it is not a point that was taken below, no doubt because neither party was legally represented. Second, there is no cross-appeal against the judge's decision; and no respondent's notice. Third, there was no application to set aside the grant of permission to appeal by Kitchin LJ. In my judgment, we should decide the appeal on its merits.

36. The argument about the right of way takes as its starting point the so-called rule in *Harris*. The basic principle is not in doubt:

"If a right of way be granted for the enjoyment of Close A, the grantee, because he owns or acquires Close B, cannot use the way in substance for K passing over Close A to Close B."

37. The principle is not without its critics. It was criticised by Ewan Paton and Gwen Seabourne in *Can't get there from here? Permissible use of easements after Das* [2003] Conv 127; and again by the Law Commission in its consultation paper *Easements Covenants and Profits à Prendre* (Consultation Paper No 186 § 5.70). Short of legislation or the Supreme Court, we must accept the principle for what it is; but I do not consider that we should be keen to extend it.

38. Mr Hill-Smith argues that the terms of the right of way do not permit Mr Tarry to drive his sheep across the Paddock and exit onto Forge Lane; and that Forge Lane is "Close B" for the purposes of the L principle in *Harris*.

39. The extent of a right of way granted by deed is a question of interpretation of the grant. In interpreting a grant, even where the grant is the grant of an easement which will in due course be registered at HM Land Registry, the court can and should take into account the physical features of the land in question at the date of the grant. In many of the cases in which the rule in *Harris* has been considered, the question was whether the right of way could be used to access land which was not in the grantee's ownership at the date of the grant. Cases like that raise different questions. Moreover, in our case it is not a question of Mr Tarry using the driveway to gain direct access from the Paddock into the M Green Land. He is using the driveway to gain access to the Paddock and thence onto the highway. In *Finch* v *Great Western Railway* (1879) LR 5 Ex D 254 an inclosure award set out a road as a carriage road and drift way from a highway to some of the inclosed lands. A railway company acquired some of the land, and built a cattle pen adjoining their railway. They used the road for the passage to and from the highway of cattle that were to be or had been conveyed on their railway. The Divisional Court held that that was a lawful use of the road. Stephen J (giving the judgment of the court) said:

"To use a private road into one close merely in order to pass over it into an adjacent close is not quite the same thing as to use a private road into a

A close in order there to make use of a public highway carried through the close subsequently to the grant."

40. He added, at p263:

> "it appears to us that it would be impossible to maintain that, if people were to have free access by a private way to new houses built on an allotment, they should be bound as often as they used that way to return by it. instead of using any other road which might happen to be open to them."

41. In our case, at the date of the grant (and indeed today) the Paddock had two means of access. One was via the driveway and the other was
B from Forge Lane. Unless there is clear contrary wording, a right of way is exercisable in both directions. In other words Mr Tarry was entitled to use the driveway to drive his sheep from Charwelton Road, up the driveway and into the Paddock. He was also entitled to drive his sheep from the Paddock, down the driveway and into Charwelton Road. Bearing in mind that the parties knew or must be taken to have known that the Paddock could be accessed from Forge Lane, it must have been contemplated that sheep could enter the Paddock from Forge Lane, graze in the Paddock, and exit down the driveway into Charwelton Road. To hold otherwise would be absurd. If sheep can be driven in that direction, I can see no reason why they cannot be driven in the
C other direction: that is to say up the driveway, into the Paddock where they graze and then into Forge Lane. Does it make any difference if the sheep do not graze in the Paddock before they exit into Forge Lane? In my judgment it does not. In *Alvis* v *Harrison* (1990) 62 P&CR 10 the House of Lords considered the scope of a Scottish servitude. Mr Harrison owned land on both sides of a driveway. On the west side was his house, and on the east side were his woodlands. The driveway began at the A73 ran alongside the woodlands and ultimately turned west to the house. Mr Harrison constructed a bellmouth on the east side of the drive so as to enable him "to traverse the woodland so as to obtain access to the A73". Although the House confirmed the principle
D in *Harris* (by reference to the law of Scotland), they held that the Inner House were wrong to treat the A73 as being "Close B" for the purposes of the principle. Mr Harrison was thus entitled to construct the new bellmouth. Lord Jauncey said:

> "it is quite wrong to treat the A73 as though it were another tenement contiguous to the woodlands. The underlying reason for restricting the benefit of a servitude right of access to the dominant tenement alone is that to use it for the benefit of a second or third tenement is likely to generate more traffic and so increase the burden. In this case, the appellant already has a right of access to the A73 over the driveway. The new road merely provides a substitute means of access without altering the volume of traffic."

E

42. In this case, Mr Tarry has a right of access to Forge Lane by virtue of his ownership of the Paddock and his rights of ownership as frontager. Whether his sheep graze in the field or not does not affect that right. I therefore reject Mr Hill-Smith's contention that it is impermissible for Mr Tarry to drive his sheep across the Paddock exiting by Forge Lane.

43. Mr Hill-Smith's next submission was that on the facts found by the judge Mr Tarry's real purpose in driving his sheep across the Paddock to Forge Lane was not to exploit the Paddock's access to the highway, but to enable him to graze both the Paddock and the Green
F Land together. He accepted that if the sheep arrived at the gate from Forge Lane into the Paddock from the highway itself (rather than from their previous crossing of the Paddock) without use of the right of way to get them there, then Mr Giles could have no objection to the sheep being allowed to graze both the Paddock and the Green Land together. He also accepted that if the sheep were driven up the driveway onto the Paddock in order to graze on the Paddock and, having grazed on the Paddock, then spread to the Green Land, Mr Giles could have no objection. So Mr Giles is not entitled to object to the Paddock and the Green Land being grazed together, either in parallel or in sequence; depending on how the sheep get onto the Paddock in the first place. This shows, to my mind, how narrow and to some extent artificial the dispute between the parties is.

44. However, Mr Hill-Smith submitted that once the judge had G
found, as he did, that:

(i) Mr Tarry's objective of "has been to graze his sheep both on the Paddock and on the adjacent green land"; and

(ii) The route taken by the sheep was a somewhat artificial device or expedient,

he ought to have concluded that use of the right of way for that composite purpose was impermissible. The use of the highway as a transient stopping place was itself colourable and it was not possible to dissect the various stages in one overall operation. That is a much more formidable submission.

45. In *Skull*, a right of way was appurtenant to a piece of land H
("Wheeler's land"). Glenister acquired an adjoining piece of land on which he wished to build a house. He carted building materials along the way and deposited them on Wheeler's land. The materials were used to build the house on his own land. The question posed to the jury and their answer were:

> "Did the defendants really use the way with carts and waggons as a way to Wheeler's land, or did they really use it as a way to the houses they were building? And was the going first to Wheeler's a mere colourable use. Answer. It was a mere colourable use."

46. On appeal the judge's direction was upheld. Erle CJ said, at J
p103:

> "The question which the learned judge left was whether the defendants used the way as a way to Wheeler's land, or was it a mere colourable use of it for the purpose of getting at their own land. That seems to me to be in substance what the summing-up amounts to. Did the defendants use the way merely for the purpose of carrying the building-materials through Wheeler's close to their own land? I think that was the correct way to leave the question…"

47. Williams J referred to authority and said, at p106:

> "These authorities appear to establish the principle that, if the defendants here had directly used the road in question as a way over the grantor's land
> through Wheeler's close to Glenister's, that would have been an excess of the K
> right. The question was whether they had not substantially done so. The jury must be taken to have found that they had."

48. In *Williams* v *James* (1867) LR 2 CP 577, Mr Jenkins was the occupier of a field called Nine Acres Field. He was entitled to a prescriptive right of way to that field. He also occupied an adjoining field called Parrott's Land. Mr Jenkins created a hay stack on Nine Acres Field, but the hay had been grown partly on Nine Acres Field and partly on Parrott's Land. He then carted away the hay along the right of way. It was held that he was entitled to do so. Bovill CJ said, at p580:

L

> "In most cases of this sort the question has been whether there was a bonâ fide or a mere colourable use of the right of way."

49. He added, at p581:

> "The question, therefore, of what was the ordinary and reasonable use of the land, was practically left to the jury. They found that Jenkins acted honestly, and that is equivalent to finding that what had been done was done in the ordinary and reasonable use of the land to which the right of way was claimed, and in the ordinary and reasonable use of the right of way itself. It was for the plaintiff to shew that there had been some excess of user on the part of the defendant, as by shewing that the user of the right of way was only colourable, or that
> the Nine Acre Field was used for purposes other than those included in the M
> ordinary and reasonable use of the land. The finding of the jury excludes both these questions."

50. Willes J said, at p582:

> "Honest user of the Nine Acre Field must have been understood by the jury in the large sense of bona fide and reasonable, not a use in order to get an advantage to which the defendant was not entitled. The finding of the jury was, that the land was used honestly, and not in order to get a right of way further on. This is equivalent to finding that the stacking of the hay on the Nine Acre Field was in the reasonable and ordinary use of it as a field; also that the carting was from the Nine Acre Field and not from Parrott's land."

51. Montagu Smith J said, at pp582-583:

"It is an admitted fact that some of the hay carried from the Nine Acre Field was grown on Parrott's land, and the carrying away of this hay is the excess in the user of the right of way which is complained of. This alone, however, does not determine the question for the plaintiff. The question is for the jury, whether the stacking of the hay in question and the carrying of it away was in the ordinary and reasonable use of the Nine Acre Field."

52. The principles established by these cases were applied by the Court of Appeal in *Harris* itself. Having referred to these cases, Vaughan Williams LJ said, at pp130-131:

"we have to consider here what was in substance and intention the user claimed by the defendant in the present case. ... The question of the user is a question of fact ... [H]ere the question of fact does not depend on any conflict of evidence, but is a question of the proper inference to be drawn from the facts, which are not in dispute."

53. Romer LJ agreed, saying, at p133:

"what he is really doing here is, under guise of the enjoyment of the dominant tenement, to try and make the right of way become a right of way for the enjoyment of both lands, the pink and the white, and using the land coloured pink as a mere continuation of the right of passage from the pink to the white."

54. Cozens Hardy LJ also agreed.

55. In *Peacock*, Schiemann LJ put the question thus, in [22]:

"However, in some circumstances a person who uses the way to access the dominant land but then goes off the dominant land, for instance to picnic on the neighbouring land, is not going outside the scope of the grant. The crucial question in the present case is whether those circumstances include a case where one of the essential purposes of the use of the way is to cultivate land other than the dominant land for whose benefit the grant was made."

56. This court answered that crucial question in the negative. I agree with Norris J that although the language in which the critical question is posed varies from case to case, the underlying thrust of the question is the same: what in substance and intention is the dominant owner's use? Since "intention" or "essential purpose" is one of the components of the question, I agree with Norris J that the judge fell into error in excluding Mr Tarry's "ultimate intention" from consideration.

57. We cannot go behind the judge's findings that Mr Tarry's "objective" was to graze the Paddock and the Green Land together, and that he was using the highway as an "artificial device" to attain that objective. Mr Johnson submitted that we should break down the movement of sheep into its constituent elements. If (as I think) Mr Tarry was entitled to use the right of way in order to access Forge Lane from the Paddock, then Mr Johnson submitted that what happened to them thereafter was irrelevant to the question whether the use of the right of way was lawful. I regret that I cannot accept that submission. In my judgment we are dealing with one continuous operation, the object of which as the judge found, is to enable the sheep to graze the Green Land via the right of way. The use of the highway as a transient stopover is itself a "colourable" use of the highway. The whole sequence is, as the judge found, an artificial device. It therefore falls foul of the principle in *Harris*.

58. There are of course other ways by which Mr Tarry can achieve his objective. One is to transport the sheep by road up to the gate into the Paddock from Forge Lane. Another is to ensure that the sheep are turned onto the Paddock to graze it first before proceeding on to the Green Land. That might necessitate the erection of some gated barrier between the Paddock and the Green Land (which Mr Tarry was obliged to do by the terms of the conveyance to him of the Paddock). But the fact that there may exist legitimate ways in which Mr Tarry can achieve his objective does not make an illegitimate one lawful. With regret, therefore, because the dispute between the parties is so narrow, I agree that the appeal on the right of way point must be allowed.

59. That leaves the question of the covenant. This turns solely on the question whether the facts found by the judge can properly be described as "temporary" use. Whereas the judge contrasted "temporary" with "permanent", Mr Hill-Smith invited us to contract "temporary" with "regular". Mr Hill-Smith referred us to judicial observations about the meaning of "temporary" in different contexts which, he said, justified the contrast that he urged. But the context in this case is a covenant restrictive of the use of an owner's right to do what he pleases with his own land. Where a seller of land wishes to impose fetters on his buyer's right to use the land being sold, he must, in my judgment, do so clearly; and in case of ambiguity or uncertainty the court should lean in favour of greater rather than lesser freedom on the part of the buyer.

60. Moreover, the word "temporary" is an ordinary English word. The judge directed himself in accordance with one of the ordinary meanings of the word as found in the dictionary. In *Brutus* v *Cozens* [1973] AC 854 an appeal concerned the meaning of the word "insulting" as applied to the phrase "insulting behaviour" in the Public Order Act 1936. Lord Reid said, at p861B-D:

"It is not clear to me what precisely is the point of law which we have to decide. The question in the case stated for the opinion of the court is 'Whether, on the above statement of facts, we came to a correct determination and decision in point of law.' This seems to assume that the meaning of the word 'insulting' in section 5 is a matter of law. And the Divisional Court appear to have proceeded on that footing.

In my judgment, that is not right. The meaning of an ordinary word of the English language is not a question of law. The proper construction of a statute is a question of law. If the context shows that a word is used in an unusual sense the court will determine in other words what that unusual sense is. But here there is in my opinion no question of the word 'insulting' being used in any unusual sense."

61. The same principle applies to the interpretation of a written contract. There are many examples, but I will refer to only one. In *Fitzroy House Epworth Street (No 1) Ltd* v *Financial Times Ltd* [2006] EWCA Civ 329; [2006] 1 WLR 2207†, the question was whether a tenant had "materially" complied with its covenants so as to be able to exercise a break clause in a lease. Sir Andrew Morritt C referred to *Brutus*, saying, in [36]:

"The application of an ordinary English word to a set of primary facts is itself a question of fact..."

62. Viewed as a question of fact, I have not been persuaded that the judge's factual conclusion was wrong.

63. I too would dismiss the appeal on the covenant point.

Ward LJ said:

64. My Lords have so carefully covered the ground that there is nothing I can usefully add save to record my agreement.

Appeal allowed in part.

† Editor's note: Also reported at [2006] 2 EGLR 13; [2006] 19 EG 174

A
Re Clarke and another's appeal

UPPER TRIBUNAL: LANDS CHAMBER

10 July 2012

HH JUDGE NICHOLAS HUSKINSON

[2012] UKUT 226 (LC)

Estates Gazette 3 November 2012

B [2012] 44 EG 96

Easements – Right of way – Extent of right – Vehicular access to certain leasehold land gained across adjoining freehold land in same ownership – Whether right of way existing across freehold land for that purpose – Effect of express right of way in favour of freehold land contained in 1946 conveyance of both freehold and leasehold land – Whether right of way in favour of leasehold land to be assumed when determining price payable on leasehold enfranchisement claim

The leasehold valuation tribunal was asked to determine the price payable by the appellants, under the leasehold enfranchisement provisions of the Leasehold Reform Act 1967, to acquire the freehold of
C a house and land of which they were the long leaseholders. The relevant lease dated from 1641 and the freeholder was unknown; consequently, the court was to execute the relevant transfer of the freehold on payment into court of the appropriate purchase price. An issue arose as to whether the property enjoyed a vehicular right of way to the south, permitting access from the highway across adjoining freehold land that the appellants also owned. The appellants contended that the property enjoyed no right of way for that purpose, was therefore effectively landlocked and should be valued as such.

The LVT certified that a price of £70,620 was payable under section 9 of the Act, based on a freehold value of £1.25m, deferred for 80.5
D years at 5%. In reaching that figure, it found the property did enjoy a vehicular right of way over the adjoining freehold land. It based that conclusion on a conveyance of 1946, by which both the leasehold and freehold land had been conveyed together. The sale of the freehold land had expressly included a vehicular right of way over a track that crossed it; however, the vendor had not owned the soil of the track but only the land to either side, and the track had since become a public bridleway. Moreover, the right of way shown in the 1946 conveyance did not extend all the way to the leasehold land. The LVT none the less considered that the parties must have intended it to be used as a substitute for a former access way to the leasehold property from the
E north that had since fallen into disuse. The appellant appealed.

On appeal, the appellant sought a valuation according to the LVT's alternative finding that, had no right of way existed, an adjoining owner would have been willing to grant such a right for half of the value of the property in the sum of £625,000, producing a purchase price of £35,300.

Decision: **The appeal was allowed. (1) An intention in the minds of the parties to the 1946 conveyance that access to the leasehold property should be gained across the freehold land and the track was insufficient, of itself, to give rise to an easement of way in fee simple. Easements may be granted by statute or express grant,
F by implication under a grant or by prescription. There was no question of any easement arising by statute or express grant in the instant case. No easement could have arisen by prescription since the freehold land had been in the same ownership as the freehold property, such that any use of it for access purposes would have been use with the permission of the owner. No easement could be implied under the 1946 conveyance since no sufficient intention to grant the easement could be inferred from that document. So far as the 1946 conveyance intended to grant a right of way, it dealt with the matter expressly: it granted a right of way, so far as the transferor could lawfully grant one, in favour of the freehold land but did not grant any right of way in favour of the leasehold property and the right of way that was granted did not extend as far as that property. Moreover, the vendor had not owned the** G
track and appeared to have had no right to grant a right of way over it to the leasehold property. There was no scope for implying any easement of necessity in favour of the leasehold property since the 1946 conveyance had conveyed the freehold and leasehold land together; accordingly, an adequate access over the freehold land was available for the transferee by reason of his ownership of it and there was no need for the grant of an easement. (2) The valuation under section 9(1) of the 1967 Act should proceed on the assumption that the property enjoyed no right of way over the freehold land and the track. Although a right of way still existed to the north as a matter of law, since there was no longer any trace of a road or track H
the hypothetical purchaser would realise that he faced problems with that route, including the prospect of litigation to establish the right and difficulties in obtaining planning permission for the construction of an access. Consequently, the hypothetical purchaser would not consider that the right of way to the north added value to the property and would cast his bid on the assumption that the property was effectively landlocked.

The following cases are referred to in this report.

Benn v *Hardinge* (1993) 66 P&CR 246
Earl Cadogan v *2 Herbert Crescent* LRA/91/2007 J
Kutchukian v *Trustees of John Lyon's Charity* [2012] UKUT 53 (LC)
Lester v *Ridd* [1990] 2 QB 430; [1989] 1 EGLR 114; [1989] 22 EG 101
Wheeldon v *Burrows* (1879) 12 Ch D 31

This was an appeal by the appellants, Timothy Clarke and Rosealind Clarke, from a decision of the leasehold valuation tribunal determining the price payable to acquire the freehold of property, under the leasehold enfranchisement provisions of the Leasehold Reform Act 1967, on the assumption that the property benefited from a right of way across adjoining land.

Nicola Muir (instructed by DC Kaye & Co of Prestwood) appeared for the appellants; there was no respondent to the appeal. K

Giving judgment, **HH Judge Nicholas Huskinson** said:

Introduction

1. This is an appeal from the decision of the leasehold valuation tribunal for the Eastern Region Residential Property Tribunal ("the LVT") dated 22 February 2011 whereby the LVT decided that the price to be paid by the appellants for the acquisition by them under the Leasehold Reform Act 1967 of the freehold of Ballinger Hill House, Great Missenden, Bucks ("the Property") was £70,620.

2. The Property is subject to a 400-year lease granted in 1641. The freeholder is missing. The appellants applied to the County Court which L
made an order under section 27 of the 1967 Act to the effect that the appellants should obtain a certificate of valuation pursuant to section 27(5) from the LVT and should pay the appropriate sum into court and that thereafter the court would execute a transfer of the freehold in favour of the appellants.

3. In summary, the case concerns the question of whether, when assessing under section 9 the price payable for the Property on the statutory assumptions, it should be assumed that the Property was landlocked or at least was without any proper vehicular access to a highway. The LVT decided that it should value the Property on the basis that there existed a right of way over some freehold land to the M
south of the Property (being freehold land which was in fact owned by the appellants themselves). It is against this decision that the present appeal is brought.

Facts

4. The facts regarding the previous history of the Property and its surrounding land are only known to a limited extent, principally from the conveyancing documents. Neither the Property nor the land to the south is registered land. So far as it is possible to extract the facts from the evidence available, the position is as follows:

5.1 On 9 April in the 17th year of the reign of King Charles I (which it is agreed was therefore 9 April 1641) William Elwes granted a lease to Thomas Elwes and Thomas Fountain of certain lands (which included the Property) for a term of 400 years at a peppercorn rent. It is unclear how much land was included within this lease, but at the least it included not merely the Property but also the northerly green land referred to below.

5.2 Soon thereafter some buildings were erected on the land – there is evidence that some part of the Property was constructed in the 17th century. There have been various further additions since then with the result that the Property is now a substantial building.

5.3 There is an abstract of title in the papers. The earliest conveyance abstracted is a document of 1 October 1910, with attached plan. This shows that prior to October 1910 the Property together with further land to the north (the northerly green land), all of which were coloured green on the plan and all of which were the subject of the lease, had become vested for the remainder of the term in a Mr Hart. The abstract also shows that Mr Hart owned the freehold of a parcel of land coloured pink on the plan which lay to the south of the Property and fronted onto a public highway. The pink land was crossed diagonally by a track (coloured brown on the plan and hereafter referred to as the brown track) which it appears was not in the ownership of Mr Hart.

5.4 It is not entirely clear how the Property was accessed as at 1910. However the plan attached to the 1910 conveyance showed a track running north from the Property across the northerly green land and effectively making a T-junction with another track running east-west along the northern boundary of this northerly green land. This east-west track appears eventually, anyhow so far as the westerly part of the track is concerned, to have reached a highway after crossing certain other lands. The plan does not show any track running across any part of the pink land – thus no connection between the Property and the brown track (which crosses the pink land) is shown and in fact no access to the south from the Property is shown at all. It therefore appears that as at 1910 the access to the Property was obtained along the tracks to the north – ie northwards from the Property across the northerly green land to the T-junction and then either east or west leading eventually to a highway.

5.5 By a lease dated 20 August 1931, Mr Hart demised to ET Hibbert the Property together with the pink land for a term of 21 years. In 1940 this lease became assigned to Squadron Leader Carslaw.

5.6 By a conveyance and assignment dated 12 April 1946, Mrs EE Hart (in whom had previously become vested various properties of her deceased husband including the lease and the pink land) conveyed the pink land to Group Capt Carslaw in fee simple and assigned to him the residue of the 400 year term in respect of the Property. The plan attached to the conveyance shows a track (which is labelled FP for footpath) running north from the northern boundary of Property to the T-junction with the east-west track previously mentioned. It is unclear whether at this date Mrs Hart still possessed the long lease of the northerly green land (ie over which these tracks ran) or whether she had already assigned this away. The 1946 conveyance and assignment did not grant any right of way over any of these tracks. The conveyance and assignment did however, in that part of the document dealing with the conveyance of the freehold pink land, grant a right of way over the brown track in the following terms:

> "Together with full and free right and liberty so far as the vendor can lawfully grant the same for the purchaser his servants and all others authorised by him and with or without horses carts and vehicles of all kinds in common with other persons entitled to the like right at all times and for all purposes to pass and repass over across and along the road or drive between the public road approximately at the point marked 'A' on the said plan drawn hereon and the point marked 'B' on the said plan."

Thus this conveyance and assignment conveyed the pink land (which was in two parts, one part on each side of the brown track over which the right of way was granted) but it did not convey the brown track itself – Mrs Hart and her predecessor did not own the soil of the brown track. It appears that the brown track is in fact now a public bridleway and it continues on beyond point B. It is not known who owns the brown track.

5.7 It appears that by 1946 the situation on the ground was that access to the Property was obtained by going south from the Property across the pink land to the brown track and then along the brown track to the highway at point A. It is not known when this access arrangement commenced. It appears in 1946 there was still a track to the north to a T-junction with an east-west track, see the markings on the 1946 plan, although this track was now marked FP.

5.8 By a conveyance and assignment dated 18 December 1961, the successor in title to Group Capt Carslaw conveyed the pink land together with the above-mentioned right of way over the brown track to Mr Clarke (one of the present appellants) in fee simple and also assigned to him the residue of the 400 year lease in respect of the Property. By 1961 the position to the north of the Property was that there was no visible path or track in existence – in effect whatever track had existed had been ploughed over or had grown over. This remains the position. There does at present remain the east-west track across the north of the northerly green land (ie the east-west track with which the track which used to run north from the Property made a T-junction) but it is not known whether the easterly limb of this track reaches any highway without impediment. The westerly limb of the track runs not to a highway but into the gardens of properties lying between the track and any highway.

The LVT's decision

6. There is attached hereto marked Appendix 1 a copy of the LVT's calculation of the price to be paid. It will be seen that the value of the rent for the unexpired term was £nil, because only a peppercorn rent was payable; that the LVT took the value of the Property at £1.25m; it took the site value at 30% of this; the LVT calculated a section 15 rent which it appropriately deferred; and it then valued the freehold reversion at a value of £1.25m deferred for 80.5 years at 5%. The total was £70,620.

7. Originally Mr Michael Carr FRICS, who appeared as the expert witness on behalf of the appellants before the LVT, argued that as the Property enjoyed no right of way to the highway and was therefore landlocked the appropriate value to attribute to the Property for the purpose of calculating the price to be paid was £nil. The LVT rejected that argument and it has not been suggested before this Tribunal that the LVT was wrong to do so. The LVT concluded (see [42] of its decision) that if no access to the Property existed then an adjoining owner would be willing to grant a right of access for no more than half of the value of the Property, namely £625,000. The LVT therefore stated that, had it not reached the conclusion it did in relation to the access issue (ie a conclusion to the effect that there existed an access to the Property to the south) it would have assessed the value of the freehold at £625,000. As it was the LVT concluded this access to the south existed and the LVT therefore calculated the price to be paid on the basis that the value of the Property was the full £1.25m. The appellants do not seek to argue that a value lower than £625,000 should be adopted as the value of the Property when calculating the price. However the appellants argue that the LVT was wrong in concluding that there existed an access to the south over the pink land and the brown track and that in consequence the LVT should have adopted the value which it said it would have adopted in the absence of an access to the south, namely £625,000.

8. The question in the present appeal therefore is whether the LVT was correct in concluding that there existed a right of way giving access to the Property over the pink land (and also the brown track) to the south so as to join a highway.

9. In [30] of its decision, the LVT concluded that there were three possibilities to be considered namely:

(A) that the Property was landlocked,

(B) that the original rear access over the fields to the north had not been abandoned, and

(C) that the original access (ie over the fields to the north) was replaced with the current access (ie over the pink land and the brown track to the south) by a former tenant of the property with the intention that the current access should be permanent.

A 10. In [43] of its decision, the LVT observed it was not persuaded that the right of access at the rear (ie to the north) had been abandoned by cessation of user. However, in [45], the LVT concluded:

> "Whether or not the rear right of access has been abandoned so as to leave the Property landlocked is a complex legal question which cannot properly be determined in these proceedings. In view of our finding on the third proposition the question can remain unanswered."

11. The LVT observed that there was no rear (ie northerly) access now. The LVT stated that it must consider what was in the mind of the leaseholder when the division of the leasehold interest was effected. B The LVT examined the terms of the 1946 conveyance and assignment and concluded on the balance of probabilities that it was Mrs Hart's intention to replace the rear access with the existing access through the front land (ie the pink land and the brown track) as a permanent arrangement. The LVT observed that the creation of this new access was an alteration unauthorised by the landlord but one which did not adversely affect the use of the Property or the value of the reversion. Accordingly, the LVT concluded it could be assumed that the landlord had accepted this permanent change by the adoption of the new front access in place of the rear access.

12. Part of the LVT's analysis which led it to this conclusion C involved a consideration of the terms of the 1946 conveyance. The LVT asked itself why it was necessary for the right of way to be granted over the diagonal track (ie the brown track). The LVT stated in [49] of its decision:

> "That right is granted with the front land but why was it necessary? If the purpose was to give access to the parcels of the front land on either side of the roadway, the rights need only have been granted over the western part of the roadway, not all the way up to point B. The two likely explanations are that either at some time before 1946 the front access had been created and was the only access being used by then or that it was intended to create the front access on completion of the conveyance. Either way, we find that this illustrates D an intention by the tenant at the time that the access to the Property would from 1946 (if not before) be through the front land. The plan attached to that conveyance indicates that the rear track had become a footpath and so it is a reasonable assumption that the intention was that the original rear access was replaced by the front access."

13. The LVT also considered that the original lease was subject to an implied (if not express) covenant that the tenant should not do anything that would render the Property unusable and should not make any alterations which would adversely affect the landlord's reversion. The LVT also considered that the removal of all rights of access to the Property would amount to waste. The LVT concluded that Mrs Hart E would not have wished to put herself in a position of breaking such a covenant or committing waste and that in consequence this confirms the conclusion that she must have intended to substitute permanently the access to the south in place of the previous access to the north. The LVT also considered there was a real possibility that the current right of access (ie to the south) could be enforced through the doctrine of necessity; and also that it would offend against equity if a tenant could make an alteration to a property which resulted in a substantial decrease in its value and if then he or his successor could enfranchise at a reduced price because of this alteration.

F *Statutory provisions*

14. The present valuation falls to be made under section 9 of the Leasehold Reform Act 1967 as amended. This provides that the price payable for a house and premises is to be the amount which at the relevant time the house and premises "if sold in the open market by a willing seller (with the tenant and members of his family not buying or seeking to buy) might be expected to realise" on certain assumptions. One of these assumptions is that the vendor was selling with and subject to the rights and burdens with and subject to which the conveyance to the tenant is to be made, and in particular with and subject to such permanent or extended rights and burdens as are to be created in order to give effect to section 10.

15. Section 10(3) provides that the conveyance shall include:

> "(a) such provisions (if any) as the tenant may require for the purpose of G securing to him rights of way over property and not conveyed, *so far as the landlord is capable of granting them*, being rights of way which are necessary for the reasonable enjoyment of the house and premises as they have been enjoyed during the tenancy and in accordance with its provisions" (Counsel's emphasis).

The hearing

16. The grant of permission to appeal to the tribunal stated that the appeal would proceed by way of review, with a view to a re-hearing if it was concluded at the review stage that the LVT's decision could not stand. At the hearing the appellants were represented by Ms Muir who H advanced the legal submissions as summarised below. She then called Mr Carr to give brief evidence upon which the appellants relied in so far as the ultimate decision was that the LVT's decision could not stand.

17. Mr Carr gave evidence confirming the contents of his written report. He agreed that the only point in the LVT's decision with which he now disagreed was the conclusion that the Property should be valued on the basis that there existed an entirely secure access for vehicles to the south.

18. Mr Carr said that, even if at the valuation date there still existed as a matter of law a right of way northwards from the Property so as to meet the east-west track at the T-junction, a hypothetical purchaser J would not conclude that this right of way was adequate or that it added any value the Property. Instead the hypothetical purchaser would treat the Property as effectively landlocked. This is because as a matter of fact, as at the valuation date, there existed on the ground no trace of any track running northwards from the property, nor does there exist any clear and obviously unobstructed access along the east-west track either to the east or to the west to a highway. Even assuming that as a matter of law the right of way to the north still existed, the hypothetical purchaser would note that he would face litigation in seeking to enforce this right of way, involving the construction on the ground of a track over a third party's land where none at present exists and involving the obtaining K (which might well be problematic in this area of green belt and area of outstanding natural beauty) of the relevant planning permission for the construction of a hard surfaced and properly engineered access way.

19. Ms Muir produced a helpful written skeleton argument which she developed further in oral submissions. In summary she advanced the following arguments:

(1) Ms Muir submitted that the LVT's analysis in [49] of its decision was wrong in two particular respect, namely: (A) it was wrong to reason that the granting a right of way up to point B was indicative of an intention to create a right of way to the Property across the pink land – this was wrong because a right of way over the entirety of the brown track, right L up to point B, was needed so as to enable the transferee to get from one part of the pink land across the brown track to the other part; and (B) it was wrong to proceed (as it apparently did) on the assumption that Mrs Hart was the freehold owner of the brown track and had power to grant a right of way over it.

(2) Ms Muir submitted that the hypothetical purchaser would be advised that it was extremely unlikely that a right of way permitting access via the pink land could be established because: (A) the conveyances of the pink land contain a legal easement over a roadway (namely the brown track) but this bypasses the Property; (B) prior to 1910 (and perhaps later) there was no need of an access over the pink M land because there was access from the north; (C) since 1910 there has been no reason to grant or imply a right of way because the pink land and the green land have been in single ownership; and (D) there is no documentation whatsoever indicating the existence of a right of way over the pink land.

(3) Ms Muir submitted that such right of way as had ever existed to the north over the northerly green land either no longer existed (through abandonment) or, if it did still exist, was not such as to be of any significant value in the mind of a hypothetical purchaser who would treat the Property as effectively landlocked unless it was established that there existed a secure right of way to the south – and in connection with this submission she relied upon Mr Carr's evidence.

(4) Ms Muir drew attention to the highlighted wording in section 10(3) as set out in [15] above which makes reference to the conveyance including such provisions as may be required for the purpose of securing rights of way over property not conveyed "so far as the landlord is capable of granting them". The absent landlord in the present case is not capable of granting any right of way over the pink land or the brown track.

Conclusions

20. This is an unusual and difficult case. I have ultimately come to the conclusion that the LVT's decision was wrong. However, before giving my conclusions I wish to express respect for the careful and detailed way in which the LVT has analysed the case.

21. I conclude from the plan attached to the 1910 conveyance that as at 1910 there existed no right of way to the Property to the south. As at 1910 the Property was accessed by a track running north over the northerly green land to a T-junction with the east-west track. I assume that ultimately this east-west track, either in its westerly or its easterly limb or both, gave access to a highway. It seems that this was no more than an agricultural track.

22. The LVT appears to have concluded that what was important was to consider "what was in the mind of the leaseholder when the division of the leasehold interest was affected", see [46] of the decision. It is not entirely clear when this division was effected, but the LVT appear to have proceeded on the basis, for the purposes of its analysis, that this division occurred on the occasion of the conveyance and assignment in 1946. I agree that as a matter of fact it was no doubt in the mind of Mrs Hart as transferor and Group Capt Carslaw as transferee that the access to the Property would indeed be to the south across the pink land and the brown track rather than to the north. This intention in their minds is not however sufficient of itself to give rise to the creation of an easement of way in fee simple across the pink land and the brown track in favour of the Property. It is necessary to ask whether what was intended was (A) merely an arrangement which suited Mrs Hart and her successors being an arrangement to use their own front land to access the property for so long as was convenient (e.g. until the end of the 400-year lease) or (B) something which could amount in law to the creation of an easement in fee simple burdening the pink land and the brown track in favour of the Property.

23. Bearing in mind that what is being considered is whether an easement in the nature of a right of way has been created over the pink land and the brown track in favour of Property, it is appropriate to have regard to the ways recognised by the law in which an easement can be created. An easement can be created by statute or by express grant, but clearly no such consideration arises here. An easement can also be created by implication under a grant if an intention to grant the easement can properly be inferred. *Gale on Easements* (17th ed), paras 3-17, summarises the circumstances in which such an intention may be inferred:

> "1. Where the grant contains particular words of description. Alternatively, in such a case, the easement may be created by estoppel.
>
> 2. Where the circumstances indicate that it was contemplated that the land would be used in some particular manner. The easement may be implied by the necessity of the case.
>
> 3. Under the doctrine of non-derogation from grant, by virtue of which, as already noticed, there may be acquired not only easements but also immunities of a special kind not recognised as easements.
>
> 4. Under the rule in *Wheeldon* v *Burrows* (1879) 12 Ch D 31, which is a branch of the general rule against derogation from grant, but which is commonly considered under a separate head.
>
> 5. Where without a means of access or other right, the land granted or retained would be rendered completely inaccessible or unusable."

Easements can also be established by prescription.

24. In the present case there cannot have become established by prescription any right of way over the pink land in favour of the Property because the pink land has been in the same ownership as the Property and any use for access purposes of the pink land will have been use with the permission of the owner of the pink land.

25. The question arises as to whether upon the occasion of the 1946 conveyance and assignment an easement was created by inference over the pink land and brown track in favour of the Property. In my judgment no such creation of an easement can be inferred for the following reasons:

(1) The 1946 document expressly grants (so far as the transferor lawfully could grant) a right of way over the brown track between points A and B in favour of the pink land. Thus the draftsman of the document has dealt expressly with such grant of a right of way as the parties intended to effect. The document conspicuously does not grant any right of way in favour of Property. The right of way which is granted extends only to point B and does not connect to the Property.

(2) The plan attached to the 1946 document does not mark any track running to the Property across the pink land – but the plan does still mark a track (labelled FP) to the north of the Property.

(3) Mrs Hart did not own the brown track and would appear to have had no right to grant a right of way over the brown track to the Property.

(4) The effect of the 1946 document was to transfer simultaneously to the transferee both the Property and the pink land. There was no need to grant a right of way to the transferee over the pink land because the pink land was by that same document to be the transferee's own freehold property.

26. I do not see any basis on which it could be contended that either the rule in *Wheeldon* or section 62 of the Law of Property Act 1925 operated in relation to the 1946 conveyance and assignment to grant some right of way over the pink land and the brown track in favour of the Property.

27. There is no scope for some inference on the basis of necessity of a grant of a right of way over the pink land and brown track in favour of the Property. An adequate access to the Property was available for the transferee over the pink land by reason of the transferee's freehold ownership thereof.

28. The LVT did not decide the question of whether a right of way to the north continued to exist, but the LVT appears to have reached its conclusion (namely that a permanent right of way was created to the south) partly on the basis that the right of way to the north was lost, through abandonment or otherwise, and that the access to the south was created in substitution.

29. In so far as it is necessary, for the purpose of analysing whether some easement has arisen over pink land and the brown track in favour of the Property, to decide whether the right of way to the north has ceased, I conclude that it has not ceased. My reasons for so concluding are these:

(1) It is not known in what order the lease of the Property became separated from the lease of the northerly green lands. If as at the date that Mrs Hart assigned the residue of the lease in respect of the Property she still retained the lease of the northerly green land, the grant of a right of way in favour of the Property over the track (marked FP on the plan attached to the 1946 document) would in my judgment have been inferred under the rule in *Wheeldon*.

(2) If Mrs Hart had assigned away the residue of the lease in respect of the northerly green land before she assigned the lease of the Property, then we do not know the terms of any such assignment. However, even if no right of way was reserved across the northerly green land in favour of the Property the following point arises. We are here concerned with the assignment of part of some demised premises, because the northerly green land and the Property were all originally demised under the same 400-year lease. So far as the landlord was concerned when he granted this lease all parts of the demised property could be accessed from other parts of the demised property. *Woodfall: Landlord and Tenant*, para 16 p135 states in relation to the effect of the severance of the term:

> "Severance of a leasehold interest by assignment of parts of the demised premises to several assignees does not without more create, as against the landlord, two or more separate holdings with separate tenants for each holding."

The Court of Appeal decision in *Lester* v *Ridd* [1990] 2 QB 430* is cited which decides (see per Dillon LJ at p439) that there would be an injustice if by a transaction to which he was not a party a landlord had thrust upon him two separate tenancies. Accordingly, I conclude that, so far as concerns the freeholder who is the landlord of the Property and of the northerly green land, the fact that they have been assigned separately should not operate to allow the holder of the residue of the term in respect of the northerly green land to argue as against the freeholder that the freeholder has no right of access from the Property over the northerly green land.

(3) In so far as cases concerning abandonment of easements are relevant I conclude that the fact that the access to the north over the northerly green land has not been used for over 50 years is insufficient to constitute an abandonment of that right of way bearing in mind that the tenant of the Property has not had occasion to use this northerly right of way there being an alternative means of access over his own freehold land to the south, see *Gale* at paras 12-64 and *Benn* v *Hardinge* (1993) 66 P&CR 246.

30. For the reasons set out above, I have concluded that no right of way arose in favour of the Property over the pink land or the brown track by virtue of the 1946 conveyance and assignment. Also I do not see how any such right of way can have arisen from the conveyancing documentation or the conduct of the relevant persons at any other time since 1910 (at which date I have concluded that no such right of way existed).

31. Returning to the question of the value of the freehold of the Property on a hypothetical sale under section 9(1) of the 1967 Act, I conclude that the Property should be valued on the basis that it enjoyed no right of way to the south over the pink land and the brown track. I have concluded that, as a matter of law, a right of access still existed to the north. However despite this conclusion of law, the position as a matter of fact as at the valuation date was that no trace of a road or track to the north existed. Thus the hypothetical purchaser who was minded to rely upon the access to the north would realise he faced problems including:

(1) prospective litigation against the leaseholder of the northerly green land to establish and (assuming established as a matter of right) to secure the construction of an adequate access way to the north;

(2) prospective litigation against the owner(s) of property across which the east-west track ran to establish and (assuming established as a matter of right) to secure the construction of an adequate access way to a highway; and

(3) difficulties in obtaining planning permission for the construction of this access in what is an area of outstanding natural beauty and greenbelt.

32. In these circumstances I accept the evidence given by Mr Carr, summarised in [18] above, to the effect that the hypothetical purchaser of the Property would not consider that the right of way to the north (even assuming it existed as a matter of law) added value to the Property and he would treat the Property as effectively landlocked and would cast his bid upon that basis.

33. There was discussion at the hearing as to whether, having regard to the Lands Tribunal's decision in *Earl Cadogan* v *2 Herbert Crescent* LRA/91/2007 and to this tribunal's decision in *Kutchukian* v *Trustees of John Lyon's Charity* [2012] UKUT 53 (LC), it was necessary for the tribunal to decide whether the Property enjoyed a right of way either to the south or to the north, or whether instead the tribunal should merely examine how a prudent and properly advised hypothetical purchaser would assess the value of the Property in the light of the potential disputes and uncertainties. I conclude that in the present case it is necessary to reach a decision as to whether these rights of way exist. For the reasons set out above I have decided that no right of way exists over the pink land and the brown track so as to give access from the south, but I conclude that as a matter of law a right of way does still exist to the north. However this northerly right of way is so problematic as to bring no value to the Property as explained in Mr Carr's evidence.

34. I do not consider that the conclusion I have reached would "offend against equity" as the LVT was concerned it might. There is no evidence that the freeholder ever provided any better access to the Property than an apparently lengthy access by agricultural track. I see nothing inequitable in a conclusion that this freeholder should not be entitled to the benefit of a shorter and properly constructed access which the tenant of the Property has created at his own expense for his own convenience over his own freehold land to the south.

35. I do not consider that the appellants or their predecessors can be shown to have committed waste or to have broken some covenant to be implied in the (lost) 400-year lease.

36. In the result therefore I allow the appellants' appeal and I determine that the price payable in accordance with section 9 (and hence the appropriate sum for the purposes of section 27(5)) is £35,300 as calculated in Appendix 2 hereto, which is a calculation provided by Mr Carr.

Appeal allowed.

Appendix 1

Unexpired Term			
30.5 years at a peppercorn			£ nil
Statutory Term			
Entirety value		£1,250,000	
Site value 30%		£375,000	
S.15 Ground rent @ 7%		£26,250 pa	
For 50 years @ 7%	x 13,8007		
Deferred 30.5 years @ 7%	x	0.1270	£46,008
Freehold Reversion			
Freehold value		£1,250,000	
Deferred 80.5 years at 5%	x	0.01969	
			£24,612
			£70,620

* Editor's note: Also reported at [1989] 1 EGLR 114; [1989] 22 EG 101

Appendix 2

Ballinger Hill House, Ballinger 10.03.2011

Input Information					
Ground Rent	Peppercorn	£	-		
Term	400 yrs from 9 April 1641		30 50 yrs		
Yield – Capitalisation			7%		
Yield – Deferment – (per Sportelli + 0.25%)			5%		
Market Value – if freehold		£1,250,000			
BUT No access to highway, thus Stokes v Cambridge 50% rule		£625,000			
Site Value at 30% of Gross Development Value (MV)		£187,500			
S 15 Rent 7% of Site Value		£13,125			
	Statutory Term		50 yrs		
Valuation					
Unexpired Term					
Ground Rent (£'s pa)		£	-	£ -	
YP for term of			30 50 years	12.4715	
@			7.00%		£ -
Statutory Term					
S 15 Ground Rent (£'s pa)		£13,125.00		£ 13,125.00	
YP for terms of			50.00 years	13,8007	
@			7.00%	£ 181,134.80	
Defer until end of First Term PV of £1 @			7.00%		
after			30 50 years	0.1270	
					£23,003.65
Reversion to an Estate in Fee Simple					
Capital Value of Reversion		£625,000			£625,000
Defer until end of lease term PV of £1 @			5.00%		
after			80.50 years	0.019691	
					£12,306.70
THUS: TENANT					£35,310.34
				Say	**£35,300.00**

A
Pick v Chief Land Registrar

CHANCERY DIVISION

21 January 2011

PROUDMAN J

[2011] EWHC 206 (Ch)

Estates Gazette 29 September 2012

[2012] 39 EG 125

B ***Insolvency – Bankruptcy – Sale of property – Land Registration Act 2002 – Bankrupt transferring property before bankruptcy restriction entered against title – Restriction subsequently entered – Transferee applying to register title – Effect of section 86(5) and (6) of 2002 Act – Whether transferee entitled to registration as purchaser for valuable consideration in good faith without notice of bankruptcy under disposition made at time when no bankruptcy restriction entered – Whether title of trustee saved where bankruptcy restriction entered before transferee applying for registration – Whether registration in favour of transferee a mistake entitling trustee in bankruptcy to claim statutory indemnity***

C The sole registered proprietor of a property was adjudged bankrupt, whereupon his estate, including that property, vested in the claimant trustee in bankruptcy pursuant to section 306 of the Insolvency Act 1986. The bankrupt subsequently purported to transfer the property to a purchaser. At the time of the transfer, no bankruptcy restriction had yet been entered against the title to the property. The purchaser did not effect registration of her purchase within the priority period. The claimant then entered a bankruptcy restriction against the title. The purchaser later applied to the defendant registrar to be registered as proprietor of the property and to have the restriction cancelled. The defendant acceded to that application.

D The claimant contended that the registration of title and the cancellation of the restriction were a mistake, for which he was entitled to a statutory indemnity under the Land Registration Act 2002. A preliminary issue was tried as to whether the purchaser took free of the claimant's rights by virtue of section 86(5) of the Land Registration Act 2002, as a purchaser for valuable consideration in good faith and without notice of the bankruptcy, under a disposition made at a time when no restriction was entered. The claimant contended that the purchaser could not rely on section 86(5) where the restriction had been entered before the purchaser sought to register her title. He cited section 86(6) of the 2002 Act, pursuant to which subsection (5) only
E applied if the relevant registration requirements were met in relation to the disposition.

Held: The preliminary issue was determined in favour of the defendant. Section 86(5) of the 2002 Act focuses on the time of the disposition, not the time of registration. The trustee's title will be avoided if no restriction is registered at the date of disposition. The trustee's title cannot subsequently be saved by entering a restriction before registration. The reason why the protection afforded by section 86(5) only takes effect in favour of a proprietor who becomes registered is to ensure that the protection is confined to those purchasers who are entitled to be registered; however, such registration can be effected at any time. Section 86(5) and (6) are modelled on the provision that they replace, namely section F **61(6) of the Land Registration Act 1972. The new provisions are not different in effect from the old but merely clarify them; there is nothing in the 2002 Act to suggest a shift in policy. The statute cannot both protect the innocent disponee from the bankrupt and, at the same time, protect innocent unsecured creditors; to the extent that the 1972 Act favoured one or the other, the same applies to the 2002 Act. The registrar is specifically empowered to disapply a restriction where the circumstances warrant it: see section 41(2) of the 2002 Act. It had been appropriate to do so in the instant case.**

The following cases are referred to in this report. G

Re Smith (a Bankrupt) ex parte Braintree District Council [1990] 2 AC 215
St John Poulton's Trustee in Bankruptcy v *Ministry of Justice* [2010] EWCA Civ 392; [2010] 3 WLR 1237
Thompson v *Foy* [2009] EWHC 1076 (Ch); [2010] 1 P&CR 16

This was the hearing of a preliminary issue as to the effect of section 86(5) and (6) of the Land Registration Act 2002 on a claim by the claimant trustee in bankruptcy, Robert Pick, against the defendant, the chief land registrar, for a statutory indemnity in respect of an alleged mistake in registering the title of a purchaser from a bankrupt and removing a bankruptcy restriction.

James Dawson (instructed by DWF LLP) appeared for the claimant; H Timothy Morshead (instructed by the Treasury Solicitor) represented the defendant.

Giving judgment, **Proudman J** said:

1. I am asked to decide a point of statutory construction. It is comprised in a preliminary issue ordered by Master Bowles on 13 April 2010 as follows:

> "Whether, apart from and without prejudice to para 14 of the defence, the cancellation of the restriction in favour of the claimant on 13 August 2008 and the restriction [sic]" (by which it is agreed was meant "registration") "on that date of Ms Awokiyesi in place of the bankrupt was a mistake for the purpose of J Schedule 8 to the Land Registration Act 2002."

Schedule 8 entitles the claimant to an indemnity against the Land Registry in specified in circumstances. I am not concerned with para 14 of the defence, which is an allegation that any loss suffered by the claimant was caused through his own fault.

2. In brief, the point is whether the rights of a trustee in bankruptcy are avoided pursuant to section 86(5) of the Land Registration Act 2002 on a sale of land by a bankrupt to a bona fide purchaser at a time when no bankruptcy restriction appears on the register, notwithstanding that the purchaser sought to register her title after the restriction had been K entered.

Facts

3. The following facts are agreed:

(i) Stella Edughele was formerly the sole registered proprietor of a property at Sydenham Hill in London.

(ii) She was adjudged bankrupt on 29 March 2007, the Official Receiver being the trustee. The date of the petition is not in evidence.

(iii) Despite the bankruptcy, the bankrupt completed a transfer of the property to Tinuke Awokiyesi on 14 November 2007. Ms Awokiyesi allowed her priority period in which to affect registration to lapse.

A (iv) On 14 December 2007 the Land Registry entered a restriction on the title to the property.

(v) On 17 December 2007 the claimant was appointed trustee in bankruptcy in place of the Official Receiver. (See Chapter III of the Insolvency Act 1986.)

(vi) On 17 January 2008 the defendant entered a bankruptcy restriction against the title.

(vii) On 13 August 2008 the defendant (i) registered Ms Awokiyesi as the registered proprietor of the property in place of the bankrupt and (ii) cancelled the restriction.

4. The proper procedure is common ground: namely, the bankruptcy
B petition should have been notified to the defendant by the court under the Insolvency Rules 1986, rule 6.13. On receipt of notification the registrar was required to enter the petition as a pending land action under the Land Charges Act 1972 and then, as soon as reasonably practicable thereafter, to enter a notice against the registered title under section 86(1) of the 2002 Act so as to make anyone dealing with the bankrupt aware of the bankruptcy. Although there is no requirement to make a search under the Land Charges Act 1972 (see section 86(7) of the Land Registration Act 2002), anyone who did so would have relevant actual notice of the petition. It is not alleged for present purposes that the absence of such a notice or restriction involved any
C error or omission by the defendant.

5. Further, in order for Ms Awokiyesi to take free of the trustee's rights, she must (a) have given valuable consideration for the transfer; (b) have acted in good faith; and (c) have had no notice of the bankruptcy petition. It is not alleged for present purposes that Ms Awokiyesi fell foul of any of these requirements.

6. Accordingly, the preliminary issue is one of statutory construction only.

Section 86 of the Land Registration Act 2002

7. Section 306 of the Insolvency Act 1986 vests the estate of the
D bankrupt in the trustee, and that section has effect without the necessity to register the trustee's interest (see section 27(5)(a)). Section 284 of that Act deals with the position between the date of presentation of the petition and the adjudication. It must however be remembered that the Land Registration Act 2002 derogates from the general rule that a person may not dispose of property which he has ceased to own, in that section 58 of that Act works what has been described as "the statutory magic" by vesting land in a person merely by virtue of the fact that he is the registered proprietor.

8. In general, the Land Registration Act does not recognise the doctrine of notice in determining priorities. However, the effect of a
E disposition of property after the owner has become bankrupt depends on principles of notice and good faith, and these have been reflected in the Land Registration Act 2002 in bankruptcy cases, in effect protecting a bona fide purchaser for value and without notice, notwithstanding that the purchase was made from a bankrupt who had ceased to own the property.

9. To cope with the different principles applicable to insolvency and to land registration, the Land Registration Act 2002 has imposed a special regime for bankruptcy. Thus neither a bankruptcy petition nor a bankruptcy order is, for the purposes of the Act, an interest affecting an estate or charge (section 86(1)) and by section 86(5) the Act makes
F express provision for the protection of disponees from the bankrupt in relation to such petitions and orders.

10. Section 86(5), (6) and (7) provide as follows:

> "(5) Where the proprietor of a registered estate or charge is adjudged bankrupt, the title of his trustee in bankruptcy is void as against a person to whom a registrable disposition of the estate or charge is made if—
>
> (a) the disposition is made for valuable consideration,
>
> (b) the person to whom the disposition is made acts in good faith, and
>
> (c) at the time of the disposition—
>
> (i) no notice or restriction is entered under this section in relation to the registered estate or charge, and
>
> (ii) the person to whom the disposition is made has no notice of the bankruptcy petition or the adjudication.
>
> (6) Subsection (5) only applies if the relevant registration requirements are met in relation to the disposition, but, when they are met, has effect as from the date of the disposition. G
>
> (7) Nothing in this section requires a person to whom a registrable disposition is made to make any search under the Land Charges Act 1972."

11. A registrable disposition is defined by section 132 of the Act as, in effect, a disposition required by section 27 to be registered. An interest can be registrable notwithstanding that it has been sold to the person seeking registration after the date of presentation of a bankruptcy petition or the making of a bankruptcy order. That is plain from section 86(1) consistently with section 58. It is common ground in this case that the disposition referred to in section 86 was the transfer H to Ms Awokiyesi (see the judgment of Lewison J in *Thompson* v *Foy* [2009] EWHC 1076 (Ch); [2010] 1 P&CR 16 at [121]-[124]) and that it was effected on 14 November 2007. It is also common ground that it was a registrable transfer; that is to say, a transfer which had to be registered in order to be effective by virtue of section 27.

12. The issue I have to decide is whether section 86(5) applies so as to invalidate the claimant's title as trustee in bankruptcy as against Ms Awokiyesi. Mr Dawson on the claimant's behalf says that it does not, because the bankruptcy restriction was entered on the title before Ms Awokiyesi sought to register her title. He submits that the effect of section 86(6), which requires registration of the registrable interest, is J that registration must have been effected before entry of the bankruptcy restriction. While it is true that section 86(5) postulates the date of the disposition as the date at which the restriction must have been entered for the purpose of fixing the disponees with notice, he submits that the requirements of section 86(6) are additional and mean that registration also has to be effected before the restriction is entered.

13. He relies on Schedule 2 to the 2002 Act, which provides that:

> "In the case of a transfer… the transferee… must be entered in the register as the proprietor."

14. In support, he also relies on the wording of the restriction itself, K which is prescribed in obligatory form by the Land Registration Rules 2003, r 166:

> "'BANKRUPTCY RESTRICTION entered under section 86(4) of the Land Registration Act 2002, as the title of [the proprietor of the registered estate] or [the proprietor of the charge dated … … … … referred to above] appears to be affected by a bankruptcy order made by the [name] Court (Court Reference Number … … … …) against [name of debtor] (Land Charges Reference Number WO … … … …).
>
> [No disposition of the registered estate] or [No disposition of the charge] is to be registered until the trustee in bankruptcy of the property of the bankrupt is registered as proprietor of the [registered estate] or [charge].'" L

I note that r 166(3) goes on to define "bankruptcy restriction" as the restriction which the registrar must enter in the register under section 86(4) of the Act.

15. Mr Dawson's argument is simple. It is that in order to fall within section 86(5) a disponee must not only satisfy the provisions of section 86(5) but must also go on to satisfy section 86(6). Ms Awokiyesi could only do so by applying to register her interest. The Land Registrar was then faced with an application for registration by the disponee. It is at that time that he had to determine whether the requirements of section 86(5) were met so as to avoid the trustee's title. At the time he considered the matter there was *ex hypothesi* no registration, so that the M requirements were not met and section 86(5) could not apply. He was therefore bound by the wording of the obligatory restriction. He could not register without cancelling the restriction and he could not cancel the restriction unless there had been registration. Therefore it must follow, runs the argument, that registration had to have been effected prior to entry of the restriction. The registrar could not look forward to future registration to justify cancellation of the restriction.

16. Mr Dawson went on to demonstrate that this result is not unfair to the disponee since, if the restriction is registered after the transfer, the disponee still has the priority period in which to register and obtain good title. The fact that she failed to do so in this case was her fault or that of her advisers.

A 17. The difficulty with the appellant's argument is that the whole focus of section 86(5) is on the time of the disposition rather than the time of registration. The mind of the reader is directed to the avoidance of the trustee's title if no restriction is registered at that date rather than any later date. I can see no point in specifying that date so precisely for the purposes of notice if the restriction could be registered at any time before registration in order for the trustee's title to be saved.

18. By contrast, the date when the "relevant registration requirements" for the purposes of section 86(6) have to be met is not specified in that subsection, as one would expect if a different date from the precisely formulated date in section 86(5) were to be applicable.

B 19. It therefore seems to me that the defendant is right in saying that the reason why the protection afforded by section 86(5) only takes effect in favour of a proprietor who becomes registered is to ensure that it only extends to persons who are entitled to be registered. If for some unrelated reason the disponee does not become registered, the protection does not come into operation. However, that registration can be effected at any time.

20. This interpretation is accentuated by consideration of the subsection which sections 86(5) and 86(6) replace; namely, section 61(6) of the Land Registration Act 1972. Section 61(6) provided:

> C "Where under a disposition to a purchaser in good faith for money or money's worth such purchaser is registered as proprietor of an estate or a charge, then, notwithstanding that the person making the disposition is adjudged bankrupt, the title of his trustee in bankruptcy acquired after the commencement of this Act shall, as from the date of such disposition, be void as against such purchaser unless at the date of such disposition, either a creditors' notice or a bankruptcy inhibition has been registered, but a purchaser who, at the date of the execution of the registered disposition, has notice of the bankruptcy petition or the adjudication, shall not be deemed to take in good faith."

21. In this connection I bear in mind the observations of Lloyd LJ in *St John Poulton's Trustee in Bankruptcy* v *Ministry of Justice* [2010] EWCA Civ 392; [2010] 3 WLR 1237. He said at [68], having cited
D Hoffmann J's approach to construction expressed in In *Re A Debtor* (No 784 of 1991) [1992] Ch 554 at 558-559, (himself citing the decision of the House of Lords in: *Re Smith (a Bankrupt) ex parte Braintree District Council* [1990] 2 AC 215):

> "It is therefore a question of considering both how similar the current text is to that of the previous legislation and whether there are policy reasons, having regard to the terms of the present legislation, which should lead to any earlier construction being ignored and to taking a fresh view of the content of the relevant provision."

22. For present purposes, there is, as far as I know, no judicial
E authority on the meaning of section 61(6). However, section 61(6) was described by the Law Commission (in *Land Registration for the Twenty-first Century, Land Registration Bill and Commentary*, Law Commission No 271) as the model for the new provisions. It seems to me unlikely in the extreme that section 86(5) and 86(6) were intended to F have a different effect from section 61(6) of the earlier Act, as opposed merely to clarifying its provisions. Certainly there is nothing in the *Law Commission Report*, or indeed the Explanatory Notes to the Act, to suggest that it did.

23. Even more pertinently, I can find nothing in the terms of the 2002 Act to suggest a shift in policy. As with the putative mothers of the baby in the judgment of Solomon, the statute cannot protect both parties; namely, an innocent disponee from the bankrupt on the one hand, and the innocent unsecured creditors on the other. To the extent that the 1972 Act favoured one or the other it seems to me that the same applies to the 2002 Act. G

24. I therefore asked Mr Dawson whether he contended that section 61(6) of the 1972 Act had the same effect that he contends for in construing section 86(5) and 86(6) of the 2002 Act. At first he would not be drawn on the issue, but eventually he submitted that the position is the same under the two statutes. In other words, that for the trustee's title to be avoided under section 61(6) of the 1972 Act, registration had to have taken place before the registration of the bankruptcy inhibition. Again, the submission is founded on the inability for the disponee to be registered as proprietor while the restriction is in place.

25. I simply cannot read section 61(6) in this way. Since the requirement for registration is imposed in the same subsection the H focus on the date of disposition is even more pronounced than under the 2002 Act.

26. Mr Morshead relied on views expressed in *Ruoff & Roper on the Law and Practice of Registered Conveyancing* (2006) at 34.009 and 34.001. Mr Dawson valiantly tried to persuade me that these passages are at best equivocal, (applying only to notice) and at worst confusing. I accept that they are not binding on me but I derive some comfort from the fact that my view of the construction of section 86(5) is apparently shared by the editors. I also derive comfort from the views expressed by Lloyd LJ in the *St John Poulton* case at [22], [35] and [50], although I accept that they were stated in a different context and were not J addressed to the registration requirements of section 86(6).

27. The one matter that gives me pause is the wording of the restriction itself, which, as I have said, is obligatory under the Regulations. However, I accept Mr Morshead's explanation (a) that the present situation, where the requirements of section 86(4) were apparently not met, is unusual, (b) that section 41(2) of the 2002 Act specifically empowers the registrar by order to disapply a restriction where circumstances warrant and (c) that it was appropriate and correct for him to do so on an application by Ms Awokiyesi to register on the basis that section 86(5) applied.

28. I therefore determine the preliminary issue in favour of the Chief K Land Registrar.

The preliminary issue was determined in favour of the defendant.

LANDLORD AND TENANT

Beitov Properties Ltd v Martin

UPPER TRIBUNAL: LANDS CHAMBER

8 May 2012

MR GEORGE BARTLETT QC, PRESIDENT

[2012] UKUT 133 (LC)

Estates Gazette 1 September 2012

[2012] 35 EG 70

Landlord and tenant – Service charge – Written demand – Landlord and Tenant Act 1987 – Managing agent serving demands on tenant – Demands giving name of landlord but address of agent – Section 47(1) of 1987 Act – Whether demands complying with requirement in section 47(1) to give name and address of landlord – Whether that requirement met by provision of any address with which landlord having sufficient connection and at which able to receive communications

The appellant landlord brought proceedings against the respondent tenant in the county court to recover unpaid service charges in respect of a residential flat. The matter was transferred to the leasehold valuation tribunal (LVT), which determined the matter in the exercise of its powers under section 27A of the Landlord and Tenant Act 1985. The respondent did not attend the hearing. The LVT raised an issue, of its own motion, as to whether the sums that the appellant sought to recover had been the subject of a written demand complying with the requirements of section 47(1) of the Landlord and Tenant Act 1987. It concluded that they had not, since the relevant demands served by the appellant's managing agent had contained the address of the agent rather than that of the landlord itself. The LVT accordingly determined that the sums demanded were not due by reason of section 47(2).

On appeal from that decision, the appellant contended that where the name of the landlord was given correctly, the address of the landlord's agent was sufficient to comply with section 47(1). It submitted that since the 1987 Act did not prescribe or limit the particular address to be used by the landlord, it was entitled to specify any address with which it had a sufficient connection and at which it could receive communications, such as the address of its managing agent.

Decision: **The appeal was dismissed. The statutory requirement to provide the address of the landlord, in addition to its name, is not imposed solely for the purpose of providing the tenant with an address at which he can communicate with the landlord but serves the wider purpose of enabling a tenant to know who his landlord is. A name might not be sufficient for that purpose and the provision of an address at which the landlord can be found assists in the process of identification. Accordingly, the address of the landlord, for the purpose of section 47(1), is the place where the landlord can be found. In the case of an individual, that will be his place of residence or the place from which he carries on business. In the case of a company, it will be the company's registered office or the place from which it carries on business. If there is more than one place of residence, or place from which the landlord carries on business, then, depending on the facts, any one such address might suffice. The address of the appellant's managing agent was not the landlord's address because it was not the registered office of the appellant company or an address from which it carried on business. The fact that the records, files and accounts relating to the property were kept at that address did not make it an address from which the appellant itself carried on business. It followed that the LVT had decided the issue correctly.**

Per curiam: **It is generally inappropriate for a tribunal to take, on behalf of one side of a dispute between parties, a purely technical point that does not go to the merits or justice of the case. There was nothing to suggest that the tenant in the instant case had wished to know, or been prejudiced by not knowing, the address of the landlord. No purpose had been served by imposing on the landlord the need to deal with the issue raised, to serve a fresh demand and, quite possibly, to take further proceedings for recovery.**

This was an appeal by the appellant, Beitov Properties Ltd, from a decision of the leasehold valuation tribunal dismissing a claim against the respondent, Elliston Martin, for unpaid service charges on the ground that no valid written demand had been served pursuant to the requirements of the Landlord and Tenant Act 1987.

The appeal was determined on the written representations of the parties.

Giving judgment, **Mr George Bartlett QC, President** said:

Introduction

1. This is an appeal by the landlord in a service charge dispute against a decision of a leasehold valuation tribunal. It raises a single issue: whether the requirement in section 47(1) of the Landlord and Tenant Act 1987 that any written demand to a tenant for rent and other sums must contain the name and address of the landlord is satisfied by giving the name of the landlord and the address of its agent. The LVT held that the requirement was not satisfied in this way. The landlord now appeals with permission granted by His Honour Judge Huskinson.

2. Section 47(1) provides as follows:

> (1) Where any written demand is given to a tenant of premises to which this part applies, the demand must contain the following information, namely
> (a) the name and address of the landlord, and
> (b) if that address is not in England and Wales, an address in England and Wales at which notices (including notices in proceedings) may be served on the landlord by the tenant.

Subsection (4) provides that in this section "demand" means a demand for rent or other sums payable to the landlord under the terms of the tenancy (including, therefore, a service charge). Under subsection (2) where any demand for a service charge does not contain the information required by subsection (1) the amount demanded is to be treated as not being due from the tenant at any time before the information is furnished to him.

3. The LVT proceedings arose from a claim by the landlord in the county court for payment of arrears of service charges. It was transferred to the LVT, which thus exercised its powers under section 27A of the Landlord and Tenant Act 1985. The tenant failed to attend the hearing before the LVT. At the hearing, the landlord abandoned one element of the amount claimed, and the LVT concluded that all other elements of the claim were made out: that the service charge for 2008 and the balancing charge for that year were reasonable in amount and that the on-account payments for 2009 and 2010 and amounts

A demanded for payments into a reserve fund were reasonable and payable. However, the LVT raised with the landlord the question of the address given in the demand for service charges, although it does not appear that this had been a matter of concern to the tenant. It acceded to the request of the solicitor who was appearing for the landlord that she be given the opportunity to make written submissions on this matter following the hearing, and, having received these submissions, it held that the amounts demanded were not payable because the demands had not complied with section 47 in that they had not contained the address of the landlord.

4. The demands, which had been prepared and sent by the landlord's B managing agents, BLR Property Management (BLR), contained the following statement:

> "Notice is hereby given pursuant to the Landlord and Tenant Act 1987 section 48 that all notices (including notices in proceedings) may be served upon the Landlord, Beitov Properties Ltd, Hyde House, The Hyde, London NW9 6LH."

The address given was that of BLR. The LVT accepted the evidence of Mr E Hughes, a property manager with BLR, that the address of Beitov Properties Ltd was 266a Regents Park Road, London N3 3HN. It said that this was reinforced by the Land Registry entries, C which recorded that as the company's address; while a company search revealed that its registered office was at Sapphire House, 73 St Margarets Avenue, Whetstone, London N20 9LD.

5. The LVT set out its reasons fully and clearly. It contrasted the requirements of section 47 with those of section 48, which provides for notification by a landlord of an address for service of notices, and it noted that in section 47 there was one piece of information that must always be contained in the demand, namely the name and address of the landlord. It said that the words used in section 47(1)(a) were to be construed as imposing an obligation to give the name and the address of the landlord and not the name and an address; but that D where a landlord had more than one address (for instance a company's registered office address and a different trading address) the landlord could choose which address to use in the demands. It accepted that there was no need to refer expressly in the demand to section 47 and that it was sufficient if it contained the specified information. It rejected the contentions advanced on behalf of the landlord that the address given in the demands was the landlord's address because it was the address of its managing agents and its accountants and because it had some papers there. It also rejected the further contention that it was sufficient to give an address at which the landlord could receive communications, pointing out that, if that was all that was required, section 47 would be E otiose since such requirement was contained in section 48.

6. In their written submissions in this appeal the appellant's solicitors, Conway & Co, say that the landlord's registered office address, Sapphire House, 73 St Margarets Avenue, Whetstone, London N20 9LD, is an address of accountants, Lecome & Co. The landlord owns, it is said, a portfolio of properties, some of which, including the subject property, are managed by BLR from the address at Hyde House, while other properties are managed by VA Property Management Ltd, which is located at 266a Regents Park Road. There are two directors of Beitov. One of these is also company secretary and is a director and part-time employee of VA Property Management. The other is a director F but not an employee of VA Property Management.

7. It is submitted that the demands in question complied with the requirements of section 47 as they specified the name of the landlord and gave Hyde House as the address. It is said that the Act does not prescribe or limit the particular address to be used by the landlord, so that the landlord is entitled to specify Hyde House as the address. Provided the address is one with sufficient connection with the landlord and at which it could receive communications, it is said, section 47 is satisfied. The landlord receives mail addressed to Beitov Properties at the Hyde House address. It is there that the managing agent instructed by the landlord to manage and administer the portfolio of properties operates the business of administration of the property. The landlord's records, files and accounts relating to the property are held at this address, and it is at this address that leaseholders are notified that G inspection of invoices may be carried out.

8. The submissions go on to draw attention to certain provisions in the Companies Act 2006 relating to the service of documents. They refer to section 1142, which provides that any obligation under the Companies Acts to give a person's address is, unless otherwise expressly provided, to give a service address for that person (and "service address" is defined in section 1141(1) to mean an address at which documents may be effectively served on the person). They also refer to section 1139, to section 1143 and, in particular, to Schedule 4, which provides (at paras 1(1) and 4) that documents or information in hard copy form may be sent or supplied to a company (a) to an address specified by the H company for that purpose, (b) to the company's registered office and (c) to an address to which any provision of the Companies Acts authorises the document or information to be sent or supplied.

9. It is, in my judgment, clear from the wording of section 47(1) that the purpose of the requirement in (a) to provide the address (as well as the name) of the landlord is not solely for the purpose of providing the tenant with an address at or through which he can communicate with the landlord. That is clear because (b) provides that, if the landlord's address is not in England and Wales, an address in England and Wales must be given at which notices may be served on the landlord by the tenant. Thus even if the landlord's address is not in England and Wales J it still has to be given (and a further address provided for the service of notices). The purpose of the requirement in section 47 to include in any demand the name and address of the landlord, in my judgment, is to enable a tenant to know who his landlord is, and a name alone may not be sufficient for this purpose. To provide an address at which the landlord can be found assists in the process of identification.

10. That this is the purpose of the requirement to provide the landlord's address is in my view clear from section 47 alone. It is, however, to be noted that section 48 makes separate provision for "Notification by landlord of address for service of notices" (as the section is headed), so that that provision carries the implication K that the requirement in section 47 is not solely for the purpose of providing the tenant with an address at or through which he can communicate with the landlord but has a wider purpose. For this reason the provisions of the Companies Act 2006, relied on by the appellant, are of no assistance because they are concerned with the service of documents or other information.

11. The address of the landlord for the purpose of section 47(1) thus seems to me to be the place where the landlord is to be found. In the case of an individual this would be his place of residence or the place from which he carries on business. In the case of a company it would be the company's registered office or the place from which it carries L on business. If there is more than one place of residence or place from which business is carried on, then, depending on the facts, it may be that any one of such addresses will do. I do not think that it is useful to say any more than this. Of course, in many cases, providing the address of a company landlord may for purposes of identification add nothing of practical value and is unlikely to be of any interest to the tenant, who will be more concerned about having an address at or through which he can communicate with the landlord. (In the present case there is nothing to suggest that the tenant was concerned to establish the company's address). But there will be cases in which provision of an address does assist in the process of identification. M

12. In the present case the address given, Hyde House, The Hyde, London NW9 6LH, was not the landlord's address because it was not the registered office of the company or an address from which it carried on business. It was the address of the agents of the company who managed the subject property, and the fact that they kept there the records, files and accounts relating to the property did not make it an address from which the company itself carried on business. The fact that it was an address to which communications to the company could be sent is nothing to the point. The LVT in my judgment determined the matter correctly in point of law, and the appeal must accordingly be dismissed.

13. Having said that, I should add that it is in my view generally inappropriate for a tribunal to take on behalf of one side in what is a party and party dispute a purely technical point, by which I mean a point that does not go to the merits or justice of the case. Here there is nothing to suggest that the tenant wished to know the address of the landlord or was concerned that the address given in the demands might not be the right one or that he was prejudiced in any way by not knowing the address. The LVT said that if the landlord were now to serve a demand that gave the address required by section 47 the service charges would be payable. No purpose will in the circumstances have been served in imposing on the landlord the need to deal with the issue raised, to serve a fresh demand and, quite possibly, to take further proceedings for recovery.

Appeal dismissed.

E.ON UK plc v Gilesports Ltd

CHANCERY DIVISION

31 JULY 2012

ARNOLD J

[2012] EWHC 2172 (Ch)

Estates Gazette 24 November 2012

[2012] 47 EG 128

Landlord and tenant – Assignment – Liability for rent – Sublease of commercial premises – Sublease assigned without consent of landlord – Transfer of sublease not registered – Arrears of rent accruing – Whether consent to assignment required on proper construction of sublease – Landlord and Tenant (Covenants) Act 1995 – Land Registration Act 2002 – Whether assignor having continuing liability for rent under tenant's covenants – Whether recovery from assignor limited to amounts specified in notices served under section 17 of 1995 Act indicating intention to recover rent from former tenant

The defendant company held a sublease of shop premises in Nuneaton, granted out of a headlease of the same premises. The terms of the sublease required the defendant to observe and perform the tenant's covenants in the headlease "so far as the same are not inconsistent with the terms hereof as if the same had been herein repeated and set out in full mutatis mutandis". The headlease contained a prohibition on assignment or underletting without the previous written consent of the landlord, not to be unreasonably withheld. The sublease expressly incorporated the provisions of section 196 of the Law of Property Act 1925 regarding the service of notices.

In 2007, the defendant carried out an assignment of the sublease. Although licence to assign was requested from the headlessee by email, the assignment was in fact completed before any such licence was granted, without the headlessee's knowledge. The transfer of the sublease was not registered at the Land Registry.

The assignee went into administration leaving arrears of rent. The headlessee served notices, under section 17 of the Landlord and Tenant (Covenants) Act 1995, of its intention to claim the rent from the defendant as the former tenant. Those notices were served without prejudice to its further claim that the assignment was ineffective and that the defendant remained the tenant. The headlesee assigned its headlease and cause of action to the claimant.

Issues arose as to whether: (i) any consent was required for an assignment of the sublease on a proper construction of its terms, and, if so, whether the headlessee was precluded from unreasonably delaying the giving of such consent; (ii) the benefit and burden of the tenant's covenants in the sublease had been effectively assigned notwithstanding the failure to register the transfer of the sublease; and (iii) the claimant could recover the full arrears from the defendant as the current tenant or was confined to recovering the sums specified in the section 17 notices.

Held: The claim was allowed. (1) The words "so far as the same are not inconsistent with the terms hereof as if the same had been herein repeated and set out in full mutatis mutandis", as they appeared in the sublease, required the defendant, as tenant under the sublease, to observe and perform the same covenants with regard to the headlessee as the headlessee was required to observe and perform with respect to the head lessor. The purpose of the relevant clause was to enable the sublease to be expressed in short form. It was evident that the headlessee would want to be protected by tenant's covenants from the defendant corresponding to those that it had given to the head lessor. Any other construction would produce an uncommercial result. Consequently, the sublease contained an obligation on the defendant, imported from the headlease, not to assign the property without the previous written consent of the headlessee. (2) Although consent was not to be unreasonably withheld, the sublease, by reference to the headlease, imposed no requirement that consent should not be unreasonably delayed. Nor did any requirement to give consent within a reasonable time arise under section 1(3) of the Landlord and Tenant Act 1988. That provision applies only where a written application for consent has been "served" on the landlord and, by section 5(2)(a), such service must be effected in a manner provided in the tenancy. The sublease required service to be effected in a manner permitted by section 196 of the 1925 Act, either by delivery to the landlord's last known place of abode or business or by registered post. Neither method had been used and the request sent by email in May 2008 did not suffice for that purpose. Accordingly, the application for consent had not been served on the headlessee and section 1(3) did not apply. In any event, a reasonable time had not elapsed on the facts of the case. The defendant had therefore executed the transfer in breach of covenant and, since the transfer was an excluded assignment under section 11(1) of the 1995 Act, the defendant therefore remained liable under the tenant's covenants in the sublease. (3) The defendant also remained liable by reason of the failure to register the transfer of the sublease. Although the sublease had not been registrable when first granted, its assignment triggered the requirement for compulsory first registration as a qualifying estate under section 4 of the Land Registration Act 2002. The failure to register had resulted in the assignment of the legal estate becoming void under section 7 of the 2002 Act, with the result that the legal estate reverted back to the defendant, which held it on a bare trust for the assignee. However, on the date when the assignment was executed, it had been an assignment of the entire legal and beneficial interest under the sublease to the assignee, which had accordingly become the tenant. The reversion of the legal estate to the defendant, once the transfer became void, amounted to an assignment of the legal estate by operation of law from the then tenant to the defendant. Accordingly, the defendant, as assignee, became bound by the tenant's covenants by virtue of section 3(2)(a) of the 1995 Act. (4) For similar reasons, section 17 of the 1995 Act did not apply to limit the defendant's liability to the sums claimed in the section 17 notices. Section 17 applied where a former tenant was, as the result of an assignment, no longer the tenant. As from July 2008, when the legal estate reverted to it, the defendant was not a former tenant but the present tenant.

The following cases are referred to in this report.

Brown & Root Technology Ltd v *Sun Alliance & London Assurance Co Ltd* [2001] Ch 733

Go West Ltd v *Spigarolo* [2003] EWCA Civ 17; [2003] QB 1140; [2003] 1 EGLR 133

NCR Ltd v *Riverland Portfolio (No 1) Ltd (No 2)* [2005] EWCA Civ 312; [2005] L&TR 25; [2005] 2 EGLR 42; [2005] 22 EG 134

Norwich Union Life Insurance Society v *Shopmoor Ltd* [1999] 1 WLR 531

This was the hearing of a claim by the claimant, E.ON UK plc, against the defendant, Gilesport Ltd, to recover arrears of rent under a sublease.

A Jonathan Seitler QC (instructed by Wragge & Co LLP, of Birmingham) appeared for the claimant; Charles Harpum (instructed by Cobbetts LLP, of Manchester) represented the defendant.

Giving judgment, **Arnold J** said:

Introduction

1. In these proceedings the claimant ("E.ON") seeks a declaration that the defendant ("Gilesports") remains the tenant of the property known as Unit 17, Abbeygate, 132 Abbey Street, Nuneaton ("the Property") under a lease dated 28 October 2008 ("the sublease") and B arrears of rent, alternatively damages. E.ON is the leasehold owner of the property under a lease dated 12 December 1991 ("the headlease"). At the times relevant to this claim, the owner of the leasehold interest was a company then called East Midlands Electricity plc and later Central Networks East plc ("Central Networks"). E.ON was substituted as claimant following an assignment by Central Networks to E.ON of the leasehold interest and its cause of action.

Relevant terms of the headlease

2. The headlease and its schedules runs to 42 pages. Clause 2.3 provides for the tenant to pay the rent, the insurance rent and service C rent (all as defined in clause 1.1). The headlease includes the following tenant's covenants:

"3. <u>TENANT'S COVENANTS</u>

<u>THE TENANT COVENANTS</u> with the Landlord as follows – that is to say:-

…

3.12 <u>User</u>

3.12.1 Not to use the Demised Premises otherwise than as a shop for the retail sale of electrical products and ancillary thereto the servicing of customer accounts or any other use within Class A1 of the Town and Country Planning (Use classes) Order 1987 with the prior consent of the Landlord but:-

3.12.1.1 the Demised Premises shall not in any case be used for the sale of D hot or cold food or beverages nor in connection with the business of a sex shop betting shop nor for the playing of amusement machines;

3.12.1.2 subject to the foregoing the Landlord's consent shall not be unreasonably withheld or delayed

3.14 <u>Alienation</u>

3.14.1 Not to assign or charge part only of the Demised Premises nor to underlet the whole or part of the Demised Premises in consideration of a fine or premium

3.14.2 Not otherwise than as may be permitted under clause 3.14.9 to share possession or occupation of the whole or any part of the Demised Premises

3.14.3 Not otherwise than by assignment or underletting permitted under E the following provisions of this Clause 3.14 nor otherwise than as provided in clause 3.14.9 to:-

3.14.3.1 part with or share possession or occupation of the whole or any part of the Demised Premises;

3.14.3.2 grant to third parties any rights over the Demised Premises

3.14.4 Not to assign or charge the whole of the Demised Premises without the previous written consent of the Landlord which shall not unreasonably be withheld

3.14.5 Upon any assignment to obtain (if the Landlord shall reasonably so require) a guarantor reasonably acceptable to the Landlord for any assignee and to obtain a direct covenant by the assignee with the Landlord to observe and perform the covenants and conditions on the part of the Tenant contained F in this lease and a direct covenant by any such guarantor in the terms set out in Schedule 5

3.14.6 Not to underlet the whole or any part of the Demised Premises without the previous written consent of the Landlord which shall not be unreasonably withheld

3.14.7 Upon the grant of any underlease to obtain covenants on the part of the underlessee that the underlessee will:-

3.14.7.1 not assign underlet or charge part only of the premises comprised in the underlease nor underlet the whole of the premises comprised in the underlease

3.14.7.2 not part with or share possession or occupation of the whole or any part of the premises comprised in the underlease nor grant to third parties rights over them otherwise than by an assignment permitted hereunder or in accordance with clause 3.14.9 (but substituting 'underlessee' for 'Tenant')

3.14.7.3 not assign or charge the whole of the premises comprised in the underlease without obtaining the written consent of the Landlord under this lease G

3.14.7.4 provide for the inclusion in any sub-underleases granted out of such underlease (whether immediate or mediate) covenants to the same effect as those contained in this clause 3.14

3.14.8 Upon the grant of any underlease:-

…

3.14.8.3 to include such covenants on the part of the underlessee as shall secure the due performance and observance of the covenants on the part of the Lessee herein contained

3.14.8.4 to obtain (if the Landlord shall reasonably so require) a guarantor reasonably acceptable to the Landlord for any underlessee the guarantee to be H in the form set out in Schedule 5 (mutatis mutandis) or such other form as the Landlord shall approve such approval not to be unreasonably withheld

3.14.9 Not to part with possession or occupation or share possession or occupation of the Demised Premises otherwise than to or with a member of a group of companies of which the Tenant is itself a member of having no interest in the Demised Premises other than a licence at will (which shall forthwith be determined if the Tenant and the relevant member shall cease for any reason whatsoever to be members of the same group of companies) ...

…

3.15 <u>Registration of assignments etc</u>

Within one month after the execution of any assignment charge transfer or upon any transmission by reason of a death or otherwise affecting the Demised J Premises to produce to the solicitors for the time being of the Landlord the deed instrument or other document (and in each case a certified copy thereof for retention by the Landlord) evidencing or effecting such transmission and on each occasion to pay to such solicitors a reasonable registration fee (being not less than <u>TWENTY POUNDS</u> (£20))".

Relevant terms of the sublease

3. The sublease runs to just seven pages. Clause 1.1.5 defines the headlease in the same manner as I have. The other relevant terms are as follows:

"**2.3 Subjections:**

The Demise is made subject to and with the benefit of all the rights K easements exceptions reservations covenants and other matters contained mentioned or referred to in the headlease as if the same were repeated herein in full mutatis mutandis save where the same are amended by the provisions hereof

…

4. TENANT'S COVENANTS

The Tenant covenants with the Landlord in the terms set out in this Clause :

…

4.2 Alienation:

Not to assign underlet or charge part only of the Demised Premises nor underlet the whole of the Demised Premises. L

…

4.4 To Observe and Perform Covenants in headlease

At all times during the term to observe and perform the covenants and conditions contained in the headlease on the part of the tenant thereof to be observed and performed with the exception of the covenant to pay rent (but including the obligation to pay insurance rent and service charge rent) and so far as the same are not inconsistent with the terms hereof as if the same had been herein repeated and set out in full mutatis mutandis and to indemnify the Landlord and by way of separate indemnity the Head Landlord against all costs claims damages and demand whatsoever and howsoever arising as a result of any breach non-performance and non-observance of such covenants and conditions as aforesaid M

…

5. LANDLORD'S COVENANTS

The Landlord covenants with the Tenant in the terms set out in this Clause :

…

5.4 To Obtain Consents

At the reasonable cost of the Tenant to take all reasonable steps to obtain the consent and/or approval of the Head Landlord whenever the Tenant makes an application for any such consent and/or approval required hereunder where the consent of both the Landlord and the Head Landlord is required by virtue of this lease and the headlease

…

6.6 Notices:

This lease shall incorporate the provisions as to notices contained in Section 196 of the Law of Property Act 1925 as amended by the Recorded Delivery Service Act 1962."

Factual background

4. Abbeygate is a shopping centre in Nuneaton. As can be seen from clause 3.12.1 of the headlease, at the time of the headlease, the Property was to be used as a shop for the retail sale of electrical products. Gilesports is, and has at all material times been, a subsidiary of Sports Direct International plc ("Sports Direct"). It used the Property as a shop selling sports clothing and equipment.

5. In March 2006, Sports Direct acquired Original Shoe Company Ltd ("OSC"). It appears that in late October 2006 OSC moved into occupation of the Premises. E.ON's case is that this was permissible by virtue of clause 3.14.9 of the headlease as incorporated into the sublease by clause 4.4. It is not clear to me on what basis Gilesports contends that it was permissible.

6. On 17 December 2007 Sports Direct contracted to sell OSC to JJB Sports plc ("JJB") for £5m, the sale to be completed on 28 January 2008. This entailed a transfer of eight leases to OSC, including the Property, since occupation would no longer be permissible under group sharing arrangements such as clause 3.14.9/clause 4.4. Sports Direct instructed Cobbetts as its solicitors to deal with this, while JJB instructed Halliwells.

7. On 11 January 2008, Cobbetts sent Sports Direct an email requesting the name and address of the landlord and/or managing agent of the Property, together with any contact details so that they could make an application for a licence to assign. Sports Direct replied on the same day to say that the landlord was The Royal London Mutual Insurance Society Ltd ("Royal London") and that the managing agent was NB Real Estate ("NBRE"). In fact, Royal London was the head lessor under the headlease.

8. On 14 January 2008, Cobbetts emailed NBRE a letter seeking Royal London's consent to assign the sublease to OSC and for a change of use of the Property to the sale of menswear, ladieswear, childrenswear and men's, ladies' and children's footwear and accessories. The letter enclosed a copy of OSC's accounts for the year ended 31 October 2005. On the same day, NBRE asked Cobbetts for confirmation that Gilesports was seeking an assignment of the sublease, not the headlease. Cobbetts confirmed this the next day.

9. On 6 March 2008, Halliwells confirmed to Cobbetts that JJB would provide guarantees for the leasehold properties to be assigned where licences had not been forthcoming, one of which was the Property. On 7 March 2008, Cobbetts relayed this to NBRE.

10. On 3 April 1998, Royal London's in-house solicitor Susan Guy requested a copy of the sublease from Cobbetts. Cobbetts sent a copy the same day. On 7 April 2008, Ms Guy sent Cobbetts a draft licence to assign the sublease and for change of use. One of the parties to the draft licence was Central Networks. On the same day Cobbetts sent Ms Guy proposed amendments to the draft licence, with copies to Halliwells and NBRE. On 8 April 2008, Halliwells agreed the amendments and on 21 April 2008 Ms Guy did. Later the same day, Ms Guy sent Cobbetts a clean copy of the licence for engrossing. On 22 April 2008, Cobbetts sent the licence to assign to Sports Direct for signature. This was duly returned on 23 April 2008, and sent to Halliwells on 24 April 2008. Halliwells returned it on 30 April 2008.

11. On 1 May 2008, Ms Guy sent Cobbetts an email saying that she was holding a signed licence. The email also stated:

> "I will email to you a [rent review memorandum] for signature by [Central Networks] and would you please send it to them for signature. I assume that you have a contact as they have signed the licence to assign."

12. This email prompted a telephone conversation between Cobbetts and Ms Guy during which it transpired that each had assumed that the other was liaising with Central Networks. Thus, contrary to Ms Guy's assumption, Cobbetts had had no communication with Central Networks, which had not signed the licence. Cobbetts reported this to Sports Direct the same day, and asked for contact details for Central Networks.

13. On 2 May 2008, Sports Direct replied by providing contact details for Gary Woodward of Innes England, Central Networks's managing agent. On the same day Angela Bhaseen of Cobbetts telephoned Mr Woodward and discovered that he was out of the office until 6 May 2008 (5 May 2008 being a bank holiday). She left a telephone message. Shortly afterwards, she sent him an email which explained that Gilesports was seeking a licence to assign to OSC and change of use, that application had originally been made to Royal London on 14 January 2008, that Royal London had given its consent and that Cobbetts had assumed that Royal London had contacted Innes England. She attached copies of the licence as signed by Gilesports, OSC and JJB and of the letter dated 14 January 2008 (but not the accounts enclosed with that letter). She asked Mr Woodward to contact her on 6 May 2008. Her email did not state that the matter was urgent, let alone give any reason for urgency or stipulate any deadline for completion of the licence.

14. On 7 May 2008, Alexandra Plane of Cobbetts telephoned Mr Woodward, who said that solicitors would be instructed and he would forward details that afternoon. On 8 May 2008, she sent Mr Woodward an email confirming the conversation and saying "I look forward to hearing from you as soon as possible".

15. On 9 May 2008, Mr Woodward replied asking for copies OSC's accounts "so that I can provide our due diligence to our client". He went on to say that he understood that Cobbetts had the superior landlord's consent, "but clearly from our point of view I need to ensure our client is happy". Ms Plane sent Mr Woodward OSC's accounts for the 18 month period ended 29 April 2007 by email later that day. These showed that OSC had made a loss of £7.6m in that period, against a loss of £529,000 in the year to 31 October 2005, and a deficiency in shareholders' funds as at 29 April 2007 of £6.8m. The directors' report revealed that during this period OSC was financed by a non-interest bearing loan from Sports Direct and that OSC had been sold to a third party (ie JJB) with effect from 28 January 2008.

16. On 12 May 2008, Mr Woodward sent a memo to his colleague Matthew Hannah, reporting on Gilesports' application for consent to assign. He said that he had carried out Experian credit checks on OSC, which was reported as high risk for 2008, and on JJB, which was shown as very low risk. He concluded that:

> "subject to them meeting our costs and E-ON's costs in respect of the transaction, there seems no particular reason to look at withholding consent to the under-lease. This is subject just to my general quick review to make sure everything is paid up to date etc".

17. On 16 May 2008, Ms Plane sent Mr Woodward an email asking whether he had been able to consider the licence yet. She went on, "We are hoping to get the licence signed as soon as possible and complete this matter," but did not give any reason for urgency or stipulate any deadline. She concluded by noting that she would be away for the next two weeks and Mr Woodward should contact Ms Bhaseen during that time.

18. On 19 May 2008, Mr Woodward's personal assistant sent Ms Bhaseen an email saying that he had submitted the documentation to his client and would chase during the course of the week to seek further instructions. He asked for confirmation that Gilesports would meet Central Networks' legal and surveyor's costs. The next day Ms Bhaseen asked Mr Woodward for confirmation of the amount of those costs.

19. On 23 May 2008, Mr Woodward sent a memo to Mr Hannah with an update asking him to identify the solicitor from whom to request an estimate. He went on to discuss the rent review memorandum received from Royal London and the fact that the sublease had not been assigned to E.ON.

20. Monday 26 May 2008 was a bank holiday. On Tuesday 27 May 2008, Ms Bhaseen telephoned Innes England and he was told that Mr Woodward was away until 2 June 2008 (ie for the Whitsun school half-term week).

21. Ms Bhaseen gave evidence that she had been instructed by Sports Direct to complete any outstanding assignments of the eight properties by the end of May 2008 whether or not the licences had been

A completed by then. She was unable to be specific as to precisely when she received this instruction, since she apparently had no documentary record of it, but thought that it might have been in mid May 2008. There is no evidence as to the reason for this deadline, if any.

22. Accordingly, Ms Bhaseen telephoned Steve Peel at Halliwells on Wednesday 28 May 2008 and they completed the assignment of the sublease from Gilesports to OSC ("the transfer"). She wrote to him the same day recording completion on that date and enclosing her client's part of the transfer together with the deeds for the Property. Mr Peel replied on 2 June 2008 acknowledging receipt. For reasons that are unclear, both counterparts of the form TR1 executed by Gilesports and B OSC are dated 29 May 2008, even though completion in fact took place on 28 May 2008.

23. Ms Bhaseen gave evidence that, as at 28 May 2008, she assumed that Halliwells would give notice of the transfer to Central Networks. It appears from documents obtained and disclosed in these proceedings that Halliwells prepared draft letters giving notice of the transfer to both Central Networks and the Land Registry dated 11 June 2008, but these were never sent. There is no evidence which explains why not.

24. On 4 June 2008, Mr Hannah sent Geldards an email asking for an estimate of their fees for acting as Central Networks' solicitors on the transaction.

C 25. On 12 June 2008, Mr Woodward wrote to Richard Rogers of Central Networks seeking his approval to the assignment of the sublease and recommending that the application should proceed in view of the guarantee by JJB. On the same day, Geldards sent Mr Hannah a fee estimate.

26. On 13 June 2008, Mr Woodward relayed the fee estimate to Cobbetts together with a figure for Innes England's fees and asked for Gilesports' undertaking to pay these. On the same day he confirmed to Geldards that he had recommended consent to Mr Rogers, subject to Geldards' review of the documentation. On 16 June 2008, Cobbetts emailed Mr Woodward a letter containing the necessary undertaking.

D 27. On 17 June 2008, Mr Rogers sent Mr Woodward an email agreeing to the assignment subject to preparation of the necessary legal documentation. Innes England relayed this to Cobbetts the same day.

28. After a certain amount of toing and froing, Cobbetts sent Geldards the draft licence agreed with Royal London on 30 June 2008.

29. On 3 July 2008, Halliwells informed Geldards that they had been instructed to withdraw the offer of a guarantee from JJB in respect of the Property and another property.

30. After more toing and froing, on 28 July 2008 Geldards sent Cobbetts proposed amendments to the draft licence by email. One of the E amendments proposed was to add the underlined words to clause 3.1:

> "The landlord at the request of the guarantor grants to the tenant and the tenant grants to the undertenant its licence to assign the whole of its estate and interest in the premises to the assignee on condition that the transfer will completed within three months from the date of this licence <u>otherwise the provision of this licence shall immediately determine and cease to have effect</u>..."

Another amendment added a new clause 4.2.5 to the effect that, if the guarantor (ie JJB) became insolvent, then the assignee (ie OSC) would procure a substitute guarantor acceptable to the Landlord (ie Royal London) and the tenant (ie Central Networks).

F 31. On 29 July 2008, Ms Plane, having discussed the matter with her superior Ms Bhaseen, sent Geldards an email refusing to agree to these changes and trying to persuade Geldards to drop them. Ms Bhaseen accepted that it must have been apparent to her by this time that Geldards did not know that the transfer had been completed, yet Ms Plane's email did not reveal the real reason why Gilesports could not agree to the proposed amendment to clause 3.1. Ms Bhaseen's explanation for this was that she was concentrating on trying to complete the licence. Furthermore, Ms Plane's email did not reveal that JJB had withdrawn its offer, but on the contrary sought to justify the deletion of clause 4.2.5 on the basis of "the covenant strength of the guarantor". Ms Bhaseen's explanation for this was that her instructions G were that Sports Direct hoped to persuade JJB to change its mind. Ms Plane reiterated this point in an email dated 8 August 2008.

32. By 12 August 2008, the only point not agreed was clause 4.2.5. After that, the negotiations proceeded in a desultory manner. By 26 September 2008, it was Geldards who were chasing Cobbetts rather than the other way around, noting in communications with Mr Rogers and Mr Woodward that (i) JJB had lost half of its market share value in the previous week and (ii) the June quarter rent and service charge were outstanding.

33. On 2 October 2008, Ms Plane, with Ms Bhaseen's express approval, again tried to persuade Geldards to drop clause 4.2.5, saying H "you are being offered JJB Sports plc as guarantor and I would ask, in the light of your suggested clause at 4.2.5, that you consider the covenant strength of this plc". Geldards did not reply to this email, and Cobbetts only chased for a reply once, by telephone on 5 November 2008.

34. On 8 October 2008, Ms Plane asked Halliwells whether they had instructions for JJB to act as guarantor. On 13 October 2008, Halliwells informed Ms Plane and Ms Bhaseen that JJB had "dug their heels in" with regard to the guarantee. Ms Bhaseen relayed this to Sports Directs' head of property on 20 October 2008, who replied "Leave it to Dave [Forsey, Sports Directs' CEO] to 'sort out' JJB". On 23 December 2008 Halliwells confirmed to Cobbetts that no JJB guarantee would be given J in respect of the Property. On 29 December 2008 Cobbetts asked Mr Forsey to take the matter up again with JJB. On 15 January 2009, Mr Forsey replied that no progress had been made.

35. On 19 February 2009, OSC went into administration. Cobbetts did not inform Geldards of this. Indeed, on 30 March 2009, Innes England accepted advice which Geldards had given on 24 December 2008 to drop clause 4.2.5. Before Geldards could communicate this to Cobbetts, however, Innes England discovered that OSC had gone into administration.

36. On 26 August 2009, Geldards wrote to Cobbetts noting that K Gilesports had not paid any rent since April 2008 and asking for its proposals for discharging arrears of rent which stood at £111,2333.67 as at the March quarter day. Not having received any reply, Geldards wrote again on 15 September 2009 with an up to date schedule of arrears as at 2 September 2009 totalling £139,869.48.

37. On 5 October 2009, Cobbetts informed Geldards for the first time that the transfer had taken place on 29 May 2008. Geldards registered their surprise at this news in a letter dated 20 October 2009.

38. On 23 June 2010, Central Networks' (now E.ON's) present solicitors served on Gilesports a notice under section 17 of the Landlord and Tenant (Covenants) Act 1995 ("LTCA 1995") without prejudice to Central Networks' contention that Gilesports remained the tenant of the L Property. Further notices have been served every quarter since then.

39. These proceedings were commenced on 10 February 2011.

The issues

40. The following issues arise for decision:

(i) As a matter of construction of the sublease, was Central Networks' consent to the transfer required?

(ii) If consent was required, did Central Networks unreasonably delay in giving its consent to the transfer?

(iii) Was the transfer effective to assign the benefit and burden of the tenant's covenants in the sublease given that the transfer was not M registered at the Land Registry?

(iv) Does section 17 of the LTCA 1995 apply so as to limit Gilesports' liability to the amounts claimed in the notices? If it does, does E.ON have a claim for damages for the remaining arrears of rent?

Issue 1: Construction of clause 4.4

41. E.ON contends that clause 4.4 of the sublease imported clauses 3.14 and 3.15 of the headlease into the sublease with the effect that the obligations imposed by those provisions were owed by Gilesports to Central Networks. Gilesports disputes this. This is a short point of construction.

42. It is common ground that the language of clause 4.4 can be broken down into three parts:

(i) The first part consists of the words "At all times during the term to observe and perform the covenants and conditions contained in the headlease on the part of the tenant thereof to be observed and performed with the exception of the covenant to pay rent (but including the obligation to pay insurance rent and service charge rent)".

(ii) The second part consists of the words "and so far as the same are not inconsistent with the terms hereof as if the same had been herein repeated and set out in full mutatis mutandis". It is these words which give rise to the dispute.

(iii) The third part consists of the words "and to indemnify the landlord and by way of separate indemnity the head landlord against all costs claims damages and demand whatsoever and howsoever arising as a result of any breach non-performance and non-observance of such covenants and conditions as aforesaid".

43. It is agreed that the first part requires the tenant under the sublease to observe and perform the tenant's covenants and conditions in the headlease (except the covenant to pay rent). This shifts the onus of performing the tenant's covenants under the headlease to the sublessee.

44. Counsel for E.ON submitted that the effect of the disputed words, and in particular the words "mutatis mutandis", was to require the tenant under the sublease to observe and perform the same covenants with respect to the sublessor as the sublessor was required to observe and perform respect to the head lessor. Counsel for Gilesports submitted that the effect of the disputed words was to ensure that all the obligations that were cast on the tenant under the headlease were complied with, by reading into the sublease such terms as were necessary to achieve this.

45. In my judgment, E.ON's construction is the correct one for the following reasons. First, I consider that it is the more natural reading of the disputed words, and in particular the words "mutatis mutandis".

46. Secondly, as counsel for Gilesports accepted, Gilesports' construction means that the disputed words add nothing to the first part of clause 4.4. He submitted that this was not a point against Gilesports' construction, because on E.ON's construction clause 4.2 of the sublease was redundant, since clause 4.4 of the sublease would import clause 3.14.1 of the headlease. As he accepted, however, clause 4.2 of the sublease gives effect to clause 3.14.7.1 of the headlease. That cannot be achieved merely by importing clause 3.14.1 of the headlease. Thus I do not agree that clause 4.2 of the sublease is redundant on E.ON's construction.

47. Thirdly, I agree with counsel for E.ON that E.ON's construction is supported by clause 5.4 of the sublease. Counsel for Gilesports had no answer to this point.

48. Fourthly, I agree with counsel for E.ON that Gilesports' construction is uncommercial. It is clear that the purpose of clause 4.4, and in particular the disputed words, is to enable the sublease to be expressed in short form, and that the sublessor would want to be protected by tenant's covenants from the sublessee corresponding to those he had given to the head lessor.

Issue 2: Unreasonable delay

49. It follows from my conclusion on issue 1 that clause 3.14.4 of the headlease as imported by clause 4.4 of the sublease obliged Gilesports not to assign or charge the whole of the Property without the previous written consent of Central Networks "which shall not be unreasonably withheld". By contrast with clause 3.12.1.2, there is no requirement in clause 3.14.4 that consent shall not be unreasonably delayed. Gilesports contends that that requirement is supplied by section 1(3) of the Landlord Tenant Act 1988 ("LTA 1988"). This gives rise to two sub-issues. First, does section 1(3) of the LTA 1988 apply? Secondly, if it does, did Central Networks fail to give its consent to the transfer within a reasonable time?

50. *Does section 1(3) apply?* LTA 1988 provides so far as relevant as follows:

"**Qualified duty to consent to assigning, underletting etc of premises**

1.(1) This section applies in any case where—

(a) a tenancy includes a covenant on the part of the tenant not to enter into one or more of the following transactions, that is—

(i) assigning,

…

the premises comprised in the tenancy or any part of the premises without the consent of the landlord or some other person, but

(b) the covenant is subject to the qualification that the consent is not to be unreasonably withheld (whether or not it is also subject to any other qualification).

…

(3) Where there is served on the person who may consent to a proposed transaction a written application by the tenant for consent to the transaction, he owes a duty to the tenant within a reasonable time—

(a) to give consent, except in a case where it is reasonable not to give consent,

(b) to serve on the tenant written notice of his decision whether or not to give consent specifying in addition—

(i) if the consent is given subject to conditions, the conditions,

(ii) if the consent is withheld, the reasons for withholding it.

…

(5) For the purposes of this Act it is reasonable for a person not to give consent to a proposed transaction only in a case where, if he withheld consent and the tenant completed the transaction, the tenant would be in breach of a covenant.

(6) It is for the person who owed any duty under subsection (3) above—

…

(c) if he did not give consent and the question arises whether it was reasonable for him not to do so, to show that it was reasonable,

and, if the question arises whether he served notice under that subsection within a reasonable time, to show that he did.

…

Interpretation

5. …

(2) An application or notice is to be treated as served for the purposes of this Act if—

(a) served in any manner provided in the tenancy,

…"

51. It is common ground that the conditions in section 1(1)(a)(i) and (b) are satisfied. The issue is whether Gilesports "served" on Central Networks a written application for consent to the transaction. Gilesports relies for this purpose on Ms Bhaseen's email to Mr Woodward dated 2 May 2008. There is no dispute that that was a written application for consent. E.ON disputes that the sending of the email to Innes England constituted good service for the purpose of section 1(3).

52. It is common ground that (i) section 5(2)(a) of the LTA 1988 requires the application to be served in a manner provided in the sublease (Gilesports does not rely upon section 5(2)(b)) and (ii) clause 6.6 of the sublease incorporates section 196 of the Law of Property Act 1925 ("LPA 1925").

53. Section 196(2), (3) and (4) of the LPA 1925 provide:

"(2) Any notice required or authorised by this Act to be served on a lessee or mortgagor shall be sufficient, although only addressed to the lessee or mortgagor by that designation, without his name, or generally to the persons interested, without any name, and notwithstanding that any person to be affected by the notice is absent, under disability, unborn, or unascertained.

(3) Any notice required or authorised by this Act to be served shall be sufficiently served if it is left at the last-known place of abode or business in the United Kingdom of the lessee, lessor, mortgagee, mortgagor, or other person to be served, or, in case of a notice required or authorised to be served on a lessee or mortgagor, is affixed or left for him on the land or any house or building comprised in the lease or mortgage, or, in case of a mining lease, is left for the lessee at the office or counting-house of the mine.

(4) Any notice required or authorised by this Act to be served shall also be sufficiently served, if it is sent by post in a registered letter addressed to the lessee, lessor, mortgagee, mortgagor, or other person to be served, by name, at the aforesaid place of abode or business, office, or counting-house, and if that letter is not returned by the postal operator (within the meaning of Part 3 of the Postal Services Act 2011) concerned undelivered; and that service shall be deemed to be made at the time at which the registered letter would in the ordinary course be delivered."

54. Counsel for E.ON submitted that section 196(3) required physical delivery of the application to Central Networks' last known

A place of business, and that the email dated 2 May 2008 was neither physically delivered nor sent to Central Networks. Counsel for Gilesports submitted that section 196(3) merely stipulated that such service was "sufficient", not that it was required, and that there was good service because the application was in fact received by Central Networks' agent. I do not accept that submission. In my judgment section 196 requires service by one of two methods: either delivery to (in the case of a landlord) the landlord's last known place of abode or business or by registered post. In the present case neither method was used. Accordingly, the application was not "served" on Central Networks.

B 55. In the alternative, counsel for Gilesports submitted that Central Networks was bound by an estoppel by convention from denying that the application had been served. In support of this he relied upon the fact that all parties proceeded on the basis that Gilesports had made an application for consent. In my judgment that does not establish an estoppel by convention. There was no agreement, mutual understanding or mutual assumption that section 1 of the LTA 1988 was engaged.

56. *Did Central Networks fail to give its consent within a reasonable time?* If, contrary to my previous conclusion, section 1(3) of the LTA 1998 does apply, Gilesports contends that Central Networks failed to give its consent within a reasonable time. It is not open to a landlord
C to put forward reasons justifying the withholding of consent if those reasons are not put forward within a reasonable time: see *Go West Ltd* v *Spigarolo* [2003] EWCA Civ 17; [2003] QB 1140* at [22] and [80(ii)]. By virtue of section 1(6)(c) of the LTA 1988, the onus lies on E.ON to show that Central Networks' failure to give consent by 28 May 2008 was reasonable.

57. The correct approach to deciding what constitutes a reasonable time has been considered in two decisions of the Court of Appeal. In the *Go West* case, Munby J (as he then was), with whom Potter LJ agreed and Pill LJ concurred, said:

> D "35. If one asks what is the reference point by which one answers the question posed in section 1(3) — Has the landlord done 'within a reasonable time' what he should have done? — the answer can only be that the reference point is the date when the tenant made his written application. Section 1(3) invites a question — 'Within a reasonable time' of what? — to which there can be only one answer: within a reasonable time after the service by the tenant of his written application. But this does not mean that what is a 'reasonable time' thereafter is something that can be determined a priori as at the date when the application is made. Nor does it mean that only the circumstances at that date can be taken into account in determining what is 'a reasonable time'. That has to be assessed, in my judgment, having regard to all the circumstances of the particular case. Those circumstances will obviously include, but will not be
> E confined to, the circumstances known to the landlord and the tenant at the date when the tenant makes his application. I entirely agree with Mr Dutton that subsequent events also have to be taken into account. This, after all, accords with Sir Richard Scott V-C's view in *Norwich Union Life Insurance Society* v *Shopmoor Ltd* [1999] 1 WLR 531 — a view with which I entirely agree, though the point was a slightly different one — that the position must be tested by reference to the state of affairs at the expiry of the reasonable time.
>
> …
>
> 39. This leads on to the second point. In one important sense the 'reasonable time' requirement in section 1(3) is there to protect the tenant: the landlord must perform his duty 'within a reasonable time'. But in another and equally important sense it is there to protect the landlord: the landlord
> F is given a reasonable time within which to do what the 1988 Act requires of him. Moreover, the 'reasonable time' referred to in section 1(3) is the time within which the *landlord* has to do something, not the time within which the tenant has to do anything. In other words, and as section 1(3) makes clear, the 'reasonable time' is the time reasonably required by the *landlord* to do the things which the Act requires of him.
>
> …
>
> 73. … I repeat, and for my own part would wish to emphasise, Sir Richard Scott V-C's references in the *Norwich Union* case to the landlord dealing with his tenant's application "expeditiously" and "at the earliest sensible moment". It may be that the reasonable time referred to in section 1(3) will sometimes
> have to be measured in weeks rather than days; but even in complicated cases, G
> it should in my view be measured in weeks rather than months."

58. In *NCR Ltd* v *Riverland Portfolio (No 1) Ltd (No 2)* [2005] EWCA Civ 312; [2005] L&TR 25**, Carnwath LJ (as he then was), with whom Lord Slynn of Hadley and Ward LJ agreed, said:

> "21. … In my view, whatever earlier discussions there had been, Riverland was entitled to adequate time following receipt of the completed application to consider the serious financial and legal implications of a refusal with its advisers, and if necessary to report to the relevant Board. In the absence of special exceptional circumstances, a period of less than three weeks (particularly in the holiday period) cannot in my view be categorised as inherently unreasonable H
> for that process.
>
> …
>
> 23. … it is in neither side's interests, at least where a refusal is being contemplated, for the decision to be rushed. The lessor is properly concerned to protect himself against the possible consequences of a finding of unreasonableness, while the lessee's primary objective is to achieve an underletting, rather than an uncertain cause of action under the Act."

59. In the present case, counsel for Gilesports conceded that time should not be regarded as having started to run until 16:14 on Friday 9 May 2008, when Cobbetts sent OSC's accounts to Innes England
by email. As counsel for E.ON pointed out, 11 working days elapsed J
between then and 28 May 2008. I do not regard that as a reasonable time in which to require Central Networks to either consent or give reasons for refusing consent for the following reasons:

(i) 11 working days is on any view a fairly short time in which to require Central Networks to consider the financial and legal implications of the transaction with its advisors, to consider the matter internally and to formulate its response.

(ii) One of those working days was during the Whitsun half-term break. Not only was that a period during which it was inherently likely that relevant personnel would be on holiday, but also Gilesports
knew (through Cobbetts) prior to completion of the transfer that Mr K
Woodward was in fact away that week (like Ms Plane).

(iii) As noted above, Ms Bhaseen's email dated 2 May 2008 did not state that the matter was urgent, let alone give any reason for urgency or stipulate a deadline for response. A landlord cannot be blamed for not treating a request for consent as urgent if it is not informed that the request is urgent or why it is urgent.

(iv) In my view, it would have been reasonable for Innes England (and hence Central Networks) to conclude from the email dated 2 May 2008 that the matter was far from urgent, because Royal London's consent had been sought as long ago as 14 January 2008 and only
now was Gilesports seeking Central Networks' consent. Furthermore, L
the letter dated 14 January 2008 did not suggest that the matter was urgent.

(v) It took Royal London roughly three and a half months to give its consent, yet Gilesports did not suggest at any stage that it was unreasonably delaying the matter. As counsel for E.ON pointed out, Central Networks had a more direct interest in the identity and financial security of the sublessee than Royal London did.

(vi) The only chaser which Cobbetts sent between 9 May 2008 and 28 May 2008 was the email dated 16 May 2008. This did not state that the matter was urgent or why or stipulate any deadline for response. Nor
did Cobbetts ever assert to Innes England that Central Networks was M
under a duty to consider the matter expeditiously by virtue of section 1(3) of the LTA 1988.

(vii) Cobbetts did not inform Central Networks of the deadline of the end of May 2008 imposed by Sports Direct. Even now it is not clear what the reason for the deadline was given that the transaction under which OSC was sold by Sports Direct to JJB had completed on 28 January 2008.

(viii) While the transaction was not particularly complicated, it did require Central Networks to consider both the adverse financial position

* Editor's note: Also reported at [2003] 1 EGLR 133

** Editor's note: Also reported at [2005] 2 EGLR 42; [2005] 22 EG 134

A of OSC and JJB's position as guarantor. Furthermore, it also involved a change of use.

60. I therefore conclude that Gilesports executed the transfer in breach of covenant. It is common ground that it follows that the transfer was an excluded assignment within section 11(1) of the LTCA 1995, and accordingly Gilesports remains liable under the tenant's covenants in the sublease.

Issue 3: Effect of the transfer

61. As is common ground, the sublease was not registrable when it was first granted in 1998, because it was not for a term exceeding 21
B years and such leases were not then registrable: Land Registration Act 1925, section 19(2). The Land Registration Act 2002 ("LRA 2002") was brought into force on 13 October 2003. The assignment of the sublease by the transfer triggered the requirement of compulsory first registration, because it was a transfer of a qualifying estate (within LRA 2002, section 4(2)(b)) for valuable or other consideration: LRA 2002, section 4(1)(a)(i).

62. Section 6 of the LRA 2002 provides as follows:

> "**Duty to apply for registration of title**
>
> (1) If the requirement of registration applies, the responsible estate owner, or his successor in title, must, before the end of the period for registration, apply to
> C the registrar to be registered as the proprietor of the registrable estate.
>
> (2) If the requirement of registration applies because of section 4(1)(g)—
>
> (a) the registrable estate is the estate charged by the mortgage, and
>
> (b) the responsible estate owner is the owner of that estate.
>
> (3) If the requirement of registration applies otherwise than because of section 4(1)(g)—
>
> (a) the registrable estate is the estate which is transferred or granted, and
>
> (b) the responsible estate owner is the transferee or grantee of that estate.
>
> (4) The period for registration is 2 months beginning with the date on which the relevant event occurs, or such longer period as the registrar may provide under subsection (5).
>
> (5) If on the application of any interested person the registrar is satisfied that
> D there is good reason for doing so, he may by order provide that the period for registration ends on such later date as he may specify in the order.
>
> (6) Rules may make provision enabling the mortgagee under any mortgage falling within section 4(1)(g) to require the estate charged by the mortgage to be registered whether or not the mortgagor consents."

63. The obligation to register the transfer was on the "responsible estate owner": section 6(1). The responsible estate owner was the transferee ie OSC: section 6(3)(b). The period for registration was two months from 28 May 2008: section 6(4). As noted above, Halliwells did not register the transfer on behalf of OSC within that period. There has been no application under section 6(5).

E 64. Section 7 of the LRA 2002 provides so far as relevant:

> "**Effect of non-compliance with section 6**
>
> (1) If the requirement of registration is not complied with, the transfer, grant or creation becomes void as regards the transfer, grant or creation of a legal estate.
>
> (2) On the application of subsection (1)-
>
> (a) in a case falling within section 4(1)(a) or (b), the title to the legal estate reverts to the transferor who holds it on a bare trust for the transferee
>
> …"

65. It is common ground that, because the transfer was not registered,
F by virtue of section 7 of the LRA 2002 the assignment of the legal estate under the sublease became void and on 28 July 2008 the legal estate reverted to the transferor (Gilesports), who thereafter held it on a bare trust for the transferee (OSC).

66. The issue is as to the effect in these circumstances of section 3(2)(a) of the LTCA 1995. It is common ground that the sublease was a new tenancy for the purposes of the LTCA 1995, and so section 3 applies to it. Section 3 of the LTCA 1995 provides so far as relevant:

> "**Transmission of benefit and burden of covenants**
>
> (1) The benefit and burden of all landlord and tenant covenants of a tenancy—
>
> (a) shall be annexed and incident to the whole, and to each and every part, of the premises demised by the tenancy and of the reversion in them, and
>
> (b) shall in accordance with this section pass on an assignment of the whole G
> or any part of those premises or of the reversion in them.
>
> (2) Where the assignment is by the tenant under the tenancy, then as from the assignment the assignee—
>
> (a) becomes bound by the tenant covenants of the tenancy except to the extent that—
>
> (i) immediately before the assignment they did not bind the assignor, or
>
> (ii) they fall to be complied with in relation to any demised premises not comprised in the assignment;
>
> …"

67. E.ON contends that, if (contrary to my previous conclusions) the transfer was not in breach of the covenant not to assign without previous H
consent, then (i) from 28 May 2008 to 28 July 2008, OSC was the tenant under the sublease, (ii) on 28 July 2008 there was an assignment of the sublease from OSC back to Gilesports by virtue of section 7(2)(a) of the LRA 2002 and thus (iii) Gilesports became bound by the tenant's covenants by virtue of section 3(2)(a) of the LTCA 1995.

68. In answer to this, Gilesports relies upon section 28(1) of the LTCA 1995, which provides:

> "In this Act (unless the context otherwise requires)-
>
> 'assignment' includes equitable assignment and in addition (subject to section 11) assignment in breach of a covenant of a tenancy or by operation of law". J

69. Gilesports contends that (i) despite the transfer of the legal estate being void for non-registration, the transfer operated as an equitable assignment and hence an "assignment" within section 28(1), and thus (ii) OSC became bound by the tenant's covenants by virtue of section 3(2)(a) of the LTCA 1995.

70. E.ON replies that, even if this definition applies in the present context (which it disputes), this does not assist Gilesports, because section 28(1) also includes an assignment by operation of law within the definition of "assignment". E.ON contends that (i) reversion of the legal estate under section K
7(2)(a) of the LRA 2002 is an assignment by operation of law, and thus (ii) Gilesports became bound from 28 July 2008 as set out above.

71. While I have not found this question easy, I find E.ON's analysis more persuasive. In my view the transfer was not, on the date it was executed, an equitable assignment: it was an assignment of the entire legal and beneficial interest under the sublease from Gilesports to OSC, and OSC thus became the tenant. The effect of section 7 of the LRA 2002 was that, on 28 July 2008, the transfer of the legal estate became void and the legal estate reverted to Gilesports. In my view that amounts to an assignment of the legal estate by operation of law from the then tenant, OSC, to Gilesports. Accordingly, the assignee, L
Gilesports, became bound by the tenant's covenants by virtue of section 3(2)(a) of the LTCA 1995.

72. Counsel for Gilesports submitted that this result was unfair, since it imposed liability on the transferor as a result of a default on the part of the transferee. Counsel for E.ON gave three answers to that submission.

73. First, he submitted that the result was the same as in the pre-LTCA 1995 and pre-LRA 2002 decision of the Court of Appeal in *Brown & Root Technology Ltd* v *Sun Alliance & London Assurance Co Ltd* [2001] Ch 733. I do not regard that as a satisfactory answer, however. M

74. Secondly, he submitted that the result could only be regarded as unfair if the transferor was unable to register the transfer, as counsel for Gilesports suggested, but disputed that that was correct. It is not necessary for me to reach a conclusion on this point, but my provisional view is that counsel for Gilesports is correct. While I find it surprising that the transferor should not be able to register the transfer, that appears to be implied by section 6(6) of the LRA 2002.

75. Thirdly, he submitted that, even if the transferor was unable to register the transfer, the transferor did have a remedy, which was to apply to the registrar for an extension of time under section 6(5) of the LRA 2002 and to apply to the court for an order compelling the transferee to register the transfer. In my view this is the most satisfactory answer of

the three, short of an amendment to the LRA 2002 to make it clear that the transferor can register a transfer.

Issue 4: Section 17

76. Section 17 of the LTCA 1995 provides so far as relevant:

> "Restriction on liability of former tenant or his guarantor for rent or service charge etc
>
> (1) This section applies where a person ('the former tenant') is as a result of an assignment no longer a tenant under a tenancy but-
>
> (a) (in the case of a tenancy which is a new tenancy) he has under an authorised guarantee agreement guaranteed the performance by his assignee of a tenant covenant of the tenancy under which any fixed charge is payable; or
>
> (b) (in the case of any tenancy) he remains bound by such a covenant.
>
> (2) The former tenant shall not be liable under that agreement or (as the case may be) the covenant to pay any amount in respect of any fixed charge payable under the covenant unless, within the period of six months beginning with the date when the charge becomes due, the landlord serves on the former tenant a notice informing him-
>
> (a) that the charge is now due; and
>
> (b) that in respect of the charge the landlord intends to recover from the former tenant such amount as is specified in the notice and (where payable) interest calculated on such basis as is so specified."

77. If the transfer was made in breach of covenant (as I have concluded), Gilesports contends that section 17 applies so as to limit its liability to the amounts claimed in the section 17 notices served by Central Networks and E.ON. Accordingly, Gilesports disputes liability for the rent due under the sublease during the period between 28 May 2008 and 25 December 2009, the latter being the earliest date covered by the first section 17 notice.

78. E.ON contends that section 17 does not apply for similar reasons to those advanced by it in relation to the previous issue. Section 17(1) provides that section 17 applies where "a person ('the former tenant') is as a result of an assignment no longer a tenant under a tenancy". E.ON argues that, as from 28 July 2008, Gilesports was not the former tenant, but the present tenant, since the effect of section 7(2)(a) of the LRA 2002 was that the legal estate under the sublease reverted to it. I accept that argument.

79. It follows that it is not necessary for me to deal with E.ON's alternative claim in respect of the period from 28 July 2008 to 25 December 2009, which is for damages for breach of clause 3.15 of the headlease as imported into the sublease. As for the period from 28 May to 28 July 2008, Gilesports accepts that, in the light on my previous conclusions, it is liable to pay the rent for this period.

Conclusions

80. For the reasons given above, I conclude that:

(i) On the true construction of the sublease, Gilesports required Central Networks' consent to the transfer.

(ii) Section 1(3) of the LTA 1988 did not apply, and so Central Networks was not under a duty to give consent or reasons for refusal within a reasonable time; but if it was, a reasonable time had not expired by 28 May 2008 when the transfer was executed.

(iii) Even if the transfer was not in breach of covenant, by virtue of section 7(2)(a) of the LRA 2002 and section 3(2)(a) of the LTCA 1995 Gilesports was bound by the tenant's covenants in the sublease from 28 July 2008.

(iv) Section 17 of the LTCA 1995 does not apply so as to limit Gilesports' liability to the sums claimed in the section 17 notices.

Claim allowed.

Re Teathers Ltd (in liquidation); Baroque Investments Ltd v Heis and another

CHANCERY DIVISION

22 October 2012

SIR ANDREW MORRITT, C

[2012] EWHC 2886 (Ch)

Estates Gazette 15 December 2012

[2012] 50 EG 102

Landlord and tenant – Dilapidations – Damages – Landlord and Tenant Act 1927 – Liquidators of tenant company surrendering leases of commercial premises – Landlord submitting proofs of debt to liquidators in respect of dilapidations and alleged liability of tenant to reinstate demised premises – Whether debts provable – Whether liability on reinstatement obligation arising prior to end of lease term – Section 18(1) of 1927 Act – Whether claim for breach of covenant to keep in good repair properly taking into account rent-free period and reduced rent granted to new tenant

The claimant landlord demised the fifth floor of a commercial property to a tenant under four leases for terms expiring in March 2014. The tenant's obligations under the leases included: (i) a covenant to keep the demised premises in good and substantial repair and yield them up in such condition at the expiration or sooner determination of the term; and (ii) an obligation, contained in two licences for alterations granted by the claimant in December 2005, to remove the works of alteration and reinstate the premises "before the end of the lease". The tenant company went into creditors' voluntary winding up and the defendants were appointed as its liquidators. The defendants surrendered the four leases in November 2009 on terms that the landlord and tenant released each other from their respective rights and obligations under the leases and from all liability in respect of any breach of those rights and obligations arising on or after, but not before, the date of surrender. The claimant immediately re-let the premises for a term of four years at a rent of £706,550, but with an initial rent-free period of 19 months.

The claimant submitted various proofs of debt to the defendants for sums said to be owed under the leases. The defendants admitted claims in respect of rent, service charges and utilities, but disputed a claim of more than £1.212m for dilapidations. On an appeal against that decision, under r 4.83 of the Insolvency Rules 1986, a preliminary issue was tried on the extent to which the alleged dilapidations liability had been released by the terms of the surrenders. Questions arose as to: (i) whether the liability to reinstate, under the terms of the licences for alterations, arose before or after the date of the surrenders; and (ii) how damages for breach of the covenant to keep the premises in good repair were to be quantified, given that the landlord's loss was limited to the diminution in the value of the reversion by virtue of section 18(1) of the Landlord and Tenant Act 1927. The claimant's calculation of damages under the latter head sought to take into account the fact that it had given a rent-free period and reduced rent to the new tenant because of the disrepair.

Held: **The preliminary issue was decided in favour of the defendants. (1) Under the terms of the licence for alterations, the tenant was given the full period of the term created by the leases within which to carry out the requisite reinstatement. Accordingly, as at November 2009, there was no breach of the obligation to reinstate. Consequently, the potential liability under that obligation was released by the terms of the surrenders. (2) The covenant to repair imposed two distinct obligations: an obligation to keep the demised premises in good repair during the term and an obligation to yield them up in good repair at the expiration or sooner determination of the term. The latter obligation was released by the surrenders. As to the former, the normal date for**

the assessment of damages is the date of the breach; that was the day before the surrenders in November 2009. Although the usual rule may be displaced if some other date for assessment would more accurately reflect the loss sustained, that did not entitle the court to have regard to the later events of surrender and re-letting when assessing damages for breach of the covenant to keep the demised premises in good repair. When calculating such damages under section 18(1) of the 1927 Act, by reference to the diminution in the value of the reversion, it must necessarily be assumed that the lease to which the reversion is subject is continuing. A valuation of the reversion assumes that a purchaser will take that reversion subject to the lease, with the benefit and burden of all the covenants and other stipulations that it contains, for the remainder of the term. The calculation of damages requires a comparison between the value of the reversion with the property in its actual state and in its repaired state. The claimant's computation of its claim was therefore inconsistent with both the terms of section 18(1) and the fact that the covenant to yield up in good repair had been released by the surrenders. The defendants had correctly rejected the proof of debt so calculated.

The following cases are referred to in this report.

Bwllfa & Merthyr Dare Steam Collieries (1891) Ltd v *Pontypridd Waterworks Co* [1903] AC 426
County Personnel Ltd v *Pulver* [1987] 1 WLR 916, 926A
Ebbetts v *Conquest* (1900) 82 LT 560
Gagner Property Ltd v *Canturi Corporation Pty Ltd* [2009] NSWCA 413; (2009) 262 ALR 691
Gaze v *Holden* [1983] 1 EGLR 147; [1982] 266 EG 998
Hanson v *Newman* [1934] Ch 298
Matthey v *Curling* [1922] 2 AC 180, 240

This was the hearing of a preliminary issue on an appeal, under r 4.83 of the Insolvency Rules 1986, from a decision of the defendants, as liquidators of an insolvent company, rejecting proofs of debt submitted by the claimant landlord in respect of sums alleged to be owing under four leases of commercial premises.

Daniel Margolin (instructed by Eversheds LLP) appeared for the claimant; Mark Sefton (instructed by Nabarro LLP) represented the defendants.

Giving judgment, **Sir Andrew Morritt, C** said:

Introduction

1. The applicant, Baroque Investments Ltd ("Baroque"), is the owner of the freehold property situated and known as Beaufort House, 15 St Botolph Street, London, EC3. At the material time, the fifth floor was let to Landsbanki Securities UK Ltd, now Teathers Ltd ("Teathers"), on the terms of four leases dated respectively 31 July 1991, 7 May 1996, 30 July 1997 and 14 December 2005 made between them or their respective predecessors in title. Each lease was for a term of years expiring on 30 March 2014 and each contained a covenant on the part of the tenant in the terms or to the effect of the following:

> "to keep the demised premises... in good and substantial repair and condition... and in such repair and condition... to yield up the same at the expiration or sooner determination of the term".

2. On 14 December 2005, Baroque granted Teathers two licences to make certain alterations to the fifth floor, including the infill of an atrium area on terms, in each licence, that:

> "Before the end of the lease the tenant is to dismantle and remove the works and reinstate the premises to the same plan and design as before the carrying out of the works and as if the works had not been carried out (unless and to the extent that the landlord requests that it does not do so)."

3. On 23 October 2008, the defendants were appointed administrators of Teathers. On 8 October 2009, Teathers went into creditors voluntary winding up and the defendants were appointed joint liquidators ("the liquidators"). On 13 November 2009, Teathers, by its liquidators, surrendered each of the leases under which it held the fifth floor of Beaufort House. The terms of the surrender included the following:

> "[Baroque] and [Teathers] respectively release each other from the rights and obligations contained in the lease and from all liability in respect of any breach of those rights and obligations whether arising on or after, but not before, the date of this surrender."

On the same day, Baroque relet the fifth floor to Instant Offices Management LLP for a term of four years at an annual rent of £706,550, but with an initial rent-free period of 19 months.

4. Thereafter Baroque submitted various proofs of debt to the liquidators of which the highest was for £7,013,964. By a notice dated 30 November 2011, the liquidators rejected them. On 19 December 2011, Baroque applied under the Insolvency Rules 1986 r4.83 for an order to vary that decision so as to admit to proof £3,844,831. Subsequently agreement was reached between Baroque and the liquidators for the admission of the claim in respect of rent, service charges and electricity, leaving outstanding a dilapidations claim of £1,212,395 and a minor issue with which I am not concerned. That claim has two components. The first is the liability to reinstate as provided in the licences referred to in [2] above. The second is the liability to keep the demised premises... in good and substantial repair and condition as provided in the repairing covenant contained in each lease and quoted in [1] above. The issue between the parties is whether those liabilities were released by the terms of the surrender quoted in [3] above.

5. To facilitate the determination of that issue, on 15 June 2012, Registrar Derrett directed the trial of a preliminary issue in the following terms:

> "Whether (or to what extent) the liability of Teathers Ltd (in liquidation) averred in paras 2.1 to 3.7 of the second witness statement of Jeremy Howard Grey dated 23 April 2012 is a liability which was released by clause 2.1 of the deed of surrender dated 13 November 2009, entered into between Baroque Investments Ltd and Teathers Ltd (in liquidation), and by para 2.4 of section 11 of the Land Registry Form TR1 entered into between the same parties on the same date."

I will consider the reinstatement and repair liabilities separately and in that order.

Reinstatement liability

6. The obligation is "Before the end of the lease... to dismantle and remove the Works and reinstate the premises...". Counsel for Baroque accepts that he cannot point to a particular time when the liability arose, but he contends that the liability to reinstate must, in accordance with its terms, have arisen before, not on or after the date of the surrenders. Accordingly, as he submits, the liability was not released by the surrenders. Consequently, in relation to that liability Baroque is entitled to prove in the liquidation of Teathers.

7. Counsel for the liquidators disputes this construction. He points out that, absent the words "before the end of the lease..." the tenant would have a reasonable time after its termination to carry out the reinstatement works, see *Matthey* v *Curling* [1922] 2 AC 180, 240. The inclusion of these words in the licences was evidently designed to limit the time within which the tenant might carry out the works to the term of the lease. Accordingly, so he submits, there was no breach of that obligation at the time of the surrenders on 13 November 2009 because the term of the leases did not expire until 30 March 2014.

8. I prefer the submissions of counsel for the liquidators. The licences gave to the tenant the full period of the term created by the leases within which to carry out the requisite reinstatement. Accordingly, as at 13 November 2009, there was no breach of the obligation to reinstate. If the works were not carried out before 30 March 2014 then the tenant would have been in breach but that would have been after, not before, the date of the surrenders. In my view, the inescapable consequence is that that potential liability was released by the surrenders. It follows that there can be no provable liability under this head.

The repairing liability

9. It is common ground that, as held in *Ebbetts* v *Conquest* (1900) 82 LT 560, the covenant to repair quoted in [1] above imposes two distinct obligations. The first is to keep the demised premises in good repair during the term. The second is to yield up the demised premises

in good repair at the expiration or sooner determination of the term. It is not disputed that the second was released by the surrenders. The issue between the parties is the extent of the provable liability of Teathers under the first.

10. Paras 2.1 to 3.7 of the witness statement of Mr Grey referred to in the preliminary issue explain how the claim for £1,212,395 is made up. (Seemingly an unquantified part of it relates to the liability to reinstate. To that extent there is some duplication.) In those paragraphs Mr Grey explains that in order to mitigate Baroque's loss it immediately relet the fifth floor of Beaufort House to Instant Offices Management LLP, but due to the poor state of repair it was obliged to give the tenant a rent-free period of 19 months and accept a reduced rent for the remainder of the term. He attributes 15 of the 19 months of the rent-free period to the want of repair and reinstatement and quantifies it as £882,812 (para 3.3). In respect of the rent reduction, he quantifies the loss incurred by Baroque as £329,583 (para 3.7).

11. The liquidators accept that, whether or not that was a proper way to quantify losses arising from a breach of the covenant to yield up the demised premises in good repair, that liability was released by the surrenders. In the case of the covenant to keep in good repair, the measure of the landlord's loss is limited by section 18(1) of the Landlord and Tenant Act 1927. That subsection, so far as material, provides:

> "Damages for a breach of a covenant or agreement to keep… premises in repair during the currency of a lease… shall in no case exceed the amount (if any) by which the value of the reversion (whether immediate or not) in the premises is diminished owing to the breach of such covenant or agreement as aforesaid…"

Accordingly, as the liquidators maintain, the calculation of the damages necessarily involves the assumption that the lease to which the reversion is subject is continuing. Such claims are relatively rare because their availability is restricted by section 1(2) of the Leasehold Property (Repairs) Act 1938, as amended, to, in effect, the last three years of the term.

12. Counsel for Baroque accepts that the normal date for the assessment of damages is the date of the breach, namely the day before the surrenders on 13 November 2009. He points out that that rule may be displaced if some other date for assessment would more accurately reflect the loss sustained, see *County Personnel Ltd* v *Pulver* [1987] 1 WLR 916, 926A. In addition, he contends that, in accordance with the principle established in *Bwllfa & Merthyr Dare Steam Collieries (1891) Ltd* v *Pontypridd Waterworks Co* [1903] AC 426, the court is entitled to have regard to the facts that occurred after the relevant date rather than speculate.

13. Counsel for Baroque relies on the decision of the Court of Appeal of New South Wales in *Gagner Property Ltd* v *Canturi Corporation Pty Ltd* [2009] NSWCA 413; (2009) 262 ALR 691. It is sufficient to refer to [54] of the judgment of Campbell JA, with whom Macfarlan JA and Sackville AJA agreed. Having reviewed a number of authorities on the compensatory purposes of damages, he said:

> "It follows that, even though a cause of action for breach of contract has accrued at the time the breach occurs, it cannot now be said that there is an accrued right at that time to receive any particular sum of damages. That is because it must await the trial to decide what is the most appropriate way, in light of events then known, to give effect to the compensatory principle of damages."

In the light of these principles, counsel for Baroque contends that the court is not bound to assume the continuation of the leases in assessing the amount of damage to the reversion but is entitled, indeed bound, to have regard to the subsequent events of surrender and reletting.

14. This is disputed by counsel for the liquidators. He submits that the terms of section 18(1) predicate the continuation of the lease in the valuation of the damage to the reversion during its currency. He relies on the decision of the Court of Appeal in *Hanson* v *Newman* [1934] Ch 298, in which Romer LJ said, at p306:

> "Whatever may be the strict meaning at law of the word 'reversion', there can, I think, be no doubt as to the meaning that that word bears in the section of the Landlord and Tenant Act, 1927, with which we are concerned here. The reversion in that section, if it be a freehold reversion, means the freehold subject to the lease, and valuing the reversion for the purposes of the section you must value the freehold subject to so much, if any, of the term of the lease as remains in existence."

In that case, a similar covenant to yield up in good repair was also relied on and the Court of Appeal took account of the fact that the tenancy was forfeited the following day. In the instant case any such liability to yield up in good repair has been released.

15. Counsel for the liquidators contends that in this case the computation of the claim is inconsistent with both the terms of section 18(1) and the fact of the release of the covenant to yield up in good repair. In relation to the *Bwllfa* principle, he submits that it cannot apply so as to undermine the express terms of section 18(1) requiring a comparison of the value of the reversion with the property in its actual state and in its repaired state. He relies on the decision of HH Judge Finlay QC in *Gaze* v *Holden* [1983] 1 EGLR 147* to the effect that the *Bwllfa* principle does not apply to valuations as at a particular date.

16. Once again, I prefer the submissions of counsel for the liquidators. It is clear from Mr Grey's description of how he arrived at the figure of £1,212,395 that he did not follow the processes required by section 18(1). He made no attempt to value the reversion to the fifth floor of Beaufort House in either its actual or repaired state but assumed that the difference between those valuations would be equal to the loss sustained by Baroque on reletting the fifth floor to Instant Office Management Ltd. It may or may not be. Accordingly, the liquidators were right to reject the proof of debt so calculated.

17. That is not to say that there may not be some provable debt if the proper processes are followed. Given the argument I have heard it is appropriate for me to say what they are. First, section 18(1) imposes a limit on what may be recovered to "the amount (if any) by which the value of the reversion is diminished owing to the breach of" the covenant to keep the demised property in repair. Second, the ascertainment of that amount necessarily requires the valuation of the reversion to the property in its actual state and in its repaired state. Third, a valuation of the reversion necessarily assumes that a purchaser will take it subject to the lease with the benefit and burden of all the covenants and other stipulations it contains for the remainder of the term.

18. Given those statutory requirements, there is no room for the *Bwllfa* principle to be applied in the manner for which Baroque contends. In addition, the date as at which the valuations are to be made, whether the date of breach or the date of judgment cannot alter the requirements imposed by section 18(1). As the editors of *Dilapidations, the Modern Law and Practice* 3rd ed, observe in para 28.21:

> "In these circumstances it may be difficult to prove that the effect of the disrepair has been to cause any diminution in the value of the reversion."

Finally, by insisting that the debt for which Baroque seeks to prove is calculated in the manner section 18(1) requires the liquidators are not relying on an "accrued right", such as Campbell JA referred to in the passage from his judgment in *Gagner Property* referred to in [13] above. They are relying on their statutory right to have the damage sustained in consequence of the breach of the covenant to keep in good repair limited in the way section 18(1) requires.

Summary

19. In these circumstances I will declare that:

(1) Baroque is not entitled to prove for damage sustained in consequence of the failure of Teathers to reinstate the fifth floor of Beaufort House as required by the Licences; and

(2) The alleged debt of £1,212,395 referred to by Mr Grey in his witness statement identified in the preliminary issue has not been calculated in accordance with section 18(1) of the Landlord and Tenant Act 1927 and, in consequence, is not provable.

I invite counsel to consider what other orders or directions I should make and to produce a minute my order.

The preliminary ruling was made in favour of the defendants.

---

* Editor's note: Also reported at [1982] 266 EG 998

LEASEHOLD ENFRANCHISEMENT

Day and another v Hosebay Ltd; Howard de Walden Estates Ltd v Lexgorge Ltd

SUPREME COURT

10 October 2012

LORD PHILLIPS, LORD WALKER, LORD MANCE, LORD CLARKE, LORD WILSON, LORD SUMPTION AND LORD CARNWATH

[2012] UKSC 41

Estates Gazette 27 October 2012

[2012] 43 EG 116

Leasehold enfranchisement – Leasehold Reform Act 1967 – Enfranchisement of house – Terraced properties originally built as single residences – Some properties furnished and let as bedsits for short-term holiday lets – One property used entirely as office accommodation – Whether properties qualifying as "houses" for purposes of section 2(1) of 1967 Act – Whether leaseholders entitled to enfranchise

The two conjoined appeals concerned claims by the respondents, as long leaseholders of certain properties, to acquire the freehold of those buildings from the appellants pursuant to the leasehold enfranchisement provisions of the Leasehold Reform Act 1967. In the courts below, the claims were allowed on the ground that each of the properties was a "house" qualifying for enfranchisement within the definition in section 2(1) of the 1967 Act. The Court of Appeal, upholding the decision of the county court, held that each property fulfilled both limbs of the section 2(1) definition since each was a building that had originally been designed and subsequently adapted for living in and could reasonably be called a house: see [2010] EWCA Civ 748; [2010] 3 EGLR 66; [2010] 38 EG 106.

The buildings in question were all large terraced properties that had originally been constructed and occupied as houses. The first appeal related to three properties in South Kensington, London SW7; the leases of two of the properties provided that they were to be used as 16 high-class, self-contained, private residential flatlets, while the lease of the third specified use as a single family residence or high-class furnished property for up to 20 occupiers. In fact, the individual rooms in all three properties had been converted into self-catering units, each with bathroom and cooking facilities, and were let as short-term accommodation for tourists and other visitors to London.

The second appeal related to a property in Marylebone, London W1; the lease described the demised premises as a "messuage or residential or professional premises" and stipulated use as professional offices on the lower two floors and self-contained flats or maisonettes on the upper two, with associated storage in the basement. In fact, all four upper floors had long been used as professional offices.

Held: **The appeals were allowed. (1) The 1967 Act is not intended to confer statutory rights on lessees of buildings that are used for purely non-residential purposes; it is concerned with houses as places to live in. In that context, it would be inappropriate to take a literalist approach to the definition in section 2(1), treating the words "designed" and "adapted" for living in as alternative qualifying requirements. The fact that a building was originally designed for living in is not sufficient to bring it within the first part of the definition regardless of any subsequent adaptation to other uses. Instead, the first part of the definition should be regarded as a simple way of defining the present identity or function of a building as a house, by reference to its current physical character, regardless of whether that character was derived from its original design or from subsequent adaptation. The word "adapted" in that context means no more than "made suitable" and, although it applies to the building itself rather than its contents, it does not imply any particular degree of structural change. Where a building is in active and settled use for a particular purpose, it is likely to have undergone at least some physical adaptation to make it suitable for that purpose. In most cases, that purpose may be taken as the use for which the building is currently "adapted" and in most case it will be unnecessary to look further. The second part of the definition, asking whether the property is a house "reasonably so called", ties the definition to the primary meaning of "house" as a single residence, as opposed, for example, to a hostel or block of flats. (2) None of the appeal properties qualified for enfranchisement. A property that was wholly used as a self-catering hotel was not a house "reasonably so called" within the meaning of the 1967 Act; nor was a building that was wholly used for offices, whatever its original design or current appearance. The fact that the buildings looked like houses, and might be referred to as houses for some purposes, was insufficient to displace the fact that their use was entirely commercial.**

The following cases are referred to in this report.

Ashbridge Investments Ltd v *Minister of Housing and Local Government* [1965] 1 WLR 1320

Boss Holdings Ltd v *Grosvenor West End Properties Ltd* [2008] UKHL 5; [2008] 1 WLR 289; [2008] 1 EGLR 51; [2008] 15 EG 174

Cadogan v *McGirk* [1996] 4 All ER 643; [1996] 2 EGLR 75; [1996] 39 EG 175

Moyna v *Secretary of State for Work and Pensions* [2003] UKHL 44; [2003] 1 WLR 1929

Prospect Estates Ltd v *Grosvenor Estate Belgravia* [2008] EWCA Civ 1281; [2009] 1 WLR 1313; [2009] 1 EGLR 47; [2009] 02 EG 86

Tandon v *Trustees of Spurgeons Homes* [1982] AC 755; [1981] 1 EGLR 66; [1981] 258 EG 553

These were conjoined appeals by Hugo Day and Lady Hilary Greenslade Day, to which Hosebay Ltd was the respondent; and by Howard de Walden Estates Ltd, to which Lexgorge Ltd was the respondent, from two judgments of the Court of Appeal upholding decisions, by HH Judge Dight and HH Judge Hazel Marshall QC respectively, to allow the respondents' claims to acquire the freehold of certain properties under the leasehold enfranchisement provisions of the Leasehold Reform Act 1967.

Edwin Johnson QC and Oliver Phillips (instructed by Pemberton Greenish LLP) appeared for the appellants in the first appeal; Stephen Jourdan QC and Anthony Radevsky (instructed by Bircham Dyson Bell LLP) appeared for the respondents in the first appeal; Jonathan Gaunt QC and Katharine Holland QC (instructed by Speechly Bircham LLP) appeared for the appellant in the second appeal; Anthony Radevsky and Mark Sefton (instructed by Wallace LLP) represented the respondent in the second appeal.

Giving the judgment of the court, **Lord Carnwath** said:

Introduction

1. The Leasehold Reform Act 1967 is on its face a statute about houses, not commercial buildings. The buildings with which we are concerned were originally designed and used as houses, but at the relevant date were used entirely for commercial purposes, one for offices, the other (in the judge's words) as a "self-catering hotel". In both cases the courts below felt constrained to hold that they were "houses" within the meaning of the 1967 Act, with the consequence that the lessees were entitled to "enfranchise", that is, to acquire the freeholds compulsorily from their lessors on the terms fixed by the Act.

2. In the Court of Appeal [2010] EWCA Civ 748; [2010] 1 WLR 2317, Lord Neuberger MR regretted this result. He saw it as the probably unintended consequence of amendments made by the Commonhold and Leasehold Reform Act 2002, removing the previous residence requirements. However, he felt bound to apply his view of the relevant provisions as they stood after those amendments, rather than to decide what "the legislature would have said if it had fully appreciated the consequences…" [57].

3. From the material we have been shown, he was clearly right to think that his interpretation did not reflect Parliament's intentions. The thinking behind the 2002 legislation is apparent from the preceding Draft Bill and Consultation Paper "Commonhold and Leasehold Reform" (Cm 4843), published by the Lord Chancellor in 2000. It included proposals for the introduction of an entirely new form of tenure, known as "commonhold", and for amendment of the existing provisions relating to leases of flats (under the Leasehold Reform, Housing and Urban Development Act 1993) and of houses (under the 1967 Act). The first paragraph of the introduction leaves no doubt that its purpose was to address perceived flaws in the "residential leasehold system" (p107), not in the leasehold system more generally.

4. In relation to flats, the government's view was that the residence tests under the 1993 Act were too restrictive, for example, in excluding someone subletting a flat, or occupying a flat as a second home. The residence requirement would therefore be abolished; but, to "restrict the scope for short-term speculative gains", it would be replaced by a rule requiring the qualifying tenant to have held the lease for at least two years (pp155-156).

5. A similar approach was proposed for leases of houses under the 1967 Act, at p189:

> "This would bring the residence test for houses in line with the proposals for flats. It would allow long leaseholders of second homes to benefit and would also enable leaseholders who lease houses through a company to enfranchise. Furthermore, as in the case of flats, it would restrict the scope for short-term speculative gains."

There is no evidence then or thereafter of any ministerial or parliamentary intention to extend the scope of the Act more generally, or in particular to confer statutory rights on lessees of buildings used for purely non-residential purposes.

6. Although the 1967 Act, like the 1993 Act, is in a sense expropriatory, in that it confers rights on lessees to acquire rights compulsorily from their lessors, this has been held not to give rise to any interpretative presumption in favour of the latter. As Millett LJ said of the 1993 Act in *Cadogan* v *McGirk* [1996] 4 All ER 643*, at p648a-b:

> "It would, in my opinion, be wrong to disregard the fact that, while the 1993 Act may to some extent be regarded as expropriatory of the landlord's interest, nevertheless it was passed for the benefit of tenants. It is the duty of the court to construe the 1993 Act fairly and with a view, if possible, to making it effective to confer on tenants those advantages which Parliament must have intended them to enjoy."

By the same token, the court should avoid as far as possible an interpretation which has the effect of conferring rights going beyond those which Parliament intended.

Statutory definition

7. Section 2(1) defines "house" in the following terms:

> "… 'house' includes any building designed or adapted for living in and reasonably so called, notwithstanding that the building is not structurally detached, or was or is not solely designed or adapted for living in, or is divided horizontally into flats or maisonettes; and-
>
> (a) where a building is divided horizontally, the flats or other units into which it is so divided are not separate 'houses', though the building as a whole may be; and
>
> (b) where a building is divided vertically the building as a whole is not a 'house' though any of the units into which it is divided may be."

8. In the present cases, nothing turns directly on the qualifications introduced by the word "notwithstanding" (which I shall refer to as "the proviso"). We are concerned with the main part of the definition, which raises two separate but overlapping questions: (i) is the building one "designed or adapted for living in"? (ii) is it a "house … reasonably so called"? Both questions remain live in *Hosebay*; in *Lexgorge* the first has been conceded in favour of the lessees.

9. The two parts of the definition are in a sense "belt and braces": complementary and overlapping, but both needing to be satisfied. The first looks to the identity or function of the building based on its physical characteristics. The second ties the definition to the primary meaning of "house" as a single residence, as opposed to say a hostel or a block of flats; but that in turn is qualified by the specific provision relating to houses divided horizontally. Both parts need to be read in the context of a statute which is about houses as places to live in, not about houses as pieces of architecture, or features in a street scene, or names in an address book.

The facts

10. The first case (*Hosebay*) concerns three properties, 29, 31, and 39 Rosary Gardens, South Kensington, London SW7. They were originally built as separate houses as part of a late Victorian terrace forming the west side of Rosary Gardens. The current leases of Nos 29 and 39 were granted in 1966 for terms expiring in December 2020, subject to covenants for their use as "16 high-class self-contained private residential flatlets". The current lease for No 31 was granted in 1971 for a term expiring in December 2030, subject to a covenant restricting its use to that of "a single family residence" or "a high-class furnished property for accommodating not more than 20 persons". It was common ground that the current use, which had begun some time before 1981, was not in accordance with the covenants.

11. It was unclear from the evidence when the premises had been converted to their present layout. The judge, at [83], proceeded on the basis that the conversions "may well" have been carried out substantially before the current leases were granted in 1966 and 1971. Although there was no evidence as to the actual purpose of the conversions, the Master of the Rolls "on the balance of probabilities" inferred (principally from the lack of documentation in the hands of the landlords to indicate otherwise) that they had been for the uses described in the leases [37].

12. Hosebay Ltd acquired all three leases in 1996. On 23 April 2007, it served notices on its landlords under section 8 of the 1967 Act to acquire the freeholds of the three properties.

13. Judge Marshall QC found that the three properties were at the relevant date being used together to provide "short-term accommodation for tourists and other visitors to London", or what she described as a "self-catering hotel" [8] and [19]. Each of the three properties had been "fully adapted to provide individual rooms for letting out" [9], with the exception of two rooms in No 31, one of which was used for office and reception purposes, and the other for storage. The great majority of the rooms could be described as "rooms with self-catering facilities". Each room had between one and four beds, furniture, and limited storage space, cooking facilities, and small "wet rooms" with shower, basin and

---

* Editor's note: Also reported at [1996] 2 EGLR 75; [1996] 39 EG 175

WC. Fresh bed linen and room cleaning, but no other services, were provided to those staying in the rooms.

14. On these facts, the judge concluded that each of the three properties was physically "adapted for living in" even though the current use was itself too transient to qualify as such. The Court of Appeal agreed. I quote the Master of the Rolls:

> "33. My primary reason for that conclusion is that, in order to determine whether premises are adapted for living in, one looks at the most recent works of adaptation, and assesses objectively, whether they resulted in the property being adapted for living in.
>
> ...
>
> 36. ... In this case, I consider that the effect of the most recent works of conversion to the three properties, if they were works of adaptation, adapted those properties for living in. Ignoring one or two rooms, each room in the three properties is a self-contained unit of accommodation, with its own basic small shower room/WC, and its own even smaller and more basic cooking facilities. As Moore-Bick LJ pointed out in argument, the rooms are entirely appropriate for letting to students on three year degree courses, and, as Mr Johnson rightly accepted, if they had been, all the rooms, and therefore the three buildings, would have been used for living in."

Even if, as Mr Johnson QC argued and I am prepared to assume without deciding, the current use of the three properties is not for living in, that certainly does not mean that, viewed objectively, the three properties were not adapted for living in."

15. The judge and the Court of Appeal held also that the properties were "houses reasonably so called", as the Master of the Rolls explained, at [38]:

> "externally, each of the three properties has the appearance of being a relatively large town house; internally, each of the three properties has been converted so that almost every room can be used as a self-contained unit for one or more individuals, with cooking and toilet facilities. ... I find it hard to see how the judge could be faulted for concluding that, even if each of the three properties might be called something else as well, they could each reasonably be called a house."

16. The other case (*Lexgorge*) relates to 48 Queen Anne Street, in Marylebone, London W1. It was built in the early 18th century as a house comprising five floors including basement, in a terrace of substantial houses. It was occupied for that purpose for many years until 1888, when it began to be used for commercial purposes.

17. Coming to more recent times, planning permission was granted in December 1949 for conversion of the second and third floors into a self-contained maisonette, and there is some evidence that it was implemented. However, from about 1961, all four upper floors were used as offices, and they were so used when the notice was served under the Act on 4 March 2005. The whole building was still in office use in June 2005. However, by the time of the trial in October 2009, when the judge inspected the property, the upper two floors were in use for residential purposes. The office use of the lower floors continued.

18. The current lease was granted in 1951 for a term of 110 years. The lease described the property as a "messuage or residential and professional premises", and restricted its use (subject to landlords' consent) to "self-contained flats or maisonettes" on the upper two floors, professional offices on the first and ground floors, and in the basement "storage ... and lavatory ... in connection with other parts of the demised premises". In 1978, the lease was acquired by Lexgorge Ltd. At the time of the notice the office use of all floors had become "established", and therefore lawful for planning purposes, although in breach of the lease as respects the upper floors. The building is listed as a building of special architectural or historic interest (grade 2); English Heritage's records describe it as a "terraced house".

19. In this case, as already noted, it is conceded by the lessors that at the material date the premises, although used for offices, were still at least in part "designed or adapted for living in". It was held by the judge (Judge Dight) and by the Court of Appeal that it was a "house reasonably so called", and therefore within the definition. The Master of the Rolls said:

> "53. If the upper two floors of the property had been empty, I have little doubt but that the property could reasonably have been called a house, bearing in mind its external character and appearance (a classic town house in London's West End), its internal character and appearance at least on the upper two floors (which were, as I understand it, substantially as constructed), the description of the property in the lease as 'messuage or residential or professional premises', and, to the extent that it is relevant, the terms of the lease (restricting the use of the upper two floors to residential). I find it hard to see why the fact that the upper two floors had been used (even for many years) as offices (in contravention of the terms of the lease) should wreak such a change that the property could no longer reasonably be called a house."

The authorities

20. The first relevant case under the Act was *Lake* v *Bennett* [1970] 1 QB 663. However, I find it helpful to start from an authority in a different statutory context, Lord Denning MR's judgment in *Ashbridge Investments Ltd* v *Minister of Housing and Local Government* [1965] 1 WLR 1320. The case related to compulsory acquisition of two properties for the purpose of slum-clearance under the Housing Acts. The level of compensation would vary significantly depending on whether the property was or was not a "house". In the absence of a statutory definition of "house", Lord Denning adopted the following formula: "a building which is constructed or adapted for use as, or for the purposes of, a dwelling" (p1324). In *Lake*, he suggested that the draftsman of the 1967 Act definition had "adopted these words, but added the limitation 'reasonably so called'" (p670G).

21. *Ashbridge* itself concerned two adjoining buildings in the same terrace, which had been designated for compulsory purchase, the first (No 17) as an unfit "house", the second (No 19) as a building other than a house. The buildings were very similar in appearance; both had been designed as shops with rear living rooms and living quarters above, but neither was in current use for living purposes. No 17, which had undergone no structural alterations, was held by the minister to have "retained its identity as a dwelling". No 19, by contrast, was held to have "lost its identity as a dwelling", following structural alterations involving the extension of the shop into the rear living area (p1325). The latter decision was described in the Court of Appeal as "extraordinary" (p1327F, per Harman LJ), but that did not undermine the validity of the decision in relation to No 17. Lord Denning's formula can be seen as his way of expressing the present "identity" (in the inspector's words), or perhaps function, of a building not currently in use, defined by reference to the purpose of its construction or subsequent adaptation.

22. *Lake* itself concerned a three-storey house, the ground floor of which had been converted into a shop. There was no issue as to the first part of the definition, as it was clear that the building was at least in part adapted for use for living in. The Court of Appeal held that notwithstanding the commercial element, the building as a whole was a house "reasonably so called" and was therefore within the scope of the 1967 Act.

23. The reasoning of *Lake* was adopted and extended by the House of Lords in *Tandon* v *Trustees of Spurgeons Homes* [1982] AC 755**, which remains the leading House of Lords authority on this part of the definition. Unfortunately, the reasoning of the single majority speech of Lord Roskill, although carrying the unqualified support of Lord Scarman and Lord Bridge, is not without difficulty. Further, the case needs to be read in its factual context. As in *Lake*, the main problem was to reconcile the statutory recognition (under the "proviso") that the building need not be "solely" designed or adapted for living in, with the need for the building as a whole to be a house "reasonably so called". This is not a problem in the present cases.

24. At the end of his judgment Lord Roskill referred with approval to *Lake*, which, at p767B-C, he welcomed as "stating a principle and [confining] the question of fact to a narrow area", and from which he "deduced" the following three "propositions of law" :

> "(1) as long as a building of mixed use can reasonably be called a house, it is within the statutory meaning of 'house', even though it may also reasonably be called something else; (2) it is a question of law whether it is reasonable to

---

** Editor's note: Also reported at [1981] 1 EGLR 66; [1981] 258 EG 553

call a building a 'house'; (3) if the building is designed or adapted for living in, by which, as is plain from section 1(1) of the Act of 1967, is meant designed or adapted for occupation as a residence, only exceptional circumstances, which I find it hard to envisage, would justify a judge in holding that it could not reasonably be called a house. They would have to be such that nobody could reasonably call the building a house."

25. Although expressed as "propositions of law", they do not in my view offer much assistance as such, at least beyond the facts of the case. The first proposition was in terms directed to a building in mixed residential and commercial use. Such a building could plausibly be described either as a house with a shop below, or as a shop with a dwelling above. That was enough to show that it could "reasonably" be called a house. That proposition cannot in my view be applied more generally. The mere fact that a building may be described as a "house" for other purposes (for example, in the English Heritage list) is not enough to bring it within this part of the definition.

26. The second proposition, that what is a house "reasonably so called" is a question of law, is not easy to extract from the judgments in *Lake*. Lord Denning described the judge's negative answer to that question as "an inference from primary facts" depending "in part at least on the true interpretation of the words "reasonably so called", and one with which the court could interfere if it was a conclusion to which the judge "could not reasonably come" ([1970] 1 QB 663, 671F). Salmon LJ described it as "partly a question of fact but also a question of law as to the true construction of the meaning of the word 'house' in this Act." (p672F). Elsewhere, Lord Roskill himself had accepted counsel's submission that the definition of "house" was "a mixed question of fact and law" ([1982] AC 755, 765E), but he saw it as one in which, in the interests of consistency, the question of fact should be confined "within narrow limits": p767A. More modern authorities have leant against such conceptual debates (see, for example, Lord Hoffmann, in *Moyna* v *Secretary of State for Work and Pensions* [2003] UKHL 44; [2003] 1 WLR 1929, [25]-[27]). In any event, none of these formulations throws much light on how the question should be answered in any particular case.

27. The third proposition is again in terms hard to extract from *Lake*. Lord Denning described the case before them as a "typical case", but thought that difficult issues might arise in other cases: [1970] 1 QB 663, 671. He did not suggest that, in such cases, an affirmative answer to the first question would lead to any presumption in respect of the second. The examples given in the judgments (pp671-672) of cases that would not satisfy the second test – a block of flats, the Ritz Hotel or Rowton House (a working men's hostel) – can hardly be described as "exceptional". Rather than a free-standing proposition of law, deduced from *Lake*, this proposition seems more an expression of Lord Roskill's own view as to the correct policy approach to a building of the kind before him, which was adapted at least in part for occupation as a residence.

28. It may be that the real difference between the majority and the minority in *Tandon* came down to one of policy. Lord Wilberforce (in the minority) thought it clear that the building could not reasonably be called a "house"; it was rather a "mixed unit consisting in part of a shop and in part of a dwelling", and as such was not within the policy of the Act: [1982] AC 755, 760G. For Lord Roskill (in the majority), Parliament had made clear that such mixed units were not in principle to be excluded. He noted that such small shops combined with living accommodation were a familiar feature of towns and villages across the country (p766). In this he echoed the view of Salmon LJ (*Lake*, p672G), who thought that a tenant living above a shop in the circumstances of that case was "obviously the sort of person to whom the legislature intended to give security of tenure". Such policy considerations do not assist the lessees in this case. For the reasons already given, policy if anything points the other way.

29. Of more significance for present purposes is the relative lack of weight given by the majority to the appearance of the buildings as a factor in answering the second question. Lord Fraser of Tullybelton (in the minority) had regarded appearance as "the main element in the character of a building": [1982] AC 755, 762B. He attached particular weight to the photograph which showed "a shop in a row of shops", in contrast with the "converted house" in *Lake*; to him it was "obvious from the photograph" that the building could not reasonably be called a house (p763A). That, however, was not the approach of the majority. Lord Roskill had apparently accepted that in determining the "character" of the building for these purposes, physical appearance could be relevant, as also its history and the terms of the lease (p766B). However, those factors played no detectable part in the final decision. The determinative points were that the proportion of residential use, even if only 25%, was "substantial" (p766H), and that a tenant occupying such a building as his residence was within what was perceived to be the scope of the protection intended by Parliament (p766). Those factors were enough to bring the case within the "principle" established by *Lake* notwithstanding the differences from that case in relation to the original design and physical appearance of the respective buildings.

30. The only other relevant authority at the highest level is the much more recent decision in *Boss Holdings Ltd* v *Grosvenor West End Properties Ltd* [2008] UKHL 5; [2008] 1 WLR 289†. The House of Lords held that a building previously designed or adapted for living in remained a house, even though at the material time it was not only disused but in parts "stripped out to the basic structural shell" [24]. In contrast to *Tandon*, this case was concerned solely with the first question. It was not in dispute that if that question was answered in the affirmative the building qualified as a house "reasonably so called".

31. As will be seen, I do not regard the case as determinative in either of the present appeals. However, some comment is desirable, in view of the change of view of Lord Neuberger on one aspect of his leading speech. He had proposed the following grammatical analysis of the relevant words of the statutory definition:

"18. In my judgment, the words 'designed or adapted for living in', as a matter of ordinary English, require one first to consider the property as it was initially built: for what purpose was it originally designed? That is the natural meaning of the word 'designed', which is a past participle. One then goes on to consider whether work has subsequently been done to the property so that the original 'design' has been changed: has it been adapted for another purpose, and if so what purpose? When asking either question, one is ultimately concerned to decide whether the purpose for which the property has been designed or adapted, was 'for living in'.

19. The notion that the word 'designed' in section 2(1) is concerned with the past is reinforced by the later words in the same section 'was or is solely designed or adapted'. The use of the past tense is striking in a section which contains a number of verbs only in the present tense. In my judgment, the expression is to be construed distributively: thus, the word 'was' governs 'designed', and the word 'is' governs 'adapted'. The present tense is appropriate for 'adapted' because, as Lord Scott of Foscote pointed out in argument, there could have been several successive adaptations, and it is only the most recent which is relevant. The word 'was' is in any event difficult to reconcile with Grosvenor's case (as accepted by the judge and the Court of Appeal), as it would be irrelevant whether the property could have been fit for residential occupation at any time in the past."

32. Later in his speech, he considered the implications of this analysis for other cases, including how the definition should apply to a property which had been designed for living in, but had subsequently been adapted to another use. As a matter of "literal language", he thought such a property would be within the definition. If, as appeared, "designed" and "adapted" were alternative qualifying requirements, a building which had been designed as a house would remain within the definition in spite of its adaptation to other uses. Such a conclusion, he accepted, might seem surprising, but it could have been more readily understandable when taken with the residence requirement in the original Act [26].

33. It was on this latter point that, as Master of the Rolls in the present case, he has had second thoughts. It had been put directly in issue by the tenants in *Hosebay*, who argued (as they have in this court) that because the buildings were originally designed for "living in", that was sufficient to bring them within the definition, regardless

† Editor's note: Also reported at [2008] 1 EGLR 51; [2008] 15 EG 174

of any subsequent adaptation to other uses. On reconsideration, Lord Neuberger felt bound to reject the argument. Although the "literalist meaning" of "designed or adapted" was that either alternative would do, that was "not by any means what the words naturally convey". His earlier thoughts had been based on "an over-literalist approach to the language used by the legislature": [2010] 1 WLR 2317, [31]. In his revised view, a building originally designed for living in, but adapted for some other purpose, was not "designed or adapted for living in", unless subsequently re-adapted for that purpose [40].

34. I have no doubt, with respect, that Lord Neuberger's second thoughts on this point were correct. Context and common sense argue strongly against a definition turning principally on historic design, if that has long since been superseded by adaptation to some other use. However, that approach may also have implications for the earlier part of his grammatical analysis in *Boss Holdings* (see [31] above). The expression "was or is designed or adapted" is, as he says, to be read "distributively": that is, as equivalent to "was designed or is adapted". While that may support the view that the word "designed" is directed to the past, the same cannot be said of the expression "is adapted". Nor (pace Lord Scott) is that grammatically the same as "was most recently adapted". Logically that expression can only be taken as directed to the present state of the building.

35. Once it is accepted that a "literalist" approach to the definition is inappropriate, I find myself drawn back to a reading which accords more closely to what I have suggested was in Lord Denning's mind in *Ashbridge* [1965] 1 WLR 1320, that is a simple way of defining the present identity or function of a building as a house, by reference to its current physical character, whether derived from its original design or from subsequent adaptation. Furthermore, I would not give any special weight in that context to the word "adapted". In ordinary language it means no more than "made suitable". It is true that the word is applied to the building, rather than its contents, so that a mere change of furniture is not enough. However, the word does not imply any particular degree of structural change. Where a building is in active and settled use for a particular purpose, the likelihood is that it has undergone at least some physical adaptation to make it suitable for that purpose. That in most cases can be taken as the use for which it is currently "adapted", and in most cases it will be unnecessary to look further.

36. That interpretation does not of course call into question the actual decision in *Boss Holdings*. The basis of the decision, as I understand it, was that the upper floors, which had been designed or last adapted for residential purposes, and had not been put to any other use, had not lost their identity as such, merely because at the material time they were disused and dilapidated. It was enough that the building was partially "adapted for living in", and it was unnecessary to look beyond that: see [2008] 1 WLR 289, [25]. That reasoning cannot be extended to a building in which the residential use has not merely ceased, but has been wholly replaced by a new, non-residential use.

37. Finally I must refer to *Prospect Estates Ltd* v *Grosvenor Estate Belgravia* [2008] EWCA Civ 1281; [2009] 1 WLR 1313††. The Court of Appeal held that a building which had been designed and built as a house, but which for many years had been used almost wholly as offices, was not a house within the definition. As in *Tandon*, the case turned ultimately only on the second question, whether the building was a "house reasonably so called". The facts were much closer to those of the present cases. The leading judgment was given by Mummery LJ.

38. The building had been built in the 1850s as a house for residential occupation, but since 1958 it had been used substantially (88.5% of the floorspace) for office purposes. Under the most recent lease granted in 1972 the use was restricted to offices on all floors, except the top floor which was limited to use as a flat for a director or senior employee of a business occupying the offices below. It was accepted by the lessors that there had been insufficient works of adaptation to conclude that it had ceased to be designed for living in (Mummery LJ, [9]), but they challenged the judge's conclusion that it was "a house reasonably so called". That had been based, as the "overwhelmingly significant factor", on the fact that the building was "designed for living in" and that "its structure and appearance have (largely) remained unchanged" [8]. Mummery LJ held that the judge had given too much weight to those factors, and insufficient weight to "the prescriptive terms of the lease, the actual uses of the building and the relative proportions of the mixed use at the relevant date" [20]. Goldring LJ, agreeing, found it impossible to accept that a "building can reasonably be called a house although no one can lawfully live in virtually 90% of it" [23].

39. In the present case, the Master of the Rolls ([2010] 1 WLR 2317, [43]) questioned the weight placed on that case by counsel for the present appellants in *Hosebay*:

> "There can be no doubt that the external and internal appearance of the properties are highly relevant factors on this issue, and it is clear from the *Prospect Estates* case [2009] 1 WLR 1313 that, in so far as user is significant, the permitted use under the lease is a relevant factor. In those circumstances, even assuming that actual use is also relevant, I find it hard to see how it can be sensibly said that each of the three properties cannot 'reasonably [be] called' a 'house'. To hold otherwise would involve concluding that the actual user, even where it involved people occupying virtually all the rooms in the building for relaxing, sleeping, cooking and washing, albeit on a short term basis, trumped all the other factors to the extent of disabling the building from being able to be a 'house ... reasonably so called'."

40. He also doubted the decisive weight placed by Goldring LJ on the terms of the lease. He thought the thrust of the judgments in *Lake* and the opinion of Lord Roskill in *Tandon* was that the question was to be determined "essentially by reference to [the building's] external and internal physical character and appearance" [46]. He was not convinced that it would occur to most people, asked whether a building could reasonably be called a house, "to ask about the permitted use under any lease, or that they would be influenced if told what the permitted use was" [47]. He suggested that the ratio of *Prospect Estates* should be treated as being "limited to a case where residential use is either prohibited entirely, or restricted to a very small part of the building, and the actual use accords with that" [49].

41. As will be apparent from my earlier analysis of *Tandon*, I cannot agree that Lord Roskill regarded "external and internal physical character and appearance" as the determining factors. I agree with the Master of the Rolls that the terms of the lease as such should not have been treated as the major factor. However, in so far as Mummery LJ treated the use of the building, rather than its physical appearance, as determinative, his approach was in my view entirely consistent with the reasoning of the majority in *Tandon* as I have explained it. I consider that *Prospect Estates* was rightly decided, and that the ratio need not be limited in the way the Master of the Rolls proposed.

The present cases

42. I turn to consider the application of these principles to the present appeals, which I can deal with briefly.

43. I would allow the appeal in *Hosebay* on the grounds that a building which is wholly used as a "self-catering hotel" is not "a house reasonably so called" within the meaning of this statute. As appears from [38] of their judgment (quoted above), the contrary view of the Court of Appeal turned on two main points: (i) the external appearance of each property as a town house; (ii) the internal conversion to self-contained units, with cooking and toilet facilities. I find it difficult with respect to see the relevance of the second point to this part of the definition, which only arises in relation to a building which is in some sense adapted for living in under the first part. It is not suggested that the building is divided in a way which comes within the proviso. The first point, for the reasons given in my analysis of *Tandon*, should not have been given determinative weight. The fact that the buildings might look like houses, and might be referred to as houses for some purposes, is not in my view sufficient to displace the fact that their use was entirely commercial.

44. In these circumstances I find it unnecessary to reach a concluded view on the application of the first part of the definition in this appeal. I agree with the appellants (and the judge) that "living in" means

†† Editor's note: Also reported at [2009] 1 EGLR 47; [2009] 02 EG 85

A something more settled than "staying in"; and that the present use does not qualify as such. There is more room for debate, however, whether the premises are to be taken as "adapted" solely for such use, to the exclusion of longer term occupation. The Court of Appeal, as I understand it, was influenced not only by the consideration that the rooms might be used (for example) for longer-term student occupation, but also that their current layout probably dates from earlier adaptation to the uses described in the leases, which could well be regarded as sufficiently settled to qualify as "living in". One of the values of the two-part definition is that it becomes unnecessary to resolve such narrow factual issues.

B 45. In *Lexgorge*, I would also allow the appeal on similar grounds. A building wholly used for offices, whatever its original design or current appearance, is not a house reasonably so called. The fact that it was designed as a house, and is still described as a house for many purposes, including in architectural histories, is beside the point. In this case no issue arises under the first part of the definition. It is unnecessary to consider whether the concession in that respect was rightly made, although it is possible that it was based on a wider interpretation of *Boss Holdings* [2008] 1 WLR 289 than my own analysis would have supported.

46. In summary, I would allow both appeals, and hold that neither
C building was on the relevant date a "house" within the meaning of section 2 of the 1967 Act.

Appeals allowed.

D

Carey-Morgan and another v Trustees of Sloane Stanley Estate

COURT OF APPEAL

[2012] EWCA Civ 1181

3 September 2012

PILL AND RIMER LJJ AND MORGAN J

Estates Gazette 1 December 2012

E **[2012] 48 EG 64**

Leasehold enfranchisement – Leasehold Reform, Housing and Urban Development Act 1993 – Collective enfranchisement – Premium payable by nominee purchaser to reversioners – Deferment rate for reversions of less than five years – Upper Tribunal giving guidance that deferment rate for flats with reversions of less than five years to be appropriate net rental yield for the property with end allowance of 5% – Whether guidance vitiated by failure to consider possibility of tenants having right to remain at end of lease term by virtue of Schedule 10 to Local Government and Housing Act 1989 – Whether
F ***hope value properly awarded in respect of non-participating flats with reversions of less than 80 years***

On an appeal from the leasehold valuation tribunal, the Lands Chamber of the Upper Tribunal (UT) determined that the appellant nominee purchaser should pay a premium of £2,961,613 to the respondent freeholders on a collective enfranchisement, under the Leasehold Reform, Housing and Urban Development Act 1993, of a 1920s mansion block within a conservation area in Kensington, London SW3: see [2011] UKUT 415 (UT); [2011] PLSCS 270; [2012] RVR 92.

In reaching that decision, the UT had to determine the appropriate deferment rate to apply when valuing the freehold reversion to six flats in the block that were held on leases with less than five years unexpired. It decided that the formula in *Earl Cadogan* v *Sportelli* [2007] 1 EGLR 153, based on a long-term, risk-free rate derived from government G securities, should not be applied to the valuation of very short-term reversions; instead, the deferment rate for reversions of less than five years should be the appropriate net rental yield for the property, with an end allowance of 5%, or such other percentage as the evidence established, to take into account the lack of control over the premises when compared to an owner in possession. It applied a deferment rate of 3.25%, rounded down from 3.31%, based on evidence of net rental yields derived from evidence of comparable lettings.

The UT's valuation also included hope value for the prospect of selling new leases to the tenants of non-participating flats. The UT considered that hope value could be taken into account in respect of H non-participating flats so far as it was attributable to the lessees seeking new leases on the open market, rather than pursuant to any statutory right under the Act, and that it should be calculated as a percentage of marriage value, derived from evidence as to the likelihood of non-participating tenants coming forward before the expiry of the current lease to negotiate for an extended lease. It applied percentages of 20% for four non-participating flats with unexpired terms of 4.74 years and 10% for a flat with 70.25 years unexpired.

The appellants appealed. On the deferment rate, they raised a new point that the freeholder should not be assumed to be entitled to vacant possession of the relevant flat at the end of the unexpired term of the J lease, since the tenant might be entitled to remain in occupation under Schedule 10 to the Local Government and Housing Act 1989. They submitted that the *Sportelli* deferment rate should therefore be applied.

Held: **The appeal was dismissed. (1) The appellants' new point concerning the relevance of the 1989 Act was rejected. The UT had been entitled to calculate the deferment rate in the manner it had since the appellants had accepted its approach at the time. The court was not able to deal with the issue now as a pure point of law, without the benefit of valuation evidence from the parties. It involved questions of fact and opinion, including as to the valuation consequences of taking the 1989 Act into account. These might not K be restricted to the choice of deferment rate; nor was it inevitable that the generic *Sportelli* deferment rate should be adopted, since that rate applied to unexpired terms of 20 years or more, whereas the instant case concerned unexpired terms of 4.74 years. Although the UT had been entitled to offer guidance as to the deferment rate in the case of reversions of less than five years, the guidance might have to be qualified in a case where a party raised an issue concerning the possible application of the 1989 Act at the end of the unexpired terms of the leases. (2) The UT had correctly valued the freehold interest in the block as a whole, notwithstanding that it had arrived at that value by building it up from valuations L of individual elements. It had been necessary to consider the deferment rate to be applied to different parts of the interest being acquired and the tribunal had been entitled to proceed by assessing the contribution that the individual components made to the value of the single freehold interest. (3) It is permissible to take hope value into account, as part of the freehold value pursuant to para 3 of Schedule 6 to the 1993 Act, so far as it is attributable to the possibility of non-participating tenants seeking new leases of their respective flats by negotiation, rather than as of right under the Act. On that issue, the UT had been entitled to derive assistance from cases where tenants had exercised their statutory rights to M extended leases shortly before the end of their current terms; the UT had not fallen into the error of regarding those cases as identical to those where no statutory right existed. It had made no error of law in choosing the percentages of marriage value that it had applied to reflect hope value. The weight to be attached to the various considerations and the ultimate selection of the percentage was a matter for the UT. The percentages it had selected for hope value were open to it on the evidence.**

The following cases are referred to in this report.

Cadogan Square Properties Ltd v *Earl Cadogan* [2010] UKUT 427 (LC); [2011] 1 EGLR 155

A *Culley* v *Daejan Properties Ltd* [2009] UKUT 168 (LC); [2009] 3 EGLR 165
Earl Cadogan v *Sportelli* [2007] 1 EGLR 153
Ex parte Firth, In re Cowburn (1882) 19 Ch D 419, 429
Macdougall v *Knight* (1889) 14 App Cas 184
Pittalis v *Grant* [1989] QB 605; [1989] 2 EGLR 90; [1989] 28 EG 126
The Tasmania (1890) 15 App Cas 223

This was an appeal by the appellants, Charles Carey-Morgan and John Stephenson, from a decision of the Lands Chamber of the Upper Tribunal, on appeal form the leasehold valuation tribunal, determining the premium to be paid to the respondents, the trustees of the Sloane
B Stanley Estate, on a collection enfranchisement of a property under the Leasehold Reform, Housing and Urban Development Act 1993.

Kenneth Munro (instructed by Pemberton Greenish LLP) appeared for the appellants; Timothy Dutton (instructed by Bircham Dyson Bell LLP) represented the respondents.

Giving judgment, **Morgan J** said:

Introduction

1. This is an appeal against a decision of the Upper Tribunal (Lands
C Chamber) dated 10 October 2011 ([2011] UKUT 415 (LC)); the decision is now reported at [2012] RVR 92. The tribunal judges were the President of the Lands Chamber (Mr George Bartlett QC) and Mr Paul Francis FRICS. The decision of the Upper Tribunal was in respect of an appeal from a decision and a supplementary decision, dated respectively 16 December 2008 and 24 April 2009, of the Leasehold Valuation Tribunal for the London Rent Assessment Panel ("the LVT"). There were two hearings before the LVT, together extending to five days. The appeal to the Upper Tribunal was by way of a re-hearing in relation to the issues which remained in dispute and the Upper Tribunal itself heard evidence of fact and expert evidence. The hearing before the Upper Tribunal lasted three days. Permission for a second appeal to this
D court was granted by Arden LJ on 27 February 2012.

2. The application to the LVT was made, pursuant to section 24 of the Leasehold Reform, Housing and Urban Development Act 1993 ("the 1993 Act"), by the Trustees of the Sloane Stanley Estate as "the reversioner" within section 9 of the 1993 Act. The respondents to that application were Mr Carey-Morgan and Mr Stephenson who were together "the nominee purchaser" within section 15 of the 1993 Act. The reversioner asked the LVT to determine the terms on which the nominee purchaser was to be entitled to acquire the freehold of a block of flats known as Vale Court, Mallord Street, London, SW3. The principal term in dispute was the price to be paid for the freehold which
E was to be determined pursuant to section 32 of the 1993 Act (giving effect to the detailed provisions contained in Schedule 6 to the 1993 Act).

3. Following the LVT's determination of the price to be paid for the freehold of the block, the reversioner appealed to the Upper Tribunal against certain parts of that determination. The issues which came before the Upper Tribunal in relation to the price to be paid were as follows: (1) the vacant possession values of the leases of some of the flats in the block; (2) whether there was potential for the freehold owner of the block to undertake a development of the roof of the block and,
F if so, the value of that potential; (3) hope value in relation to flats let on leases with unexpired terms of less than 80 years and where the leases were held by tenants who were not participating tenants within section 14 of the 1993 Act; (4) the deferment rate to be used in valuing the freehold interest in relation to flats subject to leases with unexpired terms of less than five years.

4. The only parts of the decision of the Upper Tribunal which have been challenged on this appeal relate to the deferment rate and hope value. The nominee purchaser submits that the Upper Tribunal was wrong in law in relation to its treatment of both the deferment rate and hope value and it further contends that the conclusions reached separately as to the deferment rate and hope value reveal inconsistencies or even double counting and cannot stand with each other.

5. On this appeal, Mr Dutton appeared on behalf of the nominee G
purchaser; he did not appear in the Upper Tribunal. Mr Munro appeared for the reversioner on this appeal, as he did in the Upper Tribunal.

The statutory provisions

6. Section 32(1) of the 1993 Act gives effect to Schedule 6, where the detailed provisions as to the price payable for the freehold are to be found. The provisions which matter for present purposes are those in paras 2 and 3 of Schedule 6, which are in these terms:

"Part II
Freehold of specified premises H
Price payable for freehold of specified premises

2. (1) Subject to the provisions of this para where the freehold of the whole of the specified premises is owned by the same person, the price payable by the nominee purchaser for the freehold of those premises shall be the aggregate of—

(a) the value of the freeholder's interest in the premises as determined in accordance with para 3,

(b) the freeholder's share of the marriage value as determined in accordance with para 4, and

(c) any amount of compensation payable to the freeholder under para 5.

(2) Where the amount arrived at in accordance with sub-para (1) is a negative amount, the price payable by the nominee purchaser for the freehold J
shall be nil.

Value of freeholder's interest

3.- (1) Subject to the provisions of this para, the value of the freeholder's interest in the specified premises is the amount which at the relevant date that interest might be expected to realise if sold on the open market by a willing seller (with no person who falls with sub-para (1A) buying or seeking to buy) on the following assumptions—

(a) on the assumption that the vendor is selling for an estate in fee simple—

(i) subject to any leases subject to which the freeholder's interest in the premises is to be acquired by the nominee purchaser, but

(ii) subject also to any intermediate or other leasehold interests in the K
premises which are to be acquired by the nominee purchaser;

(b) on the assumption that this Chapter and Chapter II confer no right to acquire any interest in the specified premises or to acquire any new lease (except that this shall not preclude the taking into account of a notice given under section 42 with respect to a flat contained in the specified premises where it is given by a person other than a participating tenant);

(c) on the assumption that any increase in the value of any flat held by a participating tenant which is attributable to an improvement carried out at his own expense by the tenant or by any predecessor in title is to be disregarded; and

(d) on the assumption that (subject to paras (a) and (b)) the vendor is selling with and subject to the rights and burdens with and subject to which the L
conveyance to the nominee purchaser of the freeholder's interest is to be made, and in particular with and subject to such permanent or extended rights and burdens as are to be created in order to give effect to Schedule 7.

(1A) A person falls within this sub-para if he is—

(a) the nominee purchaser, or

(b) a tenant of premises contained in the specified premises, or

(ba) an owner of an interest which the nominee purchaser is to acquire in pursuance of section 1(2)(a), or

(c) an owner of an interest which the nominee purchaser is to acquire in pursuance of section 2(1)(b).]

(2) It is hereby declared that the fact that sub-para (1) requires assumptions to be made as to the matters specified in paras (a) to (d) of that sub-para M
does not preclude the making of assumptions as to other matters where those assumptions are appropriate for determining the amount which at the relevant date the freeholder's interest in the specified premises might be expected to realise if sold as mentioned in that sub-para.

(3) In determining that amount there shall be made such deduction (if any) in respect of any defect in title as on a sale of the interest on the open market might be expected to be allowed between a willing seller and a willing buyer.

(4) Where a lease of any flat or other unit contained in the specified premises is to be granted to the freeholder in accordance with section 36 and Schedule 9, the value of his interest in those premises at the relevant date so far as relating to that flat or other unit shall be taken to be the difference as at that date between—

(a) the value of his freehold interest in it, and

A (b) the value of his interest in it under that lease, assuming it to have been granted to him at that date;

and each of those values shall, so far as is appropriate, be determined in like manner as the value of the freeholder's interest in the whole of the specified premises is determined for the purposes of para 2(1)(a).

(5) The value of the freeholder's interest in the specified premises shall not be increased by reason of—

(a) any transaction which—

(i) is entered into on or after the date of the passing of this Act (otherwise than in pursuance of a contract entered into before that date), and

(ii) involves the creation or transfer of an interest superior to (whether or not preceding) any interest held by a qualifying tenant of a flat contained in the
B specified premises; or

(b) any alteration on or after that date of the terms on which any such superior interest is held.

(6) Sub-para (5) shall not have the effect of preventing an increase in value of the freeholder's interest in the specified premises in a case where the increase is attributable to any such leasehold interest with a negative value as is mentioned in para 14(2).

7. Para 4 of Schedule 6 deals with the freeholder's share of marriage value and para 5 of Schedule 6 deals with compensation for loss resulting from enfranchisement. It is not necessary to set out those provisions although I will refer to the concept of marriage value later
C in this judgment.

Comment on the statutory provisions

8. Para 3 of Schedule 6 directs that certain valuation assumptions are to be made for the purpose of determining the value of the freeholder's interest. One assumption that is relevant in the present case is that specified in para 3(1)(b), namely, that Chapter I (dealing with collective enfranchisement by tenants of flats) and Chapter II (dealing with the right of a tenant of a flat to a new lease) of the 1993 Act do not confer a right to acquire any interest in the specified premises or to acquire any new lease. This is often referred to as a "no-Act world" assumption, but it can also be described as a "no-Act building" assumption, as it
D is an assumption that the rights under Chapters I and II do not apply to the tenants in the building that is the subject of the collective enfranchisement claim (referred to as "the specified premises": see section 13).

9. Para 3(1) of Schedule 6 defines the value of the freeholder's interest as "the amount which at the valuation date that interest might be expected to realise if sold on the open market by a willing seller…". Where the freeholder's interest is subject to a lease or leases, there will be a value to a purchaser of the freehold in acquiring, first, the right to receive the rent under those lease(s) during the term(s) of those lease(s) and, secondly, the right to possession of the property when the term(s)
E of the lease(s) expire. It is conventional for valuers to refer to the first element as the value of "the term" and to the second as the value of "the reversion". The value of the "term" is calculated by capitalising the rent. The value of the "reversion" is normally calculated by taking the estimated vacant possession value of the relevant premises and then discounting it at an appropriate rate of interest to arrive at the value today of the freeholder's right to possession at some point in the future. This rate of interest is called the deferment rate.

10. Before considering the decision of the Upper Tribunal which is under appeal in this case, it is necessary to discuss the issue of deferment rate in a little more detail. It is also helpful to refer to three
F other topics, namely marriage value, hope value and the possible application of the Local Government and Housing Act 1989, Schedule 10, before addressing the decision under appeal and the grounds of the appeal.

Deferment rate

11. The deferment rate was described by the Lands Tribunal in *Earl Cadogan* v *Sportelli* [2007] 1 EGLR 153 at [2] as "the annual discount applied, on a compound basis, to an anticipated future receipt (assessed at current prices) to arrive at its market value at an earlier date". The purpose of selecting a particular deferment rate is to enable one to assess how much someone would pay at the valuation date for the right to receive a sum of money at a future date. This depends on the rate of compound interest which he wants in return for parting with his money G
at the valuation date and getting his money back only at a future date. The formula for calculating the present value of a right to receive a future payment, using a given compound interest rate is:

Future payment/(1+i)n

(where i is the compound rate of interest and n is the number of years until deferment).

12. The use of this formula shows that an investor who wishes to purchase a right to a sum of £1m in five years' time and who wants a compound rate of return of 5%, would be prepared to pay £783,526, because £1m/(1 + 0.05)5 = £783,526. If he had invested that sum at 5% compound interest, then at the end of five years he would have £1m. H
If one applies this approach to the case of a purchaser of a freehold reversion, where the value of that freehold with vacant possession "assessed at current prices" is £1m, using this formula and a deferment rate of 5%, one would conclude that the purchaser would be prepared to pay £783,526 for the reversion (ignoring the value of the right to receive the rent(s) under the lease(s) in the intervening five years). On this basis the present value of such a reversion would be considered to be £783,526. (The analysis of a purchase of a property reversion is not as straightforward as the example of an investor buying a right to a fixed sum of money payable at a future date. The value of the property may change between the date of the purchase and the later J
date when the leases expire, when the purchaser becomes entitled to vacant possession. In valuation terms, this is dealt with by taking the current value of the freehold with vacant possession and reflecting the possibility of changes in property values in the deferment rate.)

13. The amount of the deferment rate can have a very important bearing on the value of the reversion. Over the years, many cases concerning the deferment rate have been fought in the LVT and in the Lands Tribunal. The Lands Tribunal considered that it would be helpful to give guidance as to the right approach to the determination of a deferment rate. Eventually, a number of cases were collected together to enable the Lands Tribunal to give such guidance. The decision of the K
Lands Tribunal in those combined cases is reported as *Earl Cadogan* v *Sportelli* [2007] 1 EGLR 153. Certain issues in that case were considered by the Court of Appeal (see [2008] 1 WLR 2142) and later by the House of Lords (see [2010] 1 AC 226). I will refer to this case at the different levels of decision as "*Sportelli*".

14. For present purposes, it is not necessary to describe in detail the many considerations which are discussed by the Lands Tribunal in its decision in *Sportelli*. The Lands Tribunal concluded, following earlier authority, that in the case of leases with unexpired terms of 20 years or more the right approach was to use a formula expressed as follows: Deferment Rate = Risk Free Rate ("RFR") + Risk Premium ("RP") L
– Real Growth Rate ("RGR"). Those terms are explained in detail in that case and need not be explained again here. The Lands Tribunal further concluded in *Sportelli* that the components in the formula were to be given the following values: RFR = 2.25%, RP = 4.5% and RGR = 2% giving a Deferment Rate of 4.75%. It was held that this rate was appropriate for cases where the property was a house. In a case where the property consisted of a flat or flats, a further 0.25% should be added giving a Deferment Rate of 5.0%. The Lands Tribunal concluded that these rates should be regarded as generic deferment rates which could be adopted for leases with unexpired terms of 20 years and above but which may need to be adjusted by reference to the facts of individual M
cases.

15. Some parts of the Lands Tribunal's decision in *Sportelli* were the subject of an appeal to the Court of Appeal. In particular, the nominee purchaser appealed in relation to the deferment rate. That appeal was dismissed: [2008] 1 WLR 2142. The Court of Appeal agreed with the Tribunal that it was an important part of its role to promote consistent practice in land valuation matters and that it was entirely appropriate for the Tribunal to offer the guidance which it had done, although that guidance should be confined to properties in prime central London.

16. The Lands Tribunal in *Sportelli* had not sought to identify a generic deferment rate for leases with unexpired terms of below 20 years. Instead, it indicated that with unexpired terms of below 20 years

the rate would need to have regard to the property cycle at the date of the valuation. Before long, a case where the unexpired terms were below 20 years came before the Upper Tribunal. That case was *Cadogan Square Properties Ltd* v *Earl Cadogan* [2010] UKUT 427 (LC); [2011] 1 EGLR 155. The Upper Tribunal followed the guidance given by the Lands Tribunal in *Sportelli* in adopting the formula of Deferment Rate = RFR + RP – RGR. In view of its findings as to the property cycle at the dates of the relevant valuations, the Upper Tribunal adopted values for RGR of 1.75% (for two valuation dates in 2005) and 1.5% (for three valuation dates in 2007) in place of the 2% adopted in *Sportelli*. The Upper Tribunal adopted the same values for RFR and RP as were adopted in *Sportelli*. These conclusions produced deferment rates of 5.25% (for the two valuation dates in 2005) and 5.5% (for the three valuation dates in 2007). The Upper Tribunal stated that in relation to unexpired terms from 10 years up to (but not including 20 years) certain parts of its decision should have the same effect in future cases as the guidelines in *Sportelli* itself. The Upper Tribunal added that as regards unexpired terms of less than 10 years, the method used in its decision should be a starting point for the determination of the deferment rate but that for very short unexpired terms the possibility of using net rental yields as a guide to the deferment rate should be examined more closely. As will be seen, the Upper Tribunal in the present case had to grapple with that question.

Marriage value

17. There is no issue as to marriage value which needs to be addressed in this appeal. However, a brief statement of what marriage value is may assist when I later describe the issue which does arise in relation to hope value. The concept of marriage value is relatively straightforward in the case of a long lessee acquiring the freehold reversion on his lease (eg where the premises comprise a house). If the value of the freehold reversion (ignoring any bid for it by the existing lessee) is £X, if the value of the existing lease is £Y and if the value of the freehold with vacant possession is £Z, then it will often be the case that X + Y is less than Z. The degree of difference between X + Y on the one hand and Z on the other will depend on a number of factors, in particular, the length of the unexpired term of the lease. The difference is called the marriage value; so called because the extra value is released by marrying the freehold reversion with the existing lease. Thus, if the existing lessee, who owns a lease with a value of £Y, were to be able to buy the freehold reversion for £X, he would obtain a freehold with vacant possession with a value of £Z and would secure all of the marriage value for himself. In the open market, free from statutory assumptions, it is to be expected that the freeholder would only agree to sell the freehold reversion to the existing lessee on the basis that the marriage value is shared between the freeholder vendor and the lessee purchaser. Marriage value was described in *Sportelli* by Carnwath LJ in the Court of Appeal ([2008] 1 WLR 2142 at [4]) and by Lords Hoffmann and Neuberger in the House of Lords ([2010] AC 226 at [3] and [58]-[63] respectively).

18. The concept of marriage value is a little more complex where one is concerned not with the purchase of a freehold reversion but instead with the grant of a new long lease or an extended lease to the existing lessee. In the latter type of case, marriage value is the subject of an elaborate definition in para 4 of Schedule 6 to the 1993 Act. Where the unexpired term of the existing lease exceeds 80 years, marriage value is ignored. Para 4(1) provides that the freeholder's share of marriage value is 50%.

Hope value

19. A detailed description of the concept of hope value and the approach to be adopted in relation to it is to be found in the decision of the House of Lords in *Sportelli* (reported at [2010] 1 AC 226), in particular, in the speech of Lord Neuberger of Abbotsbury. Although Lord Hoffmann dissented in the result as to hope value, his description of the concept at [2]-[4] is very clear and helpful. Happily, for the purposes of determining the present appeal, it is not necessary to do more than provide a brief summary of this subject.

20. The statutory provisions require one to assume that no tenant of any premises contained in the building, which is the subject of the collective enfranchisement, is seeking to buy the freehold of the building. Further, one must assume that no tenant is entitled under Chapters I or II of the 1993 Act to acquire any interest in the building or to acquire any new lease. Nonetheless, where the freeholder is selling his interest, even when the tenants are not then in the market, a potential purchaser might think that, in addition to its investment value, the freehold interest carries with it the potential benefit of being able to charge a premium for the possible future grant of a long lease to a tenant of a flat in the building, such lease being granted prior to the end of the term of the existing lease. In accordance with the statutory assumptions, these possible future grants will be a matter for negotiation and not a matter of entitlement under Chapter II of the 1993 Act. This potential benefit to a purchaser means that the value of the freehold might be greater than the aggregate of the capitalised rental stream and the deferred right to possession at the end of the term(s). Accordingly, in assessing the value of the freehold, something should be added for the possibility of the freeholder purchaser benefiting in this way. This something is known as "hope value".

21. On a collective enfranchisement, it is often the case that not all of the qualifying tenants (defined in section 5 of the 1993 Act) in the building join in the giving of the notice to purchase the freehold. The qualifying tenants who do join in are called the "participating tenants" (defined in section 14 of the 1993 Act). Other tenants in the building can be referred to as the "non-participating tenants". As explained earlier, para 2 of Schedule 6 to the 1993 Act provides that the price payable for the freehold includes not only the value of the freeholder's interest, assessed in accordance with para 3 of Schedule 6, but also the freeholder's share of certain marriage value, assessed in accordance with para 4 of Schedule 6. It is not necessary to refer to the detail of para 4 of Schedule 6 but it is relevant to explain that the marriage value calculation is limited to considering the release of marriage value by reference to the leasehold interests under the control of the participating tenants. The result is that there is no consideration in para 4 of Schedule 6 of any hope value in relation to the interests of the non-participating tenants. Against this statutory background, the House of Lords in *Sportelli* held that hope value could be taken into account, as part of the value of the freehold pursuant to para 3 of Schedule 6, in so far as it is attributable to the possibility of non-participating tenants seeking new leases of their respective flats by negotiation but not, of course, as of right under Chapter II (because the 1993 Act required one to assume that they did not have such a right under Chapter II).

Local Government and Housing Act 1989

22. Section 186 of the Local Government and Housing Act 1989 ("the 1989 Act") gives effect to Schedule 10 to that Act. Schedule 10 confers a form of security of tenure on certain tenants of dwelling houses (including flats). The tenancy in question must be a long tenancy at a low rent which satisfies "the qualifying condition". It is not necessary to discuss what is meant by "a long tenancy". In order to determine whether a tenancy is at a low rent, in the case of a tenancy entered into before 1 April 1990 (where the dwelling house had a rateable value on 31 March 1990), one asks whether the maximum rent payable at any time under the tenancy was less than two-thirds of the rateable value on 31 March 1990. The qualifying condition is that the circumstances as respects the property let under the tenancy, the use of that property and all other relevant matters are such that, if the tenancy were not at a low rent, it would at that time be an assured tenancy within the meaning of Part I of the Housing Act 1988. For this purpose, in relation to a tenancy which was entered into before 1 April 1980, the rateable value of the dwelling house as at 31 March 1990 must not exceed £1,500. Further, the tenant must occupy the dwelling house as his only or principal home.

23. A tenancy which, immediately before its term date, is within Schedule 10 to the 1989 Act does not come to an end on that date except by being terminated under the provisions of Schedule 10. The tenancy continues until it is so terminated. While a tenancy continues

A pursuant to the 1989 Act, it continues at the same rent and in all other respects on the same terms as before the term date save that, in some circumstances, the landlord may apply for an interim rent. Where such a tenancy is terminated by a landlord's notice proposing an assured tenancy within Part I of the Housing Act 1988, then eventually when the statutory procedures are gone through the tenant will be entitled, subject to one qualification, to remain as an assured periodic tenant at a rent established in accordance with the statutory procedures. The qualification which needs to be mentioned arises from the fact that, at the time relevant to this appeal, a tenancy entered into on or after 1 April 1990 under which the rent payable exceeded £25,000 per year could not B be an assured tenancy: see Housing Act 1988, section 1(2) and Schedule 1, para 2. Thus, if the operation of the statutory procedures produced a tenancy under which the rent exceeded £25,000 per year, the tenancy so produced would not be an assured tenancy and would lack statutory security of tenure under Part 1 of the Housing Act 1988. (In his written submissions, Mr Dutton referred to the possibility of the application of the Rent Act 1977 or the Housing Act 1988 as well as the possibility of the application of the 1989 Act. In his oral submissions, Mr Dutton focussed on the position under the 1989 Act. I do not consider that it is necessary in this case to describe the operation of the Rent Act 1977 or the Housing Act 1988. In this case, those statutes do not give rise to C any additional matters over and above those which I will consider in relation to the 1989 Act.)

Deferment rate: the decision under appeal

24. In the present case, the Upper Tribunal adopted a conventional valuation approach. It sought to build up to the aggregate value of the freeholder's interest in accordance with para 3 of Schedule 6 to the 1993 Act by including the value of the terms (ie the right to receive the rent payable under the lease during the unexpired terms of the leases) and the value of the reversions to vacant possession at the end of the terms, as well as other valuation inputs. The valuation of the term in relation to each flat assumed that the rent payable during the unexpired term would D cease at the end of that term. In relation to the value of the reversion at the end of the unexpired term, the valuation proceeded on the basis that the freeholder would be entitled to possession of the relevant flat at the end of the unexpired term. Accordingly, the Upper Tribunal proceeded to assess the value, at the valuation date, of the relevant flat with vacant possession and then determined the appropriate deferment rate. The deferment rate was then applied for the period of the unexpired term; this was consistent with the basis of the valuation which was that the vacant possession value of the flat would be available to the freeholder at that point.

25. The Upper Tribunal then considered the deferment rate which E they should adopt. As already explained, two earlier cases (*Sportelli* and *Cadogan Square Properties Ltd*) had discussed the approach to be adopted when determining the deferment rate for (1) unexpired terms of 20 years or more; and (2) for unexpired terms of between 10 and 20 years. In the present case, the unexpired lease terms were of various lengths. Where the unexpired lease terms exceeded 20 years, the Upper Tribunal applied the *Sportelli* guidance. However, there were six leases where the unexpired lease term was 4.74 years. Those terms were not within the guidance in the earlier cases and the Upper Tribunal adopted an approach which was different from that adopted earlier. In summary, it held that the deferment rate for reversions of less than five years should F be the net rental yield which the evidence showed to be appropriate for the property in question and, in addition, that there should be an end allowance which, in the absence of evidence establishing some other percentage, should be 5%. In the present case, the Upper Tribunal held that the appropriate net rental yield was 3.25%.

Deferment rate: the appeal

26. The principal point raised by the nominee purchaser on this appeal in relation to deferment rate is to question the assumption made by the Upper Tribunal that the freeholder would be entitled to vacant possession of the relevant flat at the end of the unexpired term of the relevant lease. It is submitted that this assumption ignores the possible application of Schedule 10 to the 1989 Act. If it should emerge that a particular flat was within the rateable value limits of the 1989 Act G and if the lessee under the lease were (at the end of the term) to be in occupation of the flat as his only or principal home, then because the lease in question would have been a long lease at a low rent, that lessee would be entitled to remain in occupation under the 1989 Act. If the freeholder were to take steps to terminate those rights under the 1989 Act there would eventually be a tenancy which would be an assured periodic tenancy, if the rent payable under that tenancy did not exceed £25,000 per year. Accordingly, it was submitted that the Upper Tribunal approached the case on an incorrect legal basis.

27. The submission as to the possible application of the 1989 Act was put forward for the purpose of challenging the deferment H rate adopted by the Upper Tribunal in relation to the six leases with unexpired terms of 4.74 years. The nominee purchaser appeared ready to accept that the possible application of the 1989 Act was not material in relation to the other leases. As regards the six leases in question, it was submitted that this court should simply substitute the deferment rate arrived at by the Lands Tribunal in *Sportelli* for the rate adopted in the decision under appeal. It was not suggested that, when valuing the term, one should capitalise the rent payable during the unexpired term granted by the lease and also the rent payable after the end of that term, whether during a period of continuation under the 1989 Act or under any subsequent assured tenancy. It seemed to be accepted that if this J challenge based on the possible application of the 1989 Act failed, then the general approach adopted by the Upper Tribunal was not open to challenge on this appeal.

28. The nominee purchaser accepts that the point taken by it on this appeal, as to the possible application of the 1989 Act, was not taken before the LVT nor the Upper Tribunal. In order to consider whether the point can be effectively taken on this appeal, I need to describe the course which was taken before the Upper Tribunal.

29. The valuers who were called by the parties appear to have adopted a conventional term and reversion method of valuation. They therefore capitalised the rent payable during the unexpired terms. They K did not seek to capitalise any rent payable after the end of the unexpired terms. They assessed the vacant possession value of the relevant flats and then applied a deferment rate to that vacant possession value for the period of the unexpired terms. At the hearing before the Upper Tribunal there was considerable disagreement between the valuers as to the correct approach to be adopted to determine the deferment rate.

30. Following the hearing, the Upper Tribunal sent a detailed note to the parties in which it set out the approach which it favoured in relation to the deferment rate for the reversions with unexpired terms of less than five years. It invited submissions and/or evidence in response. Para 3 of the note, referring to these reversions, stated that L "their key characteristic is that they contain the guarantee of the early enjoyment of possession" (ie at the end of the fixed term of less than five years). Other parts of the note commented on the significance of this characteristic of the reversions being considered.

31. Both sides replied to the invitation to make submissions and put in evidence. It is relevant to refer to the response from the nominee purchaser's side. The advocate for the nominee purchaser who was then acting (not Mr Dutton, who appeared on the appeal to this court) expressly replied to para 3 of the Upper Tribunal's note. He wrote:

> "We would certainly agree with the statement made herein that there is a M short term certainty both as to realisation and cost of holding until the term ends… "

He added that he agreed with a further para in the Upper Tribunal's note where the Upper Tribunal had described how the "short-term reversions" were akin to freehold interests in possession, subject only to a period of deferment. He then commented on the choice of deferment rate in those circumstances.

32. Accordingly, the Upper Tribunal having clearly stated to the parties its provisional view as to how the deferment rate should be determined and explaining the significance, as the Upper Tribunal saw it, that the freeholder had the certainty of being entitled to possession

at the end of the term, the nominee purchaser expressly agreed that that was a correct statement of the position.

33. On the appeal to this court, it was submitted on behalf of the reversioner that the advocate for the nominee purchaser had gone even further, in his response to the note from the Upper Tribunal, to concede that the possibility of assured tenancies following the end of the term was irrelevant. The advocate's note stated that the valuer for the nominee purchaser had raised "the issue of assured tenancies". The court asked counsel for the parties to show the court where the valuer had raised this issue and how he had dealt with it. We were told that the valuer had not raised it in his written evidence but that he had made some reference to it when giving oral evidence. The advocate's note continued by referring to the possibility that the £25,000 ceiling as regards rent might be lifted. Although the note is not wholly clear, the point seems to have been that tenancies of any flats which were let at market rents above £25,000 might, as a result of a change in the law, come within the Housing Act 1988. The advocate then added:

> "However, this does not seem to have any relevance to the specific property but should be borne in mind for any formula or methodology."

34. In my judgment, the various responses from the advocate for the nominee purchaser make it clear that the nominee purchaser accepted that the Upper Tribunal could act on the basis identified by it that the freeholder had the guarantee of the enjoyment of possession at the end of the terms of the leases. As regards the further comment referring to assured tenancies, this seems a clear acceptance that whatever point the advocate had in mind was irrelevant. There is a lack of clarity as to what the advocate meant when he said that something "should be borne in mind for any formula or methodology". Overall, my interpretation of the nominee purchaser's position was that it accepted the Upper Tribunal's approach based on a guarantee of the early enjoyment of possession.

35. Mr Dutton submitted to us that the Upper Tribunal should not have acted upon the nominee purchaser's acceptance of its approach. He suggested that the reason that the nominee purchaser had accepted that there was a guarantee of early possession was because the valuer for the nominee purchaser, when he was considering the value of the unexpired leasehold interests, had given evidence that the flats had a high rental value which would have taken the market rents of the flats above the £25,000 ceiling for the purposes of the Housing Act 1988. It was in the interests of the nominee purchaser to put forward evidence which supported higher values for the leasehold interests. If the valuer had not given that evidence but had instead stated that the market rental value of the flats was below £25,000 per year, then it would have been open to the nominee purchaser to raise the point that the 1989 Act might potentially apply at the end of the unexpired terms. It was further submitted that if one analysed the valuations carried out by the Upper Tribunal and focussed on its figures for the value of the leasehold interests (which were not arrived at by considering rental values) one could deduce that the Upper Tribunal would have found (if it had been asked to do so) that the market rental value of some (but not all) of the six flats with unexpired terms of 4.74 years was, after all, below £25,000. Based on these considerations, it was submitted that the right thing for the Upper Tribunal to have done was to reject the concession made by the nominee purchaser as to the guarantee of early possession and instead to work out that there might have been a possibility of the 1989 Act applying to some (although probably not all) of the six flats and then to abandon its approach to the determination of the deferment rate.

36. I am not able to accept the suggested criticism of the Upper Tribunal which I have set out in the last para. It involves considerable speculation as to the case which the nominee purchaser might have put forward in other circumstances. It also involves some speculation as to what the Upper Tribunal would have concluded were the market rental values of some of the six flats. I consider that the Upper Tribunal was fully entitled to act upon the clear stance adopted by the nominee purchaser in its response to the note from the Upper Tribunal. If the nominee purchaser had wanted to put forward an alternative argument along the lines advanced by Mr Dutton on this appeal, it should have done so and should not have expressed an unqualified acceptance of the Upper Tribunal's approach.

37. It follows that the Upper Tribunal was entitled to decide the case in the way in which it did. Nonetheless, Mr Dutton submits that the possible application of the 1989 Act is a pure point of law, that the Upper Tribunal left out of account that possibility, that it thereby (however excusably) erred in law and this court should allow an appeal against that decision.

38. Mr Dutton accepted that the approach which should be adopted to the new point in relation to the 1989 Act is the approach described by Nourse LJ giving the judgment of the court in *Pittalis* v *Grant* [1989] QB 605* at p611C-F, as follows:

> "The stance which an appellate court should take towards a point not raised at the trial is in general well settled: see *Macdougall* v *Knight* (1889) 14 App Cas 184 and *The Tasmania* (1890) 15 App Cas 223. It is perhaps best stated in *Ex parte Firth, In re Cowburn* (1882) 19 Ch D 419, 429, per Sir George Jessel MR:
>
> "the rule is that, if a point was not taken before the tribunal which hears the evidence, and evidence could have been adduced which by any possibility would prevent the point from succeeding, it cannot be taken afterwards. You are bound to take the point in the first instance, so as to enable the other party to give evidence."
>
> Even if the point is a pure point of law, the appellate court retains a discretion to exclude it. But where we can be confident, first, that the other party has had opportunity enough to meet it, secondly, that he has not acted to his detriment on the faith of the earlier omission to raise it and, thirdly, that he can be adequately protected in costs, our usual practice is to allow a pure point of law not raised below to be taken in this court. Otherwise, in the name of doing justice to the other party, we might, through visiting the sins of the adviser on the client, do an injustice to the party who seeks to raise it."

39. Mr Dutton submitted that the point as to the possible application of the 1989 Act is a pure point of law. This court should allow the nominee purchaser to take the point and should further hold that there was a possibility that a tenant of one or more of the relevant six flats would have been able to claim the benefit of Schedule 10 to the 1989 Act. This court should therefore hold that the Upper Tribunal was wrong in law. He then submitted that because the Upper Tribunal gave the guarantee of early possession as its reason for not adopting the generic deferment rate laid down by the Lands Tribunal in *Sportelli*, and the decision of this court would invalidate that reason, it would inevitably follow that the correct approach should have been to adopt the *Sportelli* rate.

40. I am unable to accept Mr Dutton's submission. If the nominee purchaser had suggested that it was appropriate to take into account the possibility that the 1989 Act might have applied at the end of the terms, then a number of questions of fact and opinion would have arisen. Without attempting an exhaustive statement of all of the implications of the point being taken it seems to me to be obvious that the valuers and the Upper Tribunal would have had to consider matters such as: (1) did the rateable values of the relevant flats come within the limits laid down in para 2A of Schedule 1 to the Housing Act 1988? (2) what was the likelihood of the rack rental value of the relevant flats bringing them within the £25,000 limit? (3) what was the likelihood that the tenants of the relevant flats would be in residence at the term dates? (4) what was the likelihood of the long lessees of such flats relying on their rights to become rack rented tenants and staying in possession? (5) what were the valuation consequences of the possibility that the 1989 Act might apply at the end of the terms; (these consequences might not have been restricted to the choice of deferment rate)?

41. I would not attach equal significance to all of the matters I have identified in the last para. As to (1), I think it is quite likely that the flats were within the relevant rateable value limits but we were not shown any evidence which was before the Upper Tribunal on the point and the passage in *Ex parte Firth, In re Cowburn* cited in *Pittalis* referred to "any possibility [which] would prevent the point from succeeding".

* Editor's note: Also reported at [1989] 2 EGLR 90; [1989] 28 EG 126

A As to (2), (3) and (4), Mr Dutton submits that the relevant possibility need not be measured or assessed; he says that it suffices if it exists as a possibility because that would negative the "guarantee" of early possession referred to by the Upper Tribunal. Whilst that is logically correct, his submission misses the point that it is likely, or at any rate it is possible, that an assessment of the risk would impact on the valuation consequences of the possibility. In any case, I consider that there is no answer to the difficulty created by the point in (5). I am unable to accept that the difficulty can be side-stepped (as Mr Dutton seeks to do) by holding that the inevitable valuation consequence of taking into account the possibility of the application of the 1989 Act is that one B must adopt the generic deferment rate identified by the Lands Tribunal in *Sportelli*. That case gave guidance as to the deferment rate for unexpired terms of 20 years or more. The Upper Tribunal in *Cadogan Square Properties* gave slightly different guidance for unexpired terms of 10 years up to, but not including, 20 years. That decision also indicated that, depending upon the valuation evidence called, a different approach might be appropriate for unexpired terms of under 10 years. The present case is concerned with unexpired terms of 4.74 years. The Upper Tribunal adopted an approach different from that identified in the two earlier cases. It is impossible for a court to hold as a matter of law, without permitting the parties to call valuation evidence on the point, C that the only possible valuation consequence of taking into account the possibility of the application of the 1989 Act is that the deferment rate in this case reverts to the generic rate identified in *Sportelli*.

42. If this court were to allow the nominee purchaser to take the point as to the possible application of the 1989 Act and if that point were to be held to be legally correct, the inevitable result would be that the court could not itself determine the appropriate deferment rate and would have to remit the matter to the Upper Tribunal for it to give directions as to the further preparation needed for a yet further hearing at which the new point could be considered and dealt with. In accordance with the established approach identified in the authorities, I D consider that the nominee purchaser ought not to be allowed to raise the new point as to the 1989 Act and that the ground of appeal which raises that point ought to be dismissed.

43. The nominee purchaser had a further criticism of the decision of the Upper Tribunal in relation to the deferment rate. The nominee purchaser pointed out that para 3 of Schedule 6 to the 1993 Act states that "the value of the freeholder's interest in the specified premises is the amount which at the relevant date that interest might be expected to realise if sold on the open market by a willing seller". It submitted that the valuation in this case should have been on the basis that the whole of Vale Court was notionally sold in the open market. It suggested that the E Upper Tribunal did not value on that basis but instead valued a property interest which consisted only of the reversions on the six leases which had unexpired terms of 4.74 years. It was submitted that the Upper Tribunal considered that the notional purchaser of that more restricted property interest would have been keen to acquire the reversion on those six leases and would have bid accordingly. The attitude of such a purchaser was not to be equated with the different attitude of a purchaser of the whole of Vale Court.

44. I agree that the interest to be valued is the freehold interest in the whole of the specified premises. In the present case, the specified F premises comprised the block at Vale Court. That block consisted of 25 flats. Six of those flats were let on leases with unexpired terms of 4.74 years; 17 were let on leases with unexpired terms of between 70 and 96 years and two were to be the subject of leasebacks with terms of 999 years. I agree that it would not be right to assess a value for the reversions on the six flats with unexpired terms of 4.74 years as if the reversions on those flats were the entire subject matter of the notional sale. In particular, such an approach would produce the wrong valuation answer if one took into account the attitude of a notional purchaser who would be keen to buy a block of property consisting of the reversions on six leases with unexpired terms of 4.74 years but who would adopt a very different attitude to a possible purchase of the whole of Vale Court comprising 25 flats with a mixture of lease terms.

45. Where I part company from the submissions for the nominee G purchaser is that I do not consider that the Upper Tribunal at any time lost sight of the need to value the freehold interest in the block as a whole. It arrived at the value of the freehold in the whole block by placing values on individual elements in the valuation. Thus, as earlier explained, it adopted the term and reversion method of valuation which is conventional. That involves placing separate values on the term and on the reversion. In doing so, the Upper Tribunal (and all valuers who adopt this conventional method) do not make the elementary mistake of assuming that there is a separate sale of the right to receive rent during the term and of the right to vacant possession at the end of term.

46. In this case, it was necessary to consider the deferment rate to H be applied to different parts of the interest being acquired. Both valuers who gave evidence before the Upper Tribunal considered whether to adopt different deferment rates for the flats with unexpired terms of 4.74 years from the rates for the other flats. To consider whether to adopt different deferment rates does not mean that the valuer is making the elementary mistake that the reversions on the short leases are being sold by way of a sale separate from the sale of the block. In those parts of its decision where the Upper Tribunal considered how a purchaser would consider the advantages of the flats with short unexpired terms, I do not detect any sign that the Upper Tribunal was making the elementary mistake that the reversions on those flats were being sold separately J from the remainder of the block. In any case, there is no suggestion in its decision that the Upper Tribunal thought that the reversions on those flats would have had a particular attraction to a purchaser if those reversions were sold separately from the remainder of the block and would be viewed differently by a purchaser of the whole block.

47. I consider that this criticism of the Upper Tribunal is without foundation. What the Tribunal was doing was to build up its value for the single freehold interest by assessing the contribution to that single value made by the value of individual components.

Hope value: the decision under appeal

48. In this case, there were five non-participating tenants. In K accordance with the decision of the House of Lords in *Sportelli*, it was necessary to consider whether the value of the freeholder's interest in the block should include hope value in relation to those five flats. The statutory provisions make it clear that the valuation is to be carried out on the assumption that all the tenants in the block (including therefore the non-participating tenants) do not have a statutory right to an extended lease under Chapter II of the 1993 Act. Therefore, the relevant hope is that a non-participating tenant will come forward, seeking to negotiate with the freeholder for the grant of an extended lease, otherwise than pursuant to a statutory entitlement to acquire an extended lease. L

49. The Upper Tribunal's decision in relation to hope value is lengthy and detailed. At the beginning of its discussion of hope value, the Upper Tribunal referred to the speech of Lord Neuberger in *Sportelli* at [66] and then (at [77]) directed itself as follows:

> "The conclusion of the House of Lords (Lord Hoffmann dissenting) was that hope value can be taken into account under para 3 of Schedule 6, in so far as it is attributable to non-participating tenants wishing to obtain new leases of their flats in the open market (and not pursuant to Schedule 13)."

50. Thus, the Upper Tribunal correctly directed itself that hope value was to be assessed on the assumption that the tenants did not have a statutory right to an extended lease. The decision then M referred at length to the evidence of the valuers for the parties. Those references all proceed on the correct basis that the tenants did not have a statutory right to an extended lease. The Upper Tribunal was faced with competing evidence and submissions as to how likely it was that a non-participating tenant would come forward, before the expiry of the current lease, to negotiate for an extended lease. One of the five non-participating tenants had a lease with an unexpired term of 70.25 years; the other four had leases with unexpired terms of 4.74 years. Therefore, it was necessary to form an assessment of the relevant likelihood for these different lengths of terms. There was evidence before the Upper Tribunal as to this likelihood based on the history of lease extensions

in Vale Court and the Upper Tribunal was also referred to the evidence on this topic in other cases decided by the Upper Tribunal. Because this evidence related to transactions which actually happened (ie in the real world where tenants had rights under Chapter II of the 1993 Act) much of it (but not all) involved lease extensions being sought pursuant to Chapter II of the 1993 Act (by service of a notice under section 42 of the 1993 Act). The Upper Tribunal considered that the evidence showed a pattern which allowed it to form an assessment of the likelihood of a non-participating tenant with a short unexpired term coming forward, before the expiry of his term, to negotiate for an extended lease.

Hope value: the appeal

51. The nominee purchaser submitted that towards the end of its lengthy reasoning on the subject of hope value the Upper Tribunal took a false step. It is submitted that instead of the Upper Tribunal assessing the hope of a tenant coming forward to negotiate for an extended lease (without any statutory entitlement to an extended lease) the Upper Tribunal assessed a quite different hope, namely, that a tenant who had an entitlement to an extended lease under Chapter II of the 1993 Act would seek to exercise that entitlement.

52. If the Upper Tribunal had done that which the nominee purchaser contends it did, then it would have made an elementary error. Moreover, it would have been an elementary error contrary to its own self-direction at [77] of the decision and contrary to the basis on which it had discussed in detail the evidence of the valuers. I do not consider that the Upper Tribunal made any such error. In the course of its decision, the Upper Tribunal repeatedly referred to the need to assess the chance of the tenant "coming forward" to seek an extended lease. In context, it is clear that the Upper Tribunal was referring to a tenant coming forward without a statutory entitlement to an extended lease.

53. The nominee purchaser criticises the Upper Tribunal's reasoning in [115] and [116] of its decision. The Upper Tribunal there stated:

> "115. Turning now to the level of hope value that should be applied to each non-participating flat, we accept, as we have said, Mr Roberts's evidence as to the likelihood of lessees coming forward in respect of the four flats with 4.74 years remaining. We agree that the potential imperatives for coming forward prior to lease expiry are many and that each lessee will have his own reasons for doing so. The fact that, with only a short time to go before expiry, the lessee has not served a notice does not mean he will not do so. Despite Mr McDonald's initial arguments that the market was declining at the valuation date, and was anticipated to continue in that vein, it seems to us on the facts that that was not the case. Mr McDonald also accepted in cross-examination that the market was closer to static than the impression he had initially given. Therefore, whilst he mounted a cogent argument as to why in a falling market the chances of a lessee serving a section 42 notice would be significantly diminished, that was not the situation that existed here.
>
> 116. The statistical analysis by Mr Roberts as to the percentages of lessees who had come forward, and the lease lengths that would remain when they did so, adds weight to our conclusion that a purchaser would be justified in anticipating the receipt of such a notice and would build in hope value to reflect that view. For instance, in Erkman there was evidence [72] that a third of the applications for lease extensions came when there were less than 5.75 years unexpired."

54. In [115], the Upper Tribunal referred to the evidence of the valuer for the nominee purchaser as to "the chances of a lessee serving a section 42 notice". That is a reference to the evidence which had been given as to what happened in the block and elsewhere as to the service of section 42 notices. I do not read that reference to the evidence as revealing that the Upper Tribunal was making the elementary error now suggested by the nominee purchaser.

55. In [116], the Upper Tribunal referred to "the statistical analysis" carried out by the valuer for the reversioner. It was explained to us that this was a reference to the evidence as to the tenants who had sought lease extensions in the block (and possibly elsewhere). Much of that evidence related to tenants exercising their statutory rights to extended leases. It was in the context of discussing that evidence that the Upper Tribunal referred to a purchaser anticipating "receipt of such a notice". It is not clear that this phrase is a reference to a notice under section 42 but even if it were, such a reference was not inappropriate as the Upper Tribunal was discussing the evidence as to the service of such notices.

56. Following the comments in [114] and [115] of the decision, the Upper Tribunal then posed for itself in [117], [118] and [119], and then answered, the question as to the prospect of a non-participating tenant coming forward to negotiate for an extended lease. It is clear to me that the Upper Tribunal was considering that question on the correct basis that the tenant did not have a statutory entitlement to an extended lease.

57. The nominee purchaser next submitted that it would not be right to equate the position of a tenant with a short unexpired term, but who has a statutory right to an extended lease, with the position of a tenant with a short unexpired term, but who does not have such a statutory right. It was submitted that there were important differences between the two cases. Conversely, the reversioner submitted to us that the two cases had many similarities. I do not think that the Upper Tribunal made any error of law in its handling of the evidence as to the cases where tenants had exercised their statutory rights to extended leases shortly before the end of their current terms. Both valuers commented on this evidence. The Upper Tribunal plainly thought that the evidence was of some help in forming its assessment of hope value. However, the Tribunal did not at any time say that the two cases were identical.

58. The nominee purchaser next submitted that the Upper Tribunal had erred in law in that it had ignored the possibility that the lessees might have security of tenure under Schedule 10 to the 1989 Act. Neither party had asked the Upper Tribunal to take this possibility into account. Neither party had called any evidence which would have enabled the Upper Tribunal to assess that possibility, nor any evidence as to the valuation consequences of doing so. The parties had expressly agreed with the note from the Upper Tribunal, to which I referred when considering the submissions as to the deferment rate. The parties therefore agreed that there was certainty as to the freeholder's right to vacant possession at the end of the terms. The Upper Tribunal did not err in law in acting on that basis. The new point as to the possible application of the 1989 Act is not open to the nominee purchaser on this appeal. It is not a pure point of law. If the point had been taken before the Upper Tribunal, there would have had to have been further or different evidence to enable the Upper Tribunal to deal with it.

59. The nominee purchaser then submitted that the Upper Tribunal was wrong to quantify the hope value in relation to the four non-participating tenants with unexpired terms of 4.74 years at 20% of the marriage value in relation to those four flats. I do not consider that the Upper Tribunal made any error of law in its assessment. It considered the evidence which had been given by both valuers. The evidence from the reversioner's valuer supported a figure of 20% of marriage value. He contended that 20% was the right figure not only for the four flats with unexpired terms of 4.74 but also for the fifth non-participating tenant who had an unexpired term of 70.25 years. The Upper Tribunal explained in detail why it was unconvinced that it should apply a "blanket rate" of 20%. In its decision, it assessed the hope value for this fifth flat at 10% of marriage value.

60. At [117] of its decision, the Upper Tribunal stated that there was nothing in the evidence "to dissuade us" from accepting 20% of marriage value for the four flats with short unexpired terms. The nominee purchaser submitted that the Upper Tribunal was directing itself that there was a presumption in favour of a figure of 20% and that there was a legal or evidential burden on the nominee purchaser to rebut that presumption. I do not agree. Read in context, the statement in [117] of the decision, coming after a detailed analysis of all of the evidence before it, is to be understood as saying that the evidence supported the figure of 20% and that there was nothing in the evidence to justify the Upper Tribunal in adopting a lower figure.

61. Finally, on the subject of hope value, it was submitted that the Upper Tribunal had gone wrong in its treatment of an earlier decision of the Upper Tribunal in *Culley* v *Daejan Properties Ltd* [2009] UKUT 168 (LC); [2009] 3 EGLR 165. The tribunal judges in *Culley* were the same as in the present case. The nominee purchaser submitted that the Upper Tribunal in the present case was entitled to take account of

A the reasoning in *Culley* which commented on certain matters which affected the choice of the percentage of marriage value. It was there considered that it might be appropriate to take a lower percentage where the unexpired terms were particularly long. Further, it might be appropriate to take a higher percentage where the proportion of non-participating flats was particularly large. In *Culley*, the unexpired terms were around 65 years and the percentage was 10%. In the present case, the Upper Tribunal selected 10% where the unexpired term was 70.25 years and 20% for unexpired terms of 4.74 years. In *Culley*, 50% of the leases were held by non-participating tenants; in the present case, there were five non-participating tenants in a block of 25 flats.

B 62. In the present case, the Upper Tribunal referred to *Culley* as part of its reasoning for its selection of 10% for the flat with an unexpired term of 70.25 years. The nominee purchaser does not criticise it for doing so. The Upper Tribunal did not rely on the evidence in *Culley* when it selected 20% for the other four flats; it relied on the evidence before it and referred to its own reasoning in *Culley*. Accordingly, the nominee purchaser's submission that the evidence in *Culley* was not admissible when coming to a conclusion on the evidence in the present case is not relevant to what the Upper Tribunal did in the present case. The nominee purchaser submitted to us that having regard to the considerations referred to in *Culley* (the length of the unexpired C terms and the proportion of non-participating flats) the right answer was not 20%, but should have been 10%. If one were to attach weight to the considerations identified in *Culley*, there is a strong argument that they support a figure higher than the 10% contended for by the nominee purchaser. In any event, the weight to be attached to these considerations and the ultimate selection of the percentage was for the Upper Tribunal. The percentages it selected for hope value were plainly open to it on the evidence before it and the Upper Tribunal did not commit any error of law.

The alleged inconsistency

D 63. The nominee purchaser submitted that the approach of the Upper Tribunal to the deferment rate and to hope value were inconsistent. It was also said that its approach involved double counting. It was said that when the Upper Tribunal determined the deferment rate, it assumed that the freeholder's interest was sold to a purchaser who was interested in having early vacant possession of the six flats with unexpired terms of 4.74 years but when it determined hope value, it assumed that the freeholder's interest was sold to a purchaser who would pay more in order to buy the hope of being able to grant extended leases to the five non-participating tenants, four of whom held the leases with unexpired terms of 4.74 years (the hope being of the grant of an extended lease E before the expiry of the relevant term). It was said that if a purchaser would pay more for that hope he could not be said to be interested in early vacant possession.

64. I am unable to accept this submission. There were four leases held by non-participating tenants where the unexpired terms were 4.74 years. In those cases, the Upper Tribunal assessed the vacant possession value at the valuation date and then deferred that value for 4.74 years, using its selected deferment rate. Its choice of deferment rate reflected the fact that the period of deferment was short and after 4.74 years the purchaser would be entitled to vacant possession. At that point the purchaser would have the benefit of a vacant flat and he could do with it F as he pleased. In particular, he would be able to let it on a long lease for a premium and thereby realise its vacant possession value. In the case of those four leases, the Upper Tribunal also included a sum for hope value. That sum reflected the hope that the lessees would come to the freeholder before the expiry of their terms and take extended leases at a premium. In that way, the purchaser would not have to wait 4.74 years before being able to charge a premium for the grant of a long lease of the relevant flat. The premium charged for a long lease while there was still a short unexpired term would be, or might be, different from the premium for a lease with vacant possession but that ought to be picked up in the assessment of the hope value. Accordingly, I do not consider that there was any inconsistency in the approach of the Upper Tribunal to the two questions of deferment rate and hope value.

65. There might have been an inconsistency in approach if the Upper G Tribunal had assessed the price for the reversion on one or more of the short unexpired terms on the basis that the purchaser would pay a higher price than anyone else because he was buying one or more of those flats for his own occupation. It might be said that such a purchaser would not be interested in the prospect of granting an extended lease to the existing lessee of such a flat in the period before the expiry of the term and so there should not be hope value in relation to such a flat, as well as an enhanced price paid by an intending occupier of the flat. Whatever the position would be in relation to that possibility, I do not think that the reasoning of the Upper Tribunal involved an inconsistency of that kind or, indeed, any inconsistency. H

The result

66. I do not accept any of the points raised by the nominee purchaser on this appeal and, accordingly, I would dismiss the appeal.

Future guidance

67. In [143] of its decision, the Upper Tribunal said:

> "Accordingly for future guidance we conclude that the deferment rate for reversions of less than five years should be the net rental yield that the evidence shows to be appropriate for the property in question; and that in addition there should be an end allowance, which, in the absence of evidence establishing J some other percentage, should be 5%."

68. The Upper Tribunal's decision to offer future guidance as to the deferment rate in the case of reversions of less than five years followed the approach adopted by the Lands Tribunal in *Sportelli* (an approach approved by the Court of Appeal in the same case) and by the Upper Tribunal in *Cadogan Square Properties Ltd*. The nominee purchaser invited this court to make it clear that the guidance given by the Upper Tribunal in the present case should not be regarded as definitive because the possible application of the 1989 Act had not been considered in this case (because no one had raised it) and this possibility might be raised in another case and, when raised, this K possibility might undermine that guidance. I agree that the guidance given by the Upper Tribunal in this case may have to be qualified in a case in which a party raises an issue as to the possible application of the 1989 Act (or the Rent Act 1977 or the Housing Act 1988) at the end of the unexpired terms of the leases. If such an issue is raised, then the relevant tribunal will have to consider it and the decision in the present case, where the issue was not raised, does not offer any guidance as to how it should be dealt with.

Rimer LJ said:

69. I agree. L

Pill LJ said:

70. I agree with the conclusion of Morgan J and with his reasoning and want only to underline his approach to the following arguments now raised:

(a) That the tribunal erred in assuming that the reversioner was guaranteed the early enjoyment of possession;

(b) That even if the tribunal was entitled to make that assumption, the nominee purchaser should now be permitted to challenge it, and

(c) Upon a successful challenge, the court must adopt the generic M deferment rate identified by the tribunal in *Sportelli*, where the tribunal gave guidance as to the deferment rate for unexpired terms of 20 years or more.

The argument was based on the possibility, considered by Morgan J at [22] and [23], that tenants may have claims to security of tenure under section 186 and Schedule 10 of the Local Government and Housing Act 1989.

71. At each stage of the argument, I agree with the reasoning of Morgan J at [26]-[42] of his judgment. It is clear that it was accepted on behalf of the nominee purchaser that the tribunal could approach the issues on the basis that the early enjoyment of possession was guaranteed. That is clear from the nominee purchaser's response to the

tribunal's note to the parties in advance of judgment, considered by Morgan J at [31] and [32].

72. In seeking to re-open the issue, the nominee purchaser relied on the decision of this court in *Pittalis* at p611 C-F where the court permitted the appellant to take "a pure point of law not raised below". In that case, a flat and shop had been let on a long lease and the flat was sublet for three years. The leasehold interest was surrendered to the landlords and the sub-tenancy of the flat had expired by effluxion of time but the defendants remained in occupation and a statutory tenancy was claimed.

73. At the hearing in the County Court, the landlord conceded that the flat was "premises" within section 137(3)(a) of the Rent Act 1977, as amended. The claim to possession failed on the ground that section 137(3) afforded protection and the landlord was permitted to withdraw the concession in this court. Nourse LJ, giving the judgment of the court, stated, at p611C that

> "the substantive question is a pure question of law which does not depend on evidence as to the actual state of the property at the material time."

74. The present situation is very different. Whether it would have been appropriate to take into account the possibility that the 1989 Act might have applied at the end of the terms depended on the questions of fact and opinion identified by Morgan J at [39] and [41] of his judgment.

75. Each of these issues would require an opportunity to give evidence and careful consideration by a fact-finding tribunal, including a risk assessment. The point was not taken before the tribunal and I agree that the nominee purchaser should not be allowed to take it now. It may be that the decision not to attempt to take the point before the tribunal was based on the nominee purchaser's interest in supporting higher values for the leasehold interest, as contemplated by Morgan J at [35]. The nominee purchaser should not be permitted a second run.

76. As to stage (c), I do not in any event accept the submission that the inevitable consequence of taking into account the possibility of the application of the 1989 Act was that the *Sportelli* deferment rate must be adopted. That case was concerned with unexpired terms of 20 years or more, a very different situation from the unexpired terms of 4.74 years under consideration. I agree with Morgan J, at [42], that the court could not adopt the *Sportelli* rate as a default position. It would have been necessary to give the reversioner an opportunity to call valuation evidence. Remittal to the tribunal would have been inevitable.

77. I find no merit in the arguments raised by Mr Dutton on any of the issues posed above, whether as a first or second appeal. In agreement with Morgan J on these and the other issues, I too would dismiss the appeal.

Appeal dismissed.

LEASEHOLD VALUATION TRIBUNAL

Havering London Borough Council v Macdonald

UPPER TRIBUNAL: LANDS CHAMBER

17 May 2012

HH JUDGE KAREN WALDEN-SMITH

[2012] UKUT 154 (LC)

Estates Gazette 8 September 2012

[2012] 36 EG 100

Leasehold Valuation Tribunal – Decision – Reasons – Leaseholder challenging reasonableness of service charge levied by landlord for provision of television and radio signals – LVT determining amount to be unreasonable – LVT giving further reasons for that conclusion when refusing permission to appeal – Whether LVT's reasons adequate – Whether entitled to expand on reasons in decision on permission to appeal

The respondent held a long lease of a flat in a block of which the appellant local authority was the freeholder. He brought proceedings in the Leasehold Valuation Tribunal (LVT), under section 27A of the Landlord and Tenant Act 1985, to challenge the reasonableness of the service charges levied by the appellants for the provision and maintenance of communal television and radio signals to the flat. The relevant costs arose under a contract between the appellants and a communications company; that agreement had been varied to upgrade the services with a consequent increase in the charges. The charges for the years ending in March 2009 and 2010 were £66.56 and £70.20 respectively.

The LVT determined that the costs of providing and maintaining the signals was recoverable under the terms of the lease but that the costs charged to the respondent were unreasonable in amount; it found that the maximum reasonable charge was £26 pa. In its reasons for that conclusion, the LVT indicated that the evidence before it was limited, there was no evidence as to the competitiveness of the relevant agreement and the large difference in the costs of satellite and television signals was unexplained; it further found that the 15-year length of the agreement was unreasonably long.

The appellants applied for permission to appeal. They contended that the LVT had given no adequate reasons for its decision and that its finding as to the maximum reasonable charge was an unjustified "quantum leap" from its other findings. The LVT refused permission, stating that its maximum reasonable figure of £26 pa was based on the limited evidence before it and its own general knowledge and experience. Permission to appeal was subsequently granted by the Lands Chamber of the Upper Tribunal.

Decision: **The appeal was allowed. The giving of reasons is necessary to satisfy the requirements of natural justice. A party, particularly the losing party, is entitled to know why the decision was made. The reasons given must be clear and specific, making it possible to determine whether the tribunal has misdirected itself. If the LVT does not give adequate reasons in its initial decision, it can remedy that defect by setting them out in its subsequent decision on permission to appeal, provided the reasons it gives were properly within its mind at the time when the decision was made and formed the basis, or part of the basis, for that decision. However, the LVT has no power to review its own decision on an application for permission to appeal. It can correct a clerical mistake under the limited "slip rule" power in regulation 18(7) but is not entitled to have second or additional thoughts or to alter its ultimate decision. The LVT is entitled to use its own expertise in deciding the issues before it, but it cannot use that expertise to raise and decide other issues that the parties might not have had an opportunity to consider and it cannot reach a conclusion on the basis of a point or argument that has never been raised by the parties or put to them. Applying the foregoing principles, the LVT had been entitled to expand on its reasons when refusing to grant permission to appeal but the expansion of reasoning that it had given did not cure the defect in its original decision. The amplified reasons were not matters that had been fully aired in front of the parties and on which they had been given an opportunity to comment. The matter was remitted to a differently constituted LVT for rehearing.**

The following cases are referred to in this report.

Arrowdell Ltd v *Coniston Court (North) Hove Ltd* LRA/72/2005
Bristol-Myers Squibb Company v *Baker Norton Pharmaceuticals Inc and Napro Biotherapeutics Inc* [2001] EWCA Civ 414
Chelsea Properties Ltd v *Earl Cadogan* LRA/69/2006
English v *Emery Reimbold & Strick Ltd* [2002] EWCA 605; [2002] 1 WLR 2409
Flannery v *Halifax Estate Agencies Ltd* [2000] 1 WLR 377
Lucie M v *Worcestershire County Council and Evans* [2002] EWHC 1292 (Admin)
Oxfordshire County Council v *GB* [2001] EWCA Civ 1358
R v *Croydon London Borough Council ex parte Graham* (1993) 26 HLR 286
Re Clarise Properties Ltd's appeal [2012] UKUT 4 (LC)
Regent Management v *Jones* [2010] UKUT 369 (LC)
Schilling v *Canary Riverside Development PTD Ltd* LRX/26/2005
Westbourne Ltd v *Spink and Joshua* LRX/14/2007
Yorkbrook Investments Ltd v *Batten* (1986) 19 HLR 25

This was an appeal by the appellants, Havering London Borough Council, from a decision of the leasehold valuation tribunal, under the Landlord and Tenant Act 1985, determining the reasonableness of service charges levied on the respondent, George Macdonald.

James Fieldsend (instructed by the legal department of Havering London Borough Council) appeared for the appellants; the respondent appeared in person.

Giving judgment, **HH Judge Karen Walden-Smith** said:

Introduction

1. This is an appeal by way of review from the decision of the Leasehold Valuation Tribunal ("LVT") for the London Rent Assessment Panel dated 4 January 2011. Permission was refused by the LVT on 4 February 2011. An application was made to the Upper Tribunal (Lands Chamber) for permission to appeal by way of re-hearing.

2. Permission to appeal was granted by the President on 8 April 2011 with the following observations:

> "There is a realistic prospect of success. I do not think, however, that the Lands Chamber should conduct a rehearing: if the LVT decision is shown to be erroneous, the proper course would be a remission.
>
> The appeal will be dealt with by way of review."

3. The appellant contends that the LVT erred in its decision-making and that the appeal should therefore be allowed and the matter remitted to a differently constituted LVT. The respondents seeks to uphold the decision of the LVT.

Background

4. The appellant is a local authority. The respondent is the long leaseholder of Flat 48 Parkview House, Sunrise Avenue, Hornchurch RM12 4YW ("Flat 48"). Flat 48 is one of 52 flats in Parkview House which is one of three blocks of flats on an estate comprising a total of 148 flats. The appellant is the freehold owner of the three blocks of flats including Flat 48.

5. By the lease dated 14 November 2005 the respondent was granted a term of 125 years of Flat 48 ("the lease"). The term of 125 years commenced on 18 June 1984 and the long lease was granted pursuant to the Right to Buy provisions of the Housing Act 1985, the respondent having previously been a secure tenant of the appellant residing in Flat 48. Clause 9 of the Eighth Schedule to the lease contains the service charge provisions.

The issue before the LVT

6. The matter for determination by the LVT was the reasonableness of the service charge levied by the appellant to the respondent for the provision and maintenance of communal television and radio signals to Flat 48. The costs charged to the respondent arise from a contract entered into between the appellant and Surtees Communications Ltd on 6 May 1992 for the provision and maintenance of radio and television signals. Surtees Communications Ltd changed its name to Surtees Holdings Ltd on 16 April 1993.

7. The agreement dated 6 May 1992 was subject to two subsequent variations in writing: one on 28 September 1997 and one on 10 January 2001. The purpose of these variations was purportedly to upgrade the services provided. In the first variation of September 1997, Surtees Holdings Ltd added the reception of additional satellite signals; added Channel 5 reception; and assumed the maintenance of the communal antenna systems in the borough. That variation provided for an increase in the sums payable under the agreement. The second variation of January 2001 provided for the phased upgrade of the communal antenna systems to accept digital television signals. That second variation also allowed for an increase in the sums payable under the agreement and for an extension of the original term.

8. The cost of the provision of television and radio signals is charged to the respondent as part of the service charge payable under the terms of the lease (clause 9 of the eighth schedule to the lease). The service charge year runs from 1 April to 31 March (clause 13 of the Fourth Schedule). Section 20 of the Landlord and Tenant Act 1985 and the Service Charges (Consultation Requirements) (England) Regulations 2003 do not apply as the sum payable by each tenant falls below the annual threshold of £100 so that the agreement of 1992, subject to the variations of 1997 and 2001, are not subject to the statutory consultation requirements.

9. The cost charged to the respondent for the period 1 April 2008 to 31 March 2009 was £66.56 and for the period 1 April 2009 to 31 March 2010 was £70.20.

10. The respondent challenged the reasonableness of the cost of the provision of the radio and television signals to Flat 48 by an application made on 10 October 2010 pursuant to the provisions of section 27A of the Landlord and Tenant Act 1985. The respondent is plainly extremely cross about the charges levied by Surtees Holdings Ltd and, as he sees it, the failures of the appellant to negotiate a contract which provided good value for money. The respondent, in his submissions before me, indicated that he had been challenging these figures for some six years or more.

11. The respondent did not challenge the appellant's entitlement to recover the cost through the service charge provisions, but he did contend that it was not reasonable for him to pay for the cost of a satellite signal as it was his case that he neither sought nor received such a signal and, indeed, to receive such a signal would, he contends, have been illegal as he had no agreement with a provider of satellite television services.

The procedural history

12. The application to the LVT had been made by the respondent on 10 October 2010. On 25 October 2010, a procedural chairman (Mrs TI Rabin JP) had given directions including, at para 6 of the order, a direction with respect to reimbursement of fees and to an application for an order under section 20C of the Landlord and Tenant Act 1985.

13. The respondent's statement of case was set out in a document referred as being a "background" document and a "conclusion" document, together with a series of annotated attachments. The appellant's statement of case was set out in a statement of reply, which referred to a report by the appellant's former director of finance, Rita Greenwood, with respect to the costs of the provision and maintenance of signals and the costs and charges made by other local authorities with respect to communal signals.

14. The hearing took place on 21 December 2010 and the decision was promulgated on 4 January 2011. The respondent appeared in person and the appellant was represented by counsel with both the Housing Management Organisation's director of finance and resources and the stock options manager in attendance.

The decision of the LVT

15. The decision of the LVT was promulgated on 4 January 2011. The conclusions of the LVT were that it was satisfied that the cost of providing and maintaining the signals was recoverable under the terms of the lease ([6] of the decision) but that the costs charged to the respondent were unreasonable [18] and that "the top of the range reasonable charges" for the years 2008/2009 and 2009/2010 was £26 pa [21]. The LVT further found that it was "just and equitable" to make an order under section 20C of the Landlord and Tenant Act 1985 and for the reimbursement of the respondent's fees to the LVT pursuant to the provisions of regulation 9 of the Leasehold Valuation Tribunal (Fees) (England) Regulations 2003 [21].

16. The reasoning of the LVT is set out in [14] to [21] of the decision. The LVT found that the evidence contained in the *Rita Greenwood Report* was hearsay but was accurate and they therefore found the contents therein as facts. The LVT further found that the evidence before them was limited and that there was no evidence as to the competitiveness of the 1992 agreement and that the large difference in the costs of satellite and television signals was unexplained. It was further found that a contract of significant length was reasonable in order for Surtees Holding Ltd to recoup its capital outlay, but that the original term of the agreement (of 15 years) was unreasonably long; that the variations to the agreements were unexplained and inexplicable; that the costs charged to the respondent were unreasonable; and, as set out above, that, while there was a range of reasonable charges, the top of the range of reasonable charges was £26 pa. The appellants say that this final finding was a "quantum leap" from the other findings particularly where it had recognised that the information underlying the costs charged to other local authorities, as contained in the *Rita Greenwood Report*, were unavailable.

Refusal of permission to appeal

17. Permission to appeal from the LVT was sought from the LVT and refused on 4 February 2011 and, in the body of its reasons for refusing permission, the LVT recorded that:

> "On the limited evidence before us, we determined that the maximum reasonable charge was £26 per annum for 2008/09 and 2009/10. This figure was arrived at on the basis of the (limited) evidence before us and our own general knowledge and experience."

18. As is set out above, the President gave permission to appeal by way of review and, if appropriate, remission to the LVT on 8 April 2011.

The issues

19. The appellant contends that the decision of the LVT should not be upheld and the matter remitted to a different constituted LVT on the following grounds:

(i) the LVT failed to give any, or any sufficient, reasons for its decision;

(ii) the failure to give any, or any sufficient, reasons for its decision cannot be cured on the refusal to give permission to appeal and, even if it could, the LVT failed to do so in this instance;

(iii) the findings on reasonableness were not open to the LVT on the evidence before the LVT;

(iv) the LVT came to a conclusion that no reasonable Tribunal could have come to on the evidence before it.

20. The respondent seeks to uphold the decision of the LVT on the basis of the matters set out in both the decision of the LVT and in the refusal to allow permission to appeal. He set out in oral submissions before me that he had spent more than six years challenging the figures charged for the provision of television signals; that he never sought nor received satellite television signals and, indeed, that it would have been illegal for him to have done so; that he has raised a number of Freedom of Information Act requests in order to obtain further information about the legitimacy of the agreements between the appellant and Surtees Holdings Ltd; and that he agreed with the comments of Mr Humphreys FRICS (a member of the LVT) that the terms of the contract with Surtees Holdings Ltd were "inexplicable". He expressed concern that the LVT may well have failed to properly set out the grounds for their conclusions but accused the appellants of being "very professional timewasters".

21. This appeal raises some general points about the decision-making process of the LVT and how the LVT should express its decisions.

Service charges and the Leasehold Valuation Tribunal

22. Sections 18 and 19 of the Landlord and Tenant Act 1985 (as amended by the Landlord and Tenant Act 1987 and the Commonhold and Leasehold Reform Act 2002) set out the statutory framework for service charges.

23. Section 18(1) provides that "service charge" means an amount payable by a tenant of a dwelling as part of or in addition to the rent which is payable directly or indirectly for services, repairs, maintenance, improvements or insurance or the landlord's costs of management; the whole or part of which varied or may vary according to the relevant costs. Section 18(2) provides that the relevant costs are the costs or estimated costs incurred or to be incurred by or on behalf of the landlord or a superior landlord in connection with the matters for which the service charge is payable.

24. Section 19(1) of the Landlord and Tenant Act 1985 (as amended) provides that relevant costs shall be taken into account in determining the amount of a service charge payable for a period only to the extent that they are reasonably incurred and where they are incurred on the provisions of services or the carrying out of works, only if the services or works are of a reasonable standard and the amount payable shall be limited accordingly.

Reasonableness

25. It is Mr MacDonald's contention that the costs incurred are not reasonable and the amount payable with respect to the provision of television and radio signals should be limited.

26. It is by virtue of the provisions of section 27A of the Landlord and Tenant Act 1987 (inserted by the Commonhold and Leasehold Reform Act 2002) that an application may be made to the LVT for a determination whether a service charge is payable and, if it is, as to the amount which is payable.

27. As is consistent with other decisions as to what is meant by "reasonableness", in determining the reasonableness of a service charge the LVT has to take into account all relevant circumstances as they exist at the date of the hearing in a broad, commonsense way giving weight as the LVT thinks right to the various factors in the situation in order to determine whether a charge is reasonable. The test is "whether the service charge that was made was a reasonable one; not whether there were other possible ways of charging that might have been thought better or more reasonable. There may be several different ways of dealing with a particular problem... All of them may be perfectly reasonable. Each may have its own advantages and disadvantages. Some people may favour one set of advantages and disadvantages, others another. The LVT may have its own view. If the choice had been left to the LVT it might not have chosen what the management company chose but that does not necessarily make what the management company chose unreasonable" per HH Judge Mole QC in *Regent Management* v *Jones* [2010] UKUT 369 (LC), in [43].

28. Once a tenant establishes a *prima facie* case by identifying the item of expenditure complained of and the general nature (but not the evidence) of the case it will be for the landlord to establish the reasonableness of the charge. There is no presumption for or against the reasonableness of the standard or of the costs as regards service charges and the decision will be made on all the evidence made available: see *Yorkbrook Investments Ltd* v *Batten* (1986) 19 HLR 25* (as applied in *Schilling* v *Canary Riverside Development PTD Ltd* LRX/26/2005 and *Regent Management Ltd.*

Written reasons and the permission to appeal decision

29. The LVT is a statutory body and the procedure of the LVT is governed by the Leasehold Valuation Tribunals (Procedure)(England) Regulations 2003 (made pursuant to the provisions of section 174 and Schedule 12 to the Commonhold and Leasehold Reform Act 2002).

30. The effect of regulation 18 of the 2003 Regulations is that reasons for a decision of the LVT (whether given orally at the end of the hearing or in writing) must be provided in writing.

31. The jurisprudence behind why written reasons must be given by a tribunal, including a Leasehold Valuation Tribunal, were set out succinctly by Lawrence Collins J (as he then was) in *Lucie M* v *Worcestershire County Council and Evans* [2002] EWHC 1292 (Admin) [10] and [11]. In that case, Lawrence Collins J was dealing with a decision of the Special Educational Needs Tribunal. However, he expressly found that the principles are not unique to that tribunal and set out the following principles:

(1) Proper and adequate reasons must be given, so that they are intelligible and deal with the substantial points that have been raised, and the reasons should deal, in short form, with the substantial issues raised in order that the parties can understand why the decision has been reached;

(2) As a result of that first principle, the absence of reasons to explain why a case was rejected may make the decision appear irrational;

(3) Where reasons are inadequate, it is not normally appropriate that the reasons should be amplified on the appeal to the High Court;

(4) A decision must be sufficiently specific and clear as to leave no room for doubt as to what has been decided;

(5) The lay members of a tribunal specifically appointed for their [educational] expertise may use that expertise in deciding issues before the tribunal, but they may not use it to raise and decide other issues which the parties may not have had an opportunity to consider as the tribunal must obey the rules of natural justice and members should not give evidence to themselves which the parties have had no opportunity to challenge

32. In *Flannery* v *Halifax Estate Agencies Ltd* [2000] 1 WLR 377, the Court of Appeal set out the following general rationale for the giving of reasons. First, it is a function of due process and justice. Fairness requires the parties (especially the losing party) to know the reason for the result as without such reasons, it cannot be determined whether the court has misdirected itself. The second reason is that it concentrates the mind and makes sure that the decision is soundly based on the evidence. As was put by Lord Phillips MR (as he then was) in *English* v *Emery Reimbold & Strick Ltd* [2002] EWCA 605; [2002] 1 WLR 2409, at [16]:

> "We would put the matter at its simplest by saying that justice will not be done if it is not apparent to the parties why one has won and the other has lost."

33. There is no power for the LVT to review its decision on an application for permission to appeal, but it can correct a clerical mistake (regulation 18(7)). In *Re Clarise Properties Ltd's appeal* [2012] UKUT 4 (LC)** the President, sitting with Mr Rose FRICS, concluded that the

---

* Editor's note: Also reported at [1985] 2 EGLR 100

** Editor's note: Also reported at [2012] 1 EGLR 83

LVT only has power to alter a decision by virtue of regulation 18(7) of the 2003 Regulations, which is very limited in scope:

> "It is in essence a slip rule, and the limitations of such a rule were stated in *Bristol-Myers Squibb Company* v *Baker Norton Pharmaceuticals Inc and Napro Biotherapeutics Inc* [2001] EWCA Civ 414 by Aldous LJ, with whom Laws LJ and Blackburne J agreed, where at [25] he concluded that the authorities established that:
>
> "the slip rule cannot enable a court to have second or additional thoughts. Once the order is drawn up any mistake must be corrected by the appellate court. However, it is possible under the slip rule to amend an order to give effect to the intention of the court."
>
> Further, in *Chelsea Properties Ltd* v *Earl Cadogan* LRA/69/2006, HH Judge Huskinson sitting with Mr Rose FRICS expressed the view, at [13], that a correction certificate under regulation 18(7) was valid so long as it did not "purport to alter the ultimate decision of the LVT…"

34. However, there is a clear rationale for the LVT to be able to amplify its reasons where reasons are said to be inadequate so long as those reason where properly within the mind of the LVT at the time the decision was made and formed the basis (or at least part of the basis) for the decision being reached.

35. In *English*, Lord Phillips held, at [25], that where an application for permission to appeal on the ground of lack of reasons is made to the trial judge, then the judge should consider whether his judgment is defective for lack of reasons:

> "If he concludes that it is, he should set out to remedy the defect by the provision of additional reasons refusing permission to appeal on the basis that he has adopted that course. If he concludes that he has given adequate reasons, he will no doubt refuse permission to appeal. If an application for permission to appeal on the ground of lack of reason is made to the appellate court and it appears to the appellate court that the application is founded, it should consider adjourning the application and remitting the case to the trial judge with an invitation to provide additional reasons for his decision or, where appropriate, his reasons for a specific finding or findings."

36. This decision is clear authority that a judge should properly provide additional reasons for a decision, where those detailed reasons are not given in his judgment. Similarly, an LVT should properly provide additional reasons for a decision where those detailed reasons are not given in the decision letter as there is no purpose in cases being appealed to the Upper Tribunal (Lands Chamber) simply for the Lands Chamber to remit for those lack of reasons. In service charge cases, the LVT may have a very large number of items to deal with and the LVT will not necessarily give detailed reasons on each item. If the losing party appeals on the basis that inadequate reasons are given, it should be open to the LVT, if appropriate, to expand on those reasons. This has the double advantage of potentially avoiding the need for an appeal and, if not, giving the Lands Chamber the opportunity of considering the reasons and whether the LVT has erred in its decision-making.

37. In *R* v *Croydon London Borough Council ex parte Graham* (1993) 26 HLR 286, a case dealing with the local housing authority's decision on whether someone was intentionally homeless, Steyn LJ (as he then was) set out, at p292, that:

> "In my judgment the idea that material gaps in the reasons can always be supplemented *ex post facto* by affidavit or otherwise ought not to be encouraged. That in effect is what we have been asked to do on behalf of the council. No doubt questions of the sufficiency of reasons usually involve a judgment as to matters of degree. Nevertheless it seems to me that if the reasons are insufficient to enable the court to consider the lawfulness of the decision the obligation of furnishing reasons has been breached and in that event the decision itself will be unlawful."

and, following the reasoning of Steyn LJ that the very absence of reasons may have rendered the decision itself unlawful, Sedley LJ giving the decision of the Court of Appeal in *Oxfordshire County Council* v *GB* [2001] EWCA Civ 1358 said, at [9]:

> "…we do not consider it generally appropriate that a statutory tribunal which is required to give reasoned decisions should responds to an appeal by purporting to amplify its reasons."

38. In this matter, the LVT was not reviewing its own decision (which is plainly not permissible) it was providing what it said were the reasons for its decision. That is permitted in accordance with the principles set out in *English* and does not prevent a potential appellant seeking to contend that the very existence of the gaps in reasoning makes the decision unlawful or that the reasons given on the refusal of permission cannot have been the reasons for the decision, on a proper reading of that decision.

The expertise of the LVT

39. In *Arrowdell Ltd* v *Coniston Court (North) Hove Ltd* LRA/72/2005, the President sitting with Mr Rose FRICS set out that, as an expert tribunal, an LVT should use its knowledge and experience to test and, if necessary to reject, evidence that is before it. As was said by Judge Mole in *Regent Management*, at [36]:

> "The LVT is perfectly entitled, as an expert tribunal, to raise matters of its own volition. Indeed it is an honourable part of its function, given that part of the purpose of the legislation is to protect tenants from unreasonable charges and the tenants, who may not be experts, may have no more than a vague and unfocussed feeling that they have been charged too much. But it must do so fairly, so that if it is a new point which the tribunal raise, which the respondent has not mentioned, the applicant must have a fair opportunity to deal with it."

40. In *Arrowdell*, the President set out three "inescapable requirements", at [23]:

> "Firstly, as a tribunal deciding issues between the parties, it must reach its decision on the basis of evidence that is before it. Secondly, it must not reach a conclusion on the basis of evidence that has not been exposed to the parties for comment. Thirdly, it must give reasons for its decision. In the present case the tribunal rejected the evidence of both the experts on relativity, and it was entitled to do this provided its reasons for doing so were explained. But in basing its decision on "its own knowledge and experience, particularly in relation to relativities which have been agreed between parties or their valuers in other similar cases" it was in error because those agreements on relativity had not been identified nor had the parties had the opportunity to comment on them."

41. Further, an LVT must not reach a conclusion on the basis of a point or argument which has never been raised by the parties or put to the parties by the Tribunal (per HH Judge Reid QC in *Westbourne Ltd* v *Spink and Joshua* LRX/14/2007).

Conclusion

42. The primary criticism of the LVT made by the appellant is that the decision letter dated 4 January 2011 failed to set out any, or any adequate, reasons as to how the conclusion was reached that the service charges levied by the appellant for 2008/9 and 2009/10 were unreasonable. The decision letter sets out that the evidence before the LVT "was limited"; that a contract of a significant length was appropriate in the circumstances of this matter but then, without any reason or justification given, finds that 15 years was unreasonable; that the increases in prices brought about by the two variations to the agreement with Surtees Holding Ltd were "inexplicable" despite an explanation being provided within the body of the variations to the original agreement; that it was acknowledged that it would not be fair to follow the Barking & Dagenham charges as set out in the *Rita Greenwood Report* as the basis of those charges was not known; but then comes to a conclusion, without any reasoning being expressed, that there was a range of reasonable charges and that the top of that range was £26 pa, or 50 pence per week.

43. For the reasons I have already set out, it is necessary for the LVT to give written reasons for its decision. This is both because the regulations governing the LVT require written reasons to be given, but is also to satisfy the needs of natural justice. A party (particularly the losing party) is entitled to know why the decision was made and the reasons given must be clear and specific.

44. Following the reasoning of Lord Phillips in *English*, it is appropriate for the LVT to be able to remedy any defect apparently caused by lack of reasons being given, by giving those reasons in the permission to appeal decision.

45. The LVT was, therefore, entitled to expand upon its reasons but their expansion of reasoning set out in refusal to grant permission to appeal, namely that the LVT was relying on the limited evidence before it and "our own general knowledge and experience" did not cure the defect. The amplified reasons must be matters which have been fully aired in front of the parties in order that they have been able to comment upon the same.

46. The LVT is an expert body and its expertise is an extremely important part of the decision-making process. However, the knowledge and expertise relied upon by the LVT must be raised before the parties, again in order that the parties have the opportunity to comment upon the same.

47. The appeal is therefore allowed.

48. There is no record of an application for an order pursuant to the provisions of section 20C of the Landlord and Tenant Act 1985 and on the application to the LVT, the respondent said that it was "not applicable". In any event, the order relating to section 20C and the remission of fees are both matters which fall away by reason of the appeal being allowed.

49. The final issue is that, as the appeal was by way of review rather than re-hearing, this matter must be remitted for rehearing. In all the circumstances, it seems to me highly sensible that the case is remitted to a differently constituted LVT. The respondent is able to argue all the points he wishes to argue with respect to the reasonableness of the service charge before that LVT.

Appeal allowed.

Birmingham City Council v Keddie and another

UPPER TRIBUNAL: LANDS CHAMBER

25 September 2012

[2012] UKUT 323 (LC)

HH JUDGE GERALD

Estates Gazette 8 December 2012

[2012] 49 EG 71

Leasehold valuation tribunal – Jurisdiction – Service charge – Landlord and Tenant Act 1985 – Application to LVT under section 27A of Landlord and Tenant Act 1985 – Tenants disputing amount of service charge demanded by landlords for window replacement – Tenants alleging poor workmanship – LVT determining charge unreasonable since not reasonable to replace windows – Whether LVT entitled to reach decision on point not advanced by parties – Whether exceeding jurisdiction

The respondents held a lease of a flat in a block owned by the appellant landlords. They applied to the leasehold valuation tribunal (LVT), under section 27A of the Landlord and Tenant Act 1985, for a determination that a service charge of £5,909.57, which the appellants sought to recover from them for the replacement of windows and balconies carried out as part of a programme of major works, was not payable since it was not reasonable within section 19(1). The respondents argued that they should be offered compensation, or a significant discount from the cost of the works to the windows, because the standard of workmanship was poor and ongoing repairs were needed.

The LVT disallowed the cost of the works on the ground that it had not been reasonable to replace the old windows. Although the respondents had not advanced such a contention, the LVT took the view that its jurisdiction extended to determining the entire service charge, not just the matters in dispute as pleaded or otherwise specifically identified in the application.

The appellants appealed. They contended that the LVT had erred in reaching a decision on grounds that were not raised in the respondents' application, without giving the appellants an opportunity to make submissions on the point and without having any evidence on which to base its decision. The respondents agreed that the appeal should be allowed, since their only complaint was as to the standard of workmanship, not the reasonableness of replacing the windows. The parties then reached agreement on the amount of service charge that was reasonably recoverable.

Decision: **The appeal was allowed. The LVT had had no jurisdiction to determine whether it was reasonable to replace the old windows, since it had not been asked to determine that issue and it was clear from the application that the respondents agreed or accepted that replacement was reasonable. On an application under section 27A of the 1985 Act, the LVT does not have jurisdiction to determine the entire service charge beyond the matters in dispute, as pleaded or otherwise specifically identified in the application. The application form and the parties' statements of case set out the nature and scope of the issues in dispute and serve the functions of: (i) identifying the issues on which the parties seek the resolution of the LVT; (ii) enabling the parties to know the matters to which they need to address their evidence; (iii) vesting the LVT with jurisdiction and focusing its attention on what needs to be resolved; (iv) setting the parameters of, and providing the tools for, case management of the application by the LVT; and (v) by confining the issues to what is actually, rather than might theoretically be, in dispute between the parties, assuring economical and expeditious disposal of the dispute while promoting efficient and economical use of judicial resources at first instance and appellate levels. The LVT's jurisdiction is derived both from statute and from what the applicant and the respondent ask it to resolve. The LVT has no jurisdiction to embark on its own inquisitorial process and to identify and resolve issues that neither party has asked it to resolve. Although there might be rare cases in which it is appropriate or necessary for the LVT, in order to properly determine the issues expressly in dispute, to raise issues that are not expressly raised by the parties, those must still fall within the broad scope of the application and cannot relate to something that arises outside it; the LVT must also give the opportunity to both parties to make submissions and, if appropriate, to adduce further evidence on the new issue before it reaches its decision.**

The following case is referred to in this report.

Regent Management Ltd v *Jones* [2012] UKUT 369 (LC), LRX/14/2009

This was an appeal by the appellants, Birmingham City Council, from a decision of the leasehold valuation tribunal determining the reasonableness of a service charge on an application by the respondents, Richard Keddie and Glen Hill, under section 27A of the Landlord and Tenant Act 1985.

Justin Bates (instructed by the legal department of Birmingham City Council) appeared for the appellants; the first respondent appeared in person for the respondents.

Giving judgment, **HH Judge Gerald** said:

Introduction

1. The appellant local authority landlord appeals against the 25th March 2011 decision ("the decision") of the leasehold valuation tribunal ("the LVT") disallowing recovery of the £5,909.57 cost of window replacement and balcony works because it was not reasonable to replace the old ones.

2. In 2006/07 the appellant undertook window replacement and balcony works to the respondent tenants' flat 56 at Galton Tower, Civic Close, Birmingham, as part of a programme of major works to the block. The respondents purchased the flat after those works had been

completed, and therefore had no knowledge of the condition of the old windows which had been replaced.

3. The appellant sought to recover the £5,909.57 costs of the works via the service charge. The respondents applied to the LVT for a determination that the amount claimed was not reasonable under sections 27A and 19(1) of the Landlord and Tenant Act 1985 ("the 1985 Act"). The material part of the application form stated that the question they wanted the LVT to determine was:

> "whether or not [the appellant] should offer us a) compensation or b) a significant discount owing to the fact payment is being demanded for the windows, yet on-going repairs are still needed three years on?"

Further details of the windows' problems were provided from which it was clear that the only complaint was the standard of workmanship, not the old windows did not need to be replaced or that it was not reasonable to replace them.

4. Having inspected the flat on 15 November 2010, the LVT issued is on 25 March 2011 ("the decision") the kernel of which is to be found in the following paras (italics supplied).

> "34. The applicants have not raised the issue of whether the replacement of the windows can comprise a repair (and thus maintenance) or whether the windows to the property were in such a condition in 2007 that they actually needed to be replaced. However, in meeting the allegation raised by the application, the respondent dealt with the point in its submissions to the tribunal dated 8 October 2010. the respondent is clearly of the view that the implied and express covenants and the sixth schedule of the lease are sufficient to allow the respondent to recover these sums as repairs or maintenance, *there still must have been grounds for replacing the windows, those grounds must be based on evidence and it must be within the reasonable responses available to the respondent to conclude that those grounds justified the replacement of the windows at the time and in the manner of the works they undertook.*
>
> "35. In the circumstances, it is with considerable surprise that we note from para 14 of the respondent's statement that the window replacement works were carried out under the Decent Homes Standard initiative and that properties were not selected for inclusion in this programme on an individual basis. *From which we conclude that, before carrying out the window replacement works, the respondent did not consider whether or not such works amounted to a repair or maintenance at all or even whether the existing windows were still serviceable without replacement. Therefore, we conclude that the respondent did not have such grounds for replacing windows, because no evidence has been presented to use to arrive at such a decision.* Accordingly, the respondent did not act reasonably in concluding that the window replacement works should be carried out. Therefore on the evidence available to us, we find that the cost of the window replacement of £6,909.52 was not reasonably incurred."

5. On 13 April 2011, the appellant applied to the LVT for permission to appeal the decision on the basis that the LVT had breached natural justice by reaching a decision on grounds not raised in the application without giving the appellant an opportunity to make submissions thereon and in any event that the decision was perverse, there being no evidence before it about the condition of the old windows which were replaced.

6. On 21 April 2011, the LVT refused permission to appeal because it was "patently obvious" that the question of whether it was reasonable to replace the windows at all was in issue and the LVT's jurisdiction under section 27A of the 1985 Act is "to determine the entire service charge not only the matters in dispute, pleaded or otherwise specifically identified in the service charge application" ("the refusal decision").

7. On 7 July 2011, HH Judge Huskinson, on the appellant's renewed application, granted permission to appeal to be by way of review with a view to a re-hearing stating inter alia:

> "The grounds of appeal raise reasonable arguments to the effect that the LVT erred as there alleged by reason of reaching a conclusion that the windows did not need to be replaced in 2007 when, as the [appellant] alleges, this was no part of the respondent's case and was a point that had never been raised with the [appellant]."

8. In its statement of case dated 8 August 2011, the appellant stated inter alia that "nowhere in either the application form or the detailed evidence that was provided, was it suggested that it had been unnecessary to install new windows. The entire focus of the evidence is on the allegedly poor quality of the workmanship and the manner in which Birmingham have conducted themselves in connection with this matter".

9. In their lengthy response dated 8 September 2011, the respondents agreed. They said: "What isn't being argued throughout this appeal (what is really the crux of the matter and the reason why we applied to the LVT in the first instance) is that the installation of our UPVC windows to our property is simply not of a reasonable and acceptable standard." Later on: "we had not raised the issue with the LVT of 'whether the windows to the property were in such a condition in 2007 that they actually needed to be replaced – if the works had been completed satisfactorily in the first instance, then there would be no need for us to take out a case against Birmingham City Council."

10. The hearing of the appeal took place on 12 September 2012 in Birmingham. Mr Keddie, who was representing himself and also his co-respondent Mr Hill, who was unable to attend through ill-health, frankly said that he was surprised that the LVT had made the decision it did because only the standard of work was being challenged not that the old windows needed to be replaced, that as far as he was concerned the LVT had misconstrued his argument which only related to the standard of work and finally that he felt that the LVT by their decision had put Mr Hill and himself in an "unfair position".

11. Mr Keddie was therefore content for the appeal to be allowed, the parties leaving me to give written judgment as to why the appeal should be allowed. That left the question of how much the £5,909.57 should be reduced by, the appellant not appealing that part of the decision which found that the work had not been carried out to a reasonable standard. It is, of course, unfortunate that the LVT did not answer the key question it had been asked to.

12. I rose to give the parties an opportunity to reach agreement as to the sum reasonably recoverable, which they did. They also agreed that an order under section 20C of the 1985 Act should be made to disallow the appellant from recovering the costs of the appeal via the service charge, the appellant most reasonably accepting that it would be unfair to visit those costs on the respondents when it was through no fault of their own that the LVT reached the decision it did.

13. It is regrettable that it appears to be a developing practice within some leasehold valuation tribunals to take it upon itself to identify issues which are of no concern to the parties and then reach a decision on issues they have not been asked to which then results in an appeal and all the waste of time and money and attendant general aggravation. It may therefore be helpful to set out the legislative framework and general principles applicable.

14. The jurisdiction of the LVT is founded in section 27A of the 1985 Act the material parts of which are:

> "(1) An application may be made to a leasehold valuation tribunal for a determination whether a service charge is payable and, if it is, as to-
> (a) the person by whom it is payable
> (b) the person to whom it is payable
> (c) the amount which is payable
> (d) the date at or by which it is payable and
> (e) the manner in which it is payable
> (2) Subsection (1) applies whether or not any payment has been made.
> (3) An application may also be made to a leasehold valuation tribunal for a determination whether, if costs were incurred for services, repairs, maintenance, improvements, insurance or management of any specified description, a service charge would be payable for the costs and, if it would, as to-
> (a) the person by whom it would be payable
> (b) the person to whom it would be payable
> (c) the amount which would be payable
> (d) the date at or by would be payable and
> (e) the manner in which it would be payable
> (4) No application under subpara (1) or (3) may be made in respect of a matter which-
> (a) has been agreed or admitted by the tenant…
> (5) But the tenant is not to be taken to have agreed or admitted any matter by reason only of having made any payment.
> (6) An agreement by the tenant of a dwelling (other than a post-dispute arbitration agreement) is void in so far as it purports to provide for a determination –

A (a) in a particular manner, or
(b) on particular evidence,
of any question which may be the subject of an application under subsection (1) or (3)".

15. Applications are commenced by landlord or tenant issuing a pro forma application form prescribed by the Residential Property Tribunal Service, which requires that details of the questions relating to service charge expenditure requiring resolution by the LVT be set out. If they are not sufficiently set out, as is often the case, the LVT will at the pre-trial review order that the applicant serve a statement of case giving full particulars of precisely what is in issue and why. The respondent will B be ordered to serve a statement of case setting out its case to which the applicant will usually be given an opportunity to respond if he so wishes by serving a statement of case in reply.

16. Those documents, whether they be described as pleadings or statements of case or whatever, set out the nature and scope of the issues in dispute. They operate to limit the issues in respect of which the parties must adduce evidence in support of their respective cases. They also operate to define the issues in respect of which they seek resolution by the LVT. They therefore serve five functions. First, to identify the issues. Secondly, to enable the parties to know what issues they must address their evidence to. Thirdly, to vest the LVT C with jurisdiction, and focus the LVT's attention on what needs to be resolved. Fourthly, setting the parameters of, and providing the tools within which, the LVT may case manage the application. Fifthly, by confining the issues requiring resolution to what is actually (as distinct from what might theoretically be) in dispute between the parties they will be assured economical and expeditious disposal of their dispute whilst also promoting efficient and economical use of judicial resources at first instance and appellate levels.

17. In this respect, it is important to bear in mind not just that the jurisdiction of the LVT is a creature of statute but that it is also a function of what the applicant and, by his response, the respondent D wish the LVT to resolve. It is the jurisdiction and function of the LVT to resolve issues which it is asked to resolve, provided they are within its statutory jurisdiction. It is not the function of the LVT to resolve issues which it has not been asked to resolve, in respect of which it will have no jurisdiction. Neither is it its function to embark upon its own inquisitorial process and identify issues for resolution which neither party has asked it to resolve, and neither does it have the jurisdiction to do so. To do so would be inimical to the party-and-party nature of applications to the LVT and would greatly increase the costs (frequently recoverable from the tenant through the service charge) and difficulties attendant to service charge disputes which by their nature are frequently E fractious, involving relatively small sums within a complex matrix of divers items of expenditure.

18. It follows from the above that the LVT does not have jurisdiction under section 27A "to determine the entire service charge not only the matters in dispute, pleaded or otherwise specifically identified in the service charge application" as stated in the refusal decision. It is not an inquisitorial tribunal. It is there to resolve issues it is asked to resolve, not uncover ones which do not exist or which the parties are not concerned about.

19. That said, there may of course be rare cases in which it is appropriate or necessary for the LVT to raise issues not expressly raised F by the parties but which fall within the broad scope of the application in order to properly determine the issues expressly in dispute. But even then, the issues must fall within the scope of the application, not something which arises outside of it. This no doubt is what HH Judge Mole QC had in mind when he said in *Regent Management Ltd* v *Jones* [2012] UKUT 369 (LC), LRX/14/2009 that:

> "29. The LVT is perfectly entitled, as an expert tribunal, to raise matters of its own volition. Indeed it is an honourable part of its function, given that part of the purpose of the legislation is to protect tenants from unreasonable charges and the tenants, who may not be experts, may have no more than a vague and unfocussed feeling that they have been charged too much. But it must do so fairly, so that if it is a new point which the tribunal raise, which the respondent has not mentioned, the applicant must have a fair opportunity to deal with it."

20. In those rare cases where an LVT does feel compelled of its own G volition to raise an issue not raised by the application or the parties, it must as a matter of natural justice first give both parties an opportunity of making submission and if appropriate adducing further evidence in respect of the new issue before reaching its decision. Failure to do so is not only unfair, but may give the unfortunate impression that the LVT has descended into the fray and adopted a partisan position which may well serve to undermine the confidence of the parties in the impartiality of the LVT.

21. Applying those principles to this case, it is in my judgment plain that the LVT had no jurisdiction to determine whether it was reasonable to replace the old windows because it had not been asked to do so. Far H from being "patently obvious" that this was in issue, it was clear from and implicit within the application that the respondents agreed or at any rate accepted that it was reasonable to replace the old windows, which they have since confirmed during the appellate process. Indeed, it is clear from [34] of the decision that the LVT itself recognised that the respondents had not challenged the reasonableness of replacing the old windows. It is therefore difficult to understand how, in its refusal decision, it could have reached the conclusion that it was "patently obvious" that this matter was in issue within the application.

22. In any event, by failing to notify *either* party of this novel point the LVT acted in breach of natural justice. Had it consulted the parties, J it would have learned or at any rate confirmed that it was not in issue so that the LVT need not trouble itself with this issue: it would then have applied itself and its resources to resolving that which it had been asked to resolve, namely, the reasonableness of the cost of the window replacement and balcony works.

23. As it is, this has led to the unfortunate result that the parties have not had resolution of what they asked the LVT to resolve. Both are dissatisfied not only with the result (or lack of it) but just as importantly with the process. Both have had to expend further time, money and trouble in attending to the appeal which could and should have been avoided had the LVT confined itself to the issue it was asked to resolve. Both have expressed their feeling at having been treated K unfairly by the LVT, Mr Keddie saying with some justification that the LVT has put himself and Mr Hill in an unfair position. Fortunately for the respondents, the appellant has taken the reasonable approach of not seeking to recover its costs of the appeal from the respondents via the service charge. Instead, those costs will have to be borne by the council tax payers of Birmingham.

24. I am bound to observe that it appears clear from the decision and also from the refusal decision that the LVT decided to take this issue when it inspected the premises and found the old windows still in situ in the balcony, which had been encased with new windows. The L LVT then concluded that those windows were in fair condition, from which they concluded that the other windows which had been removed form the other parts of the flat and replaced (so were no longer in evidence) must have been in the same condition and therefore did not need replacement, so that the appellant had acted unreasonably in its decision to replace at all rather than repair. The respondents could have but did not take this position. It was wrong of the LVT to do so. By doing so, the LVT descended into the fray and virtually became a party to the dispute, pitting itself against the common position of applicant and respondent.

25. I therefore allow the appeal. For the reasons already stated, is it M not necessary for me to resolve the residuary question of what would be a reasonable sum to charge for the window and balcony works. The appellant has consented to a section 20C order being made preventing it from recovering the costs of the appeal via the service charge which I hereby make.

Appeal allowed.

LICENCE

Fitzhugh v Fitzhugh

COURT OF APPEAL

1 June 2012

LONGMORE, RIMER AND PATTEN LJJ

[2012] EWCA Civ 694

Estates Gazette 1 September 2012

[2012] 35 EG 72

Licence – Termination – Notice – Agricultural land – Administrators of deceased's estate granting licence to one of them to occupy certain land – Licensee failing to pay licence fee – Claim for possession of land – Whether licence effectively terminated by notice in accordance with its terms – Whether notice ineffective where served on behalf of only one of two surviving licensors – Whether requirement for all licensors to give notice excluding licensor who was also licensee

The appellant and the respondent were brothers. They and a third brother were the administrators of their deceased father's estate. In 1998, all three administrators conveyed the principal asset of the estate, comprising a farmhouse, garden and five plots of land, to the appellant and his partner together with a licence, granted by all three administrators as "the Licensor", to graze animals on further land and take grass from it. Under the terms of the licence, an annual licence fee of £1 was payable and the licence was to terminate immediately in the event that the licensee failed to remedy any breaches of the licence within the period specified in a written notice served by "the Licensor". The third brother subsequently died and the appellant and respondent became the sole surviving administrators.

In 2006, the appellant failed to comply with a notice requesting payment of £7 in unpaid licence fees within seven working days. The respondent's consequent claim for possession of the licensed land was allowed in the court below. The judge found as a fact that the notice had been served only on behalf of the respondent, whereas the licensor comprised both the respondent and the appellant. However, he considered that the requirement for "the Licensor" to give notice to remedy any breach could be construed as referring to all persons who together were the licensor, apart from any person who was also the licensee. He concluded that the notice had been effective and that the licence had terminated according to its terms by reason of the appellant's failure to pay the arrears of licence fees specified in that notice: see [2011] EWHC 3553 (Ch). The appellant appealed.

Held: **The appeal was allowed. In order to be effective, the relevant notice had to be given by all those individuals who together made up the licensor. Since the licence expressly described the appellant as one of the individuals making up "the Licensor", that phrase plainly included him. A reasonable person, having all the background knowledge that would reasonably be available to the audience to whom the licence was addressed, would regard it as improbable that the phrase "the Licensor", in a short, simple, professionally-drawn document, was intended to mean one thing in the opening words and elsewhere and another in the clause dealing with the giving of notice. There was nothing on the face of the document to indicate that a change of meaning was intended in that clause or that anything had gone wrong with its drafting. While the reasonable person might foresee the potential for practical difficulties in the event that the respondent sought to serve a notice on the appellant, that person would also know that the appellant's duties to the estate as an administrator would require him to subordinate his own conflicting personal interests and concur in the service of the notice, without prejudice to his right, in his personal capacity, to dispute the justification for the notice in any proceedings that might be brought against him. The potential for practical difficulties did not render the notice machinery unworkable, notwithstanding that it might prove cumbersome or expensive in the event of obstruction from the appellant. It was not necessary to implying any term to make the notice provisions workable. Accordingly, the notice served by the respondent alone was ineffective and the licence had not terminated.**

The following cases are referred to in this report.

Attorney General of Belize v *Belize Telecom Ltd* [2009] UKPC 10; [2009] 1 WLR 1988
Featherstone v *Staples* [1986] 1 WLR 861; [1986] 1 EGLR 6; [1986] 27 EG 867
Hammersmith and Fulham London Borough Council v *Monk* [1992] 1 AC 478; [1992] 1 EGLR 65; [1992] 09 EG 135
Hounslow London Borough Council v *Pilling* [1993] 1 WLR 1242; [1993] 2 EGLR 59; [1993] 26 EG 123
Johnson v *Moreton* [1980] AC 37
Mediterranean Salvage & Towage Ltd v *Seamar Trading & Commerce Inc* [2009] EWCA Civ 531; [2009] 1 CLC 909
Rainy Sky SA v *Kookmin Bank* [2011] UKSC 50; [2011] 1 WLR 2900

This was an appeal by the appellant, Anthony Fitzhugh, from a decision of Morgan J, sitting in the Chancery Division in Manchester, allowing a claim for possession by the respondent, Harry Fitzhugh, on the ground that a licence had been validly terminated by notice.

James Howlett (instructed by Nelsons Solicitors, of Derby) appeared for the appellant; Timothy Clarke (instructed by Moody & Woolley, of Derby) represented the respondent.

Giving judgment, **Rimer LJ** said:

Introduction

1. The question raised by this appeal is a short one as to the interpretation of a licence agreement. In essence, it is this. If A and B (described as "the Licensor") grant a licence to occupy land to B and C (described as "the Licensee"), and the licence automatically terminates upon the failure of B and C to remedy any remediable breaches within the time specified by a notice given by "the Licensor" to "the Licensee", can such a notice validly be given by A alone?

2. That question arose for decision before Morgan J in a family dispute in which the claimant is Harry Fitzhugh and the defendant is his brother Anthony. In his judgment delivered on 11 November 2011 ([2011] EWHC 3553 (Ch)), Morgan J answered the question in the affirmative and, by his order of the same date, declared that the licence had been determined by notice with effect from 27 February 2006.

3. With the permission of the judge, the defendant appeals against that declaration. He was represented before us, as before the judge, by James Howlett. Timothy Clarke represented the claimant, as he did before the judge.

A *The facts*

4. Harry Fitzhugh Senior died intestate on 10 April 1995 survived by nine children of his first marriage. The claimant is his sixth child, and I shall call him Harry. The defendant is his youngest child, born in 1960, and I shall call him Anthony. Letters of administration of the father's estate were, on 15 June 1995, granted to his four sons, Geoffrey, Colin, Harry and Anthony. Colin died in 1997.

5. The principal asset of the father's estate was Poplars Farm, near Derby, comprising a farmhouse and garden and five other plots of land. On 3 July 1998 the three surviving administrators, Geoffrey, Harry and Anthony, conveyed the farmhouse and garden to Anthony and his
B partner, Karen Boddey ("Karen"). On the same day, they also granted the licence to Anthony and Karen. It was of farm outbuildings and various fields.

6. I must set out the material terms of the licence. It is described as a "licence" and as being between (1) Geoffrey, Anthony and Harry (described as "the Licensor") and (2) Anthony and Karen (described as "the Licensee"). It then provides:

> "1. The Licensor grants to the Licensee the right to keep sheep, cattle horses or pigs on the premises described in the First Schedule hereto for grazing purposes and the right to mow premises twice a year and to take away the grass
>
> 2. Further the Licensor permits the Licensee to enter onto the premises to
> C the extent necessary to exercise the rights and for the avoidance of doubt full occupation and possession of the premises remains with the Licensor subject only to the rights granted by this Licence to the Licensee
>
> 3. The Licensee agrees with the Licensor to pay the Licence fee of one pound a year and to use the premises for the exercise of the rights and for no other purposes and on the termination of the Agreement to remove the Licensee's stock from the premises
>
> 4. The rights granted under and this Licence itself will terminate immediately on the happening of any of the following events:-
>
> (a) The Licensee dying or becoming incapable by reason of mental or physical illness from discharging his obligations under this Agreement
>
> (b) The Licensee commits any grave or persistent breaches of this Licence
> D and the Licensor having given written notice to the Licensee of such breach or breaches the Licensee fails within such period as the Licensor may specify to rectify such breaches if capable of rectification
>
> (d) [sic: there is no (c)] Upon the Licensor obtaining planning permission for the development of all or any part of the land the subject of this licence…
>
> 6. This Licence and the rights contained in it are personal to the Licensor and Licensee and will not be capable of being assigned or otherwise dealt with."

7. Geoffrey died in 2001, upon which event Harry and Anthony became the sole surviving administrators, as they still were at the date of the judge's judgment, although his order provided for both to cease to be administrators and to be replaced by two new administrators.

E 8. Anthony and Karen did not pay the annual licence fee of £1. On 16 February 2006 Walker Terry & Wilson, solicitors, purportedly writing "on behalf of all the family members of the late Harry Fitzhugh Senior regarding his estate", made various requests of Anthony and Karen, including that, within seven working days, they paid the £7 arrears of licence fee owing to the estate. They did not do so. Over three years later, in September 2009, Harry commenced these proceedings against Anthony in the Chancery Division, seeking various heads of relief including possession of the land the subject of the licence. Anthony defended the claim, which was tried before Morgan J in 2011. Morgan J's 112-paragraph judgment reflects that he had to consider many
F issues, the one giving rise to the present appeal being but one of them.

The judge's decision

9. That issue was whether the licence had been terminated. Harry claimed that it had been terminated under clause 4(b) in consequence of the failure of Anthony and Karen to pay the £7 arrears of the licence fee within the seven day period prescribed by the notice given in February 2006. Morgan J found, and it is not challenged, that the failures to pay the licence fee from 1998 to 2006 amounted to "persistent breaches" of the licence within the meaning of clause 4(b). He did not, in those circumstances, find it necessary also to decide whether such breaches were "grave". He found that Anthony and Karen did not remedy the breaches within the seven day period specified in the notice. He explained that Mr Howlett did not invite him to assume any jurisdiction G
(if such there was) to grant Anthony relief against forfeiture.

10. In that state of play, the only remaining issue was whether the solicitors' letter of 16 February 2006 was a valid notice for the purposes of clause 4(b). That turned on the fact, as the judge found, that it had been given only on behalf of Harry, but not also on behalf, or with the authority of Anthony, whereas at the relevant time "the Licensor" was both Harry and Anthony. Did the notice meet the requirements of clause 4(b)?

11. The judge referred to the well-established common law principle that a contractual periodic tenancy granted by joint landlords to joint tenants continues only for so long as they all agree in its continuation. H
Such a tenancy can thus be determined by a notice to quit given by just one of the landlords without the concurrence of the others or (likewise) by just one of the tenants: see *Hammersmith and Fulham London Borough Council* v *Monk* [1992] 1 AC 478*. The judge, however, regarded that principle as having no application to the issue before him, since he was faced with an express contractual provision in the licence (clause 4(b)) which required the notice to be given by "the Licensor"; and the critical question was whether, upon the true interpretation of the licence, that meant both Harry and Anthony or just Harry. The distinction that the judge was there recognising was that expressly identified in the *Hammersmith* case by Lord Bridge, at [1992] 1 AC J
492, 490G to 491A, and by the later decision of the Court of Appeal in *Hounslow London Borough Council* v *Pilling* [1993] 1 WLR 1242** (see at 1246H to 1247E).

12. The question was thus whether, as a matter of interpretation, "the Licensor" in clause 4(b) meant both Harry and Anthony; or whether it meant, or could be read as meaning, a reference just to Harry. The judge's opening view, in [86] of his judgment, was that there was no easy way to read "the Licensor" as there referring to any one of the licensors. He was, however, reluctant not so to read it and considered it necessary to take into account that there was an overlap between the persons who were the licensor and licensee. In such a case, he said it K
was unlikely, if not certain, that a licensee who is also a licensor would join in giving the notice on a voluntary basis. If not, and all licensors must join in the service of the notice, then the non-licensee licensors must precede the giving of any notice either by (i) seeking the removal of the reluctant licensee as a trustee (as a co-owner of the land, he will necessarily be a trustee), or (ii) perhaps by seeking an injunction requiring him to join in the giving of a notice.

13. In the judge's view:

> "90. … to hold that no notice can be given is an unsatisfactory contractual result which the parties cannot have intended. It also seems to me that a requirement in my example of A and B that A gets B removed or A gets an L
> injunction requiring B to serve a notice on himself is equally a cumbersome, slow, expensive proceeding which I hesitate to think the parties intended. There is in my judgment a solution to these difficulties. The solution is to construe the reference to the licensor in Clause 4(b) as referring to all persons who together are the licensor apart from any person who is also the licensee. If that is the construction of the word licensor in Clause 4(b) then the notice can be given by Harry alone, and this notice, being so given, was an effective notice."

14. In concluding that that was the correct construction of "the Licensor" in clause 4(b), the judge drew on two sources of support. The first was assistance drawn from some obiter observations of Slade LJ (with whose judgment Stocker LJ and Sir Roualeyn Cumming-Bruce M
agreed) in *Featherstone* v *Staples* [1986] 1 WLR 861†. There the four plaintiffs were the trustees of a will and owned the relevant land. They let it under three tenancies to three partners as tenants. Two of the partners/tenants were individuals and the third was a company wholly owned by the landlords. The issue was as to the validity of counter-notices served only by the two individual tenants (without the authority of the company) requiring that section 2(1) of the Agricultural Holdings (Notices to Quit) Act 1977 should apply to notices to quit served by the

* Editor's note: Also reported at [1992] 1 EGLR 65; [1992] 09 EG 135

** Editor's note: Also reported at [1993] 2 EGLR 59; [1993] 26 EG 123

† Editor's note: Also reported at [1986] 1 EGLR 6; [1986] 27 EG 867

plaintiffs. Clause 11 of the partnership agreement provided, however, that no partner might serve such a counter-notice without the consent of the company. The issue in the proceedings was as to the validity of the counter-notices: section 2(1)(b) of the 1977 Act required any counter-notice to be served by "the tenant". Did that mean by all joint tenants or by all or any one or more of them?

15. Slade LJ explained that, subject to a qualification, he had come to the clear conclusion that "the tenant" in that sub-section meant, or ordinarily meant, all the joint tenants. But he also considered whether, by necessary implication, the phrase "the tenant" nevertheless excluded the landlord himself, or a company controlled by the landlord, in cases in which the landlord or such a company also happened to be one of the joint tenants. In relation to that, he said, [1986] 1 WLR 861, at pp876H-877C:

> "There are, in my opinion, quite strong grounds for saying that in the particular context where the landlord himself is one of the tenants (not the present case), the word "tenant" in section 2(1)(*b*) would be capable of bearing more than one meaning, namely, (1) all the joint tenants or (2) all the joint tenants other than the landlord himself; and that the court should prefer the second construction as leading to a "sensible and just result" complying with the statutory objective (see *Johnson* v *Moreton* [1980] AC 37, 50G, per Lord Salmon) instead of adopting a construction which would invalidate a counter-notice given in such circumstances, unless the landlord had joined in or consented to its service on himself. On the other hand, I think that the engrafting of further exceptions by a process of statutory interpretation on the ordinary meaning of the phrase "the tenant," to cover the case (such as the present) where one of the joint tenants was merely a company controlled by the landlord or was otherwise associated with the landlord, would present rather greater difficulties, if only because of problems of definition and deciding where to draw the line.
>
> However, I find it unnecessary to express any concluded view on the question whether, as a matter of construction, any exception should be engrafted on to what I would regard as the ordinary meaning of the phrase "the tenant," as used in section 2(1)(*b*), in the case where the tenants are more than one in number. Even assuming for present (as I will assume) that as a matter of statutory interpretation the phrase must mean all the joint tenants, in any case where joint tenants are involved, I am still of the opinion that, for the reasons to be given later in this judgment, the counter-notices in the present case must be treated as having been served with the authority of [the company], even though that authority was not in fact given."

The reasons referred to in the latter part of that quotation were public policy reasons relating to the efficacy of the restrictive condition in clause 11 of the partnership agreement. Slade LJ explained them as follows ([1986] 1 WLR 861, 881C-E):

> "I am driven to the conclusion that, if a land owner chooses to grant other persons a tenancy of agricultural land (whether or not including himself as a tenant), public policy (affirmatively) requires that those other tenants should have authority, or be treated as having authority, to serve an effective counter-notice under section 2(1) of the Act of 1977 on behalf of all the tenants without his concurrence, and thus (negatively) requires the avoidance of any contractual condition, whether express or implied and whether contained in the tenancy agreement itself or in a partnership agreement or elsewhere, which purports to deny those other tenants such authority. I might add that any contrary decision of this court would be likely to open the door to widespread evasion of the Act of 1977 to the detriment of the security of tenure which Parliament clearly intended to confer on agricultural tenants: see and compare *Johnson* v *Moreton* [1980] AC 37, 52-53, per Lord Salmon."

16. The decision in *Featherstone* therefore turned ultimately on a consideration of public policy. There was, rightly, no suggestion made to Morgan J or to us that like considerations have any part to play in this case. Morgan J, in coming to his own decision, drew no more from *Featherstone* than such support for his preferred construction of clause 4(b) of the licence as was to be derived from Slade LJ's *obiter* suggestion in the first quotation cited in [15] above, although Morgan J, in [92] of his judgment, also observed that such suggestion was "quite a bold one". He considered, however, that it was rather easier to imply into a contractual arrangement an appropriate qualification by applying the principles as to the implication of contractual terms discussed by Lord Hoffmann in the decision of the Privy Council in *Attorney General of Belize* v *Belize Telecom Ltd* [2009] UKPC 10; [2009] 1 WLR 1988. Morgan J brought into consideration the type of case to which he had referred in [89] of his judgment (see [13] above). Of that example, he said:

> "94. … A and B will be joint owners and necessarily trustees. In a case where the facts permit a notice to be given under 4(b), that is a case of grave or persistent breaches by the licensee, there must be a very strong argument that B's duty as a trustee, assessing him in that capacity, is to join in a notice to himself under 4(b). If he fails to join in that notice he is acting in breach of trust, he is committing a wrong to his fellow trustee A and the beneficiaries under the trust. If he then receives a notice from A alone and is able to challenge that notice on the ground that it is given by A alone and not by A plus B, he is in a real sense taking advantage of his own breach. There is a principle, whether it is a general principle of law or a principle of interpretation it matters not, that one should not take advantage, the court should not permit one to take advantage, of one's own breach. That is a powerful argument for holding that when one comes to the word licensor in clause 4(b), one should interpret it as referring to everyone who is not also a licensee.
>
> …
>
> 96. By that process of reasoning, I do reach the conclusion that the notice given on 16 February 2006 was valid and effective under Clause 4(b). The non-payment of rent was a persistent breach, it was not rectified within the period permitted. The consequence is that the licence was indeed terminated."

The appeal

17. We had succinct arguments from counsel. Mr Howlett said there is no scope for interpreting the words "the Licensor" in clause 4(b) as meaning (in the events that have happened) other than both surviving licensors and therefore as including Anthony. He is expressly described in the opening words of the licence as one of the individuals making up "the Licensor", it is obvious that "the Licensor" in clause 1 includes Anthony and it is equally obvious that it includes him in the like phrase in clause 4(b). The *obiter* observations in *Featherstone* – upon which Slade LJ expressly eschewed the expression of any final view and being observations as to the interpretation of a statutory provision – provided no support for an interpretation of "the Licensor" in clause 4(b) that involved such a basic departure from its ordinary meaning. As for the judge's reliance on the *Belize Telecom* decision, nothing in that authority weakened the requirement that it is always necessary to imply the proposed term; it is never sufficient that it should merely be reasonable to imply it: see the subsequent decision of the Court of Appeal in *Mediterranean Salvage & Towage Ltd* v *Seamar Trading & Commerce Inc* [2009] EWCA Civ 531; [2009] 1 CLC 909, at [15] to [18], per Lord Clarke of Stone-cum-Ebony MR. There was in this case no necessity to imply into clause 4(b) that "the Licensor" does not there mean what it means elsewhere in the licence. That is because if Anthony declined to join in the service upon him and Karen of a clause 4(b) notice with a view to the frustration of the operation of clause 4(b), he would be failing in his duties as an administrator of the estate and it would be open to Harry to seek to have him removed as an administrator. Once he was removed, he would no longer be a co-owner of the licensed land, or therefore a licensor, and Harry alone could serve a notice on him.

18. Mr Clarke's written argument sought to support the judge's reasoning by the proposition that the present case was not materially distinguishable from cases such as *Hammersmith*. The judge, however, rejected that approach and so any such argument strictly required a respondent's notice, which there is not. In the event, Mr Clarke did not seek to develop that argument orally and I consider he was right not to do so. The judge was correct to recognise that the case was not of a like kind as *Hammersmith* but that the relevant issue turned exclusively on the true interpretation of "the Licensor" in clause 4(b). Mr Clarke felt compelled to accept that "the Licensor" in clause 1 of the licence necessarily included Anthony but maintained that it did not do so in clause 4(b). In his submission this was a case where the meaning of "the Licensor" in clause 4(b) had two possible constructions, it was therefore ambiguous, and accordingly the judge was right to prefer the construction that was consistent with business common sense and to reject the other. By way of support for that approach, Mr Clarke referred us to the decision of the Supreme Court in *Rainy Sky SA* v *Kookmin Bank* [2011] UKSC 50; [2011] 1 WLR 2900, at [21], per Lord Clarke of

A Stone-cum-Ebony. Mr Howlett's response to that in his reply was that there was no such ambiguity as Mr Clarke suggested.

19. I prefer and accept Mr Howlett's submissions. Anthony is expressly described as one of the individuals making up "the Licensor" in the opening words of the licence. The phrase "the Licensor" in clause 1 plainly includes him. So must it in clauses 2 and 3. The critical question, however, is as to the meaning that the same defined phrase in clause 4(b) would convey to a reasonable person having all the background knowledge which would reasonably be available to the audience to whom the licence is addressed (see *Belize Telecom*, at [16], per Lord Hoffmann).

B 20. In my view, such a person would regard it as improbable that in this short, simple, professionally drawn document the defined phrase "the Licensor" was intended to be mean one thing in the opening words of the document, the same thing in clauses 1 to 3 but something else in clause 4(b). There is, on the face of the document, nothing to suggest that there should in clause 4(b) be any such change of meaning, or that something has gone wrong with its drafting so as to support the conclusion that "the Licensor" does not there mean the same as it means elsewhere. If, as is the ordinary inference, it does mean the same, the reasonable person might well also foresee the potential for practical difficulty in the event (as happened) that Harry wanted to serve a clause C 4(b) notice on Anthony: in particular, he might foresee the possibility that Anthony would refuse to join in its service. He would, however, also know that Anthony was a party to the licence in his capacity as one of the joint administrators, and was presumably intending to remain one. He would therefore recognise (i) that Anthony's duties owed to the estate as such an administrator would require him, at least for the purpose of the service of such a notice, to subordinate his own conflicting personal interests and so concur in its service; (ii) that there would be no reason why (if so advised) Anthony could not so concur expressly without prejudice to his right in his personal capacity to dispute the justification for the notice; and (iii) that it would then be open to Anthony to dispute D its justification in any proceedings that might subsequently be brought against him, in which proceedings the interests of the estate would be sufficiently represented by Harry. Alternatively, if it should turn out that Anthony were to decline to join in the giving of any notice for the E service of which there was proper justification in the interests of the estate, the reasonable man would recognise that it would in principle be open to Harry to seek his removal as an administrator, if necessary, by court proceedings, upon the successful conclusion of which Anthony would cease to have any status as a licensor and would not need to join in any clause 4(b) notice.

21. The reasonable man would therefore be likely to conclude that although an interpretation of "the Licensor" in clause 4(b) as including any licensor who was also a licensee might perhaps give rise to practical difficulties, and perhaps unwanted expense, in the future, there was no basis for a conclusion that it would render the machinery of clause 4(b) F unworkable. He might well consider that clause 4(b) could instead have been drafted in a way that would avoid any such difficulties arising in the future – for example, by providing for the relevant notice to be given by "the Licensors other than any who is for the time being a Licensee". But even if he were to be of that view, it is no part of the function of a court of construction to improve the document it is called upon to construe, nor does it have any power to do so (see again *Belize Telecom* at [16], per Lord Hoffmann). Moreover, as clause 4(b) remains workable even if its working out may, in the event of obstruction from Anthony, prove cumbersome or expensive, there is no necessity to imply the sort of term that the judge was prepared to imply. There was G therefore no justification for implying it.

22. In my judgment, the judge was in error in his interpretation of clause 4(b). I would allow Anthony's appeal, set aside the first declaration in his order and substitute for it a declaration to the effect that the licence has not terminated.

Patten LJ said:

23. I agree.

Longmore LJ said:

24. I agree also. H

Appeal allowed.

NOTICES

Ener-G Holdings plc v Hormell

COURT OF APPEAL

31 July 2012

LORD NEUBERGER MR, LONGMORE AND GROSS LJJ

[2012] EWCA Civ 1059

Estates Gazette 10 November 2012

[2012] 45 EG 96

Notices – Service – Time limit – Share sale agreement specifying that advance written notice be given of any claim for breach of warranty and that proceedings be issued and served within 12 months thereafter – Agreement specifying certain methods of service including personal delivery to recipient's address – Whether notice served when delivered to recipient's home or later on deemed receipt by recorded delivery – Whether subsequent claim served in time

In March 2010, the appellant served written notice on the respondent of its intention to bring a claim for breach of warranty under a share sale agreement. The agreement required notice of any such claim to be given to the respondent within two years of the completion of the agreement; further, proceedings in respect of the claim were to be issued and served on the respondent not later than 12 months after the date of the notice, failing which the claim would be deemed to be irrevocably withdrawn. The agreement specified that any notice "may" be served by delivering it personally or sending it by pre-paid recorded delivery post at or to the respondent's address, and that a notice served by the latter method would be deemed to be received two business days after posting. The claim form was to be served in accordance with the agreement or in any other manner allowed by law.

The appellant served its notice by two methods: (i) first, it was delivered to the respondent's home on 30 March 2010 and, since nobody was home, was left on a table in the respondent's front porch, where the respondent found it later the same day; and (ii) on the same date, a second notice was sent to the respondent by recorded delivery post, with deemed delivery occurring on 1 April 2010.

The appellant subsequently served a claim form seeking damages of nearly £2m for the alleged breaches of warranty. That document was left in the letterbox of the respondent's property on 29 March 2011, although the respondent did not see it until 2 April.

On a trial of a preliminary issue, the appellant's claim was held to be time-barred. The judge held that: (i) there had been no service of the notice or claim form by personal delivery, since that required the documents to be handed over to the respondent personally; (ii) none the less, the first notice had been served effectively on 30 March 2010 when the respondent in fact received it, since the methods of service specified in the agreement were permissive not exclusive; and (iii) since the claim form was deemed to have been served on 1 April 2011, that claim had been brought outside the applicable 12-month period. The appellant appealed.

Held (Longmore LJ dissenting): **The appeal was dismissed. (1) If personal service or delivery of a document is required, it should be handed to the intended recipient personally. Service by delivering personally, within the meaning of the agreement, therefore meant that the document in question had to be handed over to the respondent; leaving it at his home did not suffice. The requirement to serve "at or to" the respondent's address did not affect that conclusion. It was not absurd or capricious for the parties to have limited the location where personal service could be effected; although such a restriction rendered it easier for the intended recipient to avoid personal service, it was unexceptionable where there was an easy alternative of sending by recorded delivery. It followed that the appellant had not effected service of either the first notice or the claim form by personal delivery within the meaning of the agreement. (2) The first notice was none the less effective. The methods of service set out in the agreement, properly construed, were permissive rather than exclusive, as indicated by the use of the word "may". The purpose of such provisions was to shift any risk from the server to the intended recipient. Clear words will normally be required before it is possible to ascribe to the parties an intention that a recipient who actually receives and reads a notice in time should be treated as not having received the notice at all. The service of a second notice by recorded delivery could not retrospectively invalidate the first notice. Consequently, the first notice had been validly served on 30 March 2010 and, since the claim form was deemed to have been served on 1 April 2011, the claim had been brought in time.**

Per Longmore LJ. Setting out two alternative express methods of service in the agreement clearly implied that these were intended to be exclusive; the word "may" merely recognised the availability of choice.

The following cases are referred to in this report.

Allison Ltd v *Limehouse & Co* [1992] AC 105
BCCI v *Ali* [2002] 1 AC 251
Blunden v *Frogmore Investments Ltd* [2002] EWCA Civ 573, [2003] 2 P&CR 84; [2002] 2 EGLR 29; [2002] 29 EG 153
Chiswell v *Griffon Land & Estates Ltd* [1975] 1 WLR 1181; [1975] 2 EGLR 58; [1975] 236 EG 411
Galinski v *McHugh* (1988) 57 P&CR 359
Lantic Sugar Ltd v *Baffin Investments Ltd* [2009] EWHC 3325 (Comm); [2010] 2 Lloyd's Rep 141
Rainy Sky SA v *Kookmin Bank* [2011] UKSC 50; [2011] 1 WLR 2900
United Scientific Instruments Ltd v *Burnley Borough Council* [1978] AC 904
Valentines Properties Ltd v *Huntco Corporation Ltd* [2001] UKPC 14
Yates Building Co v *RJ Pulleyn & Sons (York) Ltd* [1976] 1 EGLR 157

This was an appeal by the appellant, Ener-G Holdings plc, from a decision of Burton J, sitting in the Commercial Court, dismissing a claim against the respondent, Philip Hormell, for breach of warranty under a share sale agreement on the ground that the claim was time-barred on a proper construction of the agreement.

George Bompas QC (instructed by Stevens & Bolton LLP, of Guildford) appeared for the appellant; Nicholas Lavender QC (instructed by BRM Solicitors, of Chesterfield) represented the respondent.

Giving judgment, **Lord Neuberger MR** said:

1. This appeal is brought by Ener-G Holdings plc ("Ener-G") against an order made by Burton J, following a judgment he gave on 14 December 2011, determining preliminary issues between Ener-G and Philip Hormell. The ultimate issue on this appeal is whether the judge was right to hold that Ener-G is precluded from pursuing a claim for nearly £2m based on alleged breaches of warranties given by Mr

A Hormell in an agreement dated 2 April 2008 under which he sold certain shares to Ener-G ("the agreement"). The reason the judge held Ener-G to be so precluded was because it failed to serve the claim form on Mr Hormell within the period specified by the agreement.

The relevant provisions of the agreement

2. By clause 6.3.3(b) of the agreement, Ener-G, as "the buyer", had to give written notice ("notice") of any breach of warranty claim to Mr Hormell, as "the seller", by "the second anniversary of completion", ie by 2 April 2010. By clause 6.3.7(a) of the agreement, where notice had been given in respect of any claim:

B "that claim should be deemed to have been irrevocably withdrawn and lapsed unless… proceedings in respect of that claim have been issued and served on the seller not later than the expiry of the period of 12 months after the date of that notice."

3. The centrally relevant provisions of the agreement for present purposes are clauses 13 and 14, which are concerned with service of documents under the agreement. However, it is also right to mention clause 1.5.7, which provides that "[u]nless expressly stated to the contrary… 'in writing' or 'written' includes faxes but does not include e-mails".

4. Clause 13 is headed "Notices", and it is in these terms:

C "13.1 Notice in writing

Any notice or other communication under this agreement shall be in writing and signed by or on behalf of the party giving it.

13.2 Service

Any such notice may be served by delivering it personally or by sending it by pre-paid recorded delivery post to each party (in the case of the buyer, marked "for the attention of directors") at or to the address referred in the agreement or any other address in England and Wales which he or it may from time to time notify in writing to the other party.

13.3 Deemed service

D Any notice delivered personally shall be deemed to be received when delivered (or if delivered otherwise than between 9am and 5pm on a business day, at 9am on the next business day), any notice sent by pre-paid recorded delivery post shall be deemed to be received two business days after posting and in proving the time of despatch it shall be sufficient to show that the envelope containing such notice was properly addressed, stamped and posted."

5. Clause 14 is headed "Governing law and jurisdiction". Only clause 14.2 is relevant to the present dispute; it is headed "Proceedings" and is in these terms:

E "The parties agree that the documents which start any legal proceedings relating to a dispute ("proceedings") and any other documents required to be served in relation to those proceedings may be served on any party in accordance with this agreement. These documents may, however, be served in any other manner allowed by law. This clause applies to all proceedings wherever started."

The facts

6. Having decided that it had claims against Mr Hormell for breaches of warranty, Ener-G arranged for notice under clause 6.3.3 of its claim for breach of warranties to be served on him in two different ways.

7. First, a process server was instructed to deliver a notice ("the first notice") in an envelope addressed to Mr Hormell, to his home address, being the address referred to in the agreement (no other address having been notified pursuant to clause 13.2), namely Ringerlane
F Farm, Elmton, Worksop, Nottinghamshire ("Ringerlane"), which was a farmhouse. No one was at Ringerlane when the process server attended on the afternoon of 30 March 2010, so he was unable to obtain a reply, when he rang the bell. Accordingly, he left the envelope in the front porch on a table. Before 5pm that afternoon, Mr Hormell found the envelope, opened it, and read the first notice.

8. On the same day, 30 March 2010, an identical copy of the notice ("the second notice") was sent by recorded delivery in an envelope addressed to Mr Hormell at Ringerlane. It is common ground that this copy of the notice was deemed received, by virtue of clause 13.3, on 1 April 2010.

9. On 29 March 2011, a claim form ("the claim form") seeking damages for the alleged breaches of the warranties, with the necessary response pack, was taken by a process server to Ringerlane. As nobody G was at home, the process server placed the claim form in the letter box serving the property at 4.20pm. As the judge found, the documents were not in fact received by Mr Hormell until 2 April, when he emptied the letter box. Under CPR 6.14, it is common ground that this would mean that service of the claim form occurred on 31 March 2011.

The issues between the parties

10. By virtue of clauses 6.3.3 and 6.3.7(a), it is, rightly, common ground that Ener-G has to establish that (i) the notice was served on or before 2 April 2010, and (ii) the claim form was served no more than one year after the notice was served. Mr Bompas QC, on behalf of H Ener-G, contends that his client can succeed on two alternative bases.

11. The first basis is that the first notice was validly served, because it was "deliver[ed] personally… at… the [relevant] address" on 30 March 2010 under clause 13.2, as it was left at that address by an individual, namely the process server on that date. This proposition is challenged by Mr Lavender QC on behalf of Mr Hormell: he contends that the reference to "delivering personally" in clause 13.2 means personal service, in the sense of given to the individual to be served personally, which it was not. However, if that contention is wrong, and Ener-G's interpretation is right, then it is common ground that the claim form was served in time, because it was delivered to the same address J by a process server on 29 March 2011, and that would have been valid service as clause 14.2 permits service of the claim form to be effected, inter alia, by any manner permitted by clause 13.2.

12. The second basis on which Mr Bompas QC puts Ener-G's case is that, if Ener-G is wrong on the first point, then the only way in which a notice can be validly served under the agreement is in strict compliance with clause 13.2. On behalf of Mr Hormell, Mr Lavender QC challenges this argument on the ground that the service of the first notice was valid even though it was not in accordance with clause 13.2. If Ener-G is correct on that issue, then the first notice was not served and was therefore ineffective, but the second notice was served and was therefore K effective, and, on that hypothesis, it is common ground that the claim form was served in time, as the second notice was deemed served on 1 April 2010, and the claim form was deemed served on 1 April 2011.

13. For completeness's sake, Mr Hormell's case is thus that the first notice was effectively served on 30 March 2010, because, although it was not served in accordance with clause 13.2, that clause does not set out an exclusive code, and therefore the claim form was served too late as it was deemed to have been served on 31 March 2011 (see clause 14.2 and CPR 6.14).

14. Accordingly, for Ener-G to win this appeal, the court must be satisfied either that "delivering [the notice] personally" did not require L it to be handed personally to the intended recipient, or that the two methods of service set out in clause 13.2 are exclusive and no other means of service of a notice will be effective (unless the server can invoke estoppel or the like, which does not arise here). The judge rejected both these arguments. I shall consider them in turn.

The meaning of "delivering it personally" in clause 13.2

15. The first of the two issues to be decided is whether the first notice was "deliver[ed]… personally… at… [Mr Hormell's] address", within clause 13.2. As explained in [11] above, Mr Bompas QC, for Ener-G, contends that it was, because it was left there on 30 March 2010, by a person, namely a process server instructed by Ener-G, whereas Mr Lavender QC, M for Mr Hormell, contends that it was not, as the relevant words required personal service of the notice on Mr Hormell. If Mr Bompas QC is right, it was enough for the envelope containing the first notice to have been left by the process server at Ringerlane on 30 March 2010, and the fact that Mr Hormell actually opened the envelope on that day is irrelevant. If Mr Lavender QC is right, the fact that Mr Hormell opened the envelope on 30 March 2010 is also irrelevant, as it was not handed to him.

16. Like the judge, I prefer the argument advanced on behalf of Mr Hormell. First, there is the natural meaning of clause 13.2 in its context. Although in normal social parlance, "delivering" a document "personally" would often be understood to mean service by the sender personally, I do not consider that that is the natural meaning in a

A provision such as clause 13.2. After all, in the case of a notice on behalf of Mr Hormell, this would mean that Mr Hormell would have to deliver the notice: that cannot be right, and is not suggested by Mr Bompas QC. So the normal social meaning cannot be invoked.

17. Secondly, the concept of "personal service" is well understood to mean service on the recipient personally, not service by the server (or anyone else) personally – see eg per Lord Bridge and Lord Goff in *Allison Ltd* v *Limehouse & Co* [1992] AC 105, 113 and 124 respectively. Although it is true that the well known expression "personal service" has not been used, it seems to me that the legally familiar concept of personal service is redolent in the relevant words of clause 13.2. The B reason why the word "delivering" rather than "serving" is used in clause 13.2 is that having used the word "served" as a generic term, it was then thought appropriate to distinguish between two types of service, namely, handing over ("delivering") and posting ("sending").

18. Thirdly, where parties are concerned to prescribe how a notice may be served or delivered, the identity of the server of the notice is rarely, if ever, important, whereas the identity of the recipient is, at least normally, of central importance. The law and common sense both support the notion that if "personal" service or delivery of a document is required, it should be handed to the intended recipient personally.

19. Fourthly, if Mr Bompas QC is right, the parties were concerned C enough to require any notice to be delivered by someone "personally" without caring who that person was, whereas, on Mr Lavender QC's approach, the "person" is of central importance.

20. Fifthly, the word "personally" adds very little or nothing to the relevant part of clause 13.2 on Mr Bompas QC's approach, because any document has to be served by an individual (save perhaps if it is sent by e-mail or fax). Even a notice sent by post is delivered by a postman.

21. Sixthly, my conclusion is consistent with the reasoning of this court in a case decided on a similarly worded provision. *Bottin (International) Investments Ltd* v *Venson Group plc* [2004] EWCA Civ 1368 concerned an agreement with a clause which required any notice D or other document required to be given under the agreement to be "delivered personally or sent by prepaid recorded delivery to [specified addresses]". Peter Gibson LJ, with whom Longmore LJ and Lindsay J agreed, said that the clause "did not mean that Bottin personally had to effect the delivery but that the adverb 'personally' qualified the delivery to the recipient of the notice": [45]. Peter Gibson LJ then referred to the view "that a document is personally served on a company if the document is simply left at the registered office without any need to consider into whose hands, if any, it is delivered". At [47], Peter Gibson LJ said that "personal delivery of a notice to a company is effected by delivering it to somebody authorised to receive it".

E 22. Of course, the same word or expression can have different meanings in different contexts, but there is little difference between the context of the relevant words in *Bottin* and the relevant words in this case. Mr Bompas QC rightly points out that, in this case, the notice has to be delivered personally "at" the relevant premises (sc. Ringerlane), whereas in *Bottin* there was no restriction as to where personal delivery was to be effected. More specifically, he makes the point that it would have been peculiar for the parties to restrict where personal service can be effected. I see the force of that, but it is not a powerful point, because it is far from absurd, or even capricious, for the parties to have so limited the locations where personal service could be effected. Such a restriction merely F renders it easier for the intended recipient to avoid personal service. But the recorded delivery process was an obvious alternative.

23. Mr Bompas QC also suggested that his interpretation of the personal delivery provision in clause 13.2 required a notice to be served at the stipulated premises in the same way as, and consistently with, the alternative recorded delivery provision in the clause. I do not agree: if a notice is sent by recorded delivery, in an appropriately addressed envelope, it is deemed to be served even if it is not actually received by the intended recipient, whether at the premises or elsewhere – see per Megaw LJ in *Chiswell* v *Griffon Land & Estates Ltd* [1975] 1 WLR 1181, 1188-1189*, per Slade LJ in *Galinski* v *McHugh* (1988) 57 P&CR 359, 365 and per Robert Walker LJ in *Blunden* v *Frogmore Investments Ltd* [2002] EWCA Civ 573, [2003] 2 P&CR 84, [42]-[45]**. However, even of that were not right, I do not see why one should assume that the two stipulated forms of delivery were intended to have the same physical outcome. G

24. Mr Bompas QC also relied on the words "marked for the attention of the directors" in clause 13.2, as showing that personal service was not intended. However, following the reasoning of Peter Gibson LJ in *Bottin*, service of the envelope on an authorised agent of Ener-G would do, so it would be sensible to make it clear that the envelope contained something the directors should see. Mr Bompas QC also relied on the ease with which a person could avoid service. I do not H see that as providing much help to Ener-G's case either. A requirement of personal service on the intended recipient at a specified address seems unexceptionable, where there is an easy alternative of sending by recorded delivery to the same address.

Are the means of service in clause 13.2 exclusive?

25. The second issue is whether it was open to Ener-G to serve the notice in a manner different from that described in clause 13.2. In other words: are the methods of service in clause 13.2 exclusive or permissive? Unusually, it is Ener-G, the server of the document in question, who is contending that the contractually prescribed methods J of service are exclusive, and it is the intended recipient of the document, Mr Hormell, who is contending that they are permissive.

26. The judge held that the methods of service set out in clause 13.2 were no more than permissive, and, although I have not found the point easy, I have reached the same conclusion.

27. The notion that clause 13.2 was intended to be permissive rather than exclusive, receives obvious support from the fourth word of the clause, "may", particularly when it is contrasted with the mandatory word "shall" in the immediately preceding clause 13.1. The force of the point is only slightly diluted by the fact that clause 13.2 envisages two alternative means of delivery: the word "shall" would have been quite appropriate if it had been intended to be exhaustive. K

28. I consider that some support for this conclusion is to be found in clause 1.5.7, with its statement that "[u]nless expressly stated to the contrary… 'in writing' or 'written' includes faxes …". There is nothing in clause 13.2 which appears to exclude clause 1.5.7, and, if delivery by fax is permitted, it would seem to be inconsistent with the notion that the two prescribed methods of delivery in clause 13.2 were exclusive. I accept that it can be said that, if clause 13.2 does permit delivery only by the two specified methods, then delivery by fax is "expressly" precluded, but it seems to me that, if one is looking to find the intention of the parties, clause 1.5.7 does provide mild assistance to L Mr Hormell's case.

29. The argument that it would have been pointless to spell out two methods of service in clause 13.2, unless they were intended to be exclusive, has some initial attraction. However, in my view, on closer analysis, the argument has no force. The purpose of a provision such as clause 13.2, if it is not exclusive, is to shift any risk from the server to the intended recipient – see per Robert Walker LJ in *Blunden* at [27]-[28]. Thus, if a document is served in accordance with clause 13.2, it is treated as served, or delivered, even if it does not come to the attention of, or even if it is not received by, the intended recipient (see the cases cited at para 23 above). But if a document is served or delivered in M any other way (eg by ordinary post or by being left at the intended recipient's premises rather than being handed personally to him), there is no such presumption.

30. That argument works perfectly well for the provision for sending by recorded delivery. However, at least at first sight, it is unnecessary for personal service: there would appear to have been no need for any presumption in such a case, as personal service, once established as a matter of fact, will always be effective. I consider that the answer to that point lies in the fact that personal delivery of a document on Ener-G might involve the document not coming to the attention of the directors,

* Editor's note: Also reported at [1975] 2 EGLR 58; [1975] 236 EG 411

** Editor's note: Also reported at [2002] 2 EGLR 29; [2002] 29 EG 153

as it would encompass delivery on a non-director who had actual or ostensible authority to accept service (see per Peter Gibson LJ in *Bottin* at [46], cited in [21] above). If a document was personally delivered to such a person, who then failed to pass on the document to a director, clause 13.2 would make it clear that service had been effected.

31. In my judgment, commercial common sense also assists Mr Hormell's argument. If Ener-G had not sent the second notice by recorded delivery, it would have been very unattractive for Mr Hormell to argue that Ener-G's right to claim on the warranties was lost by virtue of no notice having been delivered within the two-year period prescribed by clause 6.6.3(b). The purpose of providing for the notice was to ensure that Mr Hormell was notified within the two-year period of a warranty claim, so that he would know that he was free of any such claim if he received no such notice within the two-year period. As he saw the first notice within the two-year period, that aim would have been achieved even without the second notice.

32. In my view, clear words would normally be required before one could ascribe to the parties an intention that a recipient who actually receives a notice in time should nonetheless be treated as not having received the notice at all. In this case, the point is rather reinforced by the point mentioned in [23] above, namely that, if a notice is sent by recorded delivery to the prescribed address, it is deemed to have been served, even if it is not actually received. If that is right, it would seem a little curious to ascribe to the parties an intention that a notice sent or delivered in another way was, in the absence of clear words, deemed not to have been served, even though it was clear that, as a matter of fact, it had been received and read by the intended recipient.

33. Accordingly, there is a strong commercial argument, as well as a strong argument based on the language of the agreement, for saying that the first notice was delivered when Mr Hormell saw it. If that is right, then the fact that Ener-G delivered the second notice, cannot, as it were, retrospectively cause the first notice not to have been validly delivered.

34. Further, while reliance on previous judicial decisions on this sort of point has obvious dangers, I consider that this conclusion derives some support from the reasoning of the Court of Appeal in *Yates Building Co* v *RJ Pulleyn & Sons (York) Ltd* [1976] 1 EGLR 157. In that case, it was unanimously held that a notice exercising an option, sent by ordinary post, was valid, even though the contractual provision, after requiring the option to be exercised by notice in writing, continued "such notice to be sent by registered or recorded delivery post". Lord Denning MR, with whom Orr LJ agreed, described the latter provision as "directory" rather than "mandatory". I have no doubt that Scarman LJ's characterisation of the provision as "permissive" rather than "obligatory" (see [1976] 1 EGLR 157, 158) is to be preferred. In a sense, the present case is *a fortiori*, because that case concerned an option, and the contractual requirements of an option are generally more strictly construed than those of other contractual rights (see *United Scientific Instruments Ltd* v *Burnley Borough Council* [1978] AC 904).

35. However, there are arguments the other way, and it is necessary to address them. First, there is the fact that clause 13.3 would result in a document served in accordance with clause 13.2 being, at least potentially, served later than a document served other than in accordance with clause 13.2. Thus, at first sight at any rate, it would seem to be a little odd that a document sent by ordinary post might be delivered the day after it was sent, but one sent by recorded delivery would be treated as delivered two days after it was sent. In my view, the answer to that point is that, if the server of a document wishes to take advantage of the risk transfer, or the irrebuttable presumption of service, conferred on him by clause 13.2 (as discussed in [29]-[30] above and by Scarman LJ in *Yates*), part of the price is the consequence of the presumed date of delivery as set out in clause 13.3.

36. If the server wants to take advantage of the deemed delivery by serving in accordance with clause 13.2, he takes the consequences as to deemed date of delivery in clause 13.3; whereas if the server chooses to deliver in some other way, he has to prove service (rather than relying on deeming) and has to prove the date of service (again, rather than relying on deeming). I see nothing surprising about this, although it is fair to say that it may produce somewhat oddly contrasting results on comparing particular hypothetical sets of facts. Further, clause 13.3 can produce somewhat odd results on any view: it appears that, if a document is personally delivered before 9am on a particular day, it is deemed to be received the next day, whereas if it is personally delivered between 9am and 5pm on a particular day, it is deemed to be received on that day.

37. It can also be said that the word "however" in the second sentence of clause 14.2 implies that the methods of service described in clause 13.2 (as impliedly referred to in the preceding sentence) were envisaged as being exclusive, as, otherwise, the word would not have been included. In my view, that is not a strong point, as the word "however" is not inconsistent with the notion that clause 13.2 is permissive.

38. There is also something in the point that it accords with commercial common sense to make it clear that service of documents can only be served in certain specified ways. The point has added force in a case such as this when the date of service of a notice acts not merely to give rise to a right, but also to start time running against the server in relation to another step (namely service of a claim form). However, I find the commercial common sense argument the other way rather more telling – see [29]-[32] above. In any event, clause 13 does not only apply to a notice under clause 6.6.3(b). Further, even in the case of such a notice, it is up to the person who delivers the notice to make sure (i) that he knows when the notice was delivered, and (ii) that the claim form is served properly and in time, both being matters which are wholly or largely within his control. If, as in this case, he leaves service of the claim form to the last minute, he only has himself to blame.

Conclusion

39. Accordingly, I agree with Burton J that (i) the first notice was not "deliver[ed] ... personally" within the meaning of clause 13.2 on 30 March 2010, and neither was the claim form "deliver[ed] ... personally" on 29 March 2011, but that (ii) the first notice was nonetheless validly served on 30 March 2010 and the claim form was deemed to have been served on 1 April 2011.

40. I have considered whether it might be argued that, even though the first notice was validly served by it having been left at Ringerlane and received and read by Mr Hormell on 30 March 2011, it was nonetheless open to Ener-G to contend that, having served the second notice strictly in accordance with clause 13.2 on 1 April 2011, it was entitled to treat the second notice rather than the first notice as the effective notice under clause 6.6.3(b). That was not a point argued by Mr Bompas QC, although it was raised during the hearing by Longmore LJ. I have concluded that it cannot be sustained: if the first notice was validly served, then there was no room for a second notice, not least because a second notice cannot retrospectively invalidate, or cancel, an earlier notice, valid when it was served.

41. For these reasons, I would dismiss Ener-G's appeal.

Longmore LJ said:

42. I agree with the Master of the Rolls that the words "delivering it personally" mean that the notice must be served on the person of the recipient and not that it must be served by a personal deliverer.

43. I have not found the question of exclusivity (as it has been called) easy to decide, but I have concluded that the judge was wrong not to have construed clause 13.2 of the contract regarding the methods of service of any notice (such as that required by clause 6.3.3 of the share sale contract) as exclusive.

44. My main reason is that it is counter-intuitive to conclude, when the parties have taken the trouble to spell out (a) that a notice given under the agreement can be served either "by delivering it personally" or "by sending it by pre-paid recorded delivery post" and (b) that, in a case where a notice is to be served on the buyer (a company) it is to be marked "for the attention of directors", that a notice can be served in any other way the deliverer of the notices chooses. The contract could have made the matter clearer by inserting the word "only" before the words "be served" in the expression "may be served"; equally, the

A parties could have added the words “or in any other way” at the end of clause 13.2. But they did neither and it can therefore be said that the clause has either of those meanings. But the common sense of the matter is that by spelling out two methods by which a notice could be served, they did not intend that it be served in any other way. This conclusion is, to my mind, strongly supported by the fact that clause 14.2 provides that any legal proceedings may be served in accordance with the agreement and then says:

> “These documents may, however, be served in any other manner allowed by law.”

B 45. Mr Lavender QC for the defendant relied heavily on the use of the word “may” in clause 13.2 in contradistinction to the word “must” or “shall” as used in other clauses of the contract, notably in clause 13.1, which provides that “any notice shall be in writing”. That argument has some, but by no means conclusive, force; it is clear that a notice has to be in writing but if one is then providing for service of such a notice in two possible ways, it is a natural use of language to say that the parties “may” use one method of service or a second method of service. Such expression does not mean that it is not compulsory to use one method or the other.

46. Burton J made the powerful point that if the claimant had just
C left the notice on the table in the porch (without also sending it by recorded delivery) and if the notice had actually come to the attention of the defendant (as it did in this case) shortly after being so left and the defendant claimed that he had not been properly served with the notice so that no claim could be brought against him, “he would have a hard time of it in court”. But I respectfully wonder about that; one can always propose hard cases where, on different facts, a harsh result apparently ensues. The defendant could always say that a claimant who has two years within which to serve a contractual notice and always has the fall-back of sending it by recorded delivery has only himself to blame if he chooses to try to serve by some other method which (for whatever
D reason) does not contractually work out.

47. It seems to me that the judgment, with respect, produces the rather more perverse consequence that the actual method of service which was not contractual (because not served on the defendant personally) effectively becomes a contractual method once it has come to the notice of the defendant and therefore operates as a trigger for the second period (within which proceedings have to be served pursuant to clause 6.3.7) at a time of which the claimant cannot be aware, with the result that his claim is unexpectedly barred because the year allowed has begun before the year allowed by the second (contractual) method of service.

E 48. For myself, I am not influenced by decided cases about contractually prescribed methods of making offers or communicating acceptances or exercising options because they relate to the making (or otherwise) of contracts. I do not see why the law in that area should be transposed to notices given pursuant to contracts already made. Traditionally and understandably the law is stricter about the service of documents (or at any rate proceedings) than it is about the transmission of offers and acceptances. I would therefore distinguish *Yates Building Co*, on which the Master of the Rolls places some reliance. Indeed, I did not understand Mr Lavender QC to argue that the matter was concluded by authority which, after all, businessmen are unlikely to know much
F about, see *BCCI* v *Ali* [2002] 1 AC 251 [51] per Lord Hoffman.

49. Left to myself therefore, while acknowledging the force of my Lord’s conclusion and that of the judge, I would have allowed this appeal.

Gross LJ said:

Introduction

50. I gratefully adopt the summary of the provisions of the agreement and the facts contained in the judgment of the Master of the Rolls.

51. On the question of the meaning of the words “delivering it personally” in clause 13.2 of the agreement, I agree with both the Master of the Rolls and Longmore LJ and there is nothing I can usefully
G add. It follows that the appeal must be dismissed unless the appellant is entitled to succeed on the “exclusivity” question.

52. With regard to that question, I find myself in an invidious position, given that the Master of the Rolls and Longmore LJ disagree as to the answer. Like them, I have not found the point easy.

The rival cases

53. The rival cases were, if I may say so, clearly and attractively articulated by Mr Bompas QC for Ener-G (the appellant) and Mr Lavender QC for Mr Hormell (the respondent).

54. Mr Bompas QC submitted that the methods of service set out
H in clause 3.2 were exclusive. The consequence was that the 12-month period provided for in clause 6.3.7 started later, on 1 April, 2010, so that the claim form was served in time, on 31 March 2011. In essence, Mr Bompas QC submitted that in clause 13.2 of the agreement, the parties had carefully stipulated for two modes of service, reinforced by the provisions of clause 13.3. It was implausible that the parties would have constructed this detailed scheme if the modes of service were no more than permissive. Further, the wording of clause 14.2 lent support to the appellant’s case. The additional flexibility provided for the service of documents starting or relating to legal proceedings (eg, the claim form) and the service of notices or other communications (eg, the 2010
J first and second notices) was stark. The former, not the latter, could be served “in any other manner allowed by law” rather than “in accordance with this agreement”. It is common ground that if Mr Bompas QC is right on the exclusivity question, then the appeal must be allowed.

55. Mr Lavender QC submitted that the judge was right; the methods of service set out in clause 13.2 were no more than permissive. Accordingly, time started running on 30 March 2010, and the purported service of the claim form, on 31 March, 2011, was out of time. The word “may” in clause 13.2 told decisively against the argument for exclusivity. “May” did not mean “must”; contrast, for instance, the word “shall” in clause 13.1. It did not follow that clauses 13.2 and
K 13.3 were deprived of meaning. To the contrary, a party, following either mode of service for which express provision was made, obtained “copper bottomed” protection (or a “safe harbour”) against the risk of the document in question not actually coming to the attention of or being received by the intended recipient. Even if, which Mr Lavender QC sought to resist, clause 14.2 furnished a pointer in favour of exclusivity, it could not outweigh the impact of the permissive word “may”. Mr Lavender QC further submitted that the respondent’s construction enabled meaning to be given to the provision contained in clause 1.5.7 as to delivery by fax. Still further, although on the facts of the present dispute, the conclusion that clause 13.2 provided for exclusive modes of service would mean that the proposed claim was
L not time barred, there would be other situations when it would lead to a very different and unattractive outcome: for instance, it would mean that service of a notice could be out of time, simply because of the use of some other mode of service and even though the intended recipient had, as a matter of fact read the notice and was aware of its contents. Indeed, it was on that very footing that the appellant’s argument for a later start to the clause 6.3.7 period depended.

Discussion

56. (1) Preliminary matters: I begin with a number of preliminary matters which inform my consideration of the exclusivity issue.

M 57. First, as was common ground, the approach to the construction of the agreement was that set out in *Rainy Sky SA* v *Kookmin Bank* [2011] UKSC 50; [2011] 1 WLR 2900. It suffices to refer to two passages in the judgment of Lord Clarke of Stone-cum-Ebony:

> “14. … the ultimate aim of interpreting a provision in a contract, especially a commercial contract, is to determine what the parties meant by the language used, which involves ascertaining what a reasonable person would have understood the parties to have meant. …
>
> …
>
> 21. The language used by the parties will often have more than one potential meaning. I would accept the submission… that the exercise of construction is essentially one unitary exercise in which the court must consider the language

A used and ascertain what a reasonable person, that is a person who has all the background knowledge which would reasonably have been available to the parties in the situation in which they were at the time of the contract, would have understood the parties to have meant. In doing so, the court must have regard to all the surrounding circumstances. If there are two possible constructions, the court is entitled to prefer the construction which is consistent with business common sense and to reject the other. "

Elaboration is unnecessary, save to observe that one of the difficulties in the present case is that considerations of business common sense can properly be relied upon for each of the rival constructions.

58. Secondly, as observed by the Judge, at [9(ii)], certainty is
B important in commercial transactions, so that parties can know where they stand and act accordingly. Strict time bars are an aspect of certainty; there is certainly nothing uncommercial about them and they are not infrequently encountered in commercial contracts. When there is a time bar, then, as Lord Nicholls observed in *Valentines Properties Ltd* v *Huntco Corporation Ltd* [2001] UKPC 14, at [20]:

"Inherent in a time limit is the notion that the parties are drawing a line. Once the line is crossed, a miss is as good as a mile."

Mr Bompas QC, rightly, disclaimed any suggestion of seeking sympathy as to the operation of clause 6.3.7(a); his submissions were
C instead exclusively focused on the proposition that service of the proceedings was just in time, not just too late.

59. Thirdly, despite the desirability and importance of certainty, a good many commercial contracts are less tidy than might be desirable as a matter of strict theory. In this respect, commercial contracts reflect the realities of commercial life. It is thus no surprise to find in a commercial contract surplus language, for instance that which merely states the obvious. Likewise, it is by no means uncommon to find that whichever of two rival constructions is preferred, anomalies or apparent anomalies will remain. The present case is no exception, as the judgments of the Master of the Rolls and Longmore LJ readily
D allow. The task, accordingly, is to ascertain which construction best, if imperfectly, fits the language used by the parties in the context in which the agreement is located.

60. Fourthly, if I may be permitted to put it this way, the purpose of a notice provision is to give notice – ie to require the intended recipient to be made aware of the facts and matters in question. However, it by no means necessarily follows from the intended recipient having actual knowledge of such facts and matters that he has been properly served; thus, to take an example removed from the facts of this case:

" If a claimant is required to serve X and, mistakenly purports to serve Y, the mere fact that Y informs X of the purported service so that X knows of it, cannot
E convert Y's receipt of the documents into good service upon X."

See *Lantic Sugar Ltd* v *Baffin Investments Ltd* [2009] EWHC 3325 (Comm); [2010] 2 Lloyd's Rep 141, at [40]. The example, of course, serves to illustrate why pleas of hardship in this area are untenable. Conversely, however, there is much to be said for reducing the risk of a person, who has actually received a notice and become aware of its contents, asserting that he has nonetheless not been properly served. A construction which serves to reduce this risk may be thought to have obvious attraction as giving effect to the underlying commercial purpose of any typical notice provision.

F 61. (2) Decision: Very much on balance, respectfully parting company with Longmore LJ but in agreement with the judge and the Master of the Rolls, I have concluded that clause 13.2 is permissive rather than exclusive as to the modes of service there specified. My reasons follow.

62. First, I begin with the language of clause 13.2. Here, I do regard the use of the word "may" as striking. As clause 13.1 so clearly illustrates, the parties knew how to say "shall" when they intended to do so. While it is true that "may" foreshadows two specified modes of service and thus there is nothing unnatural about its usage, it is also true that "shall" or "must" would have served equally well. Moreover, either "shall" or "must" would have conveyed the exclusivity of the modes of service in question, had that been the parties' intention. For my part, I am unable to read "may" as meaning "must", at least unless G
persuaded by other language in the agreement or its commercial context that I should do so. At all events, I regard the use of the word "may" as a most powerful pointer to the permissive rather than exclusive nature of clause 13.2.

63. Secondly, I seek to test my provisional view of clause 13.2 as permissive having regard to the agreement as a whole.

(i) On the hypothesis that clause 13.2 is permissive, the scheme of clauses 13.2 and 13.3 has the important and sensible commercial purpose of risk allocation. If one of the specified modes of service is used, risk is transferred from the server to the recipient, as explained by the Master of the Rolls at [29] *et seq* above. The certainty involved in H
this transfer of risk may be seen as a price worth paying for curiosities such as the possibility that service by one of the specified modes of service might be achieved later than by some other method of service permitted by law: see [35] and [36] above. Accordingly, although I appreciate the attraction of the argument that the careful nature of the scheme of clauses 13.2 and 13.3 suggests that it was intended to be exclusive rather than permissive in nature, I am not persuaded by it.

(ii) Clause 14.2 must next be addressed. At first blush, its wording does lend support to the appellant's case. The point is an important one but, ultimately for me, not decisive. To begin with, there is some force in Mr Lavender QC's submission that the provision J
"Proceedings may be served … in accordance with this agreement" should be read as "neutral" on the question of whether clause 13.2 is permissive or exclusive. The next sentence, dealing with the service of documents commencing or relating to legal proceedings "in any other manner allowed by law" can be seen as no more than an example of "bootstraps" or a statement of the obvious, of the kind that (as already indicated) features in commercial contracts. Let it be assumed, however and contrary to Mr Lavender QC's argument, that clause 14.2 tends to favour the appellant's construction of the agreement. On this footing, the question which arises is whether the pointer furnished by clause 14.2 serves to outweigh the importance to be attached to the use of the K
word "may" in clause 13.2. The better view, in my judgment, is that it does not do so; the best fit is to accord "may" the priority it warrants. To my mind – and a matter to which I return – clear words would be needed to treat clause 13.2 as exclusive rather than permissive. The use of the word "may" at once undermines any such suggestion of exclusivity; the suggestion conveyed by clause 14.2 (even assuming that it does favour the appellant's case) falls well short of the clear wording needed to make good a case of exclusivity.

(iii) Though of distinctly limited significance, treating clause 13.2 as permissive enables full effect to be given to clause 1.5.7 and service by fax. L

64. Thirdly, I turn to questions of context and business common sense. On the one hand, I acknowledge the attraction of the appellant's argument as promoting certainty. If clause 13.2 is exclusive as to modes of service, there would be greater certainty as to the start and, hence, the finish of the 12-month period provided by clauses 6.3.3(b) and 6.3.7(a). As already foreshadowed, that is a point of significance. It remains so – even though it can also fairly be said that it is not asking too much of the appellant to issue and serve proceedings within a hardly ungenerous period of 12 months and to err on the side of caution having regard to the very limited range of possible dates when time could realistically start to count. M

65. On the other hand, there is, for my part, the yet more persuasive argument that exclusivity requires clear wording because of the consequences which can follow. Exclusivity entails that, regardless of the actual knowledge of the intended recipient, any mode of service other than those specified in clause 13.2 will not suffice. The court will, of course, give effect to exclusivity – notwithstanding the consequences – if indeed that is the bargain the parties have made; however, as already foreshadowed, there is no such clear wording here. Turning to the consequences, it is to be acknowledged that, on the facts of the present case, if clause 13.2 provided for exclusive modes of service, the appellant would not be time barred from proceeding with its proposed claims. But, in other, readily conceivable, factual situations,

A the consequences could be much less attractive. As illustrated by the judge, at [16], treating clause 13.2 as exclusive would facilitate a party asserting that the initial clause 6.3.3(b) notice was out of time despite actual receipt of the document and full knowledge of its contents. For reasons I have already sought to adumbrate, a construction which reduces the risk of such an outcome has much to be said for it. In my view, absent clear wording providing for exclusivity, this consideration strongly favours a permissive construction of clause 13.2 – and, especially when considered together with the very considerable significance of the use of the word "may" in clause 13.2, it serves to tilt the balance in favour of the respondent's case on the appeal.

B 66. Accordingly, treating the task of construction as a "unitary exercise" (*Rainy Sky*, above, at [21]) and broadly for the same reasons as the judge and the Master of the Rolls, the better view is that clause 13.2 of the agreement is permissive rather than exclusive with regard to modes of service. It follows that I would dismiss the appeal. I am conscious that the consequence of doing so is that the appellant's C proposed claims against the respondent are time-barred, but I fear that by leaving service until so late in the day, the appellant has been the author of its own misfortune.

67. For completeness:

(i) While, with respect, I am wary of relying on authority to aid in the construction of the agreement, the judgment of Scarman LJ (as he then was) in *Yates Building Co* at p158 (cited by the Master of the Rolls at [34] above), helpfully highlights the distinction between "obligatory" and "permissive" wording – and, in my view, reinforces the need for clear wording before concluding that clause 13.2 was exclusive in nature. D

(ii) Again with respect, I express no view on the point canvassed by the Master of the Rolls, at [40], above.

Appeal dismissed.

PRACTICE AND PROCEDURE

Odedra and another v Ball and another

TECHNOLOGY AND CONSTRUCTION COURT

4 July 2012

COULSON J

[2012] EWHC 1790 (TCC)

Estates Gazette 20 October 2012

[2012] 42 EG 134

Practice and procedure – Disclosure – Expert witnesses – Claim in nuisance and/or negligence for contamination of land by escape of oil – Damages claimed for lost sale of land or diminution in value – Claimants alleging that land now unsaleable and of no value – Each party obtaining an expert report on valuation – Whether claimants obliged to disclose second report obtained from same expert

The claimants brought an action against the defendants for nuisance and/or negligence in relation to an escape of heating oil from the defendants' oil tank onto the claimants' property. The claimants' claim for £402,500 in damages was based on the price that a potential purchaser had offered for the property when they marketed it in 2007; the offer had later been withdrawn, apparently on the ground that the contamination from the oil leakage had not been remedied.

The claimants' claim was framed primarily as the loss of an actual sale at £402,500; they offered to transfer the land to the defendant for that sum on such terms as the court might direct. In the alternative, they claimed the same sum for diminution in the value of the land, on the ground that the property had ceased to have any value because of the stigma of its previous contamination. They acknowledged that if, contrary to their contentions, the land were found to have some residual value, then the claim for diminution in value should be measured by reference to the lost £402,500 offer less the residual value.

By order of the court, the parties were permitted to call one expert valuer each, with the experts' reports to be served by a specified date and a joint meeting of experts to follow for the purpose of producing a joint statement of matters agreed and disagreed. Both parties obtained and served an expert report but the defendants' expert did not attend the scheduled joint meeting. In consequence of that failure, the claimant applied for further directions on expert evidence, pursuant to the order of the court and CPR 35.12. The defendants cross-applied for an order that the claimants' expert should be required to provide copies of a further, undisclosed report on valuation issues that he had produced for the claimants, or be debarred from giving expert evidence.

Held: **Directions were given accordingly. (1) The only expert evidence that could potentially be relevant to the claimants' primary claim for loss of an actual sale was whether the asking price of £402,500 was appropriate or reasonable in 2007. Since that matter was not in dispute, and £402,500 was agreed to be the appropriate starting point for any consideration of loss, then no further expert evidence was required to support the claimants' primary claim. That claim depended on the proposition that a property that could not be sold had no value, which was a matter of law and argument, rather than of expertise. (2) Expert evidence was required to support the claimants' alternative claim for diminution in value. If the court rejected the proposition that a property that could not be sold had no value, then it would need to know what, if anything, the residual value was. That was a matter for expert evidence. (3) It will not always be necessary to disclose an undisclosed report that was prepared by an expert at the same time as a report that has been disclosed. Given the importance of openness under the CPR, and the need for that openness to trump questions of privilege in certain instances, there might be cases in which an expert will have to disclose both reports as a condition of being permitted to give evidence at all. However, in other cases injustice would arise if an expert were required to disclose everything that he produced, regardless of privilege. There is no general principle that everything is disclosable, regardless of privilege. In the instant case, the claimants' expert should not be required to provide copies of his other report. Both parties' experts had become slightly confused as to what they should be doing and it would be wrong, and potentially unjust, to require either of them to provide privileged reports that might be based on a misunderstanding of the issues. If the claimants sought to advance their alternative case, they would need a new report from their expert, containing clear expert evidence on valuation and diminution in value.**

The following cases are referred to in this report.

Vasiliou v *Hajigeorgiou* [2005] EWCA Civ 236; [2005] 1 WLR 2195
Edwards-Tubb v *JD Wetherspoon plc* [2011] EWCA Civ 136; [2011] 1 WLR 1373

This was the hearing of an application by the claimants, Mayur Odedra and Charlotte Odedra, for directions on expert evidence in relation to their claim for damages in nuisance and/or negligence against the defendants, Richard Ball and Anna Ball; and of a cross-application by the defendants that the claimants' expert witness be required to disclose an undisclosed valuation report or be debarred from giving evidence.

Peter Susman QC (instructed by Jefferies Essex LLP, of Southend-on-Sea) appeared for the claimants; David Platt QC (instructed by Berrymans Lace Mawer) represented the defendants.

Giving judgment, **Coulson J** said:

Introduction

1. By an application dated 19 March 2012, the claimants' solicitors seek further directions from the court relating to expert evidence, as a result of the failure on the part of the defendants' expert valuer to attend a joint meeting pursuant to the order of the court and CPR 35.12. The defendants' cross-application, dated 20 April 2012, is that either the claimants' expert valuer be debarred from giving expert evidence, or he be obliged to provide copies of another (undisclosed) report which he has prepared on valuation matters. These somewhat unusual applications raise one potentially important issue under the CPR, and it was for that reason that, at the conclusion of the hearing, I reserved judgment. I should repeat now my gratitude to leading counsel for the measured way in which they dealt with these matters, which generated considerable heat (if rather less light) in the inter-solicitor correspondence.

A *The pleaded claims*

2. The claimants claim damages against the defendants for nuisance and/or negligence and/or under the doctrine of *Rylands* v *Fletcher* (1868) LR 3 HL 330, in consequence of the escape of heating oil from the defendants' oil tank into the claimants' property at Heddington, near Calne, in Wiltshire. In 2007, the claimants put their property on the market and received an offer, subject to contract, in the sum of £402,500. The offer was subsequently withdrawn, apparently on the basis that the contamination from the oil leakage had not been remedied. Thereafter, in a falling market, the property failed to sell. It is the defendants' case that the problem was fully remedied by August 2008.

B 3. The claimants' claim for damages in the particulars of claim is put in two different ways:

> "6.1. The claimants lost an actual sale of the claimants' land (rather than a prospective sale or the chance of a sale), in that:
>
> 6.1.1. In March 2007, the claimants believed that any contamination of the claimants' land had been remedied and put the claimants' land on the market for sale;
>
> 6.1.2. In the spring of 2007, the residential property market was buoyant;
>
> 6.1.3. On or about 25 April 2007, the claimants agreed (subject to contract) to sell the claimants' land to Mr and Mrs Victor Borok for £402,500;
>
> 6.1.4. By 1 July 2007, Mr and Mrs Borok would have exchanged contracts C and would have completed the purchase of the claimants' land;
>
> 6.1.5. However, on or about 25 May 2007, Mr and Mrs Borok withdrew from the purchase on learning that the claimants' land was still contaminated and that it would take a further year for the contamination to be remedied;
>
> 6.1.6. The claimants accordingly lost the benefit of receiving £402,500 for the claimants' land by 1 July 2007;
>
> 6.1.7. The claimants further claim interest on £402,500 under the Senior Courts Act, section 35A, from 1 July 2007 until judgment in this action, at the rate of 8 per cent a year;
>
> 6.1.8. The claimants recognise that they are not entitled both to damages on this basis and to retain the claimants land, and accordingly the claimants undertake to transfer the claimants land to the defendants on such terms as the D court may direct, if the claimants are awarded the damages claimed under this subpara 6.1.
>
> 6.2. Alternatively, the claimants suffered diminution in the value of the claimants' land, and:
>
> 6.2.1. The best evidence of the value of the claimants' land uncontaminated is the sale price of £402,500 referred to in para 6.1.3 above;
>
> 6.2.2. Because of the stigma of previous contamination (which the claimants are now obliged to disclose to any prospective purchaser), the claimants' land now has no value, as evidenced by the fact that when the claimants put it on the market again they received no offers;
>
> 6.2.3. The diminution in value is therefore £402,500;
>
> E 6.2.4. The claimants further claim interest on £402,500 under the Senior Courts Act, section 35A, from 1 July 2007 until judgment in this action, at the rate of 8% a year;
>
> 6.2.5. If and to the extent that the defendants contend that the claimants' land has some residual value, the risk of that contention being unfounded or exaggerated should in all the circumstances of the case and in justice be borne by the defendants rather than by the claimants, and accordingly the claimants undertake to transfer the claimants land to the defendants on such terms as the court may direct, if the claimants are awarded the damages claimed under this subpara 6.2."

4. The primary claim at para 6.1 is therefore a claim for the loss of F the sale in 2007. It is said that this was not the loss of a chance but the loss of an actual sale, and accordingly the sum of £402,500 is claimed in full. Mr Susman QC properly accepts that this is a novel claim but maintains that it is in accordance with first principles. The alternate claim, at para 6.2 of the particulars of claim, appears to be based on the proposition that, but for the oil leak, the property was worth £402,500 and that, in consequence of the stigma now attached to the property because of the contamination, the property has no value. That is more akin to a traditional diminution in value claim. Mr Susman QC acknowledged that, to the extent that the court concluded that there was a residual value, the alternative claim would have to be measured by reference to the asking price of £402,500, less such residual value as the court identified.

The expert evidence G

5. By an order dated 28 October 2011, Akenhead J allowed each party to call one expert environmental consultant and one expert valuer. Their reports were to be served by 10 February 2012. The experts were thereafter to meet and produce a joint statement of matters agreed and disagreed by 24 February 2012. Neither party chose to serve a report from an environmental consultant. Each side, however, served reports from expert valuers. On analysis, neither report is a model of clarity.

6. The report served on behalf of the claimants was prepared by Mr James Dutch, the head of professional services at Strakers, the estate agents acting for the claimants at the time of the unsuccessful sale. The report is extremely short (just a page and a half). The relevant sections H read as follows:

> "2. RECOMMENDED ASKING PRICE
>
> Like all estate agents, Strakers, when acting on behalf of a vendor recommend an 'asking price'. For this purpose it is necessary to make an assessment of the price that the property is likely to fetch between a willing vendor and a willing purchaser with no special interest in acquiring the particular property, and given time to market the property as effectively as is practicable. Estate agents and surveyors call this assessment a 'valuation', but it is no more than an estimate based on experience of the person making the assessment, aided by knowledge of the selling prices actually achieved on the sale of comparable properties in the locality. The test of the validity of the assessment is, of course, the price at J which the property ultimately sells, if it does.
>
> Within the appendix we enclose a copy of our original valuation and recommendations/marketing advice addressed to Mr and Mrs Odedra, together with information from our archive of interested parties and viewings of the property that took place between 2 October 2008 to 30 November 2009.
>
> 3. OFFERS
>
> You will see from the viewing history that NO offers were received during the marketing. I can also advise that I have spoken with Guy Straker, the manager at our Devizes office who was dealing with the property, who has confirmed that the archive history is correct and that NO offers were received.
>
> 4. COMPARABLE EVIDENCE K
>
> I enclose a number of comparable properties with information of figures achieved for the period August 2008 to February 2012."

7. At para 6.2 of his skeleton argument, Mr Susman QC submitted that, on the basis of this report, Mr Dutch was saying that the asking price was the price at which the property was expected to sell (based upon experience and knowledge of comparables) and that, in addition, there was an "express conclusion or necessary implication that a property which cannot be sold has no value".

8. It should also be noted that, in a letter that Mr Dutch wrote to the claimants' solicitors on 13 July 2011, and for which privilege has L not been claimed, he expressed the view that the property was saleable and identified a figure of £245,000. Other figures that he quoted, based on auction prices, ranged between £220,000 and £235,000. The views expressed in that letter may become important because, in the inter-solicitor correspondence, the claimants' solicitors say that in early 2012 they asked for, and have since received, another report prepared by Mr Dutch apparently considering the value of the property at various dates. That report has not been disclosed, and the claimants' solicitors have said that they do not intend to disclose it or rely on the report at the hearing. They maintain that the purpose of that report was to assist them in considering the report which they anticipated receiving from M the defendants' expert valuer, Ms Loweth.

9. On behalf of the defendants, Ms Loweth has produced a report which contains her explanations for two valuation figures: the figure of £290,000 as at January 2012, on the specific assumption that there was no contamination, and the figure of £295,000 as at August 2008, again on the same assumption. In her report, Ms Loweth does not deal with the comparator, namely the figure against which those valuations are to be measured. However, in his skeleton argument, Mr Platt QC said that it was the defendants' case that the appropriate measure of loss was diminution in value, and that "such a diminution is to be measured by taking the prospective sale price in May 2007 and comparing it with the value as at the date of remediation, namely August 2008." He

A confirmed this position in his oral submissions. Accordingly, although it was not clear from Ms Loweth's report, the defendants' case is that, subject to other arguments, the diminution in value is to be assessed by comparing the sum of £402,500 (the asking price) with one or other of the valuation figures produced by Ms Loweth of either £295,000 or £290,000.

The parties' submissions

10. The claimants say that Mr Dutch's report is a proper expression of expert opinion, although Mr Susman QC conceded that some judges might take the matters summarised in [7] above to be obvious, such

B that it was unnecessary to prove such matters by evidence at all. The claimants deny that there was ever any agreed basis on which the experts would be instructed and deny that they are under any obligation to disclose the other report prepared by Mr Dutch which was, they say, "commissioned purely for the benefit of advice on a topic which, on the position which the claimants have adopted, is of no relevance (namely the precise valuations attributed by Ms Loweth to a property which could not be sold)". They maintain that, in those circumstances, the experts should have met without prejudice as ordered by the court and that the defendants and/or their expert are therefore in breach of the court order.

C 11. The defendants say that there was an agreement between the solicitors as to the basis on which the reports would be prepared and that it appears that Mr Dutch has prepared a report (for which privilege is claimed) on that agreed basis, dealing with valuation figures. They say that, in consequence, that report is disclosable in accordance with the CPR. By contrast, they say that the report which has been disclosed ([6] above) does not contain any expression of expert opinion. They say that the expert process is being manipulated by the claimants so as to allow them to maintain the "no loss" argument, irrespective of the views of their own expert.

Is expert evidence required to support the claimants' primary claim?

D 12. As noted above, the claimants' primary claim is set out in para 6.1 of the particulars of claim ([3] above). In my view, the only expert evidence that might potentially have been required in relation to that claim was whether the asking price of £402,500 was appropriate or reasonable as at May 2007. Indeed, that was the main reason that Mr Susman QC gave in seeking to rely on the short report from Mr Dutch. However, that is no longer in dispute. The defendants do not argue that the £402,500 was and is anything other than the appropriate starting point for any consideration of loss. That is therefore not an issue between the parties.

13. In my view, now that the £402,500 has been agreed, no other

E expert evidence is required to support the claimants' primary claim. That claim (para 6.1 of the particulars of claim) depends on the proposition that a property which cannot be sold has no value, which is a matter of law and/or argument; it is not a matter of expertise. Furthermore, even if it were, Mr Dutch does no more in his report than state the obvious, namely that "the test of the validity of the assessment [of the asking price] is, of course, the price at which the property ultimately sells, if it does." For the avoidance of doubt, and contrary to the claimants' submissions, I do not read his report as an express or implied statement of opinion that a property which cannot be sold has no value. He would need to say that expressly if it was his view: experts' reports should not

F be a matter of "necessary implication". However, that omission is not fatal to the primary claim, which is instead a matter of law and fact.

14. Accordingly, now that the parties have reached a belated agreement about the asking price of £402,500, I do not regard the claimants' primary case as requiring any further expert evidence at all. In my view, that claim is not assisted one way or the other by the short report from Mr Dutch.

Is expert evidence required for the alternative way in which the claim is put?

15. By contrast, I am in no doubt at all that expert evidence is required to support the alternative claim (para 6.2 of the particulars of claim). The alternative claim only arises if the court rejects the

G argument that a property which cannot be sold has no value. In those circumstances, the court will want to know what the residual value of this property was. It is the claimants' pleaded case that (due to stigma and the like) the residual value is nil: see section 6.2 of the particulars of claim. That alternate claim therefore requires expert evidence. Mr Dutch has not addressed that alternative claim in his short report. He needs to say whether the property is worth nil (and if so, why) or, if there is a residual value (along the lines of his letter of 13 July 2011), what that is and how he has calculated it.

16. Ms Loweth's approach to the diminution in value issue is also somewhat elliptical. She merely provides two valuations for August

H 2008 and January 2012. It would now appear that she is saying that the diminution may be capable of being measured by subtracting one or other of those figures from the asking price of £402,500. But that only became apparent from Mr Platt QC's submissions; it is not referred to in her report.

Should Mr Dutch's other report be disclosed?

17. Mr Platt QC submitted that Mr Dutch's other report, which plainly exists, should be disclosed. He argued this by analogy with the decision of the Court of Appeal in *Vasiliou* v *Hajigeorgiou* [2005] EWCA Civ 236; [2005] 1 WLR 2195, where Dyson LJ (as he then

J was) emphasised that "if a party needs the permission of the court to rely on expert witness A in place of expert witness B, the court has the power to give permission on condition that A's report is disclosed to the other party or parties, and that such condition will usually be imposed." Possibly, I think that there is a typographical error, and that the reference should have been to the disclosure of B's report, which would have been the earlier report.

18. In addition, Mr Platt QC relied on the decision in *Edwards-Tubb* v *JD Wetherspoon plc* [2011] EWCA Civ 136; [2011] 1 WLR 1373, in which a claimant obtained an expert's report (from a nominated expert) for the purposes of the pre-action protocol, and then instituted

K proceedings using a different expert. The Court of Appeal recognised that the report from the first expert was privileged but said that, once a party had embarked on the pre-action protocol procedure of co-operation in the selection of experts, there was no justification for not disclosing the report of such an expert, who had been put forward as suitable, had been accepted as such by the opposing party, and had prepared a report. At [11] of his judgment in that case, Hughes LJ said:

> "The question of principle which this case raises is whether the power to impose a condition on the grant of permission to rely on expert B can properly be employed to require the disclosure of the privileged report of expert A, and
>
> L if so when. If this is proper, what is being done is not directly to override the privilege, because the claimant can elect to stand upon his right to it. Rather, it is presenting the claimant with a price which must be paid for the leave of the court to rely on expert B; that price is waiver of privilege in relation to expert A. ... Accordingly the order sought will have the effect of curtailing the operation of privilege by making waiver the price of being able to continue in reliance on expert B. The suggested basis and justification for doing this is the need to prevent expert shopping, and where it has taken place, to put before the court of trial the whole of the available evidence on the question at issue, and not only parts."

19. *Edwards-Tubb* was a case under CPR 35.4 (power to restrict

M expert evidence). There is no authority dealing with the status of an undisclosed report prepared by an expert at the same time as a report which has been disclosed. I agree with Mr Platt QC that, because the two authorities to which I have referred emphasise the importance of openness under the CPR, and make plain that, in certain instances, that openness will trump questions of privilege, there may be cases in which an expert may have to disclose both reports as a condition of being permitted to give evidence at all. However, I also acknowledge that there may be cases where requiring an expert to disclose everything that he produces, regardless of privilege, could give rise to injustice. Taking the point to its logical extreme, it might mean that all expert's draft reports become disclosable, or all documents produced by the expert for the sole purpose of assisting counsel with topics for cross-examination.

In my view, therefore, there could be no general rule that everything is disclosable, regardless of privilege.

20. In all the circumstances of this case, I am not going to require Mr Dutch to provide copies of his other report, pursuant to CPR 35.4. But that is principally because I consider that both experts have become slightly confused as to what they should be doing, and that it would be wrong (and potentially unjust) to require either of them to provide privileged reports which may be based on a misunderstanding of the issues. In any event, I doubt whether the point matters in view of my other rulings. In my view, if the claimants want to advance their alternative case, they will have to do so by reference to clear expert evidence as to valuation and diminution in value. That will require a further report from Mr Dutch. That will then allow the defendants to see precisely what Mr Dutch says as to valuation matters.

Should the experts meet?

21. In my view, whatever the confusion that may have been created as to the scope of the experts' reports, Ms Loweth was wrong not to meet Mr Dutch without prejudice in accordance with the court order. Whilst I accept that the position in relation to the scope of the expert evidence was unsatisfactory, there was an element of confusion on both sides and I consider that such matters may well have been sorted out by the experts, had they met. It is hard to think of a situation in which one expert, having been ordered to meet his or her counterpart by the court, would be justified in refusing to do so. The failure to meet was compounded by the unnecessarily aggressive tone of the defendants' solicitors' correspondence.

Conclusions and the way forward

22. Accordingly, the following things need to happen in this case. First, the parties' agreement, that the £402,500 is the starting point for any assessment of loss, should be recorded in writing. That could be in the experts' joint statement prepared under CPR 35.12(3). Secondly, the claimants should review their primary and alternative cases on damages. If they wish to maintain both claims (and in my view there is no reason in principle why they should not) then they need to serve an expert's report dealing expressly with the alternate basis of claim, set out in para 6.2 of the particulars of claim. I would propose that such a report be provided in 28 days. Thereafter, I would expect the experts to meet in the usual way.

23. I said to the parties that I would deal with all subsequent matters, including costs, in whichever way was most economic for the parties. I ought to say that, in the light of my analysis and conclusions, my preliminary view is that the costs of these applications should be costs in the case, because of the confusion that I have found on both sides, and the need for it to be resolved. However, I make plain that that preliminary view is always open to change.

Application for directions granted.

RATING

Re Woolway (VO)'s appeal

UPPER TRIBUNAL: LANDS CHAMBER

11 June 2012

MR GEORGE BARTLETT QC, PRESIDENT

[2012] UKUT 165 (LC)

Estates Gazette 15 September 2012

[2012] 37 EG 128

Rating – Non-domestic rates – Hereditament – Office premises – Ratepayer occupying second and sixth floors in modern office building – Whether each floor comprising separate hereditament – Whether two floors to be listed together as single hereditament – Whether appropriate to make end allowance to reflect inconvenience of separation within building

The ratepayer, a large accountancy firm, occupied two floors in a modern, eight-storey office block. The block was built in a U-shape around three sides of a glass-covered atrium, with a central lift shaft containing six high-speed lifts. The ratepayer had separate leases of the second and sixth floors in the building, each for a 15-year term from 2007 at a rent of approximately £1,000 pa, with five-yearly rent reviews. The appellant valuation officer entered each of the ratepayer's two floors in the rating list as a separate hereditament. He took the view that different floors in the same occupation could only be listed as a single hereditament if those floors were adjacent.

The Valuation Tribunal for England accepted the ratepayer's proposal to merge the two entries with effect from November 2007. It determined that the two floors should be listed as a single hereditament, with an end allowance of 5% to reflect the disadvantage of their separation within the building. In reaching that conclusion, the tribunal found, applying *Gilbert (VO)* v *S Hickinbottom & Sons Ltd* [1956] 2 QB 40, that the two floors were within the same curtilage and that there was an essential functional link between them. The appellant appealed.

Held: **The appeal was allowed in part. In determining whether premises in one occupation constitute separate hereditaments or a single hereditament, regard may be had to the physical and functional relationship between the parts under consideration. However, *Gilbert* does not establish an incontrovertible formula for the identification of a hereditament. Its references to curtilage or to a single geographical unit are directed towards buildings and other land as they lie on the ground; the concept of curtilage has no useful part to play in identifying hereditaments within a modern office building. Likewise, the test of an essential functional link was devised for premises that are not within the same curtilage and are separated by a highway, and should not be applied in the case of separate floors in an office building. Although each floor in a modern office building is normally the subject of a separate lease, lettings of two or more floors in the same building to the same occupier are normally negotiated on an overall basis. A feature common to most modern office blocks is that the ratepayer occupies floors within a single building, with communication between floors only through the common parts. There will usually be no significant practical difference, in travel time in the lift, between floors that are contiguous and those that are separated. In such circumstances, contiguity between floors has no practical significance. The proper approach in such a case is to treat floors within the building occupied by the same occupier as a single hereditament, regardless of whether those floors are contiguous. Since the occupier is occupying the floors as offices for the purposes of its business, it is not necessary to investigate the functional interrelationship between those floors at any particular time. Accordingly, it was appropriate to enter the second and sixth floors occupied by the ratepayer in the instant case as a single hereditament. However, the valuation tribunal should not have made an end allowance of 5% for inconvenience suffered as a result of the floors being separated. An end allowance of that kind should generally be supported by rental evidence. There was no evidence that the rents payable under the leases were reduced to reflect the separation between the floors. Moreover, there was no case for an end allowance where the lift service was fast enough to negate the problem of not having adjoining floors.**

The following cases are referred to in this report.

Baker (VO) v *Citibank NA* [2007] RA 93
British Railways Board v *Hopkins (VO)* [1981] RA 328
Coppin (VO) v *East Midlands Airport Joint Committee* [1971] RA 449
Gilbert (VO) v *S Hickinbottom & Sons Ltd* [1956] 2 QB 40
Ladies Hosiery and Underwear Ltd v *West Middlesex Assessment Committee* [1932] 2 KB 679
Trunkfield (VO) v *Camden London Borough Council* [2011] RA 1

This was an appeal by the appellant valuation officer, Peter Woolway, from a decision of the Valuation Tribunal for England accepting a proposal by a ratepayer to merge two entries in the non-domestic rating list in respect of two separate floors within the same office block.

Daniel Kolinsky (instructed by the legal department of HMRC) appeared for the appellant; the ratepayer did not respond to the appeal.

Giving his decision, **Mr George Bartlett QC, President** said:

1. This appeal by the valuation officer (VO), to which the ratepayer does not respond, concerns two floors in a modern city office block which are not next to each other but are in the same occupation. The VO had entered them in the rating list as separate hereditaments, but the Valuation Tribunal for England (VTE) determined that they should be entered as a single hereditament and that an end allowance should be given to reflect the disadvantage of their separation within the building. The VO contends that the decision was in error in both respects. He says that the two floors are two separate hereditaments and that there is no evidence to justify an end allowance.

2. Tower Bridge House is a modern eight-storey office building that was completed in 2005. Although in Tower Hamlets it is on the opposite side of the road from the Tower of London and thus effectively serves a City of London office function. In plan form it is in the shape of an extended U. The open space between the three inner sides of the building consists of a glass covered atrium with a central lift shaft housing six high speed lifts that serve six floors of the building. On the first floor there is a common reception area that provides security for the entire building. On the ground floor there are restaurants and there are seven floors of offices above. The solicitors Reynolds Porter Chamberlain have offices on levels 1, 3, 4 and 5, the accountants Mazar LLP on levels 2 and 6, which are the subject of the present appeals, and

there are two occupiers of space on level 7. The floorplate of levels 2 to 6 is about 2,400 sq m net, although there are small differences between levels. Mazar hold the two floors under separate leases dated 24 June 2007. Each is for a 15 year term with effect from May 2007. The rent for level 2 was £1,008,422.50 per annum and the rent for level 6 was £1,053,010 per annum, and there are five yearly reviews.

3. In the 2005 rating list there were separate entries for the Reynolds Porter Chamberlain offices on level 1, for each of the levels occupied by Mazar and for each of the two occupancies on level 7. Levels 3, 4 and 5 were entered as a single hereditament. A rate of £250 per sq m was applied to all levels, but the hereditament consisting of levels 3, 4 and 5 was accorded an end allowance for size of 6.25%. Level 2 was entered at £600,000 with effect from 24 August 2006 and level 6 at £605,000 with effect from 26 November 2007. On 10 February 2010, Paul Rabbette Ltd made a proposal on behalf of the ratepayer to merge the two entries with effect from 26 November 2007. It was this proposal that was the subject of the appeal before the VTE.

4. The ratepayer's case before the VTE was that, although the levels occupied by them were separated, there was a functional dependency between them and they therefore ought to be entered as a single hereditament with effect from 26 November 2007 when they had begun to occupy the two floors. It was contended that there should be an end allowance for fragmentation of 10%.

5. In dealing with the question whether the two floors should be entered as separate hereditaments or as a single hereditament the Tribunal referred to *Gilbert (VO)* v *S Hickinbottom & Sons Ltd* [1956] 2 QB 40, placing reliance in particular on the first and third "general rules" in the judgment of Denning LJ: that where two or more properties within the same curtilage are in the same occupation they are as a general rule a single hereditament; and that where two properties are separated by a public highway, they may form a single hereditament if they are so essential in use the one to the other that they should be regarded as one single hereditament. In the light of this, the tribunal considered that it should address itself to the questions whether the two floors were within the same curtilage and whether there was an essential functional link between them. In relation to the first question it said:

> "33. In seeking to establish whether the premises are two or more properties within the same curtilage and in the same occupation, the Panel considers that the Tower Bridge House development is contained within a single curtilage. It is the panel's view that the 'curtilage' contains the two floor levels. It considers that the structure of the building provides a visible enclosure around the development which included the two levels in question.
>
> 34. While the panel accepts that level 2 and level 6 are linked by the high speed lift it remains satisfied that they are separate floors. However, in terms of the occupation of both levels which form the appeal entries, the Panel is satisfied that they are in the 'same occupation'…"

6. In relation to the second question that the tribunal considered that it should address, it said:

> "42. In determining the question of an essential functional link, the panel notes the VO's analysis of the case law and contentions that Mr Rabette had not demonstrated the existence of an essential functional connection between the operations carried on by the appellant on the two floors. In particular, the panel notes the VO's opinion that the link was merely one of convenience.
>
> 43. However, the panel considers that these views and contentions are based on case law in respect of primarily manufacturing premises, which in the case of *Gilbert* is now more than 50 years old. It is the panel's view that the criteria for the test of 'functionally essential' should not be restricted to manufacturing processes but should also be adapted to reflect the modern era of service industries which also have 'processes' in creating service products. As such, the panel considers that the VO is contending for a degree in the level of necessity which is too high for non-manufacturing premises in the modern era.
>
> 44. The panel considers that the appellant has demonstrated that there is a link between the floors which is essential to the efficiency of the working of Mazars' overall service-provision business. While the panel accepts that it was presented with no evidence that Mazars would cease to operate if it did not occupy the two floors at Tower Bridge House, it considers that the function carried out between the two levels, as described earlier, are sufficiently integrated that they are essential to the effective business as a whole.
>
> 45. It is the panel's view that, while the actual layout of occupation is, to use the VO's term, 'convenient' to Mazars, the evidence presented by the appellant indicated that the occupation of the accommodation provided by the two floor levels went beyond 'convenience' but was essential to the efficient operation and functioning of the business operated by the company.
>
> 46. Therefore, the panel accepts the appellant's contention that there is a sufficient degree of propinquity and an essential functional link between the two floors. Thus, the panel also finds that on the ground the appeal premises are a single hereditament."

7. The ratepayer's valuer, Mr Rabbette, had contended for a basic price per sq m of £225, rather than £250, and an end allowance of 10%. The tribunal concluded that there was no evidence to support the price of £225. As far as the end allowance was concerned, it said simply:

> "It should only attract a 5% end allowance for the relatively small inconvenience suffered."

It gave no reasons beyond this, nor did it record any evidence or contentions that sought to justify the allowance. It did, however, record Mr Rabbette as saying:

> "5. Tower Bridge House was, to some extent, a compromise. The accommodation was not available on adjoining floors, however, it did have sufficient space for the firm to locate all the work force in one place and the lift service was fast enough to negate the problem of not having adjoining floors."

8. At the hearing before me, evidence was given by the appellant, Mr Woolway, the VO. He has been a fellow of the Royal Institution of Chartered Surveyors since 1988, having become a professional associate in 1972. He said that he had 44 years' experience of rating valuation, having joined the Valuation Office as a cadet valuer in 1967. He had worked in central London since 1982 and was currently litigation specialist for the litigation and technical policy team as part of the National Specialists Unit. Mr Woolway gave factual evidence about the building and about Mazar and their use of the premises that they occupy. He described the approach taken by valuation officers to the identification of hereditaments, particularly in relation to modern office buildings; and he gave valuation evidence.

9. Mr Woolway said that Mazar took the accommodation at Tower Bridge House following the merger of the firm in April 2007 with the London Office of MRI Moores Rowland, becoming the eighth largest accountancy firm in the United Kingdom. Mazars had said that ideally they would probably have preferred to acquire affordable accommodation elsewhere as a single floor or to occupy floors in Tower Bridge House that adjoined each other, but that was not possible. Mr Woolway said that the accommodation presumably met Mazar's cost criteria and suited the needs of the company's expanded workforce. Although the accommodation was on two floors separated from each other by intervening floors the lift service was fast enough to negate the problem of separation. It was not materially different from the logistical challenges faced by Reynolds Porter Chamberlain with their occupation of levels 1, 3, 4 and 5.

10. The structure and layout of the accommodation was designed, Mr Woolway said, to reflect the fact that all Mazar's operations in central London were under one roof. Mazar held themselves out as providing what they described as integrated financial services to their clients, which meant being in a position to offer the complete range of services in one location. On level 2 there were facilities for meeting clients. There were rooms for client consultations as well as for staff training and meetings. There was a staff canteen. At this level were accommodated the internal accounts department, the insolvency team, the forensic investigation team, a claims department, the HR and IT departments, the print team and the post room. On level 6 there was accommodation for a large corporate team dealing with banks, insurance companies, internal audit and large corporate clients. In addition it housed departments dealing with taxation advice and planning, VAT, corporate finance, PFI schemes and other specialist advice and services.

11. Mr Woolway said that it was clearly convenient for Mazar to locate all their activities in a single building and there was no doubt a flow of personnel between the two levels. But that convenience was not the same as finding that it was necessary for the parts to be located

A close to each other. Such proximity was not vital. He considered that if the floors were further apart or even in different buildings they would still be able to carry on the accountancy business providing a bespoke service to clients.

12. The approach taken by valuation officers in determining whether premises should be shown as a single hereditament or as two or more hereditaments was to exercise judgement on the facts of the case and to reach results that were practical and reasonable. The great majority of merger proposals on properties of common types such as shops, offices and industrial properties involved a degree of physical proximity between the units, and it was necessary to take account not only of the B degree of physical proximity but also the functional connection.

13. Mr Woolway said that, although in Denning LJ's general rules in *Gilbert* reference was made to cases where two or more properties were within the same curtilage, the concept of curtilage, as derived from conveyancing law, did not seem to be particularly relevant when dealing with office buildings subject to sub interests in the form of leases. Individual office floors were always capable of being let singly or in multiples of floor occupations. While the subject building formed a single geographical unit, the structure and layout facilitated the occupation of individual floors, multiple floors and parts of a single floor. His experience was that even when multiple floors were occupied C together the landlord required individual leases, although negotiations for the leasing of a number of floors would be directed towards agreeing an overall price which would then be divided up for the purpose of the individual floor leases. He said that he assumed that Mazar had said that they wanted 4,500 sq m and that negotiations had taken place on that basis.

14. If he was wrong to say that the two floors should be assessed as separate hereditaments, Mr Woolway said, he did not think that there was any justification for applying an end adjustment for fragmentation. It appeared to be contrary to the market evidence of the actual leases. There were no perceived or operational difficulties for the actual or D hypothetical occupier. There was little practical difficulty in moving from floor to floor, whether or not the floors were adjoining. The rents of the two floors let to Mazar were very specific, but the negotiations would not have proceeded in isolation, and the total rent could be presumed to reflect all the benefits and disadvantages to the extent to which they might or might not be valuation significant. There was, however, no evidence that the rent paid was reduced to reflect the separation of the floors.

15. Mr Kolinsky submitted that applying the general rules in *Gilbert*, although floors 2 and 6 were in the same occupation, they were not contiguous, they could not be ringed, and they were not functionally E essential one to the other so as to overcome their separateness. By contrast floors 3, 4 and 5 were contiguous and could be ringed. It was not relevant, he said, to look at the negotiations that might take place in the real world, since what mattered were the physical facts. While he would not suggest that the fact that there were separate leases for each floor should be ignored, their only relevance was in showing that each floor was capable of being separately let. Moreover it was not to be assumed that the hypothetical landlord of each floor was the same, and in relation to this Mr Kolinsky referred to *Coppin (VO)* v *East Midlands Airport Joint Committee* [1971] RA 449. That case concerned the valuation of an airport, which consisted of the principal airport F hereditament and let-out parts (hangars, workshops, offices, car parks and so forth), which were agreed to be separate hereditaments. Only the principal airport hereditament was in issue. The Court of Appeal held that it was not to be assumed for the purpose of valuation that the hypothetical landlord of this hereditament was also the landlord of the let-out parts.

16. Under section 42 of the Local Government Finance Act 1988 a local non-domestic rating list must show each hereditament in the billing authority's area that is a relevant non-domestic hereditament. It must be entered at a value equal to the rent at which it might reasonably be expected to let from year to year (see para 2(1) of Schedule 6). Under section 64(1) a hereditament is anything which, by virtue of section 115(1) of the General Rate Act 1967 would have been a hereditament for the purposes of that Act; and "hereditament" was defined in section G 115(1) as "property which is or may become liable to a rate, being a unit of property which is, or would fall to be, shown as a separate item in the valuation list."

17. The hereditament is a unit of occupation (see *Ryde on Rating* para B[113] "Introduction to the concept of the 'hereditament'"). What constitutes a separate hereditament is a question of fact and degree. Rating is intended to achieve the payment of rates on a basis that fairly reflects the value of the ratepayer's occupation (or such value as the property would have if it was occupied) and the relative worth of all rateable property (see, for example, *Ladies Hosiery and Underwear Ltd* v *West Middlesex Assessment Committee* [1932] 2 KB 679, per H Scrutton LJ at 688). The identification of the hereditament, as the unit of assessment, is part of the process by which this objective is achieved, and the aim must be to identify what on the facts can fairly be said to constitute the physical unit that the ratepayer occupies. While I fully accept Mr Kolinsky's submission that the identification of the hereditament must precede the task of valuation and that valuation can play no part in making the identification, it is right to bear in mind that a failure to identify what on the facts can fairly be said to constitute the physical unit that the ratepayer occupies may result in a rateable value or rateable values that does or do not properly reflect the value of the occupier's occupation and such value relative to the property of other J ratepayers.

18. *Gilbert* is the leading case on the identification of the hereditament. It establishes that in determining as a matter of fact whether premises in one occupation constitute separate hereditaments or a single hereditament regard may be had to both the physical and the functional relationship between the parts under consideration. It should not, however, be treated as establishing an incontrovertible formula for the identification of the hereditament. It is important to have in mind the words of Morris LJ ([1956] 2 QB 40 at p52) where, having noted the contention advanced by the appellant VO, that if premises in the same occupation were structurally and geographically separate and capable K of being separately let they must be held to be separate hereditaments, he said:

> "I do not feel able to accept this approach. I do not think that it is appropriate to lay down propositions of the above nature as being inevitably or invariably applicable. If, as I think, the decision as to whether premises form one hereditament is a question of fact, then I think that it is undesirable to prescribe some formula in words or to seek to define certain considerations as being relevant and to stipulate that others must be excluded. Parliament has not laid down a definition of a hereditament and difficulties might result if a rigid judicial definition were formulated. In the great majority of cases there will be no difficulty, after assessing all the considerations which apply according to the L weight they command, in deciding whether premises comprise a hereditament. In the borderline cases where difficulty arises it is better to employ a common-sense assessment of the features of the case than to seek to have recourse to some standard formula."

19. *Gilbert* itself was concerned with industrial premises, a bakery and a depot for the repair and maintenance of delivery vans, that were situated on opposite sides of a street, and the judgments, although cast in general terms, reflect the factual elements of the sort of case then under consideration. Thus Denning LJ's references to the curtilage (at pp48-49) and Parker LJ's reference to "a single geographical unit" (at p54) are directed towards buildings and other land, whether separate M from each other or contiguous, as they are upon the ground. Those references do not usefully translate to the occupation of floors within a modern office block. Nor should Denning LJ's functional test (at pp49-50), "so essentially one whole – by which I mean, so essential in use the one to another – that they should be regarded as one single hereditament", be treated as the test in the case of separate floors in an office building. As expressed, that related to the situation that "may happen when a nobleman's park, or a farm (when agricultural land was rated), or a golf course is bisected by a public road."

20. I agree with Mr Woolway, and with the submissions made by Mr Kolinsky on the point, that in identifying hereditaments within a modern office building the concept of the curtilage has no useful part

to play, and is far removed from what Denning LJ had in mind when formulating his general rules in *Gilbert*. The VTE, having concluded that the two floors were within the same curtilage, then explored whether there was an "essential functional link" between them. In so doing it was clearly misapplying Denning LJ's general rules, where the question of the essential functional link only arose in the case of premises that were not within the same curtilage but were separated by a highway, and I do not think that an essential connection should be treated as a criterion in the present case. I agree in any event with the VO that a detailed inquiry into the functional relationship between parts of a building in the same occupation is of no assistance in the present case and is positively undesirable. It seems to me inappropriate to explore the degree of functional interaction between two floors in common occupation. Any such process would tend to be detailed and time-consuming (as it was in the present case) and always liable to reassessment as the occupier made changes in the way that the space was utilised. The fact that the floors of office premises are in the same occupation for the purposes of the occupying firm is by itself, in my judgment, a significant pointer.

21. In the context of floors within a modern office block, although I accept that contiguity or the lack of it is a material consideration, I attach (as I shall say) more importance to the practical realities of the physical relationship between the floors. The functional relationship is also, in my view material, and I bear in mind Parker LJ's consideration (5) (at p54): "Whether the occupier finds it necessary or convenient to use the premises as a whole for one purpose, or whether he uses different parts of the premises for different purposes." I can, however, see no justification for requiring that the functional connection between the parts should be such that each is essential to the other.

22. The practice of VOs, according to Mr Woolway, is to enter individual floors as separate hereditaments because that is the way in which they are normally let (entering contiguous floors in the same occupation as a single hereditament being an exception to this practice). It is clear that he had regard to the way in which the floors are normally let not just for the purpose of establishing that they were capable of being separately let, but as a material consideration in identifying the hereditament. I accept that it may be a material consideration for this purpose. But it is also right, in my judgment, to have regard in the same context to the fact, referred to by Mr Woolway, that where two or more floors in the same building are let to the same occupier this is normally done on the basis of negotiations carried out on an overall basis. These considerations, however, although material, ought in my view to carry less weight than the facts relating to the physical nature of the premises and the purpose of the ratepayer's occupation.

23. An approach that treats as a single hereditament adjoining floors which communicate only through the common parts while insisting that floors that are separated even by a single floor must be treated as separate hereditaments seems to me to be somewhat unreal. Take floors 3, 4 and 5 in the present building. Personnel move between floors by using the lifts. Floors 3 and 4 on the one hand and 4 and 5 on the other are only a floor's travel on the lift apart. It is two floors between 3 and 5, but the extra time in the lift must be insignificantly small. (As I have already noted, Mr Rabbette for Mazar told the VTE that the lift service was fast enough to negate the problem of not having adjoining floors, and he was speaking about floors 2 and 6.) Since communication between floors is only through the common parts of the building contiguity between floors has no practical significance. Supposing the occupation was not floors 3, 4 and 5 but floors 3, 5 and 7 or 3, 5 and 6. The difference in travel time in the lift would be insignificant, but the VO would, on the basis of what he says is common-sense, treat the first as a single hereditament, the second as three hereditaments and the third as two hereditaments. This would be of no account if in each case the total rateable value was three times the value of an individual floor. But in the case of floors 3, 4 and 5 an end allowance for size has been agreed, yet if the ratepayer occupied instead floors 3, 5 and 7 or floors 3, 5 and 6 he would not get the benefit of an end allowance even though the total amount of rent he would be paying for the three floors could be expected to be the same. The approach adopted by VOs thus risks producing values that do not fairly reflect the value of the ratepayer's occupation and the relative worth of the different occupancies within the building.

24. I should make clear that I do not regard matters of valuation themselves as relevant to the identification of the hereditament. What has to be borne in mind, however, is that unless the hereditament is identified in a way that reasonably and fairly represents the occupier's unit of occupation unfairness may result when the valuation stage is reached.

25. Mr Kolinsky relied on *Coppin* for his contention that it was not to be assumed that the hypothetical landlord of each floor was the same. The context of that case was quite different (the valuation of the let out parts in a contractor's basis valuation), and is of no assistance. The VO in the present case took the view that floors 3, 4 and 5 constituted a single hereditament despite the fact that they were the subject of separate leases and were evidently capable of having different landlords, and I do not see that any different assumption has to be made where the floors are not contiguous. Many hereditaments are capable both of being let in parts and of being owned in parts.

26. Although I was told that the practice of VOs is to enter only adjoining floors as a single hereditament, it is to be noted that in *British Railways Board* v *Hopkins (VO)* [1981] RA 328 an entry had been made comprising "lower ground, ground, part first (west wing) and fifth-tenth floors" of a recently completed office building, reflecting the extent of the ratepayer's occupation. The rest of the building was unoccupied. The dispute in that case concerned the treatment in the valuation list of the unoccupied parts of the building. The VO had entered these as six separate hereditaments. The ratepayers contended that the whole of the building was a single hereditament and should be entered as such. The member (JH Emlyn Jones FRICS) rejected this submission. He said, at p341:

> "The unoccupied part of the building being capable of separate occupation cannot in my judgment be said to be part of a hereditament of the occupied part since it cannot be said that it was occupied for the same purpose as the rest nor was it occupied at all."

The member said later, at pp344-345:

> "I have come to the conclusion that it was open to the valuation officer in the light of circumstances existing at the date of the proposals to make proposals in respect of the separate parts. By virtue of section 69(2) [of the 1967 Act] a valuation officer may at any time make a proposal for the alteration of the valuation list and clearly he must exercise his judgment in so doing. In the instant case I think it would also have been open to him to make a single proposal in respect of the whole of the unoccupied parts and subsequently to carve portions out of that hereditament to accord with the lettings which subsequently occurred. In the same way it would have been open to the valuation officer to make two proposals in respect of the occupied part, that is to say one in respect of the lower ground, ground, and part first floor and another in respect of the fifth to tenth floors inclusive. The fact that he chose one way rather than the other does not seem to me to lead inevitably to the conclusion that on one approach effect is to be given to a proposal but on the other no effect on the grounds that the proposal is invalid. It seems to me that there are many instances where two parts of a single hereditament may appear as two separate entries in the valuation list even though they could also and more appropriately be described as one.

27. The context in which the question of the identity of the hereditament arose was different from that of the present case, and it should not be thought, in the light of what the member there said, that the VO's role in determining how to enter property in the list or how to deal with a proposal relating to the identity of the hereditament is other than to ensure that the list is correct. For this purpose he must determine, in the case of occupied property, what reasonably and fairly represents the occupier's unit of occupation; and a tribunal, hearing an appeal, must do the same. Where property is unoccupied, on the other hand, and could realistically become the subject either of a single occupancy or a number of occupancies, it may be impossible to say that there is only one correct way of entering it in the list.

28. It is also to be noted that, in *Baker (VO)* v *Citibank NA* [2007] RA 93, the VO entered as a single hereditament parts of two

A intercommunicating office blocks at Canary Wharf, the whole of number 33 Canada Water and parts of the 42 floor building number 25 (basement, part ground floor, and floors 3-14, 17-18 and 20) on the basis that that was the physical extent of the area occupied. In *Trunkfield (VO)* v *Camden London Borough Council* [2011] RA 1, by contrast, I held that two office buildings (one wholly, and the other partly, occupied by the ratepayer) that touched at one point but did not intercommunicate internally were separate hereditaments.

29. I have referred above to the features of this case which are common to most modern office blocks – the fact that these are floors within a single building; the fact that communication between floors, B whether next to each other or not, is through the common parts of the building; the lack of any significant practical difference, from the occupier's point of view, between floors that are next to each other and those that are separated; and the fact that, although each floor is normally the subject of a separate lease, where two or more floors in the same building are let to the same occupier this is normally done through negotiations carried out on an overall basis, albeit that there is a separate lease for each floor. To treat as a single hereditament floors next to each other that are in the same occupation but as separate hereditaments floors in the same occupation that do not adjoin each other does not, in my view, properly reflect the realities of occupation in a modern office C block. The proper approach in a case such as this, therefore, in my judgment, is to treat the floors occupied within the building by the same occupier as a single hereditament. Since the occupier will be occupying the floors as offices for the purposes of his business, it is not in my view necessary to investigate the functional interrelationship between D the floors at any particular time. In the present case, therefore, floors 2 and 6 are properly to be entered as a single hereditament, as the VTE determined; and the VO's appeal on this point fails.

30. I would add that the adoption of the approach that I have concluded to be appropriate does not require that all existing entries in current lists of separate floors in the same occupation need to be changed. Only if a proposal for merger were to be made would it be necessary to make an alteration.

31. The second question that arises is whether the VTE was correct in making an end allowance of 5% "for the relatively small inconvenience suffered" as a result of the floors being separated. The justification for E an end allowance of this kind ought generally to be sought in the rental evidence, but it does not appear that there was any rental evidence before the tribunal to justify the reduction. Mr Woolway says that there is no evidence that the rents payable under the leases were reduced to reflect the separation between the floors, and I see no reason to reject what he says. In any event the decision records the ratepayer's surveyor as having said that the lift service was fast enough to negate the problem of not having adjoining floors, so that on this basis no case for an end allowance had been made.

32. The appeal is accordingly allowed in part. The rateable value of the hereditament (levels 2 and 6, Tower Bridge House, St Katherine's F Way, London E1W 1AA) must be entered at £1,205,000.

Appeal allowed in part.

RIGHT TO MANAGE

Gala Unity Ltd v Ariadne Road RTM Co Ltd

COURT OF APPEAL

23 October 2012

ARDEN, SULLIVAN AND PATTEN LJJ

[2012] EWCA Civ 1372

Estates Gazette 15 December 2012

[2012] 50 EG 105

Right to manage – Commonhold and Leasehold Reform Act 2002 – Self-contained building – RTM company seeking right to manage two blocks of flats within larger development – Section 72(1)(a) and 72(2) of 2002 Act – Whether each block a "self–contained building" – Whether common parts comprising "appurtenant property" to which right to manage extending

The respondent RTM company served two notices on the appellant landlord, under section 79 of the Commonhold and Leasehold Reform Act 2002, seeking to acquire the right to manage two blocks of flats within a development on behalf of the tenants. In addition to the two blocks, the development included two freestanding "coach houses", comprising first-floor flats with parking spaces beneath, plus further residents' parking and "estate common parts" that included footpaths, roads, visitor parking, grassed areas and a bin area. The lease of each flat included the use of a car port or parking space and the right to use the common parts, with service charges payable for those facilities.

By its counternotices, the appellant asserted that the two blocks were not self-contained buildings or parts of buildings, as required by section 72(1)(a). It submitted that although the blocks were structurally detached, they could not be described as "self-contained" since they could not function independently from the rest of the development, relying as they did on shared facilities such as the access road and visitor parking spaces; moreover, the parking spaces demised with several of the flats lay underneath the coach houses.

The leasehold valuation tribunal, upheld by the Lands Chamber of the Upper Tribunal on appeal, held that the respondent was entitled to acquire the right to manage pursuant to its notices. The Upper Tribunal held that each of the two blocks was a self-contained building, being "structurally detached" within the meaning of section 72(2), and that the right to manage applied to those buildings plus such appurtenant property as they might have, which included the carport or parking space demised with each flat in those blocks and the various incorporeal rights of way and other rights granted by the tenants' leases: see [2011] UKUT 425 (LC); [2012] 1 EGLR 99. The appellant appealed.

Held: **The appeal was dismissed. (1) The 2002 Act defines a self-contained building by reference to it being "structurally detached": see section 72(2). Each of the two blocks was structurally detached and was therefore a "self-contained building" for the purpose of section 72(1)(a). There is no further requirement that the structurally detached building be able to function independently, without the need to make use of any shared facilities such as private access roads, car parking, gardens or other communal areas. (2) In order to qualify as "appurtenant property" under section 72(1)(a), the property in question does not have to be exclusively appurtenant to the self-contained building that is the subject of the right to manage. Appurtenant property includes appurtenances belonging to or usually enjoyed with the building, part of a building or flat: see section 112(1). Appurtenances such as gardens or yards are often so enjoyed in common with other buildings, parts of buildings or flats. The carports or parking spaces included in the leases of the flats in the two blocks "belonged to" those flats, while the bin area, access road and gardens were "enjoyed with" those flats. The fact that the occupiers of other property, namely the coach houses, also enjoyed those appurtenances did not mean that they fell outside the section 112(1) definition. Although that result raised the prospect of dual responsibility for the management of some of the appurtenant property, with the potential for duplication of management effort and conflict between the "old" management company and the new RTM company, those consequences were not so grave, and the result was not so manifestly absurd, as to justify adding a gloss to the words "appurtenant property", which were already defined in the Act. It is always open to parties, if they wish to avoid duplication and conflict, to reach an agreement that makes economic sense for all involved.**

The following case is referred to in this report.

Cawsand Fort Management Ltd v *Stafford* LRX/145/2005

This was an appeal by the appellant, Gala Unity Ltd, from a decision of the Upper Tribunal (Lands Chamber) upholding a declaration of the leasehold valuation tribunal that the respondent, Ariadne Road RTM Co Ltd, was entitled to acquire the right to manage certain property under Chapter 1 of Part 2 of the Commonhold and Leasehold Reform Act 2002.

Brian McGurk, director of the appellant, appeared for the appellant; Ellie Cameron-Daum, company secretary of the respondent, represented the respondent.

Giving judgment, **Sullivan LJ** said:

Introduction

1. This is an appeal against a decision dated 25 October 2011 of the President of the Upper Tribunal (Lands Chamber) ("UT"), George Bartlett QC, dismissing the appellant's appeal against a determination of a Leasehold Valuation Tribunal ("LVT") under section 84(3) of the Commonhold and Leasehold Reform Act 2002 ("the Act") that the respondent was entitled to manage two premises in Swindon: a block containing two flats, numbers 10 and 12 Ariadne Road, and another block containing 10 flats, numbers 14-32 (even) Ariadne Road, together, in each case, with appurtenant property. The background to the appeal is set out in some detail in the President's decision: [2011] UKUT 425 (LC).

Statutory framework

2. The right to manage ("RTM") provisions are contained in Chapter 1 of Part 2 of the Act. Sections 71 and 72 provide:

"71. The right to manage

(1) This chapter makes provision for the acquisition and exercise of rights in relation to the management of premises to which this chapter applies by a company which, in accordance with this chapter, may acquire and exercise those rights (referred to in this chapter as a RTM company).

(2) The rights are to be acquired and exercised subject to and in accordance with this chapter and are referred to in this chapter as the right to manage.

72. Premises to which chapter applies

(1) This chapter applies to premises if –

(a) they consist of a self-contained building or part of a building, with or without appurtenant property,

(b) they contain two or more flats held by qualifying tenants, and

(c) the total number of flats held by such tenants is not less than two-thirds of the total number of flats contained in the premises.

(2) A building is a self-contained building if it is structurally detached.

(3) A part of a building is a self-contained part of the building if –

(a) it constitutes a vertical division of the building.

(b) the structure of the building is such that it could be redeveloped independently of the rest of the building, and

(c) subsection (4) applies in relation to it.

(4) This subsection applies in relation to a part of a building if the relevant services provided for occupiers of it –

(a) are provided independently of the relevant services provided for occupiers of the rest of the building, or

(b) could be so provided without involving the carrying out of works likely to result in a significant interruption in the provisions of any relevant services for occupiers of the rest of the building.

(5) Relevant services are services provided by means of pipes, cables or other fixed installations."

"Appurtenant property" is defined in section 112(1) as follows:

"appurtenant property", in relation to a building or part of a building or a flat, means any garage, outhouse, garden, yard or appurtenances belonging to, or usually enjoyed with, the building or part or flat…"

Factual background

3. There is no dispute that the requirements in paras (b) and (c) of subsection 72(1) are met in this case. The issue before the LVT and the UT was whether the premises fell within para (a). The appellant is the freehold owner of land which forms part of an extensive modern development on the outskirts of Swindon. The President carried out a site inspection, and his description of the land is contained in [6] of his decision:

"On the land there stand the two blocks of flats to which the claim notices relate and also two free-standing "coach houses", which are first-floor flats with parking spaces underneath. The land is bounded on the north, east and west by estate roads that curve round it and on the south by other residential buildings. There is a short, brick-surfaced road that runs across the land from east to west, providing access on the north side to the 10 flat block and on the south side, where it opens out into a courtyard, to the two-flat block and the coach houses. On its western side the roadway also serves a house that is not within Gala's ownership. There are defined parking spaces on the roadway and the courtyard area and at the front of the 10-flat block. There is a free-standing dustbin store adjacent to the roadway and this serves all the flats on the land. Immediately to the north of the 10-flat block is an area of garden bounded by a wall but with open access. Between the estate roads that curve round the development and the 10-flat block and the garden area is a grassed area of varying width on which trees have been planted. There is also a small grassed area between the 2-flat block and the estate road. On the south side of the coach houses there is a courtyard accessible only on foot."

4. The two coach houses were not included in the RTM claim. We were told by Mr McGurk that there were eight car parking spaces under the two coach houses. One of the spaces under each coach house is allocated to the occupiers of the coach house, and the remaining six spaces are demised to the lessees of six of the flats. The President summarised the terms of the leases of the 14 flats on the land in [7] of his decision:

"The leases of the 14 flats on the land are in similar terms. I was provided with copies of the leases of 28 Ariadne Road (a second-floor flat in the 10-flat block) and 14 Ariadne Road (one of the two flats in the other block). Forming part of the demise of each lease is a numbered car port. (In the leases of other flats, I understand a car parking space rather than a car port forms part of the demise.) Each lease is for a term of 125 years from 1 July 2006 and is expressed to be made between the lessor, "the management company" (Hazelvine Ltd) and the lessee. The lessee pays rent to the lessor (under clause 2.1) and a service charge to the management company (under clause 2.2). The lessor covenants (clause 4.1 and para 6-2.1 of Schedule 6) to provide the services set out in Schedule 7. Schedule 7 specifies six categories of services and sets out under them the particular services that are to be provided. One of these (Category D) relates to lifts and has no application because there are no lifts in any of the buildings. The other categories are as follows:

Category A, relating to the estate common parts (excluding the car park);

Category B, relating to the building main structure;

Category C, relating to the building common parts;

Category E, relating the car park; and

Category F, relating to insurance."

5. The lease defines "the estate common parts" as follows:

"'The estate common parts' means the areas and amenities in the managed estate available for use in common by the lessee and the owners and all persons expressly or by implication authorised by them, including pavements, footpaths, forecourts, visitor car parking spaces, cycle store, roads, drives, landscaped areas, gardens and areas designated for the keeping and collecting of refuse, but not limited to them.

'The managed estate' means the land shown edged blue on the plan and includes the buildings and all other structures (including boundary walls and fences) erected on the managed estate."

Under Schedule 2 the lessee and all persons authorised with him are given rights of way, in common with the lessor and all other persons having a like right, over and along the roads, drives, forecourts and pavements on the development, the right to use appropriate areas of the estate common parts, the right to use car parking spaces available for common use and the right to use the dustbin area.

The President's decision

6. The President decided that each of the two blocks of flats was "undoubtedly self-contained since it is structurally detached (see section 72(2)); and accordingly on the relevant date the RTM company was entitled to acquire the right to manage them." [13]. He then considered the question: to what other parts, if any, of the estate currently managed by Hazelvine Ltd did RTM's right to manage extend? Having considered the terms of the claim notice in [14], the President answered this question in [15] and [16]:

"15. Thus the right to manage in the present case extends to the two blocks of flats and to appurtenant property. Property is appurtenant for this purpose, in my view, if it is appurtenant to a flat within the block. The appurtenant property attaching to each flat under the lease of it is of two sorts. Firstly there is the car port or car parking space that is included in the demise, and there can be no doubt, in my judgment, that each flat's car port or parking space is appurtenant property for the purposes of the statutory provisions. The second sort of appurtenant property consists of the incorporeal rights of way and other rights granted under Schedule 2 of each flat's lease. These are rights that are not exclusive to the particular flat but are shared with all or some of the other flats, including flats within the managed estate that are not within either of the two blocks in respect of which the claim notices were served. There is, I think, no reason why the right to manage should not extend to the maintenance of land over which tenants have incorporeal rights (cf, in relation to the appointment of a manager under Part II of the Landlord and Tenant Act 1987, *Cawsand Fort Management Ltd* v *Stafford* LRX/145/2005 at [17].

16. There is clearly an argument, however, that it is implicit that 'appurtenant property' should be construed as relating to land that appertains exclusively to the premises, excluding, therefore, land over which the tenants of the premises have rights in common with others whose flats (the two coach houses) are not the subject of the right to manage claim. The reason for such a construction would be that, if such land were included in the right to manage this could conflict with the rights in relation to that land of both the tenants of the coach houses and the landlord and any management company. Moreover, section 97(2), which removes the landlord's entitlement to do what the RTM company is required or empowered to do, might be thought to lend some support for this. I had myself reached the conclusion that this was indeed the correct approach, so that in the present case 'the premises' would only extend to the buildings themselves and the car ports and parking spaces that were included in each demise, and I inquired of the RTM company whether, in the light of this, it wished to pursue its claims. Its response was that it did wish to do so. One further consideration I do not think that 'appurtenant property' is to be so narrowly construed. There is nothing in the wording itself that would suggest this, and, although the scope for conflict of the sort that I have mentioned exists, this is insufficient reason for imposing a restriction on the meaning of the provision."

7. The President dealt with the practical implications of his conclusion that "appurtenant property" was not to be narrowly construed, so as to extend only to property that appertained exclusively to the flats (the numbered car ports or parking spaces), excluding land over which the lessees of the flats enjoyed rights in common with the occupiers of the coach houses (the bin area, roadway, gardens etc), in [17] of his decision:

> "the landlord is still required, and therefore entitled, under the leases of the coach houses to provide the services in categories A, B, C, D and F, including, therefore, maintenance of those parts of the managed estate over which those tenants have rights; and the tenants of those flats are still liable to pay to the landlord the service charge as provided under their leases. However, it would seem to me that if the landlord and management company continued to provide services in relation to those parts of the estate that the RTM company is obliged to the tenants of the 12 flats in the two blocks to maintain, the cost of such services would not be reasonably incurred and could be disallowed under section 19(1) of the Landlord and Tenant Act 1985."

8. When granting permission to appeal, the President said that the issue was of general importance for leaseholders who wish to exercise the right to manage and their landlords.

Submissions

9. The appellant was represented by one of its two directors, Mr McGurk. The respondent was represented by Ms Cameron-Daum, the company secretary.

10. Mr McGurk's principal concern was the desirability of the company's estate being managed as a whole. If the management responsibilities were shared between the respondent and Hazelvine Ltd there was the potential for at best wasteful duplication of management effort and at worst conflict as to, eg whether/when maintenance was required. Understandably, Mr McGurk's submissions focussed on his perception of the practical difficulties of shared management responsibility rather than the statutory language. In a nutshell, he contended that Chapter 1 of Part 2 of the Act should be interpreted so as to enable a RTM company to manage the whole, but not a part of an estate.

11. Mr McGurk accepted that the two blocks of flats were structurally detached, but he submitted that they were not "self-contained" because they were unable to function independently, without the use of the shared appurtenances: the access way, the gardens, bin store, etc. He accepted that all of the areas and amenities comprising "the estate common parts" (see [5] above) were "appurtenant property", as described in section 112, but he submitted that para (a) in subsection 72(1) applied only to appurtenant property that was "self-contained", in the sense that it appertained exclusively to the "self-contained building" in question. If, contrary to his submission, the blocks of flats were self-contained buildings, then their "appurtenant property" was limited to the car ports/parking spaces that were included in the leases of the flats, and it did not include the shared appurtenances.

12. Ms Cameron-Daum submitted that the President's decision was correct, for the reasons given in his decision.

Discussion

13. In my judgment, there is only one issue in this case: the issue identified in [16] of the President's decision ([6] above). Mr McGurk's wish that his company's estate should be managed as a whole is understandable, but there can be no doubt that the two blocks of flats are self-contained buildings for the purpose of section 72(1)(a). There is no challenge to the President's factual conclusion, reached after he had carried out a site visit, that the two blocks are structurally detached. The Act defines a self-contained building by reference to it being "structurally detached", and there is no justification for imposing Mr McGurk's further requirement that the structurally detached building must be able to function independently, without the need to make use of any shared facilities such as private access roads, car parking, gardens or other communal areas. Mr McGurk referred to the concept of an ability to function independently which is contained in subsections 72(3)-(5), but those subsections, which are concerned with whether a part of a building is a self-contained part, are of no assistance in the present context.

14. There is no dispute that all of the estate common parts are appurtenant property within the definition in section 112(1) ([2] above). Mr McGurk accepts that if, as I have concluded, the two blocks of flats are two self-contained buildings, then the designated car ports/parking spaces forming part of the demises for each of the flats are appurtenant property. That leaves the rest of the appurtenant property. As the President said in [16] of his decision there is nothing in the wording of the Act which suggests that appurtenant property is limited to property that is exclusively appurtenant to the self-contained building. Where Parliament wished words in the Act to have a particular meaning, it expressly said so: see eg the meaning of "self-contained building" in section 72(2), and the meaning of "appurtenant property" and "flat" in section 112(1). In these circumstances, the court should be slow to infer that Parliament must have intended to impose a further qualification.

15. Appurtenant property, as defined by section 112(1), includes appurtenances *belonging to or usually enjoyed with* the building, part of a building or flat. Appurtenances such as gardens and yards are frequently enjoyed by a building, or a part of a building or a flat, in common with other buildings, parts of buildings or flats. In ordinary language, the car parking ports/spaces included in the leases of the flats "belong to" the flats which comprise the self-contained block, whereas the bin area, access road and gardens are "enjoyed with" the flats which comprise the two blocks. The fact that the occupiers of other property, in this case the two coach houses, also enjoy those appurtenances does not mean that they fall outside the definition in section 112(1). The fact that the definition is not limited to appurtenances which belong to the building in question is a powerful indication that Parliament did not intend that appurtenant property for the purpose of section 72(1)(a) should be limited to property that is exclusively appurtenant to the self-contained building in question. In effect, Mr McGurk's approach is an attempt to substitute in section 72(1)(a) the words "self-contained premises" for premises which consist of a self-contained building together with appurtenant property.

16. In my judgment, the wording of section 72(1)(a) is clear: there is no requirement that the appurtenant property should appertain exclusively to the self-contained building which is the subject of the claim to acquire the right to manage. The prospect of dual responsibility for the management of some of the appurtenant property in this and other similar cases is not a happy one. As Mr McGurk submitted, there is the potential for duplication of management effort and for conflict between the "old" management company and the new RTM company in respect of such appurtenant property, but I am not persuaded that these consequences are so grave, or that the end result is so manifestly absurd, that we would be justified in adding a gloss to words – appurtenant property – which are already defined in the Act. It is always open to the parties, if they wish to avoid duplication and/or conflict, to reach an agreement which would make economic sense for all parties (see [18] of the President's decision); if they are unable to do so, [17] of the President's decision suggests a means of resolving disputes arising from dual responsibility for maintenance.

Conclusion

17. I would dismiss this appeal.

Patten LJ said:

18. I agree.

Arden LJ said:

19. I also agree.

Appeal dismissed.

TORT
DEFECTIVE PREMISES

Harrison and others v Shepherd Homes Ltd

COURT OF APPEAL

5 July 2012

HUGHES, AIKENS AND MCFARLANE LJJ

[2012] EWCA Civ 904

Estates Gazette 13 October 2012

[2012] 41 EG 116

Defective premises – Damages – Quantum – Developer of new residential estate found liable to purchasers of houses for negligence of contractor in design and construction of pilings – Consequent cracking to houses affecting value and mortgageability – Correct approach to assessment of damages

The respondents were the owners of 71 detached houses on an estate of executive homes built by the appellant developer. Since the estate was on a former landfill site, it had been necessary to use extensive piling in the construction of the houses. Owing to the negligence of the contractor engaged to design and install the pilings, some of the piles settled, resulting in significant cracking to many of the houses. In most cases, the cracking was not structural in nature but it was sufficiently serious to render the houses either unmortgageable or unsaleable, save at very low prices.

In a trial of 10 sample cases, the appellant was held to be liable to the respondents in contract, under the Defective Premises Act 1972 and pursuant to section 2 of the "Buildmark Cover": see [2011] EWHC 1811 (TCC). The judge held that it was unreasonable to assess damages by reference to the cost of repiling, and instead awarded the diminution in value of the houses as a result of the defective piling. For that purpose, he divided the properties into three categories, namely: (i) those that were probably mortgageable, albeit with difficulty; (ii) those that were probably unmortgageable; and (iii) an intermediate category that might or might not be mortgageable, to varying degrees. He valued the first category by the "capital comparison" method to derive a percentage of 32.5% for diminution in value. To the second category, he applied the "investment return" method, based on the return to a "buy to let" investor, while for the third category he applied a mixture of the two methods. He additionally decided to allow for the cost of necessary remedial works, which, after further consideration and agreement by the parties' experts, were agreed at one set of works immediately plus two further sets over the next 20 years. The judge refused to admit further evidence as to an actual sale of one of the properties that had occurred between the end of the hearing and the date of the judgment.

On appeal, the appellant challenged the judge's approach to the assessment of damages, including his use of the third, intermediate category of houses.

Held: **The appeal was dismissed. (1) The judge had been correct not to take a dogmatic approach to whether or not a particular house was "probably mortgageable", notwithstanding that the chances of movement were remote in eight of the 10 test cases. He had been entitled to calculate damages for diminution in value on a "gradated" basis, to take into account uncertainty arising from the need for even potentially mortgageable houses to be subjected to a structural engineer's report, the terms of which would affect the availability of a mortgage. (2) Having accepted that the capital comparison method for assessing diminution in value should be applied to the "probably mortgageable" houses, the judge had been required to make his own decision on the appropriate percentage. He had considered all the evidence about the 10 houses and had borne in mind the view of the parties' experts. The figure that he had applied fell inside the band within which he could reasonably assess the percentage. (3) Although the basic measure of damages was the diminution in value of the houses, the judge had been entitled to make an additional award for further works to remedy defects. That award involved no element of double recovery. Since the judge had not been in a position, at the date of his judgment, to decide whether such further works might be needed, the cost of those works had been excluded from his award for diminution in value. (4) The judge had been entitled not to admit further evidence as to the actual sale of one of the houses. That evidence would have added little to what he already knew, would have necessitated further expert evidence and possible cross-examination of witnesses and might have given rise to arguments as to whether the circumstances of the sale could be regarded as typical. There had to be a limit to the litigation process.**

This was an appeal by the appellant, Shepherd Homes Ltd, from a decision Ramsey J, sitting in the Technology and Construction Court in Leeds, assessing the damages to be awarded to the respondents, Andrew and Lisa Harrison and the owners of other properties on a housing estate built by the appellant, for defects in the properties caused by negligence in the construction process.

Roger Stewart QC (instructed by Weightmans LLP) appeared for the appellants; Andrew Bartlett QC and Robert Stokell (instructed by Tilly Bailey & Irvine LLP, of Hartlepool) represented the respondents.

Giving judgment, **Aikens LJ** said:

Synopsis

1. This appeal concerns the proper measure of damages to be awarded to property owners when their homes have suffered damage as a result of negligent design and construction of piling, which supported their homes. Eden Park, Hartlepool, is an estate of 94 new houses which were built by the appellants ("SHL") between 2002-2004. The estate plan provided for six different types of house. All are detached and were designed so that they could be sold as high quality "executive homes". The estate was built on a former landfill site, so that it was necessary to use extensive piling in the construction of the houses. In fact the company engaged to design and construct the piling, Encia Remediation Ltd, ("Encia"), failed to use reasonable care in the design and installation of the piling. The result was that some piles associated with some houses settled. As a result at least 57 of the 94 houses on the estate suffered significant cracking. This was in most cases not structural but it was sufficiently serious for it, broadly, to render the houses either unmortgageable or unsaleable save at very low prices. Between 2005 and 2007 there was litigation between SHL and Encia in which the latter eventually accepted that it had not exercised care and skill in the design and installation of the piling used on the estate.

2. Then, in 2008, the current proceedings were brought by the owners of 71 houses. I will call the current respondent house owners

collectively "the houseowners". Subsequently, it was decided that there should be a trial of 10 sample "lead cases", which were selected by the parties in the hope that this would provide a sufficient range of types of house, defects and consequent damage so that the result of the test cases would enable the other claims to be settled amicably. That is, in fact, what has happened in all but one case. The defendants were SHL, the builders, and also the National House Building Council ("NHBC") and NHBC Building Controlled Services Ltd. The latter two took no part in the trial or on this appeal and I do not need to spend time on explaining why they were brought into the proceedings.

3. The allegations of the houseowners were that: (i) all the houses suffered from significant cracking which was progressive and rendered the houses unmortgageable and/or unsaleable; (ii) all the houses needed underpinning and the claimants were entitled to claim that cost even though, (as was admitted), it would exceed the value of the houses in their undamaged state; and (iii) they were entitled to claim further damages based on a "diminution in value" of the houses and further costs or damage, once the properties had been underpinned.

4. There was a five-week trial before Ramsey J in July 2010. In the course of the trial the judge heard expert evidence on valuation (Mr Watson for the houseowners and Mr Lewis for SHL) and on structural engineering (Mr Taylor for the houseowners and Mr Johnson for SHL). The judge also went to view the houses concerned. He reserved his judgment and handed it down on 11 July 2011. The present appeal is against his order of 11 July 2011.

5. In his judgment, the judge considered first the evidence and principles in general; then he applied his findings and conclusions to each of the 10 houses concerned. On the general principles he concluded that SHL was liable to the houseowners (a) in contract; (b) under the Defective Premises Act 1972 and (c) pursuant to section 2 of the "Buildmark Cover" which provides that the builder, SHL, will have liability for "defects" and "damages" in the first two years after the house has been built. Those conclusions are not appealed.

6. As to the damages recoverable by the houseowners, the judge concluded: (1) it was not reasonable for the houseowners to receive damages based on the full cost of repiling the houses and the necessary costs associated with doing that work. Instead the correct measure of damages was an award for the diminution in value of the houses suffered as a result of the defective piling. (2) Two different methods for valuation of the houses were used; that used by Mr Lewis was called the "capital comparison" and that used by Mr Watson "investment return". The first is that most usually used in valuing residential houses but Mr Watson used the second because he thought there were insufficient comparables to use the usual method. That method produced a value of the property concerned based on the return to a "buy to let" investor. The judge concluded that both methods were equally valid and applicable and could appropriately be used: [282]. (3) Two out of the 10 properties were probably mortgageable albeit with difficulty; two were probably unmortgageable; the remaining six might or might not be mortgageable, to varying degrees. (4) As regards the first of those categories, Mr Lewis's "capital comparison" method was the more appropriate approach. For the second category, Mr Watson's "investment method" was the more appropriate: [292]. (Mr Lewis accepted that the investment method was appropriate if the property was unmortgageable). As regards the properties in between, they were to be valued using a mixture of both methods. (5) For the mortgageable properties the judge rejected Mr Lewis' figures on diminution in value. SHL contends on appeal that the judge also rejected those of Mr Watson and used his own. That is contested by the houseowners. (6) For the unmortgageable properties the judge accepted Mr Watson's revised inputs: [306]-[316]. (7) The houseowners could recover the cost of both current minor remedial work but also the cost of carrying out two lots of remedial work over the course of the next 20 years. The figures were set out in the judge's order of 8 September 2011.

7. In the period between the end of the hearing in July 2010 and the time when judgment was handed down, one of the houses on the estate, 47 Meadowgate Drive, (which was not one of the 10 houses the subject of the trial) was put on the market by a person who had bought the house at auction just before the end of the hearing before the judge. Both parties requested and were permitted to put in further evidence from the valuation experts on what they said were the conclusions that could be drawn from the figure at which the house was being marketed for sale: see [285] to [289] of the judgment. Subsequently, in February 2011, that house was sold, for £170,000 for which a loan of £150,000, secured by a mortgage, had been obtained. SHL wished to put before the judge a further statement from Mr Lewis relating to this actual sale. The judge refused permission to do so. This refusal is challenged in one of the grounds of appeal.

8. The grounds of appeal for which Tomlinson LJ gave permission concern four principal areas: (1) The judge's decision to create an intermediate category of houses that were neither "probably mortgageable" nor "probably not mortgageable" but were in a "twilight zone" between those two categories. (2) The figure (or percentage) used by the judge for the diminution in value award for those properties found to be "probably mortgageable" including those in the "twilight zone". (3) The judge's decision to grant an additional head of damages equal to the cost of remedial work to be done on the houses, immediately in the first place and then twice more in the course of the next 20 years, on top of the damages based on diminution in value. (4) The judge's refusal to admit the further evidence from Mr Lewis concerning the actual sale of 47 Meadowgate Drive in February 2011. I will set out in more detail the judge's findings and the arguments of the parties as I consider each head of appeal.

The creation of an intermediate category of houses between those that were "probably unmortgageable" and those that were "probably mortgageable"

9. The judge found, at [276], that the worst of the lead properties (plot 15) had a "low probability" of significant foundation movements in the future. For the second worst, plot 20, he concluded that there was only a "remote probability" of such movement. But for all the other eight plots he considered that the probability of movements in the foundations was "extremely remote". Such movement would require only cosmetic repair, in particular the repointing of brickwork and internal redecoration.

10. The judge also decided that the valuation of each property in its damaged state depended directly on whether a purchaser could obtain a loan secured by a mortgage on the property. That in turn would depend, not on what the judge found to be the case in his judgment, but on reports which gave engineering and valuation advice to a prospective purchaser: see [319] of the judgment. The judge concluded that he had to evaluate the likely advice that might be given in each case, given the circumstances of the particular property concerned, and thus determine the property's mortgageability. He did this in general terms at [320] and then performed this exercise and reached conclusions on each of the 10 houses.

11. Mr Stewart QC, who appeared on the appeal for SHL, submitted that the judge's "compromise" position in relation to six of the 10 houses was unsupported by either the valuation or the engineering evidence. He submitted that because the six properties in the "twilight zone" were all ones where the chance of movement was very remote, the judge should have concluded that all were in the same "probably mortgageable" class as the "top" two properties. He submitted that the judge should then have approached the valuation exercise accordingly, which would have produced a lower figure for the diminution in value. In the result, Mr Stewart submitted, the judge awarded damages that were too high for the houses in this middle category, particularly bearing in mind that the judge had also started from too high a percentage for the diminution in value for "mortgageable" houses in the first place.

12. Mr Bartlett QC, who appeared for the houseowners, submitted that the judge was correct to approach the question of what figure for diminution in value should be awarded by asking the question: how likely is it that the house will be mortgageable? He submitted that the judge was also correct to conclude, first, that there could be no certainty that a house would be mortgageable, even if the chance of movement was remote and, secondly, that the critical issue would be the engineer/

surveyor's report on the particular house concerned. Mr Bartlett QC pointed out that although the judge had concluded that the chance of movement in these houses was very remote, he had also concluded that the piles were not designed and constructed with the required factors of safety and that the standard of construction was unreliable; and, further, that there was always the risk that something unforeseen might happen which could give rise to movement of a house that had been built on piles: see [197] and [199].

13. In my view, the judge was correct not to be dogmatic in deciding whether a particular house was "probably mortgageable" or probably not so. The fact that, in eight out of the 10 test cases, the chances of movement were remote or very remote, is not the only relevant factor. The judge accepted, at [301], that even potentially mortgageable houses would all need to be subject to a structural engineer's report. The terms of that report would be crucial to the question of whether a loan secured by a mortgage could be obtained. Nuances in a particular engineer's report would depend on the particular house and the particular expert giving the report. The particular terms of a report may well determine the likelihood of a particular house being "mortgageable". The judge was correct to take account of this uncertainty by approaching the question of damages for diminution in value on a "gradated basis" as he put it in [320]. I would therefore reject this ground of appeal.

The percentage figure for the diminution in value award for those houses found to be "probably mortgageable"

14. SHL argued that the judge rejected both Mr Lewis' and Mr Watson's figures for the percentage of diminution in value for those properties (including those in the "twilight zone") that were "probably mortgageable" and that he should not have done so. The judge considered this issue at [293]-[305]. The judge accepted the principle of Mr Lewis' "capital comparison" method of arriving at a diminution in value percentage, but he rejected Mr Lewis' figure because Mr Lewis could only use one comparable (31 Clover Drive) and, even if he could have used the subsequent information about 47 Meadowgate Drive, that would have only given two comparables and it was normal to use three. The judge also found that, given the particular circumstances of the houses on the estate that had suffered from the defective piles, there would be a "limited market" for them. He therefore concluded that this lack of comparables and the limited market meant that he must assess a "discount" (ie between the undamaged market value and the actual value of the 10 houses concerned) that was higher than Mr Lewis' figure of 25% and that of Mr Watson of 27.5% "...which could be derived from the one comparable in such a limited market": [302]. The judge then said that he would choose a figure of 32.5% as one which "...represented the appropriate reduction from the market value of each property as it would have been without the problems". He then set out the figures for all 10 properties.

15. The houseowners submit that the point taken by SHL on this issue is a false one, based on a misreading of the judgment. I agree. I am satisfied that the discount figure of 27.5% referred to in [302] as being "Mr Watson's 27.5%" is a reference to the percentage for diminution in value of the one property, 31 Clover Drive, that would be obtained by taking Mr Watson's figure for the value of that property in good condition (less the conservatory), compared with the actual sale price obtained for it. As Mr Watson's "good condition" figure for 31 Clover Drive was higher than Mr Lewis', this meant that Mr Watson's diminution in value percentage would be higher, viz 27.5%.

16. In fact, Mr Watson disagreed with Mr Lewis' figures for obtaining a diminution in value figure for the "probably mortgageable" properties which, in Mr Lewis' case, was based solely on one (or even two) comparables. The judge effectively accepted that Mr Watson's broad view on the figures was correct. But, as the judge had also accepted, at [292], that the capital comparison method for assessing diminution in value was correct for houses that were "probably mortgageable", he had to make his own decision on what that percentage should be, using that comparison method. There is no basis on which this court could interfere with the judge's decision that the proper percentage figure should be 32.5% as set out at [302]. There must be a band of possible figures. The judge considered all the evidence about the 10 houses, collectively and individually. He had the experts' views well in mind. Mr Stewart QC has not convinced me that the figure chosen by the judge was outside the reasonable band within which the judge could have assessed the percentage for diminution in value. I would therefore reject this ground of appeal also.

The additional head of damages based on the cost of two sets of remedial work over the next 20 years, on top of the diminution in value head of damages

17. The judge considered this head of claim in principle at [329]-[332]. The two expert engineers (Mr Taylor and Mr Johnson) had disagreed in their joint statement on whether the defective piling would be the cause of cracks which would require future remedial work that was out of the ordinary for new properties. The valuers could therefore not assess what effect this might have on the diminution in value of the houses affected. Thus when the judge came to assess his diminution in value figures for the houses he was not able to decide finally whether he should take into account the fact that future remedial work would be needed. The judge decided that these two issues (defective piling causing cracks which would need remedial work and the cost) should be determined by the two quantity surveyor experts, who, he was confident, would be able to agree matters, but if not, he would deal with them subsequently: [331]. The judge decided, for the purposes of his judgment, that he should proceed on the basis that further remedial work would be necessary and that he should allow for "...two further similar sums to be expended over the next 20 years, with a discount for early receipt of the damages": [332].

18. Subsequently, the experts agreed the issues of whether those remedial works were necessitated by the defective piling and the costs of such remedial works. This agreement was embodied in an order of 8 September 2011. The figure for remedial work for each of the 10 houses was then multiplied by 2.3 times. This multiplicand, as I understand it, was intended to reflect the cost of one immediate set of remedial works, then two more within the next 20 years, but discounted for early receipt: see also [364] of the judgment.

19. Mr Stewart submitted that the award of these damages for two further sets of remedial work means that the houseowners have benefited from a measure of "double recovery". That is because, he argued, the judge must have taken into account the fact that further remedial work might be needed in arriving at his percentage figure for the diminution in value head of damage. He accepted that the houseowners were entitled to receive the cost of one set of remedial works which had to be done immediately.

20. Mr Bartlett submitted that the judge was fully aware of the need not to give double recovery and that he guarded against it. Properly understood, the judgment awarded a figure for diminution of value which excluded the figures for future remedial works. Therefore, it was legitimate for the judge to award a suitably discounted figure for such future works.

21. The judge rejected the houseowners' primary case that they were entitled to recover the cost of re-piling. The basic measure of damages was therefore bound to be that of the diminution in value of the houses affected. The judge also concluded that, ideally, the houseowners wished to sell their houses and move elsewhere: [277]. He also appears to have concluded, doubtless because of the combination of his conclusion that they could not recover the cost of re-piling and his conclusion on the difficulties in obtaining a loan secured by a mortgage on one of the houses affected, that it was likely that the owners would stay there for 20 years: see eg [332]. The judge had to arrive at a figure for diminution in value based on what he knew at the time of writing his judgment. He was clearly aware that he must not award a "double recovery". But he also decided that he was not in a position (at the time of writing his judgment) to decide whether the defective piling would give rise to the need for further remedial works or what the cost of such work might be. Given all these factors, it seems to me that the judge must have appreciated that he needed in his judgment to make an award for diminution of value, but excluding the cost of any further

necessary remedial work which would be the subject of further work by the experts. Hence the judge's conclusion in principle on the former head of damage, at [320], and then, individually, for each of the houses in the second part of his judgment. His conclusion in principle on the latter head is given in principle at [332] and then, after the experts had done their further work, in the order of 8 September 2011 as regards each of the 10 houses.

22. In the particular circumstances of this case, I consider the approach of the judge was proper and cannot be criticised. He did not award an element of "double recovery". I would therefore reject this ground of appeal also. In his written submissions, Mr Bartlett also suggested that Mr Stewart had conceded in argument before the judge that the judge was entitled to award damages for diminution in value and also something for making good the defects and Mr Bartlett relied on an excerpt from the transcript. I do not accept that the transcript discloses any particular concession. However, the judge was correct to approach the matter as he did for the reasons I have attempted to give.

The refusal to admit the further expert evidence relating to the sale of 47 Meadowgate Drive

23. After the sale of 47 Meadowgate Drive, Mr Lewis prepared a short further statement which set out the fact of this sale, but nothing more. It was this statement only that the judge refused to admit. The judge dealt with the application of SHL to adduce further evidence from Mr Lewis following the actual sale of 47 Meadowgate Drive without a further hearing, as the parties agreed. The judge correctly identified the issue on which that evidence might assist as being whether the lead properties were mortgageable and "…on that basis" what value those properties had at the date of trial: [290]. The judge explained in [291] that he had decided that he should not admit that evidence because it would necessitate further investigations and further evidence being submitted.

24. Mr Stewart QC submitted that this decision of the judge was wrong and led to serious prejudice to SHL. 47 Meadowgate Drive is a "Langdale" type property, which is similar to four of the 10 "lead properties" the subject of the trial (nos 2, 5, 7 and 8). The valuers agreed that such a property had a value, undamaged, of £230,000. No 47 had been sold at auction on 8 July 2010, ie at the end of the trial, for a figure of £128,000. The house was then put on the market at a price of £174,950, as the judge recorded at [286]. In fact, as noted above, the house was sold on 25 February 2011 for £170,000 and the purchaser obtained a loan of £150,000 secured by a mortgage on that house.

25. Mr Stewart QC submitted that Mr Lewis' proposed evidence affected two vital issues at the trial: (i) how easy it was to get a mortgage on the properties concerned; and (ii) the appropriate diminution in value of a mortgageable property. He argued that it was perverse of the judge to refuse to admit this evidence, particularly as he had admitted the earlier evidence about the price at which this house had been put on the market by those who had purchased it at auction.

26. Mr Bartlett QC submitted that the decision of the judge, which was a case-management decision, was reasonable and cannot be successfully impugned. He pointed out that the judge knew of the fact that the house had been put on the market for £174,950 and that there was a possibility of it being sold and with the aid of a loan secured by a mortgage on the property. Therefore the fact of the actual sale added nothing new on the two points identified by Mr Stewart QC.

27. In my view, the judge was entitled to refuse to admit the further evidence concerning the sale of 47 Meadowgate Drive. First, by itself Mr Lewis' evidence would have added little to what the judge already knew. Secondly, if SHL had wished to use that factual evidence as the basis of making further submissions on the two points identified by Mr Stewart and set out above, that would have necessitated further expert evidence from Mr Lewis and that would have required a response from Mr Watson, on which the parties may (or may not) have wanted to cross-examine the witnesses. The circumstances of the sale of this house were unusual and it is likely there would have been arguments about whether the circumstances of this sale could be regarded as typical. Thirdly , there has to be a limit to the litigation process and the judge had already been generous in admitting the evidence on events after the trial concerning the marketing of no 47. Lastly, the judge effectively took the matter into account anyway, as I think is clear at [300] and [301]. So I would also reject this ground of appeal.

Conclusions and disposal

28. For the reasons I have given, I would reject each of the grounds of appeal that have been pursued by SHL. I would dismiss this appeal. The houseowners did not attempt to pursue their cross-appeal so nothing more need be said about it.

McFarlane LJ said:

29. I agree

Hughes LJ said:

30. I also agree.

Appeal dismissed.

TOWN AND COUNTRY PLANNING

R (on the application of Millgate Developments Ltd) v Wokingham Borough Council

COURT OF APPEAL

[2011] EWCA Civ 1962

6 July 2011

PILL, RIMER AND MUNBY LJJ

Estates Gazette 29 September 2012

[2012] 39 EG 120

Town and country planning – Planning obligation – Unilateral undertaking – Town and Country Planning Act 1990 – Local planning authority refusing planning permission for residential development but indicating that unilateral undertaking to make financial contributions to infrastructure would overcome objections – Developer providing such undertaking under section 106(1) of 1990 Act – Planning permission granted on appeal – Planning inspector expressing view that undertaking unnecessary – Whether undertaking enforceable – Whether unreasonable to enforce

In 2006, the respondent council refused the appellant's application for planning permission for a development of 14 dwellings on land in Woodley, near Reading. The terms of the refusal indicated that the proposal did not adequately provide for services, amenities and infrastructure, but that this objection could be overcome by a unilateral undertaking, under section 106(1) of the Town and Country Planning Act 1990, to make financial contributions towards leisure, education and library facilities. The appellant therefore gave such an undertaking, to take effect only once planning permission was granted and the development was commenced; the undertaking measured the contributions by reference to tariffs contained in supplementary planning advice and local plan policies. Planning permission was subsequently granted by a planning inspector on appeal. In his decision letter, the inspector indicated that he had given little weight to the undertaking, since the respondents had not shown the necessity of the contributions.

The appellant proceeded with the development but did not make the required contributions; instead, it asked the respondents to discharge the undertaking. The respondents refused to do so and requested payment of the required sums. The appellant's claim for judicial review of that decision was dismissed in the court below. The deputy judge held that the enforceability of the undertaking was conditional on only two events, namely the grant of planning permission and the commencement of development, both of which had occurred, and that it was reasonable for the respondents to enforce the undertaking since it continued to serve a useful planning purpose: see [2011] EWHC 6 (Admin). The appellant appealed. The respondents conceded that the full amount of £200,000 was no longer needed and indicated that they would seek to recover only £140,000.

Held: **The appeal was dismissed. (1) The giving of the undertaking under section 106 of the 1990 Act had been a predictable, lawful and reasonable response to the respondents' legitimate decision to refuse planning permission on grounds relating to the absence of provision for infrastructure needs. Such an undertaking is enforceable as a contractual obligation. The appellant's undertaking had become binding according to its own terms and its enforceability could no longer be challenged on the ground that, when made, it had lacked a sufficient nexus with the proposed development. The inspector's willingness to grant planning permission without giving weight to the undertaking did not mean that the undertaking had ceased to have any legitimate planning purpose. The respondents were therefore entitled to enforce the undertaking. Their decision to do so was not a planning decision but merely a decision to enforce a contractual obligation. (2) Even if the decision to enforce were a determination under the planning Acts, in the making of which the respondents were required by statute to have regard to the development plan, the undertaking had been in accordance with the development plan when made and did not cease to be so by reason of any concession that the full amount undertaken to be paid was no longer appropriate. The respondents had not acted unlawfully in refusing to discharge the undertaking while stating that they would seek to enforce it only up to a certain amount; a reduction in their claim did not involve a finding that the remaining sum was there other than for planning purposes and did not give rise to an obligation to quash the undertaking. (3) Moreover, it was questionable whether a remedy by way of judicial review was appropriate given that the undertaking was lawful when given and would be enforced by a private law contract action. Since the points that the appellant sought to advance by way of judicial review could be employed as a defence to a private law claim, an alternative remedy was available.**

Per curiam**: The better view was that the respondents had a power to repay sums contributed under a section 106 agreement, pursuant to their power to conduct contractual negotiations and incidental to their planning powers under section 111 of the 1972 Act.**

The following cases are referred to in this report.

R (on the application of Batchelor Enterprises Ltd) v *North Dorset District Council* [2003] EWHC 3006 (Admin)

R (on the application of Renaissance Habitat Ltd) v *West Berkshire District Council* [2011] EWHC 242 (Admin); [2011] 22 EG 104

R (on the application of The Garden and Leisure Group Ltd) v *North Somerset District Council* [2003] EWHC 1605 (Admin)

Tesco Stores Ltd v *Secretary of State for the Environment and others* [1995] 1 WLR 759; [1995] 2 EGLR 147; [1995] 27 EG 154

This was an appeal by the appellant, Millgate Developments Ltd, from a decision of HH Judge Pearl QC, sitting as a deputy High Court judge in the Administrative Court, dismissing a claim for judicial review of a decision of the respondents, Wokingham Borough Council, not to discharge a unilateral undertaking given under section 106 of the Town and Country Planning Act 1990.

John Pugh-Smith (instructed by Pitmans LLP, of Reading) appeared for the appellant; Guy Williams (instructed by the legal department of Wokingham Borough Council) represented the respondents.

Giving judgment, **Pill LJ** said:

1. This is an appeal against a decision of HHJ David Pearl QC sitting as a deputy High Court judge on 14 January 2011 whereby he dismissed an application by Millgate Developments Ltd ("the appellants") to

quash a decision of Wokingham Borough Council ("the respondents"). The respondents had by letter dated 23 April 2008 refused to discharge an undertaking given by the appellants on 22 March 2007 under section 106(1) of the Town and Country Planning Act 1990 ("the 1990 Act"). On 10 October 2006, the appellants had applied to the respondents for planning permission to build 14 dwellings on land at Colemans Moor Lane, Woodley, Berkshire. The respondents' officers reported that the development was not appropriate to the character of the surrounding area. They also set out costs under the headings "highways", "leisure", "primary education", "secondary education" and "libraries" that the proposed development would attract, in line with the local plan, by way of contributions.

2. On 22 December 2006, the respondents refused permission. One of the reasons for refusal provided:

> "...the proposal fails to make satisfactory provision of adequate services, amenities and infrastructure needs and consequently would have an unacceptable adverse impact upon the amenities of the area. As such the proposal is contrary to policies DP4 and T4 of the Berkshire Structure Plan and Policies WOS4 WR7 WT1 WT3 and WET7 of the Wokingham District Local plan"

3. Under the heading "information" in the same document the respondents stated that the refusal "could be overcome through the submission of an acceptable unilateral undertaking". The appellants appealed against the refusal of permission and the appeal was considered on the basis of written submissions by an inspector appointed by the Secretary of State for Communities and Local Government ("the Secretary of State").

4. On 22 March 2007, the appellants submitted what they described as, and was, a section 106 unilateral undertaking in support of the appeal. Section 106 of the 1990 Act provides, in so far as is material:

> "1. Any person interested in land in the area of the local planning authority may, by agreement or otherwise, enter into an obligation (referred to in this section and in sections 106A and 106B as a 'planning obligation') enforceable to the extent mentioned in the subsection (3)
>
> ....
>
> (d) requiring a sum or sums to be paid to the authority...on a specified date or dates or periodically.
>
> 2. A planning obligation may (a) be unconditional or subject to conditions;
>
> ...
>
> (c) if it requires the sum or sums to be paid, require the payment of a specified amount or an amount determined in accordance with the instrument by which the obligations entered into and, if it requires the payment of periodical sums , require them to be paid indefinitely or for a specified period.
>
> 3. Subject to subsection (4) a planning obligation is enforceable by the authority identified in accordance with sub section 9(d) –
>
> (a) against the person entering into the obligation; and
>
> (b) against any person deriving title from that person.
>
> ...
>
> 9. A planning obligation may not be entered into except by instrument executed as a deed which:
>
> (a) states that the obligation is a planning obligation for the purposes of this section;
>
> (b) identifies the land to which the person entering into the obligation is interested;
>
> (c) identifies the person entering into the obligation and states what his interest in the land is; and
>
> (d) identifies the local planning authority by whom the obligation is enforceable...
>
> ...
>
> 11. A planning obligation shall be a local land charge for the purposes of the Local Land Charges Act 1975 the authority by whom the obligation is enforceable shall be treated as the originating authority as respects such a charge."

5. Section 106A provides, in so far as is material:

> "(1) A planning obligation may not be modified or discharged except—
>
> (a) by agreement between the appropriate authority (see subsection (11)) and the person or persons against whom the obligation is enforceable; or
>
> (b) in accordance with this section and section 106B.
>
> (2) An agreement falling within subsection (1)(a) shall not be entered into except by an instrument executed as a deed.
>
> (3) A person against whom a planning obligation is enforceable may, at any time after the expiry of the relevant period, apply to the appropriate authority for the obligation—
>
> (a) to have effect subject to such modifications as may be specified in the application; or
>
> (b) to be discharged.
>
> (4) In subsection (3) "the relevant period" means—
>
> (a) such period as may be prescribed; or
>
> (b) if no period is prescribed, the period of five years beginning with the date on which the obligation is entered into.
>
> Where an application is made to an authority under subsection (3), the authority may determine—
>
> (a) that the planning obligation shall continue to have effect without modification;
>
> (b) if the obligation no longer serves a useful purpose, that it shall be discharged; or
>
> (c) if the obligation continues to serve a useful purpose, but would serve that purpose equally well if it had effect subject to the modifications specified in the application, that it shall have effect subject to those modifications."

6. The provisions of section 106A, as now enacted, were inserted into the 1990 Act by the Planning and Compensation Act 1991. The present appeal turns on the section 106A(1) procedure but I have cited section 106A(3) because it may be relevant to the Act's construction that a different procedure is provided once the relevant period defined under the section has expired. Section 106B provides for an appeal to the Secretary of State against a decision of the local planning authority under section 106A.

7. Clause 2.3 of the undertaking provided:

> "The obligations contained in and created by this undertaking shall not take effect unless and until the planning permission shall have been granted and commencement of development shall have taken place."

8. The leisure and library contributions stated in the undertaking were to be paid "provided that the council shall apply them towards leisure and library facilities within the borough." Schools contributions were to be paid provided the respondents shall apply them towards the provision of school facilities "...where reasonably required by the council in the light of the likely or actual impact upon such facilities in the borough arising from the development." The highways contribution was to be made provided it would be applied towards road safety measures and sustainable transport schemes reasonably required by reason of the impact of the development.

9. The contributions were by their terms to be paid "on or before the commencement of the development". They were not so paid and the development has been completed. Presumably no attempt to enforce was made at that stage because the provisos, to which I have referred in the undertakings, had not been met. Whatever the reason, the sum mentioned was not and has not subsequently been paid. The payments to be made under each of the specified headings were itemised. The total was in the region of £170,000. No evidence of how the sums were calculated, or as to what were they would be applied, was submitted for the consideration of the inspector who heard the appeal.

10. A second undertaking was provided in April 2007 excluding the highways contribution but is not material for present purposes.

11. In an appeal decision dated 9 May 2007, the inspector stated, when summarising his decision:

> "The appeal is allowed and planning permission granted subject to conditions set out in the formal decision. None of the conditions relate to the section 106 undertaking."

12. At para 13 the inspector stated:

> "The council's request for contributions towards highways, leisure education and libraries are addressed by the Appellant through the submission of unilateral undertakings. However the council produced nothing to show that those conditions are necessary in order to satisfy the test in Structure Plan Policy DP4, Local Plan Policy WOS4 or Circular 5/05, Planning Obligations. I therefore conclude that contributions to the provision of infrastructure are unnecessary and afford the unilateral undertakings little weight. "

13. Structure Plan Policy DP4, mentioned by the inspector at para 3 of his determination, provides:

> "Contributions may be sought from developers to provide additional infrastructure or services where they are needed to enable the development to take place and are directly and reasonably related in scale to the proposed development."

14. That policy reflects annex B to circular 05/2005, to which reference has been made in the course of the hearing :

> "B5. The Secretary of State's policy requires, amongst other factors, that planning obligations are only sought where they meet *all* of the following tests. The rest of the guidance in this circular should be read in the context of these tests, which must be met by all local planning authorities in seeking planning obligations.
>
> A planning obligation must be:
>
> (i) relevant to planning;
>
> (ii) necessary to make the proposed development acceptable in planning terms;
>
> (iii) directly related to the proposed development;
>
> (iv) fairly and reasonably related in scale and kind to the proposed development; and
>
> (v) reasonable in all other respects.
>
> **B6.** The use of planning obligations must be governed by the fundamental principle that **planning permission may not be bought or sold**. It is therefore not legitimate for unacceptable development to be permitted because of benefits or inducements offered by a developer which are not necessary to make the development acceptable in planning terms (see B5(ii)).
>
> **B7.** Similarly, planning obligations should never be used purely as a means of securing for the local community a share in the profits of development, ie as a means of securing a "betterment levy".
>
> THE SECRETARY OF STATE'S POLICY TESTS
>
> **B8.** As summarised above, it will in general be reasonable to seek, or take account of, a planning obligation if what is sought or offered is **necessary** from a **planning** point of view, ie in order to bring a development in line with the objectives of sustainable development as articulated through the relevant local, regional or national planning policies."

15. Having considered para 13 of the determination, the appellant sought a correction of the order under section 56(2) of the Planning and Compulsory Purchase Act ("the 2004 Act"). They sought to add to para 13 the sentence :

> "This would therefore make it clear that no contributions are required."

16. The inspectorate declined to issue a correction notice, stating that para 13 simply acknowledged the existence of unilateral undertakings. The writer stated that he could not share the appellants' view "that it detracts from the Inspector's conclusion on their necessity."

17. The respondents declined to remove the undertakings from the register of local land charges and by e-mail of 14 January 2008 required the appellants within one month to meet the terms of the agreement dated 22 March 2007. Once the undertaking had become binding, its enforcement is determined by the provisions of the undertaking itself, it is submitted, not by the degree of nexus with the proposed development. That claim has not been followed by private law proceedings.

18. The present proceedings were commenced by the appellants on 29 May 2008. On 17 June 2008, the respondents served a witness statement including a detailed justification of each of the financial contributions claimed. The sum claimed by way of contribution was then and is at the latest calculation the sum of £140,000. The reasonableness of the figure was challenged by the appellants. They appear to accept that some payment within that amount was required.

19. The principles as to the legality of section 106 undertakings were stated in *Tesco Stores Ltd* v *Secretary of State for the Environment and others* [1995] 1 WLR 759*. There was no question in this case of auctioning a planning permission: Lord Hoffmann at p782. Lord Hoffmann stated at p779D:

> "Of course it is normal for a planning obligation to be undertaken or offered in connection with an application for planning permission and to be expressed as conditional upon the grant of that permission. But once the condition has been satisfied, the planning obligation becomes binding and cannot be challenged by the developer or his successor in title on the ground that it lacked a sufficient nexus with the proposed development."

20. At p783C, Lord Hoffmann stated:

> "the choice between a policy which emphasises the presumption in favour of development and fairness between developers… and a policy of attempting to obtain the maximum legitimate public benefit… lies within the area of discretion which Parliament has entrusted to planning authorities."

It was not a choice, he stated, which should be imposed on them by the courts.

21. For the appellants Mr Pugh-Smith seeks a declaration that the respondents' decision requiring the appellants to comply with their undertaking was unlawful. He accepts the undertaking was lawful when given and complied with the test for such undertakings expressed in the authorities and Circular 05/2005. What was unlawful, he submits, was a decision to enforce it in a sum now approaching £200,000, with the index linking provided in the undertaking, when the respondents knew that the entire sum was not required for the planning purposes stated. As I have said, the sum now claimed is £140,000, a concession having been made that the total contemplated earlier was no longer required for planning purposes.

22. I would analyse the situation as follows:

(a) A reason for refusal of permission by the respondents was the absence of provision in the application for infrastructure needs. That was a legitimate reason for refusal. The appellants were invited to give a section 106 undertaking.

(b) It was a predictable, lawful, and reasonable response by the appellants to offer such an undertaking and to do so by way of a financial contribution to highways, leisure and libraries facilities, and educational facilities.

(c) The appellants gave an unconditional section 106 undertaking.

(d) If planning permission was granted on the evidence before him the inspector stated that there was nothing to show that the contributions were necessary on planning grounds. The undertaking was given little weight by the Inspector in granting permission. It appears very likely that the Inspector would have granted planning permission even in the absence of a section 106 undertaking.

(e) On its own terms the undertaking became enforceable. Enforceability as a contractual obligation is a part of the rationale for the section 106 procedure.

(f) The inspector's approach does not cast doubt upon the lawfulness of the undertaking. Planning permission without giving weight to the undertaking does not mean that the undertaking was not given for a legitimate planning purpose. The inspector was not overruling the local planning authority's view as expressed in the reasons for refusal. I will return to that issue.

(g) The appellants did not, as they could have done when they were refused permission by the respondents, question the need for the facilities or the likely cost of the facilities provided.

(h) I understand the dismay of the appellants when it emerged that they could have got planning permission without the undertaking they had given in a substantial sum of money.

(i) In the circumstances the respondents were entitled to attempt to enforce the undertaking. That did not require analysis by them at the enforcement stage. Subject to the relevance of the development plan they were not making a planning decision but a decision to enforce a contractual undertaking. The relevance of that plan is the main point of issue in the case and I will return to it.

(j) The enforceability of the undertaking cannot now be challenged on the basis that, when made, it lacked a sufficient nexus with the proposed development (Lord Hoffmann in *Tesco*).

23. Mr Pugh-Smith relies, in support of his argument based on the development plan, on section 38(6) of the Planning and Compulsory Purchase Act 2004:

* Editor's note: Also reported at [1995] 2 EGLR 147; [1995] 27 EG 154

"If regard is to be had to the development plan for the purpose of any determination to be made under the planning Acts the determination must be made in accordance with the plan unless material considerations indicate otherwise."

24. Counsel submits that the refusal of the respondents to discharge the undertaking was a determination under the Planning Act within the meaning of section 38(6). It was a decision taken under section 106A(1)(a) of the 1990 Act. Accordingly the decision must be made, by virtue of section 38(6), in accordance with the development plan unless material considerations indicate otherwise. Under Structure Plan Policy DP4, it is submitted, the contribution stated in the undertaking were no longer needed to enable the development to take place and, that being so, it was unlawful to refuse the application to discharge under section 106A(1), because the inspector had stated in para 13 of his determination that the undertaking was unnecessary.

25. Mr Pugh-Smith referred to the decision of Sullivan J in *R (on the application of Batchelor Enterprises Ltd)* v *North Dorset District Council* [2003] EWHC 3006 (Admin). He relies on it to establish that the decision not to discharge was a decision under section 106A(1)(a) of the 1990 Act. In that case the decision of a council not to modify a section 106 agreement was quashed. There had been no need for the developer to take the section 38(6) point in that case because the decision was quashed on other grounds. Mr Pugh-Smith submits that, now for the first time, it comes up for consideration. He relies, however, on the quashing of the decision in *Batchelor Enterprises*.

26. In *Batchelor Enterprises* the developer had entered into a section 106 agreement providing that part of the appeal site be maintained as a grassed open area. On appeal to the Secretary of State, a planning permission involving encroachment onto the area specified in the section 106 agreement was granted. Sullivan J held that in those circumstances the local planning authority could not lawfully conclude that the agreement should be maintained. The inspector, and thus the Secretary of State, had squarely addressed the question whether the grassed open area still served any useful planning purpose. The Secretary of State had concluded that it was not necessary on amenity grounds to keep the whole of the area grassed and open. The council's view of the amenity had been rejected by the Secretary of State, acting as planning authority. The council failed to have regard to the implications of the Secretary of State's decision granting a planning permission which was based on a finding that it was not necessary to keep the area grassed and open. The undertaking had ceased to be relevant to the planning permission granted.

27. That is very different from the present situation, in which there is no determination of the Secretary of State which deprives the undertaking of a planning purpose. The inspector, for the Secretary of State, was prepared to grant the planning permission without giving weight to the undertaking and found he could do so without giving any such weight. It does not, in my view, follow that the undertaking did not have and did not continue to have a legitimate planning purpose. Its validity, when made, is not challenged on the grounds stated in *Tesco* and in the circular. It does not cease to be valid because of the respondents' concession that they would not seek to enforce the entire sum due. The Secretary of State in this case was not overruling the local planning authority's view expressed in reasons for refusal to which the appellants responded

28. For the respondents, Mr Williams relies on the test stated by Sullivan J in *Batchelor Enterprises* at [29]:

"It is accepted that the question to be considered by the local planning authority in each case is the same: does the obligation still serve a useful planning purpose? Since the court in judicial review proceedings may not substitute its own answer to that question for that of the local planning authority, the question in relation to an application for judicial review in respect of a local authority's decision under section 106A(1)(a) is whether a reasonable local planning authority could have concluded that the obligation still served a useful planning purpose. "

29. Mr Williams submits that the question can only be answered in the affirmative on the facts of this case. As to the argument based on section 38(6) of the 2004 Act, Mr Williams submits that the decision not to discharge under section 101A(1) (a) did not attract the provisions of section 38(6) even though the decision was made under the 1990 Act. It was a decision to enforce the undertaking and not a decision to grant planning permission. Sections in the 1990 Act which require regard to be had to the development plan do so explicitly. Mr Williams referred to sections 70(2), 77(4), 79(4), 91(2), 92(6), 97(2), 102(1), 172(1) and 172(2). Section 38(6) does not bite upon the decision and there is no need to revisit development plan policies, he submits. Even if the section does bite, it is submitted, an undertaking in accordance with the development plan when made does not cease to be in accordance with it by reason of a concession that the full amount undertaken to be paid is no longer appropriate. I agree with that submission.

30. Another factor to be considered in my judgment is whether a remedy by way of judicial review is appropriate in a case such as this. The undertaking was lawful when given. It would be enforced by a private law action contract. If and when a sum is claimed by the respondents in an action based on the undertaking, it will be open to the appellants to question whether the sum claimed under each head comes within the terms of the undertaking. Reference to those terms has been made and I need not expand on them. They reveal an area of argument that would be available to the appellants. Concessions have been made by the respondents. The points which in substance the appellants claim to make by way of judicial review can be employed as a defence to a private law claim. In this context a private law shield rather than a private law action is available but an alternative remedy is available.

31. I do not consider that the inspector's use of the word "unnecessary" in para 13 of his determination assists the appellants. The inspector was stating that he would grant planning permission without weight being given to the undertaking. He had not been assisted by any detailed statement by the respondents in relation to contributions. However, the undertaking remained a valid undertaking in *Tesco* and Circular 5/2005 terms as a valid contribution for planning purposes. The Inspector did not state that the contribution he mentioned fell outside those purposes. Had he taken that view, I would have expected him to say so expressly.

32. In my judgment, there is no unlawfulness in the respondents stating that they will seek to enforce the undertaking only up to a certain amount. A reduction in their claim does not involve a finding that the remaining sum subject to the undertaking is there other than for appropriate planning purposes. I would refuse the application for a declaration. The decision to discharge was not unlawful. Reduction in the sum to be claimed does not involve an obligation to quash the undertaking.

33. I would reach the same conclusion had the application been to modify rather than discharge. The undertaking remains valid even if the precise sum of money offered cannot now be claimed and a partial defence may be available in a private action to enforce the undertaking. A relevant planning purpose is served and is not defeated by concessions or detailed arguments about quantum. There is no principle that a concession that a part of the sum is no longer required for specific planning purposes requires the discharge of the undertaking.

34. It is not necessary to consider other issues which have been raised during the hearing, including the possible impact of section 106A(3) to (6) on the construction of section 106(A)(1), nor is it necessary to consider broader issues as to the position on occurrence of a change in planning purpose or in the nexus between the new purpose and the original purpose. Those issues were considered by Richards J in *R (on the application of The Garden and Leisure Group Ltd)* v *North Somerset District Council* [2003] EWHC 1605 (Admin), at [28]-[29] and [46], and Ouseley J in *R (on the application of Renaissance Habitat Ltd)* v *West Berkshire District Council* [2011] EWHC 242 (Admin)**. I state only that I find nothing in those judgments contrary to the conclusion I have reached.

** Editor's note: Also reported at [2011] 22 EG 104

35. As a second ground for appeal, it is submitted that the respondents have no power to repay to the appellants sums paid under a section 106 agreement. It is a hypothetical point because no sums have been paid. I see no merit in this point. Councils have powers to conduct contractual negotiations. The undertaking is enforceable by way of private law principles. In any event, without full argument, it appears to me that a repayment would be incidental in circumstances such as these to the planning powers of the council (section 111 of the Local Government Act 1972).

36. For those reasons, and in agreement with the judge, I would dismiss this appeal.

Rimer LJ said:

37. I agree.

Munby LJ said:

38. I also agree.

Appeal dismissed.

Moore v Secretary of State for Communities and Local Government and another

COURT OF APPEAL

[2012] EWCA Civ 1202

18 September 2012

LORD NEUBERGER MR AND LONGMORE AND SULLIVAN LJJ

Estates Gazette 17 November 2012

[2012] 46 EG 122

Town and country planning – Change of use – Use as dwelling – Enforcement notice alleging change of use of property from dwelling to commercial leisure accommodation – Whether permitted use as dwelling including use as holiday accommodation – Whether notice expressed too widely

The appellant's property had formerly been used as an eight-bedroom family dwelling pursuant to a planning permission granted in 1999. From 2008 onwards, the appellant, through her company, used the property for short-term holiday lets. The property was advertised as accommodating groups of up to 18 people, with two more on a sofa bed, and was generally let for short periods of three, four or seven nights.

The second respondent council issued an enforcement notice, alleging a breach of planning control consisting of a change of use of the property, without planning permission, from a Class C3 dwelling to a sui generis use as commercial leisure accommodation falling outside Class C3; the notice required such use to cease. Upholding the notice on appeal, the first respondent's planning inspector found that the use of the property resulting from the holiday letting business was quite different in character from that of a private family dwelling-house. The appellant's challenge to that decision, under section 289 of the Town and Country Planning Act 1990, was dismissed in the court below: see [2012] EWHC 1092 (Admin).

The appellant appealed. She contended, relying on the decision of the Court of Appeal in *Moore* v *Secretary of State for the Environment* [1998] 2 PLR 65, that where there was a permitted use as a "dwelling" or "dwelling-house", that use lawfully included not only occupation by an individual or family as a permanent home but also the use of the dwelling for holiday or temporary occupation, whether or not that occupation resulted from a commercial letting. She further submitted that if the current use of the property were a breach of planning control, then its description in the enforcement notice as "commercial leisure accommodation" was too wide since not all forms of leisure accommodation amounted to a change of use.

Held: **The appeal was dismissed. (1) A dwelling-house can lawfully be used for some degree of holiday letting without there being a material change of use. Whether commercial holiday letting involves a material change of use will be a question of fact and degree in each case, with the answer depending on the particular characteristics of the use in question. *Moore* is not authority for the proposition that use of a dwelling-house for commercial holiday lettings can never amount to a material change of use. Neither of the two extreme positions – that use of a dwelling-house for commercial holiday lettings will always amount to a material change of use, or that it will never do so – is correct. Having compared the character of the current use of the appellant's property with that of the previous lawful use, and carefully examined the characteristics of the lettings, the inspector had been entitled to conclude that, as a matter of fact and degree, they constituted a material change of use from the permitted use as a dwelling-house. The particular use of the appellant's property for holiday lettings was very far removed from the permitted use as a dwelling-house and a material change of use had occurred. (2) The terms of the enforcement notice, although wide, were neither misconceived nor unintelligible and they covered the particular use to which the property was being put. The inspector's primary duty was to consider the proposals before him and he had been presented with no viable alternative that would have secured the cessation of the breach of planning control that he found to have occurred. The parties had made no attempt to define when a use for commercial holiday letting would be a material change of use and had suggested no relevant criteria for the purpose of drafting such a definition. There was no "obvious alternative" on the material before the inspector. Accordingly, he had been entitled to uphold the enforcement notice in its existing form.**

The following cases are referred to in this report.

Blackpool Borough Council v *Secretary of State for the Environment* (1980) 40 P&CR 104
Gravesham Borough Council v *Secretary of State for the Environment* [1984] P&CR 142
Moore v *Secretary of State for the Environment* [1998] 2 PLR 65
Tapecrown Ltd v *First Secretary of State* [2006] EWCA Civ 1744
Taylor and Sons (Farms) v *Secretary of State* [2001] EWCA Civ 1254
Wilson v *West Sussex County Council* [1963] 2 QB 764

This was an appeal by the appellant, Sheila Moore, from a decision of Ms Frances Patterson QC, sitting as a deputy High Court judge, dismissing an appeal, under section 289 of the Town and Country Planning Act 1990, from the decision of a planning inspector appointed by the first respondent, the secretary of state for communities and local government, upholding an enforcement notice issued by the second respondents, Suffolk Coastal District Council, alleging a change of use of a property from a dwelling to commercial leisure accommodation.

Alun Alesbury (instructed by Gotelee Solicitors, of Ipswich) appeared for the appellant; Gwion Lewis (instructed by the Treasury Solicitor) appeared for the first respondent; the second respondents, Suffolk Coastal District Council, did not appear and were not represented.

Giving judgment, **Sullivan LJ** said:

1. This is the judgment of the court.

2. This is an appeal against the order dated 27 March 2012 of Frances Patterson QC (sitting as a deputy High Court judge) dismissing the appellant's appeal under section 289 of the Town and Country Planning Act 1990 ("the Act") against a decision of an inspector appointed by the first respondent dismissing the appellant's appeal under section 174 of the Act against an enforcement notice issued by the second respondent. The inspector's decision is dated 15 August 2011.

3. The breach of planning control alleged in the notice was a change of use without planning permission of the appellant's property, St Audry's House, Melton, Woodbridge, Suffolk ("the property"), "from a C3 dwelling to use as commercial leisure accommodation which does not fall within Class C3(a)-(c), and which therefore constitutes a sui generis use". The requirements of the notice were to cease the use of the property as commercial leisure accommodation.

4. It is common ground that the reference to Class C3 is a reference to Class C3 in Schedule 1 to the Town and Country Planning (Use Classes) Order 1987, as amended ("the Order"). Class C3 reads as follows:

> **"Class C3. Dwelling-houses**
> Use as a dwelling-house (whether or not as a sole or main residence) by:
> (a) a single person or by people to be regarded as forming a single household;
> (b) not more than six residents living together as a single household where care is provided for residents; or
> (c) not more than six residents living together as a single household where no care is provided to residents (other than a use within Class C4)."

5. The appellant appealed against the notice on grounds (a), (b), (c) and (f) in section 174(2) of the Act. The inspector dismissed the appeal on all four grounds and upheld the notice as issued. There is no appeal against the inspector's decision to dismiss the appeal on ground (a).

Factual background

6. The factual background is set out in some detail in the judgment below: [2012] EWHC 1092 (Admin). St Audry's at Melton was an extensive hospital complex. The hospital closed and outline planning permission was granted for part conversion and part redevelopment of the hospital site for a mix of uses, including residential.

7. In May 1999 approval was granted for the conversion of the property to an eight bedroomed dwelling (in fact there are nine bedrooms – a studio is used as an additional bedroom, but nothing turns on this). Conversion works were carried out, and from 1999-2007 the property was occupied as a dwelling by a family. Since May 2008 it has been let by the appellant, through her company, Prestige Holiday Lettings, for short-term holiday lets.

The inspector's decision

8. The material parts of the inspector's decision are set out in [6] of the judgment below. Under ground (b) the appellant argued that the alleged breach of planning control had not occurred because the allegation that there had been a change of use to "commercial leisure accommodation" was "misconceived and practically unintelligible and cannot be regarded as describing any recognised land use for planning purposes." Having said that there was ample evidence to support the contention (which was not disputed by the appellant) that the property was being commercially let, the inspector said in [6]-[8] of the decision:

> "6. I accept that the description 'leisure accommodation' might encompass a wide range of different forms of occupation of the property, some of which may constitute a material change of use, and others not. However, I consider holiday accommodation is one such purpose, since holidays are clearly leisure time.
>
> 7. I also accept that the allegation might be framed in a number of ways – and it is quite reasonably suggested for the appellant that it might be described as a "use as a holiday dwelling". However, the formulation used is clear, it covers the particular use to which the property is put, and it is obviously understood by the appellant whose business it is.
>
> 8. I do not find the formulation used in the notice either misconceived or unintelligible. The alleged breach has occurred as a matter of fact, and the appeal on ground (b) therefore fails."

9. Under ground (c) the appellant contended that there had been no breach of planning control because the use of the property for holiday letting was not materially different in character from the lawful use as a dwelling house. The inspector said that there was no dispute that the property fell

> "within the particular kind of building that should be described as a 'dwelling house' – that is, following the test set out in the case of *Gravesham Borough Council* v *Secretary of State for the Environment* [1984] P&CR 142 – it ordinarily affords the facilities required for day-to-day existence."

10. There was also no dispute that the use of the property from 1999 until 2007 had been as a dwelling house occupied by people living together as a family. As the inspector said in [10] of the decision:

> "Such use falls squarely within the definition of Use Class C3(a)."

11. The cornerstone of the appellant's appeal under ground (c) was the decision of this court in *Moore* v *Secretary of State for the Environment* [1998] 2 PLR 65 (see [20]-[24] below). The inspector said that Moore was of limited relevance. Having noted the appellant's argument that a planning permission for the construction of dwellings did not preclude a use for purposes other than as family dwelling-houses, the inspector said in [12] of the decision:

> "...in determining whether a material change of use has taken place it is necessary to look at and compare the character of the current allegedly unlawful use with that of the actual previous lawful use."

12. The inspector described the characteristics of the current use of the property in [13]-[16] of the decision.

> "13. As now used the house is advertised as accommodating up to 18 people with an additional two on a sofa-bed. It is let generally for short periods of three, four or seven nights, and it is apparent that there have been quite regular bookings in each of the years from 2008 to 2010, with occupation on as many as about 175 nights per year. In 2009 there were slightly over 40 separate bookings. I note that many of these are three-night bookings over weekends.
>
> 14. I accept that if the building is only occupied for short periods it is not disqualified from being a dwelling-house. Furthermore, even if as large a group as 20 people occupy it, they could conceivably be a family or regarded as a single household.
>
> 15. The appellant says that residents come to the house as a preformed group for a predetermined period, and have a family or other relationship which predisposes them to occupy the house as a single household. While it is possible that large family gatherings take place at the appeal property, there is no specific evidence to support the frequency of this, and I would expect such events to be rare. From what I have read, particularly from occupiers of nearby dwellings, numerous different groups visit the property for various reasons. There has been a yoga group of some 15 people, and a cycling group of some 20 people. There are probably many who come for countryside holiday activities, and it appears there are also many who come for reunions, parties, or celebrations of one sort or another. Such reasons for coming together largely arise from participants' shared interests, but do not establish these groups as single households.
>
> 16. A result of the current use is that new groups arrive at, and leave the property relatively frequently, often arriving on a Friday and leaving on a Monday. A new group may then arrive on a Monday and leave on a Friday, presumably with cleaners arriving at the changeover. Furthermore, it appears that groups often travel in a number of separate vehicles, as shown by a number of photographs and in third parties' written statements."

13. Having considered the differences between the current use of the property and its use by a family/household, the inspector summarised his findings and overall conclusion in [21] and [25] of the decision.

> "21. In summary, I consider there are a number of distinct differences between the current use and use of the appeal property as a family dwelling-house. Notably, the pattern of arrivals and departures, with associated traffic movements; the unlikelihood of occupation by family or household groups; the numbers of people constituting the visiting groups on many occasions; the likely frequency of party-type activities, and the potential lack of consideration for neighbours.
>
> …
>
> 25. Overall, as a matter of fact and degree, I consider the use of the property as part of the appellant's holiday letting business results in a use of the dwelling-house that is quite different in character from that of a private family dwelling-house. I consider this change in character has resulted in a material change of use of the property that is development requiring planning permission under the provisions of Section 55(1) of the Act. No planning permission has been obtained, nor is this any form of permitted development. There has therefore been a breach of planning control, and the appeal on ground (c) must fail."

14. Under ground (f) the appellant contended that the cessation of the use was unnecessary, and that any adverse effects on the amenity of nearby residential properties could be lessened by limiting the numbers

A of people using the property at any time, and by a noise management plan. The inspector was not impressed by these unparticularised suggestions, and dismissed the ground (f) appeal saying that lesser steps would not overcome the objections to the current use.

Use as a dwelling-house – the authorities

15. There are three relevant authorities. In *Blackpool Borough Council* v *Secretary of State for the Environment* (1980) 40 P&CR 104 (to which the inspector was not referred) a house was used by the owner as a second home for holidays by himself and his family, by members of his office staff, and by "family groups" who paid rent. There were
B lettings at a rent for 10 out of 18 weeks in the four month holiday season; for the remainder of the year the premises were left empty. The local planning authority served an enforcement notice alleging a material change of use from use as a private dwelling house to use for holiday lettings on a commercial basis. On appeal, the inspector concluded that there had been no material change of use. The Secretary of State adopted the inspector's conclusions and allowed the owner's appeal against the enforcement notice. The local planning authority's appeal was dismissed by the Divisional Court.

16. It was submitted on behalf of the local planning authority that the inspector had erroneously approached the matter on the basis that
C "if the house is occupied by one family, the house is residential and *therefore* in accordance with the permitted use as a dwelling-house". Ackner LJ said, at p111:

> "I would agree that, if that is what the inspector is saying, it would be wrong, because not every residential use is necessarily a use as a private dwelling-house. To my mind, however, what is said reads quite clearly as being merely a double description. If the house is occupied by one family, etc, the use is residential *and* in accordance with permitted use as a private dwelling-house. In my judgment, what was being found as fact here was that the character of the user from the planning point of view had not been changed by the fact that the premises were being occupied not only by the owner and his family but also by his friends or by members of his office staff or by paying tenants during the
> D periods that I have indicated.
>
> If that is right, it is a finding of fact, and all that this court has to ask itself is: was it a finding of fact that could reasonably have been made on the evidence before the inspector. It is common ground, as I understand it, that the question for determination in the context of this appeal is whether the character of the use of this dwelling-house as a private residence has been changed so substantially as to amount to a material change of use. It is a question of fact and degree. It is a decision that is based on the particular facts of this case that I have recited. This is not a case that lays down, as I see it, any principle. The inspector was not dealing with a house that was constantly being let in short holiday lettings. She was dealing with a house that was being occupied by the owner, by the owner's friends and by the owner's staff on a non-paying basis, with, superadded to
> E that, a period in the aggregate of 10 weeks in the year during which it was let as a rent to single households. I think that she was wholly entitled to reach the conclusion that the character of this dwelling-house and its use was not materially changed by the succession of occupiers over the period that I have mentioned in the categories that I have described. As I have said, I do not consider that she was seeking to propound any proposition of law in her para 42. She was making the legitimate findings of fact on which to base her conclusion that there had been no material change of use."

17. Jupp J agreed with Ackner LJ "that not every residential use is necessarily a use as a private dwelling house", and added, at p112:

> "I also agree that the method of taking the evidence in this case has left some
> F matters of fact unclear that might have come out and turned out very differently from the way they did had they been open to further probing and questioning. It is not to be thought that, in a case like this where the lettings had a different character from that concluded by the inspector here, there would not be a material change of use. It depends entirely on the facts, as Diplock LJ said in *Wilson* v *West Sussex County Council* [1963] 2 QB 764" 'Considerations which are relevant are planning considerations…' and these vary from case to case."

18. In *Gravesham*, the question was whether a building which had been erected pursuant to a planning permission for a "weekend and holiday chalet" subject to conditions that prohibited its use for human habitation between 1 November and 28 February each year was a "dwelling-house" for the purpose of Class I of Schedule 1 to the Town and Country Planning General Development Order 1977. On reconsideration, the Secretary of State concluded that the building was G
a dwelling-house. The local planning authority's appeal was dismissed by McCullough J.

19. McCullough J said that whether a building is or is not a dwelling-house is a question of fact. Having considered a number of situations where buildings would still be dwelling-houses even though they were not occupied throughout the year – second homes, houses left empty pending sale, houses unoccupied because they are flooded or undergoing extensive repair – McCullough J said, at p146:

> "Suppose that a London-based company requires a succession of employees to be based one at a time for four months in a location far distant from London. H
> Suppose that the company buys a house and makes it available to each employee and his family for his tour of duty. It would still be a dwelling-house. Take a holiday cottage subject to time-share with a number of owners each enjoying the right to occupy it for two particular weeks each year. That would still be a dwelling-house.
>
> What have these examples in common? All are buildings that ordinarily afford the facilities required for day-to-day private domestic existence."

20. In *Moore* (no relation to the appellant in this case) the outbuildings of a large country house had been converted into 10 single self-contained units of residential accommodation for the purpose of holiday lettings. Nine of the units were in use by May 1991. In May
1995 the local planning authority issued an enforcement notice alleging J
a material change of use from residential to mixed use of residential and as 10 units of holiday accommodation. If the change of use of each of the units was a change of use to a single dwelling-house, the enforcement notice was not served within the four year time-limit in section 171B(2) of the Act.

21. The material facts as found by the inspector were as follows (see p67B-D):

> "The units are all self-contained, with no apparent connection between them, and each is supplied with the facilities necessary for daily life, including living, sleeping and eating space, kitchens, bathrooms and WCs. Each unit has
> a small area of open air amenity space defined by hedges or fences, usually at K
> the front. There is a communal car park for the 10 units, apart from which there are no communal areas. Council tax is charged on the property by four separate assessments, one of which covers the main house and the 10 units. The units are available to the public on short lets, including weekend and mid-week breaks, with the longest letting being for three or four months. They are managed as one entity, the income being deposited in one account. Cleaning is provided at changeovers and a maid can be employed at an extra charge for cleaning on an hourly basis. Linen, including towels, is provided. Breakfast hampers are provided for guests at an extra charge, but other than that no meals area provided. None of the units is used for staff accommodation."

22. The inspector concluded in [106] of his report (see p68B-C): L

> "The 'cottages', apartments or whatever description is applied to them certainly have the physical attributes of self-contained dwellings now. However, they are not used in the normal sense as independent residential units. Their use for holiday accommodation is, in my opinion, materially different to a use of premises by a household as the long-term home of the person or persons comprising that household. Put in simple terms, no one lives in these cottages and has not done so since 1985. This requires no definition of legal principle; it is a common sense conclusion derived from the facts of this case..."

23. The Secretary of State accepted the inspector's conclusion and
rejected the appellant's contention that the four-year time limit applied. M
The High Court dismissed the appellant's appeal under section 289 of the Act. On appeal to this court, Mr Alesbury, who appeared on behalf of the appellant in *Moore*, relied on the decision in *Gravesham* and submitted that the inspector had erred in para 106 of his report ([22] above) in concluding that the units were not dwelling-houses, even though they had the physical attributes of self-contained dwellings, because their use as holiday accommodation was "materially different to a use of premises by a household as the long-term house of the person or persons comprising that household." The Divisional Court's decision in Blackpool was not referred to.

24. The appeal was allowed. Nourse LJ, with whom Pill and Thorpe LJJ agreed said, at p71A-C:

A "In my judgment, McCullough J's approach to the meaning of 'dwelling-house' was entirely correct. Although we were not referred to any of the many other decisions on the meaning of that word in other areas of the law, I am confident that an examination of them would reveal no requirement that before a building can be so described it must be occupied as the permanent home of one or more persons or the like. Nor do 10 self-contained units of residential accommodation which would otherwise be properly described as 10 single dwelling houses cease to be used as such because they are managed as a whole for the commercial purpose of holiday or other temporary lettings. Accordingly, I am satisfied that the Secretary of State applied an incorrect test in this case and that if he had applied the correct test, he could only have properly concluded that the 10 units are being used as 10 single dwelling-houses within section
B 171B(2) of the Act."

The grounds of appeal

25. Mr Alesbury challenged the inspector's decision on two grounds. In his skeleton argument he submitted that where there is a permitted use as a "dwelling" or "dwelling-house" (it is common ground that they are synonymous) that use lawfully includes not only occupation by an individual or family as a permanent home but also the use of the dwelling for holiday or temporary occupation, whether or not that occupation is as a result of a commercial letting. That proposition was, he submitted, the ratio of the court's decision in *Moore*, a decision which
C the inspector wrongly dismissed as being of only limited relevance. The inspector should not have asked himself whether there was any material difference between the present use – for commercial short-term holiday lets – and the previous use of the property as a "family dwelling" or "private family dwelling-house", but should instead have asked the question: is there any material difference between the present use and the use that the Court of Appeal has said in *Moore* is within the scope of a planning permission to use a building as a dwelling?

26. Secondly, Mr Alesbury submitted that, if the present use is a breach of planning control, the description of the use in the notice as "commercial leisure accommodation" was neither an accurate nor an adequate description of what is actually taking place in the property.
D The inspector had accepted that the appellant's suggestion that the present use might be described as "use as a holiday dwelling" was reasonable, (see [7] of the decision), and he had also accepted that the description "leisure accommodation" might encompass a wide range of different forms of occupation of the property, some of which may not constitute a material change of use (see [6] of the decision). In these circumstances the inspector should have allowed the appeal under ground (b), or varied the notice so as to define its requirements with greater particularity under ground (f). Mr Alesbury submitted that it was of vital importance to the appellant, who will face criminal penalties if she fails to comply with the requirements of the notice, to
E know precisely what it is that the notice prohibits. As issued and upheld by the inspector the notice closes down her business in its entirety, even though the inspector's decision acknowledges that using the property for some forms of "leisure accommodation" may not amount to a material change of use.

Discussion

Ground 1

27. Starting from first principles, without the assistance of any authority, whether the use of a dwelling-house for commercial letting
F as holiday accommodation amounts to a material change of use will be a question of fact and degree in each case, and the answer will depend upon the particular characteristics of the use as holiday accommodation. Neither of the two extreme propositions – that using a dwelling-house for commercial holiday lettings will always amount to a material change of use, or that use of a dwelling-house for commercial holiday lettings can never amount to a change of use – is correct.

28. The Divisional Court's decision in *Blackpool* supports this approach: the court concluded that the inspector would have erred if she had approached the matter on the basis that there could be no material change of use if the house was in some form of residential use – "not every use is necessarily a use as a private dwelling-house" – but she was entitled to conclude that the character of the dwelling-house was not materially changed by the particular pattern of holiday lettings in G
that case.

29. In *Gravesham* the Court was concerned with a different issue: whether it was an essential characteristic of a dwelling-house that it should be occupied residentially throughout the year. McCullough J rejected that proposition. McCullough J did not have to consider whether some forms of commercial letting for holiday purposes throughout the year might be capable of amounting to a material change of use. *Blackpool* was not relevant to that issue, and was not cited.

30. Nourse LJ's conclusion in *Moore* that the Secretary of State had applied an incorrect test is consistent with the approach of the Divisional Court in *Blackpool*. In *Moore* the inspector had adopted an H
extreme position: that any use of premises as holiday accommodation was materially different to a use as a single dwelling-house because it was not a use of the premises as "the long-term home" of persons comprising a household. The inspector did not consider whether the particular characteristics of the use of the 10 units as holiday accommodation amounted to a change of use.

31. The only possible basis for the contention that the use of a dwelling-house for commercial holiday lettings can never amount to a change of use is Nourse LJ's final conclusion, at p71C, that if the Secretary of State had applied the correct test "he could *only* have properly concluded that the 10 units are being used as 10 single J
dwelling houses" (emphasis added). The 10 units were being used for short-term holiday lets: see [21] above. The court was not referred to *Blackpool*, and it applied *Gravesham*. While *Gravesham* is authority for the proposition that a dwelling-house can be so described even if it is not occupied throughout the year as a permanent home, it is not authority for the proposition that any use of a dwelling-house for holiday lettings will not amount to a material change of use.

32. There is no discussion in *Moore* of the characteristics of the holiday lettings in that case. It appears that the Secretary of State defended the lawfulness of his decision on a point of principle: use as holiday accommodation was not a use as a single dwelling-house within K
section 171B(2) because it was not a use as a "long-term home". Once this point of principle was rejected by the court, it seem that nothing else was put forward by the Secretary of State to support the proposition that the 10 units were not being used as single dwelling-houses. The court accepted that whether the 10 units were being used as single dwelling-houses was a question of fact and degree for the Secretary of State (see p70B-C of the judgment of Nourse LJ), but in the absence of any argument, other than the erroneous point of principle, that there had been a material change of use, the court felt that it was able to conclude that there was only one conclusion properly open to the Secretary of State. L

33. For these reasons, we do not consider that *Moore* is authority for the extreme proposition: that the use of a dwelling-house for commercial holiday lettings can never amount to a material change of use. In his oral submissions, Mr Alesbury readily accepted that *Moore* was not authority for that extreme position, and made it clear that he did not submit that any kind of holiday letting was within the permitted use of the property as a dwelling-house. However, he submitted that the inspector had erred in treating *Moore* as being of limited relevance. He should have approached the question whether there had been a material change of use on the basis that a dwelling-house could lawfully be used for some degree of holiday letting without there being a material M
change of use, and then asked whether there was anything about the particular characteristics of the holiday lettings in the present case which amounted to such a change.

34. In our judgment, this was the inspector's approach. It is true that the inspector refers to the differences between the current use and the use of the property "as a family dwelling-house" ([21] of the decision) or "private family dwelling-house" ([25] of the decision), but the decision must be read as a whole. In [12] of the decision, the inspector noted the appellant's argument that a permitted use as a dwelling-house did not preclude use for purposes other than as a family dwelling-house. He did not reject that proposition. Had he done so, that would have been the end of the ground (c) appeal and the decision would have been

much shorter. Instead, he reminded himself that he had to compare the character of the current use with that of the previous lawful use, and he then proceeded to examine in some detail the particular characteristics of the current use in [13]-[20] of the decision.

35. In this context, [15] of the decision is of particular significance. The inspector rejected the appellant's contention that those who occupied the property did so in preformed groups so that their occupation was akin to occupation by a single household. One of the factors which had persuaded the inspector in the *Blackpool* case that there had been no material change of use was the fact that when the house was being let out for rent it was let to "family groups", or "single households": see [16] above. In the present case, the inspector concluded that the large groups (the property can accommodate up to 20 people, see [13] of the decision) who occupied the property came together largely as a result of their shared interests (yoga, cycling etc) but they did not occupy the property as "single households".

36. In our judgment, the inspector adopted the correct approach. Although he was not referred to *Blackpool*, he did not fall into the error of assuming that any use for holiday letting amounted to a material change of use. He carefully examined the characteristics of the lettings in the present case and concluded that, as a matter of fact and degree, they were a material change of use from the permitted use as a dwelling-house. In our judgment, he was not merely entitled to reach this conclusion; it was clearly, on the facts of this case, the correct conclusion. As a matter of common sense, this particular use for holiday lettings is very far removed from the permitted use as a dwelling-house and a material change of use has occurred.

Ground 2

37. The inspector rightly dismissed the appeal on ground (b). Even if the description of the breach of planning control was too wide it was neither "misconceived" nor "practically unintelligible." For the reasons given by the inspector the description of the current use as "commercial leisure accommodation" covered the particular use to which the property was being put. Holiday accommodation is a form of leisure accommodation. If the appellant has a valid complaint, it is not that the alleged breach of planning control did not occur, it is that the requirements of the notice are excessive, not least because the inspector concluded that "leisure accommodation" might encompass a wide range of different forms of occupation of the property, some of which may not constitute a material change of use. Moreover, the authorities make it clear (see above) that not all forms of commercial holiday letting amount to a material change of use of a dwelling-house.

38. On behalf of the first respondent, Mr Lewis submitted that this complaint should have been made in the appeal under ground (f), and it was not. In her appeal under ground (f) the appellant did not suggest that the description of the prohibited use should be narrowed down, she submitted that any adverse impact on residential amenity could be ameliorated by a limitation on the numbers using the property and by a noise management plan: see [14] above.

39. Mr Lewis referred us to a number of authorities, including *Tapecrown Ltd* v *First Secretary of State* [2006] EWCA Civ 1744 and *Taylor and Sons (Farms)* v *Secretary of State* [2001] EWCA Civ 1254, which establish the proposition that an appellant under ground (f) should state his "fall back" position because the inspector's primary duty is to consider the proposals which have been put before him, and he is not under any duty "to search around for solutions."

40. We readily accept that it is not the duty of an inspector to make an appellant's case tor him (see [46] of *Tapecrown* per Carnwath LJ, as he then was), but in the present case the appellant had made her case, albeit that she made it under ground (b) rather than ground (f), that the alleged use in breach of planning control, which the notice required her to cease, was too wide. As Carnwath LJ observed in *Tapecrown*, the enforcement procedure is intended to be remedial rather than punitive" ([46]). We accept Mr Alesbury's submission that the mere fact that this issue was raised under ground (b) rather than ground (f) is not fatal to this ground of appeal. If there was "an obvious alternative which would overcome the planning difficulties, at less cost and disruption than total [cessation]" the inspector should have considered it: *Tapecrown* (ibid).

41. Was there an "obvious alternative" which the inspector should have considered? In this respect he was not assisted by the procedure – the appeal was decided by written representations – or by the parties. The second respondent's written representations to the inspector did not engage with this issue, and it has played no part in the section 289 appeal. Although the appellant had argued that a more appropriate description of the current use would be "use as a holiday dwelling", she was contending that such a use was "not materially different in character from the lawful use as a dwelling-house." Thus, there was no "obvious alternative" on the material before the inspector. The appellant now accepts that using a dwelling-house for commercial holiday lettings may amount to a material change of use, but given the need for clarity as to the requirements of an enforcement notice (breach of the requirements is a criminal offence) it would be wholly inappropriate to require the appellant to cease the "commercial holiday letting use of the property insofar as it amounts to a material change of use from a dwelling-house." That would be a recipe for endless disputes before the magistrates.

42. Before the inspector there was no attempt to define when a use for commercial holiday letting might amount to a material change of use of the property, or even to suggest any criteria which might be relevant for the purpose of drafting such a definition. The second respondent contented itself with the proposition that there had been a material change of use, and the appellant denied that such a use could amount to a material change of use of a dwelling-house. In these circumstances we are not able to conclude that the inspector erred in law in deciding to uphold the notice in the terms in which it was issued. He was not presented with any viable alternative which would have secured the cessation of the breach of planning control which he found to have occurred. It is not without significance that in pursuing his second ground of appeal Mr Alesbury was not able to suggest any modification to the "excessive" requirement in the notice which would cut back the requirement whilst ensuring that the property is not used in breach of planning control. The notice is not perfect, but no better alternative has been suggested.

Conclusion

43. For these reasons the appeal must be dismissed.

Appeal dismissed.

Holliday v Breckland District Council

UPPER TRIBUNAL: LANDS CHAMBER

30 August 2012

[2012] UKUT 193 (LC)

MR GEORGE BARTLETT QC, PRESIDENT, AND MR PAUL R FRANCIS FRICS

Estates Gazette 8 December 2012

[2012] 49 EG 68

Town and country planning – Compensation – Interference with easement – Town and Country Planning Act 1990 – Right of way blocked by development carried out on land appropriated for planning purposes under section 122 of 1990 Act – Owner of right of way claiming compensation under section 237 – Whether entitled to compensation for "taking" of easement to be assessed by reference to ransom value – Whether compensation confined to diminution in value of dominant tenement by reason of loss of right of way

The claimant owned a furniture shop in Dereham, Norfolk, with a separate workshop building to the rear. Until 2004, he was able to gain access to the workshop by means of a right of way over adjoining

A land owned by the district council. In 2004, the council confirmed a resolution to appropriate their land to planning purposes, pursuant to section 122 of the Local Government Act 1972; the access way was subsequently blocked by a retail development carried out, with planning permission, on the council's land.

The claimant sought compensation, under section 237 of the Town and Country Planning Act 1990, for the interference with his easement; by section 237(4), such compensation was payable in accordance with sections 63 or 68 of the Lands Clauses Consolidation Act 1845 or sections 7 or 10 of the Compulsory Purchase Act 1965, to be assessed "in the same manner and subject to the same rules as in the case of other
B compensation under those sections in respect of injurious affection": see section 237(4)(b).

The council, as compensating authority, accepted that the claimant was entitled to compensation for injurious affection, consisting of the diminution in value of the shop premises as a consequence of the loss of the right of way. The claimant contended that he was additionally entitled to be compensated, under section 63 of the 1845 Act, for the "taking" of his right of way, to be calculated by reference to the market value of that right; he contended that this included a ransom value representing the amount that he could have extracted, as a proportion of the development value of the council's land, in return for agreeing to
C the extinguishment of the easement. A preliminary issue was tried as to whether such compensation was payable.

Decision: **The preliminary issue was determined in favour of the compensating authority. The claimant was not entitled to have compensation assessed on the basis that an interest in land had been acquired from him. Section 237(4)(b) prescribes the manner in which compensation is to be assessed for an interference with an interest or right resulting from works carried out with planning permission on land acquired or appropriated by a local authority for planning purposes. Compensation is not to be assessed as though the land has been taken; instead, it must be assessed as**
D it would be under sections 63 or 68 of the 1845 Act, or sections 7 or 10 of the 1965 Act, in respect of injurious affection. The basis of compensation for injurious affection under those Acts is the diminution in the value of the retained land. Since compensation for injurious affection under section 237(4) falls to be assessed on the same basis, it cannot include ransom value.

The following cases are referred to in this report.

Allen v *Gulf Oil Refining Ltd* [1981] AC 1001
Argyle Motors Ltd v *Birkenhead Corporation* [1975] AC 99
B&Q plc v *Liverpool & Lancashire Properties Ltd* [2001] 1 EGLR 92
Hammersmith & City Railway Co v *Brand* (1869) LR 4 HL 171
E *Hoveringham Gravels Ltd* v *Chiltern District Council* (1978) 35 P&CR 295; [1979] 2 EGLR 182; [1979] 252 EG 815
Kent County Council v *Union Railways (North) Ltd* [2009] EWCA Civ 363; [2009] 2 EGLR 1; [2009] 30 EG 68
Thames Water Utilities Ltd v *Oxford City Council* [1999] 1 EGLR 167

This was the hearing of a preliminary issue on a claim by the claimant, Richard Holliday, against the compensating authority, Breckland District Council, for compensation, under section 237 of the Town and Country Planning Act 1990, for interference with a right of way by reason of development on land appropriated for planning purposes.

F Christiaan Zwart (instructed by Stevensons Solicitors, of Dereham) appeared for the claimant; Barry Denyer-Green (instructed by the legal department of Breckland District Council) represented the compensating authority.

Giving judgment, **Mr George Bartlett QC, President** said:

Introduction

1. The claimant in this reference is the owner of a furniture shop at 53/55 High Street at Dereham in Norfolk, where he has carried on business since 1993. At the rear of the premises is a separate workshop building. Until 2004 there was access to the land at the rear of the shop across land to the north that was owned by the council. This access
was wide enough for a large furniture van, and the claimant had the G
right to use it under an easement benefiting the land on which the shop stands. Development carried out on the land owned by the council blocked up the access, and the claimant claims compensation for loss of the access under section 237 of the Town and Country Planning Act 1990. Compensation is sought not just for the injurious affection of the claimant's dominant tenement but also for the value of the easement, which is said to consist of the amount that the claimant would have been able to extract from the council as owner of the servient tenement in order to secure the extinguishment of the easement.

2. The council accept that the claimant is entitled to compensation
for injurious affection, consisting of the diminution in value of the shop H
premises as a consequence of the loss of the right of way, but they deny that he is entitled to compensation based on the development value of the council's land. On their application on 30 August 2011 the President ordered that the following two issues, identifying the respective contentions of the parties, be determined as preliminary issues:

(1) Whether the claimant is entitled to claim compensation under section 237 of the Town and Country Planning Act 1990 on the basis that the benefit of right of way to numbers 53 and 53a High Street, Dereham ("No 53 High Street") has been appropriated by the acquiring authority and that compensation should be payable
for that appropriation on a ransom value basis under section 63 of J
the Lands Clauses Consolidation Act 1845 for the loss of a ransom value that would otherwise have benefited No 53 High Street on the ground that the claimant was entitled to and has lost the right to grant or refuse consent to the development obstructing the said right of way.

(2) Whether the claimant is entitled to compensation under section 237 of the 1990 Act and in accordance with section 10 of the Compulsory Purchase Act 1965 (alternatively section 68 of the Lands Clauses Consolidation Act 1845) on the basis only that on a claim for compensation under these statutory provisions
the compensation is limited to the diminution in value of No 53 K
High Street by reason of the lawful obstruction of the said right of way and lawful use of the land over which the right of way was exercisable, without regard to any ransom value.

The facts

3. There is no disagreement as to the facts. The claimant is the registered freehold proprietor of land known as 53 and 53a High Street, Dereham, registered at the Land Registry under Title No NK150149. Number 53 is occupied by the claimant, together with number 55 High Street, which he also owns, as retail premises for the purposes of his
business as a retailer and restorer of furniture. High Street runs north- L
south, and the claimant's land fronts it on the east. To the rear (east) of number 53 is small open area, and beyond that is a workshop; to the east of the workshop is a smaller open area, and finally beyond that is a storage shed. All of these are within Title No NK150149. We will refer to this as "the land".

4. The land has the benefit of a right of way to pass and repass with or without vehicles over an accessway which is situated on land to the north that is owned by the council and is comprised in registered Title No NK102856 (which we will refer to as the council land). The terms of the right of way are noted at entry number 2 in the Property Register of the official copy of the registered title, and the servient land is shown
tinted blue on the plan thereto. The right of way is L-shaped and runs M
from the High Street, behind land abutting the High Street, to the open land between the shop and the workshop.

5. The council land, together with other adjoining land, was acquired by the council in 1996 by agreement, pursuant to section 120 of the Local Government Act 1972. On 13 March 2003, the council resolved under section 122 of the Act to appropriate this land, and other land, for planning purposes. The resolution was confirmed on 26 February 2004. On 14 November 2003, the council granted planning permission for the development of the council land and the other land for mixed uses. On 21 April 2004, they granted a lease to Dencora Dereham Retail Partnership LLP, which then proceeded to implement the planning

permission and develop the land. On 19 April 2004 a fence was erected around the council land, obstructing access to the servient land, and the servient land has now been built on. The claimant can no longer exercise the right of way. The building constructed on the servient land is now used for retail purposes in accordance with the planning permission. Construction was completed on or about 17 June 2005.

The statutory provisions

6. Section 237, so far as material, provides as follows:

> "(1) Subject to subsection (3), the erection, construction or carrying out or maintenance of any building or work on land which has been acquired or appropriated by a local authority for planning purposes (whether done by the local authority or by a person deriving title under them) is authorised by virtue of this section if it is done in accordance with planning permission, notwithstanding that it involves:
>
> (a) interference with an interest or right to which this section applies, or
>
> (b) a breach of a restriction as to the user of land arising by virtue of a contract.
>
> (2) Subject to subsection (3), the interests and rights to which this section applies are any easement, liberty, privilege, right or advantage annexed to land and adversely affecting other land, including any natural right to support…
>
> (4) In respect of an interference or breach in pursuance of subsection (1)…, compensation–
>
> (a) shall be payable under section 63 or 68 of the Lands Clauses Consolidation Act 1845 or under section 7 or 10 of the Compulsory Purchase Act 1965, and
>
> (b) shall be assessed in the same manner and subject to the same rules as in the case of other compensation under those sections in respect of injurious affection where–
>
> (i) the compensation is to be estimated in connection with a purchase under those Acts, or
>
> (ii) the injury arises from the execution of works on…land acquired under those Acts."

7. It should be noted that, with effect from 6 April 2009, section 237 was amended by the addition of the following subsection:

> "(1A) Subject to subsection (3), the use of any land in England which has been acquired or appropriated by a local authority for planning purposes (whether the use is by the local authority or by a person deriving title under them) is authorised by virtue of this section if it in accordance with planning permission even if the use involves–
>
> (a) interference with an interest or right to which this section applies, or
>
> (b) a breach of a restriction as to the user of land arising by virtue of a contract."

This provision was inserted following the decision in *Thames Water Utilities Ltd* v *Oxford City Council* [1999] 1 EGLR 167, in which it was held that section 237(1) did not authorise a use of land that interfered with an easement. It does not have retrospective effect, and thus does not apply for the purposes of the present case.

8. Section 63 of the Lands Clauses Consolidation Act 1845, so far as material, provides:

> "In estimating the purchase money or compensation… regard shall be had… not only to the value of the land to be purchased or taken… but also to the damage, if any, to be sustained by the owner of the lands by reason of the severing of the lands taken from the other lands of such owner, or otherwise injuriously affecting such other lands by the exercise of the powers of this or the special Act, or any Act incorporated therewith."

9. Section 68 of the 1845 Act, as amended, provides:

> "If any party shall be entitled to an compensation in respect of any lands, or of any interest therein, which shall have been taken for or injuriously affected by the execution of the works, and for which the promoters of the undertaking shall not have made satisfaction under the provisions of this or the special Act, or any Act incorporated therewith, such party may have the same settled."

10. Section 7 of the Compulsory Purchase Act 1965 provides:

> "In assessing the compensation to be paid by the acquiring authority under this Act regard shall be had not only to the value of the land to be purchased by the acquiring authority, but also to the damage, if any, to be sustained by the owner of the land by reason of the severing of the land purchased from the other land of the owner, or otherwise injuriously affecting that other land by the exercise of the powers of this or the special Act."

11. Section 10 of the 1965 Act provides:

> "(1) If any person claims compensation in respect of any land, or any interest in land, which has been taken for or injuriously affected by the execution of the works, and for which the acquiring authority have not made satisfaction under the provisions of this Act, or of the special Act, any dispute arising in relation to the compensation shall be referred to and determined by the Upper Tribunal.
>
> (2) This section shall be construed as affording in all cases a right to compensation for injurious affection to land which is the same as the right which section 68 of the Lands Clauses Consolidation Act 1845 has been construed as affording in cases where the amount claimed exceeds £50…"

The claimant's case

12. For the claimant, Mr Christiaan Zwart's contention was that compensation fell to be assessed under section 63 of the 1845 Act on the basis of the market value of the claimant's interest in the easement and that this encompassed the amount that could be negotiated for its removal. The route by which, he suggested, this result was reached was as follows. Section 237(1) distinguished between acquisition on the one hand and appropriation on the other. Here the interference with the easement had been through appropriation. Under subsection (4) compensation was payable for the interference permitted by subsection (1). Sections 7 and 10 of the 1961 Act had no application because there had been no compulsory acquisition. Section 63 distinguished between a "purchase" and a "taking", and appropriation equated in substance with taking. What was taken was land as defined in section 237(1) and (2), and this definition should be applied to section 63. Section 237(4)(b) had no application because neither of its conditions (a purchase under (i) or an acquisition under (ii)) was satisfied. Therefore compensation had to be assessed under section 63 for the taking of the easement.

The compensating authority's case

13. For the compensating authority, Mr Barry Denyer-Green summarised their case as follows:

(a) There was an interference with the right of way by the erection of a fence and building works in 2004-2005 on the land across which the right existed, and section 237 applied to this interference.

(b) But for the statutory authority of section 237 such interference would have been a private nuisance giving rise to an action for the substantial obstruction of the right of way.

(c) The right of way, although an interest in land, was not "acquired" or "appropriated" in the sense of a compulsory acquisition, by section 237 or any other provision.

(d) Section 237(4) provides for compensation for such interference in accordance with the rules of injurious affection.

(e) As the land across which the right existed (and other land) was acquired by agreement by the compensating authority in 1991 and no land was acquired from the claimant, the provisions of the 1845 Act cannot have application, nor can section 7 of the 1965 Act apply.

(f) The measure of compensation where section 10 has application is the diminution in value of the land of the claimant in consequence of the interference with the right of way by the works on the land over which the right of way existed.

Discussion

14. It is important to understand the function of section 237. In the absence of statutory powers, interference with a right of way would be actionable as a private nuisance: see *Gale on Easements* 18th ed, para 13-06, and *B&Q plc* v *Liverpool & Lancashire Properties Ltd* [2001] 1 EGLR 92. Where, however, works which are carried out under statutory powers cause damage that would otherwise give a cause of action in nuisance, there is no actionable nuisance: see *Hammersmith & City Railway Co* v *Brand* (1869) LR 4 HL 171 per Lord Cairns at 215, and *Allen* v *Gulf Oil Refining Ltd* [1981] AC 1001, where Lord Wilberforce said, at p1011E:

> "It is now well settled that where Parliament by express direction or by necessary implication has authorised the construction and use of an undertaking

A or works, that carries with it an authority to do what is authorised with immunity from any action based on nuisance."

Where land is compulsorily acquired an easement "may be overridden to the extent necessary to implement the statutory scheme, and compensation payable accordingly (under section 10 of the 1965 Act)": see *Kent County Council* v *Union Railways (North) Ltd* [2009] EWCA Civ 363*, per Carnwath LJ at [7]. There is no statutory scheme of development (or works carried out under statutory powers) in the case of the development of land which has been acquired or appropriated for planning purposes by a local authority. The function of section 237(1) is to make lawful in relation to such development interference with an B easement that would otherwise constitute a private nuisance. Subsection (4) then makes provision for the payment of compensation for such interference.

15. Mr Zwart's contention was that the right to compensation arose under section 237(4) but fell to be assessed under section 63 of the 1845 Act. The key element of his argument was that section 237(4) (b), which provides for assessment of compensation on the basis of injurious affection, has no application, so that compensation is to be assessed on the basis of the market value of the easement, which he said was "land taken" for the purposes of section 63. The fundamental flaw in this argument is that it assumes that section 237(4)(b) is laying down C alternative requirements, one of which has to be satisfied on the facts of the case, (ii) being that there has been an acquisition. That, however, is not what it says. What para (b) provides is that compensation is to be assessed "in the same manner and subject to the same rules as in the case of other compensation under those sections in respect of injurious affection where – (i) the compensation is to be estimated in connection with a purchase under [the 1845 and 1965 Acts], or (ii) the injury arises from the execution of works on…land acquired under those Acts". This is a prescription of how compensation is to be assessed, and it has nothing to do with the circumstances under which the interference with the interest or right arose. The two limbs of para (b) cover the two ways D in which compensation for injurious affection may be payable under the Acts: where some land of the claimant has been acquired and his retained land has been injuriously affected (section 63 of the 1845 Act and section 7 of the 1965 Act) and where no land of the claimant has been acquired (section 68 of the 1845 Act and section 10 of the 1965 Act). Clearly subpara (i) does not contain a requirement that has to be satisfied on the facts of the case. If compensation was to be estimated in connection with a purchase under the Acts, it would be assessed in accordance with the Acts, and this provision would have no role to play. Similarly, in relation to subpara (ii), if compensation was to be estimated in respect of injury arising from the execution of works on E land acquired under the Acts, it would be assessed under the Acts and the provision would have no role to play. Section 237 only comes into play where there has been no acquisition under the Acts.

16. Compensation, therefore, falls to be assessed, as it would be assessed under sections 63 (or 7) or 68 (or 10) in respect of injurious affection. It is not to be assessed as though the land had been taken. Where section 63 or section 7 applies the basis of compensation for injurious affection is the diminution in the value of the retained land of the claimant: see *Hoveringham Gravels Ltd* v *Chiltern District Council* (1978) 35 P&CR 295**. Where section 68 or section 10 applies, again the basis of compensation for injurious affection is the diminution in F the value of the retained land: see *Argyle Motors Ltd* v *Birkenhead Corporation* [1975] AC 99. Compensation for injurious affection under section 237(4) is to be assessed on the same basis. It cannot include ransom value.

17. That is the short answer to Mr Zwart's contention, which, based as it is on a misreading of section 237(4), is in our view simply misconceived. It becomes unnecessary therefore to deal with his submission that the interference with the easement constituted a taking G of land for the purposes of section 63. Even if it was correct that such interference constituted a taking of land it does not seem to us that that would impinge on the operation of section 237. We confine ourselves to saying that it is in our view not correct that it constituted a taking of land; and the contention that the easement had been "appropriated" (and that an appropriation means a taking) is also not correct: appropriation in section 237 is used in its established sense to refer to the act of a statutory authority in assigning to a particular statutory purpose land in its ownership. It does not relate to the taking of land from another person.

H

Conclusion

18. The claimant's contention that compensation should be assessed on the basis that an interest in land had been acquired from him is misconceived. Compensation under section 237(4) is payable in respect of injurious affection only. We accept the contentions of the compensating authority. The answers to the questions contained in the preliminary issues are: (1) No; (2) Yes.

19. The parties are now invited to make submissions on costs, and this decision will become final when the question of costs has been determined.

J

Addendum on costs

20. We have received submissions on costs. The claimant submits that there should be no order as to costs. It is said that the preliminary issue was a vital matter between the parties on which the tribunal's guidance was necessary. In addition the tribunal had ruled against the compensating authority's contention that the *Oxford City Council* case was wrongly decided. It was entirely reasonable for the claimant to seek the tribunal's guidance in relation to these two issues. As a matter of equity, in a case such as this the costs of the reference should be borne by the compensating authority. The claimant had a substantial legitimate interest to defend in that the loss of the potential to use the blue land has unarguably deprived him of an easement critical to his K business.

21. The compensating authority say that the decision disposes of the claimant's primary case against the authority, and they should therefore have their costs. They had always rejected the contention that compensation should be paid on a ransom value basis and had offered compensation on the basis of the diminution in value of the claimant's land. The claimant's contention was pursued at his own risk. It had no merit in it and was founded on a misreading of the relevant statutory provision.

22. The claimant has failed on his contention that compensation L should be paid on a ransom value basis, and we can see no reason why he should not pay the compensating authority's costs. Such loss as he may have suffered through the obstruction of an easement that he used for his business will be the subject of the substantive hearing if compensation is not agreed. We would add that we made no ruling about the *Oxford City Council* case. The contention of the authority was that, although they considered that it was wrongly decided, it should be treated as binding; but that it did not assist the claimant because it concerned a covenant against the use of land, whereas here there was physical interference with the easement. We expressed no views on the case. It was of no materiality for the reason given by the authority. M

23. The claimant must pay the compensating authority's costs of the preliminary issue, such costs if not agreed to be the subject of detailed assessment by the registrar on the standard basis.

The preliminary issue was determined in favour of the compensating authority.

---

* Editor's note: Also reported at [2009] 2 EGLR 1; [2009] 30 EG 68

** Editor's note: Also reported at [1979] 2 EGLR 182; [1979] 252 EG 815

VENDOR AND PURCHASER

A

Greenglade Estates Ltd v Chana and another

CHANCERY DIVISION

12 July 2012

MR DAVID DONALDSON QC, SITTING AS A DEPUTY HIGH COURT JUDGE

[2012] EWHC 1913 (Ch)

Estates Gazette 20 October 2012

B **[2012] 42 EG 138**

Sale of land – Warranty of authority – Damages – Auctioneers holding auction sale – Agreement reached after auction for sale and purchaser of unsold lot comprising seven flats – Owner of flats denying having authorised sale – Purchaser's claim against owner for specific performance failing – Auctioneers conceding claim for breach of warranty of authority – Quantum of damages – Appropriate valuation date

The second defendant, a firm of chartered surveyors and auctioneers, conducted an auction sale in which one of the lots comprised seven subtenanted flats in a large detached house. The first defendant was
C the registered leasehold proprietor of the flats. The lot failed to reach its reserve price of £670,000 and, immediately after the auction, a representative of the claimant approached the auctioneer with a view to purchasing it by private treaty. A sale was agreed at £670,000 and a memorandum of sale was completed, signed by the claimant and by the second defendant as agent for the vendor. The claimant paid a deposit of 10% to the second defendant.

The first defendant later denied having authorised any sale of the property. The sale did not complete on the contractual completion date and the first defendant did not respond to a notice to complete served by the claimant. The claimant brought proceedings against the first
D defendant for specific performance. By his defence, the first defendant alleged that he had not been a party to the sale agreement, which had in fact been made by a third party using his identity to obtain moneys by fraud. In view of that defence, the claimant joined the second defendant to the proceedings, claiming damages for breach of warranty of authority.

In the light of the first defendant's evidence at trial, the parties agreed that the specific performance claim should be dismissed and that judgment should be entered against the second defendant on the breach of warranty claim. An issue remained as to the quantum of damages. The parties accepted that the proper award was the value of
E the property, less the agreed price, and that the valuation date should be postponed for so long as the claimant had reasonably continued to seek completion of the sale. They differed as to the proper valuation date on the application of that approach.

Held: **Judgment was given accordingly. Damages for breach of warranty of authority should put the claimant in the same position as if the warranty had been accurate and the sale had been authorised by the owner of the property. The appropriate measure is therefore the value of the property, less the agreed price. Although the logical date for valuation is the agreed completion date, damages can be assessed at a later date where it is just to do. There is authority for the proposition that the valuation date can be postponed for so long as the claimant reasonably continued to seek completion of the sale;** F **the parties agreed on that approach in the instant case. The mere fact that completion is unlikely to be achieved does not make it unreasonable to pursue it. It had been reasonable for the claimant to continue to pursue its alternative claims against both defendants in all the circumstances. Accordingly, it was permissible to value the property as at the date of the trial, as agreed.**

Per curiam**: Where damages are awarded for breach of warranty of authority, the logical date for valuation is the agreed completion date. The claim for breach of warranty involves a hypothetical assumption that the owner of the property authorised the sale; on that assumption, the claimant would have received the property** G **on the contractual completion date, in the absence of evidence that the owner would have resisted or delayed completion for some other reason. An approach that postpones the valuation date for so long as the claimant reasonably pursued completion of the sale is problematic since it conflates the hypothetical with the actual situation, whereas it is the difference between those two scenarios that represents the measure of the claimant's loss.**

The following cases are referred to in this report.

Habton Farms v *Nimmo* [2004] QB 1
Suleman v *Shahsavari* 1988 1 WLR 1181; [1989] 1 EGLR 203; [1989] 09 EG 69

This was the hearing of an issue as to the correct date for the H assessment of damages payable by the second defendant, Strettons Ltd, to the claimant, Greenglade Estates Ltd, for breach of warranty of authority to sell property owned by the first defendant, Himat Singh Chana.

Mark Warwick (instructed by Kaye Tesler & Co) appeared for the claimant; Thomas Bell (instructed by CMS Cameron McKenna LLP, of Bristol) represented the second defendant.

Giving judgment, **Mr David Donaldson QC** said:

1. In the course of the first day of what was envisaged as a two-day hearing before me, this tripartite action collapsed to a single issue J relating to quantum as between the claimant and the second defendant. To explain the nature of that issue and how it arises, it is however necessary to describe at least briefly the proceedings prior to that collapse.

2. The second defendant (Strettons) is a large firm of London chartered surveyors and auctioneers. Lot 70 in one of its auctions held on 8 September 2009 comprised the long leases of seven subtenanted flats in a large detached house at 279 Church Road, Upper Norwood, London SE19. The registered leasehold proprietor of the flats was the first defendant (Mr Chana).

3. The claimant is the vehicle for a consortium of property investors, K including Mr Robert Steinberg, who attended the auction on the claimants behalf. Lot 70 failed to reach its reserve price of £670,000. Immediately afterwards Mr Steinberg approached the auctioneer directly with a view to a purchase by private treaty. Having referred back apparently to his principal, the latter told Mr Steinberg that £670,000 was the minimum acceptable price. The consortium took the view that the true market value was more like £1m and decided to accept the offer. A memorandum of the sale was completed, with Mr Steinberg signing for the consortium as Buyer and Strettons signing as agent for the Seller, and the deposit of 10% was paid to Strettons. The

A claimant then passed the matter to its conveyancing solicitors, Bude Nathan Iwanier (BNI).

4. On 25 September 2009, Mr Philip Waterfield, Strettons auction director, wrote to the claimant as follows:

> "We may have been in contact already but I am writing to all purchasers of a number of lots which were offered in our auction sale on 8 September. Hence my reason for contacting you as purchasers of the above property.
>
> Since the auction a third party has alleged that he owns this property (and others) and it should not have been offered for sale. This party has not provided any documentary evidence of identity or ownership but we are treating the allegations seriously and the matter has been reported to the police.
>
> B I confirm that Strettons are holding your deposit.
>
> Your solicitor, to whom we are copying this letter, may wish to con-tact the solicitor acting for the party who instructed us in the sale and whose contact details are:
>
> Elliott Costa, Taylors Legal…
>
> I am sorry that I do not have any further news at this stage..."

5. BNI was, however, unable to obtain any information from Taylors Legal. Completion not having taken place on the due date BNI served a notice to complete on 21 October 2009, but without response.

6. On 17 November 2009, Mr Waterfield wrote again to the
C claimant, saying:

> "As you will be aware there has been no resolution to the sale of the above property. Disappointingly, the solicitor acting for the seller advises that he has closed its files and the police do not seem to be interested in investigating. We hold your deposit but if the sale is not going to complete I would like to return it to you without delay.
>
> The issue I have is that it is not in Strettons domain to rescind the contract which would then place us in a position of having to return the deposit. However, bearing in mind the time which has elapsed since the September auction sale and the lack of recent involvement by the sellers solicitor, I would like to take the initiative by suggesting that, if you now consider the contract for your purchase to have been rescinded you should specifically request for the
> D deposit money to be refunded."

7. On 3 December 2009, the claimant issued the present proceedings against Mr Chana seeking an order for specific performance of the contract. No defence was served, and in July 2010 the claimant applied for summary judgment. At the hearing of the application Mr Chana appeared in person to object, and it was ordered that the case should continue as a Part 7 claim and that Mr Chana should serve a defence, which he did on 24 September 2010. The defence denied that Mr Chana was a party to any agreement for sale of the properties, and alleged that the contracted party was another party called Mr Karan Singh
E who used the identity of the defendant to obtain monies by way of fraudulent transaction. It further stated that it was Karan Singh and not the defendant who had engaged the services of Taylor Legal.

8. In reaction to these averments, the claimant joined Strettons as the second defendant to the action. Their defence was served on 28 April 2011. It took no position on whether Mr Chana was correct in his allegation that he had never authorised the sale, adding that Strettons had relied on representations by the sellers solicitors that Mr Chana was the seller and that he had provided ID to confirm his identity. They rightly did not suggest that such reliance would provide a defence to the claim for breach of warranty of authority.

F 9. At the trial before me, Mr Chana gave evidence, with some detail, that he had not authorised the sale, supported by testimony from one of his managing agents. It was a matter of indifference to the claimant whether it recovered judgment against the First or against the second defendant, and counsel for the claimant did not therefore challenge the evidence of either Mr Chana or his supporting witness in cross-examination. Nor did counsel for the second defendant. Both counsel then confirmed formally to me on behalf of their clients that they made no challenge to that evidence and did not ask the court to reject it and agreed that, accordingly, there should be judgment (1) in favour of the first defendant dismissing the claim against him, and (2) against the second defendant on the claim for breach of warranty of authority. It was also agreed that there should be an order for costs in favour of the first defendant but that the court should decide whether they should be G
paid by the claimant or the second defendant.

10. That left outstanding only the damages to be awarded against the second defendant for the breach of warranty of authority. On both principle and clear authority, such damages are for the fulfilment interest: they are intended to put the claimant in the same position as if the warranty had been accurate and the sale had therefore been authorised by Mr Chana. The appropriate measure is accordingly the value of the property less the agreed price. The claimant and Strettons are, however, in dispute as to the date at which the property is to be valued for this purpose.

11. Logic suggests that, absent evidence – not present here – that the H
true owner, if it had authorised the sale, would for some other reason have resisted and delayed completion, the relevant date should be the agreed completion date (in the present case, 30 days after the contract date). In *Suleman* v *Shahsavari* 1988 1 WLR 1181* Mr Andrew Park QC (sitting as a deputy High Court judge) did not, however, feel compelled to such a result. The vendors in *Suleman* were husband and wife, who refused to complete because they maintained that their solicitor had had no authority to sign on their behalf: in the event, the judge upheld that allegation as regards the husband. He therefore dismissed the claim for specific performance and gave judgment against the solicitor for breach of warranty of authority. The value of J
the house at the contractual completion date was £56,000, £9,500 more than the contract price; its value immediately prior to the hearing was £76,000. The judge pointed out that, on established case law, where the purchaser of a house who initially requires completion of the contract subsequently instead seeks or recovers damages for non-performance from the vendor, the damages may be assessed at a date other than the agreed completion date if it would be more just to do so, adding (at p1183D-E):

> "Where… there have been dramatic changes in property values, it may be more just to assess damages at a different date… That is particularly so where the innocent party reasonably continues to try and have the contract completed: K
> in such a case it is logical and just to assess damages as at the date when (otherwise than by his default) the contract is lost."

The judge then stated (at p1184C-D) that:

> "It would be extraordinary if the plaintiff was entitled to a lower amount of damages against the solicitor than he would have been awarded against Mr and Mrs Shahsavari, had he had a good claim against them.
>
> In *Suleman*, the attempts to obtain completion had continued, and reasonably so, up to the date of the judgment. On this basis, the judge ruled that the value of the house was to be assessed at the higher figure obtaining at the date of his judgment rejecting the claim for specific performance."

L

12. I confess to serious difficulty in subscribing to the judge's approach in *Suleman*. Since the claim for breach of warranty involves the hypothetical assumption that the sale *had been authorised*, it is on that basis that one has to consider when completion would have occurred: absent evidence that if the sale had been authorised by the owner completion would for some other reason have been resisted and delayed, the agreed completion date must represent the time when the purchaser would have received the property. No assistance can be obtained from the lapse of time during which in actual history completion was resisted and ultimately specific performance was refused because the sale was not in fact authorised. To M
do so is to conflate actual and hypothetical, whereas it is the difference between those two scenarios which represents the measure of the claimant's loss. The decision could be justified in the result if the evidence permitted the conclusion that *even if they had authorised the sale* the vendors would have refused completion up to the date of the hearing: since the first half of the judgment is not reported, it is, however, unclear whether or not this might have been the judge's view. That is not to say that some additional damages might not be recoverable where a claimant can demonstrate that in reliance on the representation of authority it refrained from investing the monies

---

* Editor's note: Also reported at [1989] 1 EGLR 203; [1989] 09 EG 69

A earmarked for payment of the purchase price in some similar property which would have benefited from the same market increase. Here again, given the abbreviated form of the report, it is unclear whether such a conclusion might have been available in *Suleman*; and no such evidence or argument was presented in the present case.

13. I am also far from convinced, notwithstanding a positive reference by Auld LJ to *Suleman*, that the point is foreclosed by *Habton Farms* v *Nimmo* [2004] QB 1, where the Court of Appeal was concerned with the position where the claimant was the vendor, and the property was a racehorse.

B 14. However that may be, in their submissions counsel for both parties concurred in inviting me to follow and apply the formula in *Suleman* by postponing the date of valuation for as long as the claimant had been acting reasonably in trying to have the contract completed. I shall proceed accordingly, despite a residual concern as to whether it is always right for the court to proceed on a particular view or formulation of the law simply because it has been agreed by the parties.

15. Counsel for Strettons, Mr Thomas Bell, placed emphasis in particular on the following matters:

(1) Strettons first notified the claimant of Mr Chana's allegation as early as 25 September 2009, in the letter from which I quoted at [4]
C above.

(2) On 7 January 2010, Strettons reported to the claimants' solicitors that Taylor Legal had advised them that the ID received from its client was counterfeit.

(3) On 18 May 2011, Mr Chana's solicitors sent to the claimants' solicitors a copy of Mr Chana's passport and of that which had been provided to Taylor Legal: manifestly these did not relate to the same individual. On this basis, they invited the claimant to confirm by return that it would now discontinue the claim against Mr Chana.

Mr Bell invited the court to rule that further pursuit of specific performance was therefore unreasonable after January 2010, or at latest
D May 2011.

16. For the claimant, Mr Mark Warwick objected that this analysis was inappropriately truncated. In particular:

(1) On 20 May 2011 the claimants' solicitors replied to the letter of 18 May 2011 pointing out that Strettons client apparently possessed detailed knowledge of the property sufficient to pro-vide auction particulars and stressing that Strettons had not accepted that their client was not the true owner. That letter was copied to Strettons E solicitors.

(2) On 5 August 2011, Strettons solicitors wrote that:

> "For the avoidance of doubt, our client has not admitted liability in this matter. As far as we are aware, it remains far from clear that first defendant (sic) was not the individual whom (sic) instructed our client."

This position, described by them later – and perhaps somewhat tendentiously – as one of neutrality, continued up to the trial.

17. While agreeing that the claimants pursuit of the claim against Mr Chana was to be assessed against the criterion of reasonableness, counsel differed in how that test was to be applied. Mr Warwick F argued that it was reasonable for the claimant to continue to seek specific performance until there was no longer any real prospect of success, in effect of proving that Mr Chana had authorised the sale. In counter, Mr Bell submitted that the pursuit of specific performance was unreasonable once it was unlikely that it would be obtained. These competing formulae are both restrictive glosses on the basic test of reasonableness, which is itself a gloss placed in *Suleman* on justice. As such, though in fairness to counsel they were advanced only in response to questioning by me, they are in my view to be viewed with reserve and arguably as illegitimate, not least because they exclude many other factors which might be relevant in a particular case. Having said that, G I consider that so far as it goes Mr Warwick's test is a closer guide to unreasonableness than Mr Bell's: mere unlikelihood without more is insufficient to render the pursuit of completion unreasonable. And in the present case no more was relied upon.

18. From the time the passport copies were disclosed in May 2011, few would have placed heavy money on the claim against Mr Chana succeeding. That outcome was not however in any way certain, indeed Strettons solicitors themselves described it as far from clear that Mr Chana had not authorised the sale, and Strettons were never prepared to concede the point and accept liability. In these circumstances, I consider that the continued pursuit of the alternative claims against both defendants was reasonable, and that accordingly – H in what I consider a preferable formula if *Suleman* is to be followed – it would be just to value the property as at the date of the trial. I am asked to refer the determination of the appropriate figure to the Master, in the event that it cannot be agreed by the parties.

Judgment given accordingly.

SUPPLEMENT

CASES NOT REPORTED IN "ESTATES GAZETTE"

DEFECTIVE PREMISES

A
Drysdale v Hedges

QUEEN'S BENCH DIVISION

[2012] EWHC B20 (QB)

27 July 2012

Mr John Leighton Williams QC, sitting as a deputy High Court judge

Defective premises – Duty of landlord – Personal injury – Defective Premises Act 1972 – Occupiers' Liability Act 1957 – Tenant slipping on steps leading to demised property and suffering back injury –
B ***Whether tenant entitled to damages from landlord – Whether landlord in breach of statutory duty or common law duty of care***

The claimant took a tenancy of a mid-terrace Victorian property that was arranged over three floors, including a basement level. The property had a concrete path with three stone steps leading up to the front door. The steps had been painted with red floor paint at the instance of the defendant landlord. To the left of the path was a small wall, which separated the path from a drop of more than 8ft where the property was stepped down to the basement level. On the first day of the tenancy, while moving her possessions into the property in wet weather conditions, the claimant fell from the painted steps into the basement
C area and suffered a severe back injury. In a claim against the defendant for damages for personal injury, she complained that the paint on the steps had made them unduly slippery in the wet weather and that the wall at the side of the steps was not high enough to prevent someone from falling down to basement level.

The claimant contended that the defendant was liable for injury resulting from those matters by reason of: (i) section 2 of the Occupiers' Liability Act 1957; (ii) a breach of the landlord's repairing covenant in the tenancy agreement and/or section 4 of the Defective Premises Act 1972; and (iii) a breach of a common law duty of care.

Held: **The claim was dismissed. (1) The general duty of care**
D **imposed by section 2 of the 1957 Act had no application to the case. A landlord's duty in tort is defined primarily by section 4 of the 1972 Act, which replaces the similar duty under section 4 of the 1957 Act; accordingly, a landlord's duty of care will normally be confined to the duty set out in that section. The section 4 duty arises by reason of the landlord's obligation or right to repair the demised premises. In order to show a breach of the landlord's repairing covenant and/or section 4, the claimant had to show that the premises were not in good repair. Repair involves making good any damage, so as to leave the property, as far as possible, as though it had not been damaged. A duty to repair cannot be equated with**
E **a duty to make safe and a duty to keep in good condition cannot encompass a duty to put in a safe condition. A tenant must take a house as she finds it; neither a landlord nor a tenant is bound to provide the other with a better house than was there to start with. There was no evidence that the wall, at the time of the accident, was any different from when the defendant had acquired the property; therefore it could not be said that the wall was not in good repair. The presence of the paint on the steps did not result in the steps not being in good repair. The stone steps themselves were in good repair. The paint did not replace the stone but was additional to it. The paint required to be removed, not repaired. It would be stretching the meaning of the word "repair" to apply it**
to the removal of the paint. Consequently, there was no breach of F
the landlord's repairing covenant or section 4 of the 1972 Act. (2) Nor was the defendant liable for breach of a common law duty of care. A landlord who lets premises in a dangerous condition owes no duty to remedy the defect and no duty of care to a third party injured as a result of the defect. The defendant was therefore under no duty to guard the drop to basement level. A landlord does owe a duty of care at common law to take reasonable care not to create an unnecessary risk of injury; that duty is owed even where there is no disrepair and section 4 of the 1972 Act does not apply. The defendant was therefore under a duty to take care that the
application of the paint to the steps, done at her instance after she G
acquired the property, did not create an unnecessary risk of injury. However, on the evidence, the defendant had not breached that duty. It was not unreasonable of her to apply paint that was said to be suitable for outdoor use and that contained no warning against use on steps.

The following cases are referred to in this report.

Alker v Collingwood Housing Association [2007] EWCA Civ 343; [2007] 1 WLR 2230
Anns v *Merton London Borough Council* [1978] AC 728
Anstruther-Gough-Calthorpe v *McOscar* [1924] 1 KB 716
Boldack v *East Lindsay District Council* 31 HLR 41 H
Cavalier v *Pope* [1906] AC 428
Donoghue v *Stevenson* [1932] AC 562
Dutton v *Bognor Regis Urban District Council* [1972] 1 QB 373
Greene v *Chelsea Borough Council* [1954] 2 QB 127
Lips v *Older* [2004] EWHC 1686 (QB); [2005] PIQR P14
Murphy v *Brentwood District Council* [1991] 1 AC 398
Quick v *Taff Ely Borough Council* [1986] QB 809
Rimmer v *Liverpool County Council* [1985] QB 1
Sowerby v *Charlton* [2005] EWCA Civ 1610; [2006] 1 WLR 568
Targett v *Torfaen Borough Council* [1992] 3 All ER
Wheat v *E Lacon & Co Ltd* [1966] AC 552

This was the hearing of a claim by the claimant tenant, Gillian J
Drysdale, against the defendant landlord, Joanne Hedges, for damages for personal injury sustained at the demised residential premises.

Niazi Fetto (instructed by Fentons Solicitors LLP) appeared for the claimant; Michael Pooles QC (instructed by Greenwoods Solicitors LLP) represented the defendant.

Giving judgment, **Mr John Leighton Williams QC** said:

1. This is a trial of the issue of liability in a claim for damages for
personal injuries arising out of a fall suffered by the claimant, then 40 K
years old, on 5 October 2008 when she was ascending a flight of three steps leading to the front door of 3 Cannon St, Harwich, Essex ("the property").

2. The property, owned by the defendant, is a mid-terraced Victorian house on three floors and consists of a basement, ground floor and a first floor. It is one of many similar properties in the terrace, all of which have steps, usually three steps, leading to the front door. Similar houses on the opposite side of the road tend to have more steps due to the different topography. Photographs show the property and neighbouring properties. Some of the photographs of the property were taken about

six weeks after the fall and some a little later after a wooden handrail had been fitted next to the steps.

3. Access to the front door was via a concrete path leading to the three steps. The steps are about 1m wide. The depth of the bottom two steps appears to be about 11.5in but the top step leading to the front door is less deep. At the instance of the defendant, who had acquired the property in 1998, the steps had been painted with red paint to improve their appearance.

4. To the left of the path a small area has been stepped down to basement level. This stepped area is separated from the concrete path by an ornamental brise soleil block concrete wall, typical, I am told, of the fashion of the 1970s. This wall has a continuous height of 950mm above path level along its length and continues alongside the left of the three steps to the front of the house. When it reaches the steps it is not raised so as to run parallel with the rise of the steps but the top level of the wall continues horizontally to the front of the house. Thus the height from the middle step to the top of the wall is 463mm, about 18in. To the left of the wall at the steps there is a drop of 2.524m (over 8ft) to basement level. The wall thus offers very limited protection against a fall from the steps down to basement level.

5. In June 2008, the claimant was living in Scotland. She was looking for premises to rent and contacted Mr Cundall of Concord Property Management ("Concord"), who managed the property for the defendant. Mr Cundall states the claimant visited the property in about late August/early September 2009. Terms of a tenancy were agreed and it was further agreed that she would sign and provide a copy of the tenancy agreement on her arrival in Suffolk. She was provided with keys to the property in advance of signing the tenancy, no doubt because the statutory services had agreed to pay the rent. The tenancy agreement was signed by the claimant on 8 October but backdated to 5 October.

6. On 4 October, the claimant arrived in Harwich and spent the night at the property. Her fiancé, Mr Stuart arrived with a van load of possessions on 5 October and she and Mr Stuart set about carrying their possessions into the house. 5 October 2008 was a wet day and it was raining when the claimant's accident happened.

7. The claimant's account of the accident has not been in issue. As she and Mr Stuart were jointly carrying a large box up the front steps to the property, the claimant leading and walking backwards and Mr Stuart taking the other end of the box, her right foot slipped when on the middle step and she fell to her right and over the wall alongside the steps down to basement level. She suffered serious back injuries.

8. On the claimant's behalf it is alleged that the steps were "unduly slippery" as a result of being painted, being wet and because of the presence of dirt. Of these allegations it is the combination of paint and wet which were originally said to be the real villains. The presence of dirt was added by amendment at the outset of the hearing.

9. In addition, it is alleged that the wall at the side of the steps was not sufficiently high and/or was not provided with a guardrail or handrail to prevent someone falling down to basement level. It is said that the presence of the inadequately guarded drop to the basement made it all the more important that the steps should not be slippery. It is not suggested that the defendant had altered the wall in any way during the period of her ownership of the property.

10. It is alleged that these failures gave rise to breach of section 2 of the Occupiers Liability Act 1957, breaches of a repairing covenant in the tenancy agreement and section 4 of the Defective Premises Act 1972, and breach of duty at common law. The claimant relied on her own evidence and that of Mr Stuart, her mother Mrs Porter, and her son Stephen, the latter of whose witness statement was agreed. She also relied on hearsay evidence of conversations between her then next door neighbours, Ms Kellie (or Kelly) Tyler and Mr Mark Tyler and a solicitor, Miss Lankester who then prepared statements for Ms Tyler and Mr Tyler, which were never signed. Miss Tyler's and Mr Tyler's whereabouts were not known.

11. The defendant relied on the evidence of the defendant and her husband, Mr Stephen Hedges, Mr Cundall the letting agent, and a witness statement taken from the same Miss Tyler, which conflicts with Ms Lankester's account of what Ms Tyler told her.

12. Both parties relied on expert evidence: the claimant on Mr Highfield, a chartered surveyor; the defendant on Mr Butcher, a structural and consulting engineer.

The claimant's lay evidence

13. In addition to giving her account of how the accident happened, the claimant in her first witness statement said it was raining heavily, the steps had become very slippery and she had been wearing trainers. She said that Mr Stuart had warned her that the steps were slippery.

14. In her second witness statement she said that she had not met Miss Tyler until after she (the claimant) had been discharged from hospital, that when Miss Tyler used to visit her she would ask Mr Stuart to carry her eldest daughter up the steps "because she said she found the front steps slippery" and that Miss Tyler mentioned that she had slipped on the front steps a few times. She also said that having seen Mr Butcher's report, which mentioned the possibility of contamination, nothing had been spilled on the step but she remembered that the step was dirty and had a black smudge by the wall which could be seen on the photographs.

15. In evidence she confirmed what she had said in her statements. She said she had been sent photographs of the house by Mr Cundall, knew what it looked like and that it was old. She had ascended the steps one foot at a time. She agreed with suggestions that as it was raining heavily the path had puddles in it and was slippery and that steps could be slippery when wet. She did not know what had caused her to slip. She accepted that in most of the neighbouring houses the height of wall in relation to the steps was roughly the same as hers. She said she had not considered the steps to be dangerous at the time – "It did not seem obviously dangerous at that point."

16. She was a very straightforward, reliable witness doing her very best to tell me the truth.

17. Mr Stuart was also a straightforward witness doing his best to tell me the truth. In his statement he had described the rain as drizzle. In evidence he said it had been dry at first, then drizzle, then heavy rain. Both in his statement and in evidence he said he had earlier slipped on the middle step when carrying a box up the stairs and had dropped the box. As a result he had warned the claimant that the steps were slippery. He said that in the aftermath of the fall, when paramedics were present, one of them said he had been to the premises before when an old lady had slipped on the same steps. This is something which Ms Tyler also mentioned to Ms Lankester, stating the lady had fallen over the wall.

18. Mr Stuart said he had continued to use the steps but the claimant, Mr Stuart and Mrs Porter told me that following the accident they tended to use the rear entrance. If Mrs Porter was going out via the front door Mr Stuart said he would accompany her down the steps. In the claimant's case because she has used and uses a walking stick or sticks I assume this may have been because using the rear access was easier.

19. He too said that prior to the accident he had not taken notice of the steps. He said you could see it was a low wall alongside the steps but there was nothing unusual about that. He had not paid much attention to other houses in the street. He said that a neighbour had built up the wall of his house at that point as he did not want his children to fall over but it had not been high enough to prevent an adult going over. The neighbour would appear to have been Mr Tyler.

20. Mrs Porter gave supporting evidence. She is now getting on in years and was willing to agree with almost everything Mr Pooles QC for the defendant put to her. I do not find that her evidence adds to what others have said. Similarly Mr Stephen Drysdale, who had witnessed the accident and whose evidence was read.

21. I found Ms Lankester a reliable witness who I am sure took care in recording what Ms Tyler and Mr Tyler told her and did not put anything into their mouths. I regard the statements she prepared for them as fairly representing what she was told by each of them.

22. Ms Tyler's account as given to Ms Lankester on 13 January 2011 and recorded in her unsigned statement differs from the statement she later provided to the defendant. She expressed the opinion to Ms Lank-

ester that the steps were unsafe, said she had gone out of 3 Cannon St and nearly fallen over the steps – "There is nothing to step on" and that she had nearly slipped and gone over in the same way as the claimant, it had happened to her about two or three times since the claimant's accident. She suggested the steps were not properly maintained because the paint was coming off and stated that Mr Tyler was forever painting the steps to make sure it was "secure". She stated they had used non-slip black paint on their steps. She said a lady had fallen over the wall eight years ago but, it seems, gained this information from a paramedic who attended the claimant.

23. By 2012 Ms Tyler was living in 3 Cannon Street, having moved from the nextdoor property in March 2011, two months after her conversation with Ms Lankester. Her statement for the defendant is signed and undated with the front page allowing for insertion of a date in 2012. So by this time she was the defendant's tenant. In her statement she says that the steps have been repainted once since she moved in, in April 2012, that no one has slipped on the steps during the time she has been in the property, that she has used the steps in very wet weather and in snow and not found the steps to be slippery and has had no difficulty using them. She makes no mention of the lack of guarding to the steps. The statement appears to me to have been drafted with great care to confine its observations on relevant matters to what has been her experience since March 2011.

24. Mr Tyler told Ms Lankester, that when he moved in to No 1 there was a picket fence to the right side of the steps, alongside a drop to basement level, which he replaced with a wall. He had also extended the steps since when double-glazed doors had been put in they had reduced the depth of the steps by half and the top step was not deep enough. He had installed the double glazing so the new door, brick wall and extended steps had all been done in one go. He was critical of the steps at No 3, but his criticism appears to have been directed to the restricted depth of the third step.

The defendant's lay evidence

25. The defendant and her husband both gave evidence. In her statement dated 15 August 2009, the defendant said the property had been purchased with a view to renting it. Prior to the claimant taking up her tenancy the property had been empty for about nine months. They had been unable to sell it. During that nine months they had inspected the property from time to time. She said her husband was a builder and did the maintenance. Over the years they had modernised the property internally and installed new windows and doors. They kept no records of inspections or repairs. They had not altered the front elevation nor the steps, apart from painting the steps "with appropriate concrete floor paint".

26. She said that when she acquired the property the step was "just concrete". She had asked her husband to paint the steps to improve the appearance. She said the steps were painted with red floor paint obtained from B&Q. "It is not a gloss paint. It is specifically a floor paint described on the tin as 'ideal for concrete floors'." She said they had always applied the same paint and Steve had repainted the steps every year and that the step had last been repainted in December 2007 or January 2008, when the last tenant left. She said that following the accident and on advice she added a handrail and the steps were repainted. She said no previous accidents had been reported, that she had not carried out written risk assessments but "obviously when periodically visiting the property, I together with my husband Steve, have ensured that it is properly maintained". Following the accident and on advice she had added a handrail and the steps were repainted.

27. Cross examination proceeded on the basis that ascending steps involved a greater risk of slipping/falling than walking on steps, which risk may increase in inclement weather because the steps were uncovered, to which she responded that she would presume people would not fall in the first place. She did not agree that the wall was not high enough for protection if someone fell but agreed that care would ensure that the application of paint did not increase the risk of someone falling but thought at the time that the paint had non-slip properties then agreed that she had not thought about the non-slip quality at the time. Its use had initially been cosmetic.

28. She did not agree that applying the paint, whose qualities as set out she was asked about, gave rise to an obvious risk that its application would have made the steps smoother, stating that having seen the paint on the steps they had not appeared slippery or smooth, agreed with the proposition that applying gloss paint would increase the risk of slipping but denied that that dictated the need to put up a guard rail to make the entrance safer. Before a new tenant they cleared out any rubbish. She could not remember whether they had ever examined or cleaned the steps. She did not agree that the risk of going over the edge of the steps was obvious and said that if she had thought that there was a risk she would have had something done but she had never thought of it. She accepted that if someone fell over the edge he might suffer serious injury.

29. Mr Hedges in his short witness statement did not describe his occupation but stated that after his wife purchased the property he did a lot of internal alteration work, replacing flooring, the kitchen and bathroom and redecorating the property. He added "internally the property was in very good order". No one has suggested to the contrary.

30. Others fitted the new doors and windows. He painted the steps with what he described as "a concrete non-slip red floor paint" and repainted on average once a year. He recalled last painting the step (prior to the accident) in December 2007.

31. He said there had been no accidents or complaints since they had owned the property (the defendant said the property had been transferred to joint names after purchase). He said that following the accident he had been asked to repaint the step and put up a wooden handrail. His statement makes no mention of the wall and the drop alongside the steps.

32. In evidence he said he maintained the property. Prior to late 2007/2008 he had painted the steps once or twice a year depending on how they had weathered. He suggested that as the letter box could be reached from the lowest step that step wore the most, as photographs suggest. When he repainted the steps he used to clear the flaked paint and then lightly sand the remaining paint to provide a key and then repaint the steps. He said the steps had not changed: he accepted that the top step was less deep then the lower two steps but said that the new door had been placed in the same "reveals" as its predecessor.

33. He had only used one paint, ie the B&Q paint which he thought was a non-slip paint, and had applied one coat on each occasion. In addition to maintaining his own properties he maintained property for others, used that paint on boat decks and farm buildings, and had never had the present problem. He accepted the paint tin label did not say "non-slip" and thought one may be less likely to slip on non-gloss paint. He denied that the paint was "semi-gloss" but had to accept that the tin said it was. He had erected the handrail at the request of the agent, otherwise he thought he would not have done so. He also stated that if he had known someone was going to fall over the steps he probably would have erected a handrail but the chance never really entered his head. He agreed that the steps had not been a problem before he painted them and said he had not seen any problem in putting paint on the steps. They had received no previous complaint about the steps whether in their painted or unpainted form. He accepted the obvious, namely, that the cost of erecting a guardrail would have been the cost of materials and his labour. He saw the purpose of the guardrail he had erected as twofold; to accommodate the claimant's disability and to protect against a fall.

34. Mr Cundall's witness statement said that when he visited the property the steps did not cause him any concern and had the surface done so he would have so told the defendant. Following the accident he recommended installation of a handrail.

35. In evidence he said that following the accident he recommended the handrail because of the claimant's condition and because it "would do no harm" to have a handrail there. He said that the risk of falling down to basement level was not a risk which he or anybody else had considered to be a risk prior to the accident. In this respect it was no

different from any other old house. He said: "My concern is to let: people take it as it is" or similar words.

36. As far as I am aware, Mr Cundall had no qualifications as a surveyor, engineer, in health and safety, or any field which involved considerations of the safety and structure of the premises he was letting. If he did, I was not told of them. He told me that this was the only house they represented in the terrace. His views were, I conclude, based on what he considered to be common sense rather than expertise.

37. It was quite evident that no formal risk assessment was ever carried out. Having heard Mrs Hedges, I consider she knew very little about the duties a landlord may have in law to her tenants. I do not mean by this to suggest that she was an irresponsible landlord. She could not be described as a professional landlord. Rather she was someone who had purchased a small property as an investment and then sought to make it acceptably habitable. It was also clear to me that until after the accident neither the defendant nor Mr Hedges had addressed their minds to either the danger resulting from the unguarded drop or any safety consequences of painting the steps.

The expert evidence

38. Mr Highfield came in for heavy criticism from Mr Pooles QC for the defendant for expressing opinions which were not based on expertise. In some respects Mr Poole's criticisms are justified namely that Mr Highfield expressed views about slipping risk and the slip potential of the B&Q paint used without having expertise in those subjects. His report appears to me to be a document suggesting lines of enquiry to his instructing solicitors, bolstered with a sympathetic view for arguments favourable to the claimant, rather than a truly objective report. Indeed he appeared to accept that that was how he perceived his function. What he highlighted appear to be more matters of common sense and experience rather than expertise.

39. In his report, he said he visited the property on 14 December 2009. He referred to building regulation requirements for fencing and similar places, the provisions of the 1972 Act, and British Standards for guarding balustrades. He commented on the general standard of maintenance of the property and that the guardrail subsequently provided was "somewhat basic and considerably lacking in refinement". He observed that the steps had been painted with what "at first sight appears to be a proprietary concrete floor paint". He said: "It was difficult to understand how it could be non-slip as alleged because there were no signs of non-slip additives in the paint and it was difficult to understand how a non-slip finish could be achieved with a chemical additive." He observed:

> "it is also possible, that when dry, the paint might by a lay person, be considered to be firm and non-slip. However, when wet, the material can easily become slippery."

He also referred to the possibility of the presence of green algae from reference to photographs.

40. He took a number of measurements, about which there is no dispute. He said it was reasonably obvious that no protective guarding was available to prevent a fall over the wall down to basement level. As to the condition of the step he observed that decorative paint had been applied to what appeared to be stone steps and:

> "The material (ie stone) usually is, if properly maintained by the householder, reasonably anti-slip. The provision of a painted surface would alter the characteristics of the step. In wet weather, the paint could easily become very slippery. It would also make the surface of the step impervious and subject to green algae colonisation. Again it would become very slippery. It is my understanding that it was raining at the time of the accident."

41. He regarded the condition of the steps and wall as "an accident waiting to happen" stating that problems relating to the step and lack of guarding should have been immediately recognised.

42. In response to instructions that he should look at neighbouring houses, he commented that few had handrails leading up to the front door from the path guarding the light well and that in almost every case the rails, walls or fences abut at a level height to the house wall.

43. In evidence in chief, Mr Highfield said that his expertise did not extend to the science of slipping. As part of his investigation Mr Butcher had commissioned RSK to assess the slipperiness or otherwise of the steps. RSK provided two reports/certificates dated 4 August 2011 and 18 April 2012, respectively. Mr Highfield said that when he had had a telephone discussion with Mr Butcher with a view to providing a joint statement, he had not seen the RSK certificate dated 18 April 2012, but saw it on the day he signed the report and took it at its face value, being heavily involved in something else at the time. He had sent an HSE information sheet entitled "Assessing the slip resistance of flooring" to the claimant's solicitors to assist their understanding. He had downloaded the software referred to in the document, entered the information requested and produced the resulting print-outs which described the risk of slipping on different surfaces, eg natural stone when wet ("medium"), natural stone when wet and no cleaning ("significant"), smooth painted lightly contaminated with water ("significant").

44. In cross-examination he accepted that the slip risk result produced by use of the software might be strongly influenced by the material put in, which was subjective. He said he did not know whether the various entries were weighted. For each surface under "surface usage" he had entered "Pedestrians carrying loads. Rushing pedestrians" as he would have expected the claimant to be rushing to get out of the rain. He did not know what weight the programme assigned to each entry.

45. He accepted he was not a paint expert but agreed with the suggestion that if you wanted to paint steps the B&Q paint was the sort of paint you would use. He said that when he prepared reports he would mention whether steps had been painted and if so that there was a risk of slipping but could not say that the general practice of surveyors. He said he was not suggesting it was a defect but that there were consequences to painting steps. Most surfaces became slippery when wet.

46. Of the B&Q paint, he said there was nothing on the tin to say it was non-slip, as far as he knew it might have been non-slip but he did not think it was and agreed it was a reasonable choice for a household. He said he could not say, as an expert, whether the steps without paint would have been more or less slippery, by which I understood him to be saying that he was not basing this view on any expertise.

47. In his joint statement with Mr Butcher they had agreed that:

> "It was not unreasonable for Joanne Hedges to apply or have applied B&Q paint that can be used externally and is stated as suitable for concrete, tile and stone floors, to the steps."

But he had qualified this statement by stating that it was sold as a floor sealant that could be used externally "but not specifically on steps" and by stating that when used on steps "of either materials" it would be desirable to incorporate a non-slip material such as sharp sand or carborundum grit and that both or either would significantly increase the slip resistance of the steps. He also said that he considered the paint used was ordinary quality paint with no specifically inherent non-slip qualities.

48. In re-examination he said he did not know whether B&Q sold non-slip paint. Although personally he would not have painted the steps, it was perfectly reasonable for the claimant to have painted them but if he had done it he would have incorporated a non-slip material.

49. Mr Butcher, in his revised report dated July 2012, stated he had visited the property on 14 July 2011. He agreed the dimensions given in the particulars of claim. He said the step showed negligible signs of wear given that it was over 100 years old. The steps had been painted and the paint surface had worn off where regularly used. The steps were wet when he visited. He stated:

> "In my opinion, there is no reason to consider that the steps of No 3 can be considered as being in anything other than a good state of repair."

50. He had photographed and exhibited the steps on other houses on both sides of the road and observed that the steps to the properties had

A over the years been maintained or replaced in a variety of materials; likewise the fences/walls leading to each upper ground floor entrance.

51. He observed that the Health and Safety Executive ("HSE") had produced a slip potential model setting out the factors relevant to risk of slipping. Controllable factors were environment, floor material and contamination. Predictable factors were use, behaviour and footwear. He did not deal in detail with the HSE publication containing this model, which I shall soon refer to.

52. The slip resistance of the steps was, on his instructions, assessed by RSK STATS, whose conclusions were before me. Lack of space prevented them using a pendulum tester but measurements of surface B roughness were taken using a surface roughness meter and the results had then been processed by RSK STATS. He commented on their findings, noting that "provided the steps had a degree of roughness (Rz) greater than 20 there was no reason to consider why the painting of the steps was unacceptable from a safety point of view" and pointed out that in this instance the surface roughness had varied between 24 and 56. He thought comparison of the slipperiness between steps in public areas and these steps could not be made because of the variables involved.

53. He noted that the B&Q paint, described as a floor paint, was said to be "suitable for interior and external use", "ideal for concrete floors", C "dries to a semi-gloss" and "hardwearing and durable". Further, that it was said to be "suitable use on properly prepared tiles, brickwork, wood, concrete and stone surfaces" and gave "an attractive sheen finish". Paint testing had confirmed that paint on the steps matched paint from the B&Q tin provided by the defendant.

54. The HSE publication was that which Mr Highfield had referred to. The version before me post dates the accident. It looks at test methods for assessing floor slip resistance and describes their use. It says it is aimed at employers who need to perform accurate measurements of floor slipperiness but will also help employers and other duty holders to assess slip risks in workplaces. It points out that D most slipping accidents occur when floor surfaces are contaminated by water.

55. The HSE methodology is based on the use of a pendulum tester which measures the coefficient of friction, HSE's preferred method of slipperiness assessment, and use of a surface microroughness meter. It refers to Rz as "a measure of total surface roughness, calculated as the mean of several peak to valley measurements". It suggests that where only roughness data is available it should be used with the Slips Assessment Tool (SAT), a software package, which can be downloaded. Using UKSRG (UK Slip Resistance Group) guidelines, it assesses Rz as follows:

E Below 10: High
10-20: Moderate
20+: Low.

The software package is intended to help users carry out assessments on horizontal surfaces.

56. The 2011 version of the UKSRG guidelines states that microroughness measurements should be used in conjunction with pendulum test results wherever possible and cannot be considered to be an independent test of slip resistance.

57. In evidence, Mr Butcher said he was not an expert in paint. He had not been instructed to comment on the presence or absence F of a guardrail but in evidence agreed that the drop was effectively unguarded, that an 18in wall would not have been adequate, opinions he expressed as a layman, although he had had training in health and safety.

58. He accepted that the presence of paint on the steps made their surface less rough but said that did not mean the surface was unsafe. He said the 20-micron figure was intended to apply to commercial properties. He knew of no equivalent for domestic properties. He suggested that what was unacceptable for horizontal surfaces would not apply to steps because gait was different on steps. On a horizontal surface slippage occurred because of application of a horizontal force – the heel landing before the sole, whereas on steps the force was vertical.

59. He was taken through the RSK findings. The 4 August 2011 G certificate contained the following table:

Location	**Condition**	**Mean roughness**	**Slip potential**
Middle step right	Degraded paint	24	Low
Middle step centre	Exposed concrete	56.3	Low
Middle step left	Degraded paint	36.1	Low

This was to be compared with the 18 April 2012 certificate, which stated:

Location	**Condition**	**Mean roughness**	**Slip potential**
Middle step right	Repainted – good condition	19.5	Moderate H
Middle step centre	Repainted – pitted and scratched	13.6	Moderate
Middle step left	Repainted – good condition	16.1	Moderate

It can easily be seen that the reading for the middle step centre in April 2005, after the application of paint but with some wear, is significantly lower than the degraded paint readings in August 2011.

60. Mr Fetto for the claimant delved more deeply into these figures. The mean figures were from readings taken from nine positions on the step. The readings for the centre, right and left of the step in April 2012 each had measurements below 10 microns in some places. What it comes down to is, not only that the presence of paint increased the J risk, as measured by these tests, but could have increased it to an unacceptable level on parts of the step. Mr Fetto then went one stage further and referred to photographs taken about six weeks after the accident which clearly show some, but not complete, wear down to stone at the centre of the middle step but that to the wall side of the step less wear appeared to have taken place. Thus inroads were made into Mr Butcher's observation that the reading was above 20 microns.

Overview of the facts

61. As I have already stated, I found the claimant and Mr Stuart reliable witnesses. I accept that Mr Stuart slipped when ascending K the steps and warned the claimant that the steps were slippery. No suggestion has been made that either of them was acting carelessly. There are no allegations that the claimant was contributorily negligent. I therefore proceed on the basis that she was taking due care and wearing acceptable footwear.

62. The defendant and Mr Hedges were cautious giving evidence, concerned, I suspect, not to say anything which would be damaging to their interests. But I am satisfied that they told me the truth. I would question Mr Hedges' evidence in one respect only and that is whether or not installation of the new front door had the effect of narrowing the step. But I heard no evidence of whether or not earlier door was itself L a replacement. The narrower top step was in any event not causative of the accident.

63. I did not find the expert evidence of Mr Highfield and Mr Butcher helpful, save in the most general way. Neither was an expert in paint. Neither, I consider, was an expert in slipping. Mr Highfield acknowledged this. Mr Butcher had some, but I consider limited, expertise in slipping. He relied heavily on the RSK findings.

64. Although it was Mr Butcher who appeared to have introduced reference to the HSE publication, criticism was advanced of Mr Highfield's reliance on the results of applying the model software he had downloaded on the basis that material inputted was subjective M in nature: yet the model required inputting of material which by its nature required subjective assessment. On the other hand Mr Butcher was criticised for applying the Rz measurements independently of the pendulum test and using them to assess slipperiness when the HSE publication advised against this.

65. What I am satisfied about is, as the RSK results suggest, that the application of the B&Q paint did reduce the Rz, and increased the slipperiness of the steps. Whilst, as I understand it, the risk of slipperiness is assessed on the assumption that the surface is wet, I cannot think that the presence of heavy rain did anything other than increase the risk further. Thus I find support for Mr Highfield's "amateur" view that painting the steps increased the risk of slipping,

especially in the wet. It fits the fact that both the claimant and Mr Stuart slipped that day.

66. I find the following.

(1) When the defendant acquired the property the brise soleil wall was already there and that it had probably been erected in the 1970s to replace either a wall or fencing which had previously been there. The wall, at the time of the claimant's accident was in effectively the same condition as when the defendant purchased the property.

(2) There was a dangerous drop from the middle step to the basement which was inadequately protected by the wall that was present.

(3) Such a drop would not have been unusual at the time the house was built but by the time of the accident a reasonable man, not least a reasonable landlord, ought to have realised known that the drop was dangerous and either raised the wall or provided a guardrail to protect those using the steps.

(4) When the defendant acquired the property the stone steps were unpainted.

(5) The defendant decided at some stage after acquiring the property that the steps should be painted to improve their appearance.

(6) Thereafter they were painted approximately once a year by Mr Hedges.

(7) Mr Hedges last painted the steps in December 2007 when the last tenant left.

(8) Thereafter the property was unoccupied until 4 October when it was first occupied by the claimant who slept there that night.

(9) Between December 2007 and 4 October 2008 the steps were little used.

(10) Painting the steps had increased the risk of slipping on the steps, particularly when the steps were wet.

(11) Neither the defendant nor Mr Hedges considered that painting the steps would increase the risk of slipping.

(12) Mr Hedges, acting on behalf of Mrs Hedges and with her authority, decided to apply the B&Q paint to the steps.

(13) There was no previous history known to the defendant or Mr Hedges of anyone slipping on the steps nor had there been any complaint to her or Mr Hedges about the slipperiness of the steps.

(14) The steps could fairly be described as "dirty" but whether or not this affected the slipperiness of the steps on 5 October 2008 is speculative.

Liability

68. The claimant invites me to hold the defendant liable for:

(i) breach of section 2 of the Occupiers Liability Act;

(ii) breach of covenant and/or breach of section 4 of the Defective Premises Act 1972;

(iii) breach of duty at common law.

Breach of the Occupiers Liability Act

69. Section 2 of the Occupiers Liability Act provides sets out the common duty of care, namely:

> "…a duty to take such care as in all the circumstances of the case is reasonable to see that the visitor will be reasonably safe in using the premises for the purposes for which he is invited or permitted by the occupier to be there."

70. The particulars of claim had alleged a breach of this duty should it be the position that at the time of the accident the tenancy had not commenced. At the outset of the case Mr Fetto for the claimant had been minded to withdraw this allegation but retained it at my suggestion. In his closing submissions he made a short submission that the defendant was an occupier of the steps on the basis that whilst technically it might be said that the claimant was the occupier of the steps, they being part of the demised premises, it was unreal to suggest that she was the sole occupier. As he put it: "She could not take control of the steps in any meaningful way." He pointed out that under clause 2.32 she could not carry out any decoration to the premises without prior consent. As a matter of fact it is clear that it is the defendant who has always painted the steps.

71. There may be more than one occupier of premises and people may be occupiers for different purposes: *Wheat* v *E Lacon & Co Ltd* [1966] AC 552. By clause 3.2 the defendant as landlord had to "provide and maintain the structure and exterior… in good repair". No doubt she would have justified her right to paint the steps by reference to this provision in the lease, as at least indicative of her authority to do so.

72. Section 4(6) of the 1957 Act had stated:

> "Nothing in this section shall relieve a landlord of any duty which he is under apart from this section."

Section 6(2) of the 1972 Act states:

> "Any duty imposed by or enforceable by virtue of any provision of this Act is in addition to any duty a person may owe apart from that provision."

It might therefore be argued that these provisions have not precluded reliance on breach of section 2 of the 1957 Act.

74. However, I am satisfied that Mr Fetto was right to decide the 1957 Act had no application, not merely because the tenancy and therefore occupation had commenced. Whilst section 2 of the 1957 Act would appear general in application, a landlord's duty was defined by section 4 of that Act. I cannot think that Parliament intended both sections to define a landlord's duty. Section 4 of the 1957 Act has been replaced by section 4 of the Defective Premises Act 1972 which is in similar, although not identical, terms to section 4 of the 1957 Act. I consider therefore that it is to section 4 of the 1972 Act that one has to look, in the first place, to find the extent of the landlord's duty in tort.

75. I am reinforced in this view by the opinions expressed in the speeches in *Murphy* v *Brentwood District Council* [1991] 1 AC 398 where the House of Lords declined to recognise a common law duty going beyond the terms of the Act. Mr Fetto accepts this but submits that the House was concerned with section 1, not section 4 and with a claim for pure economic loss, not personal injury.

76. However, underlying the reasoning in *Murphy* was the view that such duties are a matter for Parliament not the courts since, the consequence of imposing a common law duty could, especially in the case of local authorities, have such far reaching effects. Lord Mackay LC in *Murphy* at p457B-C said:

> "I am of the opinion that it is relevant to take into account that Parliament has made provisions in the Defective Premises Act 1972 imposing on builders and others undertaking work in the provision of dwellings obligations relating to the quality of their work and the fitness for habitation of the dwelling. For this house in its judicial capacity to create a large new area of responsibility on local authorities in respect of defective buildings would in my opinion not be a proper exercise of judicial power."

Mr Pooles QC submits the same reasoning should apply when considering duties applying to landlords.

77. I conclude therefore that a landlord's duty of care should normally be confined to that set out in section 4 of the Act.

Breach of the duty in contract and/or under section 4 of the Defective Premises Act 1972

The terms of the tenancy agreement

78. The premises demised was "All those premises known as 3 Canning Street…".

79. Clauses 2.1–2.11 made provision for the tenant's responsibility for the interior of the premises, fixtures and fittings, appliances and like matters. Clause 2.11 granted the landlord access on notice to enter the premises to inspect the state of repair and decoration thereof "and to paint the outside of the premises or building and generally to carry out therein or therefrom any repairs, additions, alterations or other works which may appear to the landlady… to be necessary to the premises or under the heading 'Cleaning'". Clauses 2.15–2.17 dealt with chimneys, the insides of windows and smoke alarms. By clause 2.18 the tenant was not to cause or allow obstructions/blockages concerning internal plumbing, drains downpipes and gutters. By clause 2.32 the tenant covenanted not to carry out any decoration to the premises without

A prior consent and not to "alter nor interfere with the construction or arrangement of the premises" without consent.

80. The landlord's covenants were contained in clause 3, the relevant subclause of which is 3.2 whereby the landlord covenanted:

"3.2 To provide and maintain the structure and exterior… in good repair…. except in respect of damage caused by the Tenant or any invitee or insofar as the Tenant is liable to keep the premises in repair under clause 2 of this agreement."

Section 4 of the Defective Premises Act 1972

81. This provides:

B **"Landlord's duty of care in virtue of obligation or right to repair premises demised.**

(1) **Where premises are let under a tenancy which puts on the landlord an obligation to the tenant for the maintenance or repair of the premises**, the landlord owes to all persons who might reasonably be expected to be affected by defects in the state of the premises **a duty to take such care as is reasonable in all the circumstances to see that they are reasonably safe from personal injury** or from damage to their property caused by a relevant defect.

(2) The said **duty is owed if the landlord knows** (whether as the result of being notified by the tenant or otherwise) **or if he ought in all the**
C **circumstances to have known of the relevant defect.**

(3) In this section **"relevant defect" means a defect in the state of the premises existing at or after the material time and arising from, or continuing because of, an act or omission by the landlord which constitutes or would if he had had notice of the defect, have constituted a failure by him to carry out his obligation to the tenant for the maintenance or repair of the premises**; and for the purposes of the foregoing provision **'the material time' means–**

(a) where the tenancy commenced before this Act, the commencement of this Act; and

(b) in all other cases, **the earliest of the following times, that is to say–**

(i) **the time when the tenancy commences;**

(ii) the time when the tenancy agreement is entered into;

D (iii) **the time when possession is taken of the premises in contemplation of the letting.**

(4) Where premises are let under a tenancy which expressly or impliedly gives the landlord the right to enter the premises to carry out any description of maintenance or repair of the premises, then, as from the time when he first is, or by notice or otherwise can put himself, in a position to exercise the right and so long as he is or can put himself in that position, he shall be treated for the purposes of subsections (1) to (3) above (but for no other purpose) as if he were under an obligation to the tenant for that description of maintenance or repair of the premises; but the landlord shall not owe the tenant any duty by virtue of this subsection in respect of any defect in the state of the premises arising from,
E or continuing because of, a failure to carry out an obligation expressly imposed on the tenant by the tenancy.

(5) For the purposes of this section obligations imposed or rights given by any enactment in virtue of a tenancy shall be treated as imposed or given by the tenancy.

(6) This section applies to a right of occupation given by contract or any enactment and not amounting to a tenancy as if the right were a tenancy, and 'tenancy' and cognate expressions shall be construed accordingly."

I have highlighted what I consider to be the relevant parts of the section.

82. In order to show a breach of clause 3.2 and/or section 4, the claimant has to show the premises were "not in good repair". On the
F meaning of "repair" Mr Fetto drew my attention to *Quick* v *Taff Ely Borough Council* [1986] QB 809*, where Lawton LJ at p821 stated:

"that which requires repair is in a condition worse than it was at some earlier time".

In the same case, Dillon LJ cited with approval the earlier words of Atkin LJ in *Anstruther-Gough-Calthorpe* v *McOscar* [1924] 1 KB 716 at p734 that repair:

"connotes the idea of making good damage so as to leave the subject so far as possible as though it had not been damaged".

83. In *Alker* v *Collingwood Housing Association* [2007] EWCA G
Civ 343; [2007] 1 WLR 2230**, the Court of Appeal had to consider whether a landlord was liable under section 4 of the 1972 Act for injuries caused to a tenant whose arm went through a glass panel in her front door. The panel was made of ordinary not safety glass. Breach of duty at common law had been abandoned below and the court was concerned essentially with whether or not the unstrengthened glass was "a relevant defect" for the purposes of the 1972 Act. The claimant's case was that there was a relevant defect because it was dangerous. The defendant's case was that the glass did not require repair.

84. Laws LJ, with whom Carnwath LJ agreed, stated a duty to repair could not be equated with a duty to make safe and that a duty to keep "in H
good condition" could not encompass a duty to put in a safe condition. He added this at p2236A-B:

"A house may offer many hazards: a very steep stairway with no railings; a hidden step; some other hazard inside or outside the house of the kind often found perhaps in particular older properties. I do not think it can be said that the Act requires a landlord on proof only of the conditions I have described for the application of section 4 to make safe any such dangerous feature."

His approach echoes that of Lawton LJ in *Quick* who said at p821G:

"…a tenant must take the house as he finds it; neither a landlord nor a tenant J
is bound to provide the other with a better house than there was to start with".

85. As regards the unguarded drop, Mr Fetto accepts that there is no evidence that the wall at the time of the accident was any different from when the defendant acquired the property. It cannot be said that it was out of repair.

86. As regards the steps, Mr Fetto says this: The stone steps unpainted were in good repair. Painted they became less slip-resistant and more prone to become slippery when wet or contaminated. Therefore they required repair. He accepts that context is important but submits that the function of the steps was to provide safe access and that that function was undermined when the steps became slippery because of the paint. K
He says that the court should bear in mind that those using the steps may be carrying objects of different shapes and sizes, that the steps were not protected against the elements and that they were alongside the unguarded drop.

87. This argument is semantically attractive. However, I consider it is stretching the meaning of the word "repair" to apply it to the removal of the paint. The paint did not replace the stone but was additional to it. The stone did not require repair. The paint did not require repair: what it required was removal.

88. I would therefore not find on the facts of this case that the presence of the paint caused the steps not to be "in good repair". L
Accordingly, the presence of the paint did not give rise to a breach of clause 3.2 and section 4 of the 1972 Act.

Breach of duty at common law

89. *Cavalier* v *Pope* [1906] AC 428 decided that a landlord who lets premises in a dangerous condition owes no duty to remedy the defect and no duty of care to a third party injured as a result of the defect. This decision has come in for much criticism and attempts have been made to limit its effect, especially since *Donoghue* v *Stevenson* [1932] AC 562, which defined the extent of the modern duty of care in personal injuries actions. In *Greene* v *Chelsea Borough Council* [1954] 2 QB M
127, Denning LJ said the doctrine, based on privity of contract, had received its quietus following *Donoghue*.

90. Judges in later cases have not expressed such positive views. In *Rimmer* v *Liverpool County Council* [1985] QB 1, John Stephenson LJ giving the judgment of the court dissented from what Denning LJ had said about *Cavalier* but said, at p9H, it should be "kept in close confinement". In the result the Court of Appeal distinguished it and held that a local authority landlord who had also designed and built the let premises owed a duty of care, not as landlord but as the designer/builder of the building for installing insufficiently strong glass in a

* Editor's note: Also reported at [1985] 2 EGLR 50; 276 EG 452-460

** Editor's note: Also reported at [2007] 2 EGLR 43; [2007] 25 EG 184

glass panel which caused the tenant injury. *Rimmer* was followed by the Court of Appeal in *Targett* v *Torfaen Borough Council* [1992] 3 All ER†. More recently, in *Boldack* v *East Lindsay District Council* 31 HLR 41, May LJ held that *Cavalier* was binding authority unless it could be distinguished. All these cases concern defects which pre-existed the tenancy.

91. Mr Fetto accepts that this court is bound by *Cavalier* but submits that the defendant owed the claimant a duty of care at common law, independent of the provisions of the 1972 Act. He submits that they should have guarded the drop and either not painted or removed the paint from the steps.

Does a duty exist at common law?

92. To establish the existence of the duty, Mr Fetto relies on the decision of Mackay J in *Lips* v *Older* [2004] EWHC 1686 (QB); [2005] PIQR P14, where a tenant of a bedsitter, one of many in a house owned by the defendant landlord, fell down into a basement well from a parapet wall on which he had been sitting. The premises had probably been built in about 1900 and if the wall had had railings they had probably been removed in 1940 or so as part of the war effort. The defendant was held to have owed the claimant a duty of care. Mackay J said at [40]:

> "To consider, first, the defendant's duty of care in these circumstances, this is entirely a case based on common law negligence. Mr Braithwaite accepts that it is not taken any further by any actionable breach of statutory duty. The duty of care could, therefore, be formulated for the purposes of this case in this way. The defendant had a duty to take such care as was reason-able in all the circumstances to ensure that people who he could contemplate using the path to this front door were reasonably safe to do so. The standard of care is, of course, flexible. It would be different for a private household from a children's home, or say, a clinic for people with disability of one sort or the other."

Cavalier was not, apparently, mentioned. Mr Pooles QC submits that the finding of a duty of care could be justified on the basis that the defendant retained occupation of the wall as part of the common parts of the building.

93. *Lips* was mentioned by Brooke LJ in *Sowerby* v *Charlton* [2005] EWCA Civ 1610; [2006] 1 WLR 568, a case broadly similar to the present, where the defendant was allowed to resile from an admission of liability but the court ruled that the judgment should stand. Brooke LJ, at [29]-[31], stated:

> "29. In *Lips*, Mackay J was concerned with a similar case involving a fall by a 42-year-old man (who was known by his landlord to be an alcoholic) into the basement area of a house in Hove. He held the house owner one-third liable. He was influenced by the consideration that it was foreseeable that any of the tenants, who were largely single students, would return to the premises in drink, and that the installation of a simple handrail on one side was an inexpensive and reasonable precaution. It is fair to say that he found this a finely balanced case and there were features of it, not present in the present case, that militated against the landlord.
>
> 30. The present case, too, would be finely balanced. But in all the circumstances we regard it as inconceivable that any High Court judge would fail to find the defendant at least partly liable. There is not only the fact that when confronted with a very large damages claim her insurers, their reinsurers, and their very experienced solicitors accepted liability after having had plenty of opportunity to investigate the circumstances of the accident. There is also the fact that par-ties given by the occupiers of the ground floor flat were known to be not infrequent events, with the reasonably foreseeable consequence that people might leave the flat the worse for wear for drink, with an unguarded 8ft drop onto a very hard surface on their left hand side.
>
> 31. A further relevant consideration is the fact that Judge Playford QC, who also has vast experience of personal injuries litigation, expressed the view that he thought that the claimant would succeed in this case on primary liability. It seemed to him that the danger of a fall was an obvious hazard that could be remedied at very little expense by anyone who chose to apply their mind to the danger.
>
> 32. In all the circumstances we considered that there was no real prospect of the defendants resisting a finding of primary liability."

It appears that *Cavalier* was not cited to the court.

94. Mr Fetto submits *Sowerby* is binding on the court as authority for the proposition that a duty of care was owed. I do not regard *Sowerby* as providing Court of Appeal authority for the existence of a common law duty of care in the circumstances of the present case. It is clear that the case was primarily concerned with the operation of the CPR and whether or not judgment should be set aside rather than whether or not a duty of care existed. I regard Brooke Li's words as expressing a view on what was likely to happen in the context of whether or not judgment should be set aside rather than declaring that a duty of care did exist, also perhaps what he instinctively considered would be a fair outcome.

95. Mr Fetto submits that the scope of the common law duty is not restricted by the 1972 Act and relies on section 6(2) of the Act. He accepts that in Murphy the House of Lords declined to recognise a common law duty going beyond the terms of the 1972 Act but seeks to distinguish it (see above).

96. The decision of Mackay J is, of course, very persuasive, coming as it does from a judge who is very experienced in this field but it is important to note that the case was presented to him as a case based on common law negligence. There is nothing to suggest that the existence of the duty of care was disputed. Rather it appears to have been accepted and therefore the essential question for the judge was whether or not it was breached. It is difficult therefore to regard it as deciding definitively that a duty of care exists.

97. There is an apparent conflict between *Lips* and *Cavalier*. Mr Fetto submits that I am bound by *Sowerby* on this point. For the reasons I have given I do not accept that submission. He also suggests that, at least so far as painting the steps is concerned, *Cavalier* can be distinguished on the grounds that in painting the steps the defendant owed the claimant a duty, as he put it in his skeleton argument, to take care not to cause harm by her ie the defendant's positive action. He reminds me of John Stephenson LJ's observation that *Cavalier* should be kept in close confinement and how landlords when acting in other capacities such as a contractor, as in *Rimmer*, have been held to owe a duty of care. He relies on the speech of Lord Salmon in *Anns* v *Merton London Borough Council* [1978] AC 728††, at p768E-G:

> "The immunity of a landlord who sells or lets his house which is dangerous or unfit for habitation is deeply entrenched in our law. I cannot, however, accept the proposition that a contractor who has negligently built a dangerous house can escape liability to pay damages for negligence to anyone who, eg, fall through a shoddily constructed floor and is seriously injured, just because the contractor happens to have been the owner of the land upon which the house stands. If a similar accident had happened next door in a house which the contractor had also negligently built on someone else's land, he would not be immune from liability. This does not make any sense. In each case the contractor would be sued for his negligence as a contractor and not in his capacity as a landowner: the fact that he had owned one plot of land and not the other would be wholly irrelevant I would hold that in each case he would be liable to pay damages for negligence."

However, almost immediately after these words Lord Salmon stated:

> "*Cavalier* is so far away from the present case that I express no opinion about it."

98. He also points out that in *Dutton* v *Bognor Regis Urban District Council* [1972] 1 QB 373 Sachs LJ distinguished *Cavalier* on the basis that in that case the landlord had not created the dangerous state of affairs. He acknowledges that *Anns* was departed from in, and *Dutton* was overruled by, *Murphy* but relies on the fact that the quoted words of Lord Salmon were approved by Stephenson LJ in *Rimmer* and submits they remain good law.

99. I am bound by *Cavalier*, *Rimmer* and *Targett*. In *Rimmer* and *Targett* the danger was caused by the defendant when acting in a different capacity from that of landlord, indeed, before becoming a landlord. In the present case, in having the steps painted the defendant was doing something that a landlord in her position could well do. I

† Editor's note: Also reported at [1992] 1 EGLR 275

†† Editor's note: Also reported at [1977] 2 EGLR 94; [1977] 243 EG 523 and 596

do not regard her as acting in any different capacity so as to justify distinguishing *Cavalier* on that ground.

100. In this connection it appears to me relevant to distinguish between the unguarded drop and condition of the steps due to the presence of paint. The unguarded drop pre-existed the defendant's acquisition of the property whereas the paint was applied, at her instance, after the defendant acquired the property.

101. I have no difficulty in concluding that I am bound by *Cavalier* so far as the unguarded drop is concerned and therefore conclude that the defendant had no duty to guard it. But so far as the application of the paint is concerned I see no reason why the defendant should not have a duty to take reasonable care to ensure its application did not create an unnecessary risk of injury. Otherwise a landlord would have carte blanche to act with impunity and create dangers which would not be caught by the 1972 Act.

102. I would therefore conclude that where personal injuries result from a failure to repair then the duty of care is that set out in the 1972 Act but where the Act does not apply, as is the case here, then a landlord owes a duty to take reasonable care not to create an unnecessary risk of injury.

Breach of the duty of care

103. There is always a risk of slipping on steps, usually one resulting from misplacing a foot or feet on a step. The presence of wet increases the risk. The claimant therefore would have faced an increased risk of slipping that day, even had the steps not been painted, because the steps were wet. She accepted that there were puddles on the concrete path and the path was slippery. But the presence of paint unnecessarily increased, and in my judgment significantly increased, the risk. I find that the presence of the paint materially contributed to her slipping. Having slipped there was nothing to prevent her falling as she did.

104. But that of itself is insufficient to establish liability. I accept that the paint was sold as paint suitable for stone surfaces, indoors and outdoors, by a well known supply chain, who as I suspect is also well known, supply DIY enthusiasts, jobbing builders and small building businesses. It is not described as non-slip paint and there is nothing in the paint analysis carried out by RSK to suggest it is. If it had been, had it contained sharp sand or carborundum, or had it been applied mixed with sharp sand or carborundum, I am sure the defendant would have been the first to tell me. Mr Hedges selected and then applied the paint.

105. In applying the paint to the steps the defendant was doing what I am satisfied many other people in his position would have done. Painting steps appears to have been fashionable on the claimant's side of the street, although what paint was used is not the subject of evidence. He was not applying ordinary paint but paint said to be suitable for outdoor use which contained no warning against its use on steps nor that its surface may become slippery in the wet, nor that it should be applied mixed with sharp sand or carborundum. The references to "semi-gloss" and producing a sheen may have alerted the knowledgeable to question whether it was suitable for steps, but not, in my judgment the man in the street. For this purpose I would regard Mr Hedges as "the man in the street" rather than a well-informed builder. Mr Highfield agreed with Mr Butcher that it was not unreasonable for the defendant to apply or have applied the B&Q paint to the steps, although it appears that he personally would not have done so.

106. I regret that I cannot conclude on the evidence that there was a breach of duty.

107. I have considerable sympathy for the claimant, who to some extent appears to be the victim of a decision which many consider needs to be reconsidered, but her remedy can only be in another court.

Claim dismissed.

LANDLORD AND TENANT

A Shellpoint Trustees Ltd and another v Barnett and others

UPPER TRIBUNAL LANDS CHAMBER

[2012] UKUT 375 (LC)

8 October 2012

HH JUDGE NIGEL GERALD AND MR ANDREW J TROTT FRICS

Landlord and tenant – Landlord and Tenant Act 1987 – Variation of lease – Landlord applying to vary leases of flats to permit replacement B *of communal heating and hot water system with individual boilers in each flat – Landlord further seeking "non-consequential" variations to permit recovery through service charge of costs of enforcing all lease covenants – Application made with approval of requisite majority of tenants – Sections 37 and 38 of 1987 Act – Identification of object of application – Whether object satisfactorily achieved without non-consequential variations – Whether such variations unreasonable and prejudicial to tenants*

The appellants were the landlords of three blocks of flats containing a total of 365 flats let to tenants. The respondents were the tenants of three of those flats. The appellants applied to the Leasehold Valuation C Tribunal (LVT), with the approval of the requisite number of tenants, to vary the terms of the leases pursuant to sections 37 and 38 of the Landlord and Tenant Act 1987 to permit the replacement of the communal heating and hot water system with individual boilers in each flat, with consequential provision for recovery of those costs through the service charge. The appellants additionally sought further, non-consequential variations designed to give them extensive powers to recover, through the service charge, their reasonable costs of enforcing all lease covenants. the reasonable costs of all proper steps that they deemed necessary in enforcing lease covenants. They gave evidence that the tenants had agreed to the non-consequential variations as D part of a deal in return for the new heating and hot water system. The respondents opposed the variations.

The LVT allowed the variations concerned with the heating and hot water system but refused the non-consequential variations relating to enforcement of covenants. They found that the single object of the application was to achieve the replacement of the old heating and hot water system and that the non-consequential variations were unnecessary to achieve that object, would prejudice the respondents and were unreasonable, so that the requirements of sections 37 and 38 were not met. The appellants appealed.

They considered that the proposed variations gave more to the E appellants than was justified and that it would be inappropriate to permit them substantially to improve their position as landlords in respect of the recovery of costs against recalcitrant lessees to the detriment of other lessees.

Decision: **The appeal was dismissed. (1) An applicant for a lease variation must identify the object of the application by adducing evidence of what it is trying to achieve by the variations and why. There can be either single or multiple objects. Since the appellants had presented their variations in the context of a single objective, namely to replace the communal heating and hot water system with individual boilers, the LVT had been entitled to assess the proposed variations by reference to that single object. (2) The purpose of section 37 is to enable the majority to apply to the LVT for a** F **variation to achieve a particular object. An applicant for variation must show that the object cannot satisfactorily be achieved without varying all the leases to the same effect; that involves showing both that the proposed variation will achieve the object and that all the leases need to be varied in order satisfactorily to achieve that object: see section 37(3). If those requirements are not met, there is no jurisdiction to entertain the application or consider it further, notwithstanding that the requisite majority are in favour of the variation. It is not the case that the majority view will prevail unless the section 38(6) grounds for refusing variation are made out. The LVT had properly concluded that the replacement of** G **the communal heating and hot water system could be achieved without any of the non-consequential variations; accordingly, those variations failed to satisfy the requirements of section 37(3) and had properly been refused. (3) The LVT had further been entitled to find that the non-consequential variations were likely to cause substantial prejudice to the respondents and others and were not reasonable in the circumstances, such that they should be disallowed under section 38(6)(a) and (b). It is inappropriate for landlords to use circumstances such as the replacement of a heating and hot water system as an opportunity to substantially improve their contractual position, by demanding new covenants that are unrelated to, and not consequent on, achieving the object.** H **The non-consequential variations would considerably widen the nature, scope and effect of the lease provisions concerning the costs that the appellants could recoup from the tenants through the service charge; they covered not just the normal costs to be expected in managing a block of flats but also the costs of enforcing covenants at the initiative of the landlords and/or the tenants, plus the costs of the landlords' own forfeiture proceedings and enforcement of covenants of particular interest to the landlords and their reversions. Such provisions shifted all financial risk from landlord to tenant, thereby giving the landlord an incentive to enforce, rather than reach a consensual agreement. That would** J **fundamentally alter the relationship between landlord and tenant in a manner prejudicial to the tenants.**

The following cases are referred to in this report.

Arrowdell Ltd v *Coniston Court (North) Hove Ltd* (2006) LRA/72/2005 (Lands Tribunal)
Baystone Investments Ltd v *Perkins* [2010] UKUT 70 (LC)
Curtis v *London Rent Assessment Committee* (1999) QB 92
Dixon and others v *Wellington Close Management Ltd* [2012] UKUT 95 (LC)
English v *Emery Reimbold & Strick Ltd* [2002] 1 WLR 2409
Flannery v *Halifax Estate Agencies Ltd* [2000] 1 WLR 377 K
LB Havering v *MacDonald* [2012] UKUT 154 (LC)
Lucie M v *Worcestershire County Council and Evans* [2002] EWHC 1292
Pepper and others v *Appleby Consultancy Ltd* BIR/00CN/LVL/2007/0001 (LVT)
Pittalis v *Grant* [1989] 1 QB 605
Schilling v *Canary Riverside Development PTE Ltd* LRX/26/2005 (LT)
South Buckinghamshire District Council v *Porter (No 2)* [2004] 1 WLR 1953
Thirlway v *Troy* [2012] UKUT 302 (LC); LRA/57/2012

This was an appeal by the appellant landlords, Shellpoint Trustees Ltd and Anston Investments Ltd, from a decision of the leasehold valuation tribunal refusing their application, under the Landlord and

A Tenant Act 1987, for certain variations to leases of flats of which the respondents, James Barnett, Jeremy Clyne, Katherine Zographos and Constantine Zographos were four of the tenants.

Howard Lederman (instructed by Teacher Stern LLP) appeared for the appellants; the respondents appeared in person.

Giving judgment, **HH Judge Nigel Gerald** said:

Introduction

1. The appellant landlords, Shellpoint Trustees Ltd and Anston Investments Ltd, appeal against the decision of the Leasehold Valuation B Tribunal (the "LVT") dated 13 January 2011 (the "Decision") refusing to order variation of the subject leases to allow recovery of the landlords' management and legal costs of enforcing divers covenants contained in the leases through the service charge under sections 37 and 38 of the Landlord and Tenant Act 1987 ("the 1987 Act").

2. There is no extant appeal against the principal part of the Decision ordering variations to the leases to permit replacement of the communal boiler system with individual flat boilers and a change to the insurance covenant: the appellants do not pursue their appeal against the insurance variation as they now accept that the LVT ordered what they had asked for.

C 3. The appellants are together the landlords of the 365 flats comprised within the three blocks known as Eton Hall, Eton Place and Eton Rise all situate at Eton College Road, London NW3 (respectively, "Eton Hall", "Eton Place" and "Eton Rise", collectively "the Etons"). Shellpoint is the freehold reversioner of the majority of the flats ("the Shellpoint flats"); Anston holds a headlease of the remainder ("the Anston flats").

4. The material covenants of the leases of the Shellpoint flats are in common form, as are those of the Anston flats albeit that their clause numbering and layout is slightly different and they do not have clause 6(8)(b). For the purposes of these appeals we adopt the approach of the D LVT of addressing itself to the sample lease granted on 16 April 1984 by Shellpoint's predecessor in title, Linthaven Ltd, to Jackinstant Ltd ("the Shellpoint leases") and, where necessary, the sample lease granted on 9 June 1975 by Anston's predecessor in title, Peachey Property Corporation Ltd, to Anthony Rudolf Sascha Suchy ("the Anston leases"). "Leases" refer to both Shellpoint and Anston leases.

5. Two applications, both dated 25 January 2010, were issued to reflect the different reversionary ownerships and leases of the flats by the appellants and, to demonstrate the requisite level of section 37(5)(b) support for the variations, by a total of 299 of the 365 tenants, 96 from Eton Hall, 103 from Eton Place and 100 from Eton Rise. The rest either E opposed the proposed variations, or did not respond to the ballot of all 365 flats conducted shortly before issuance, the results of which were as follows, "S" referring to Shellpoint flats and "A" to Anston flats:

	ETON HALL		**ETON PLACE**		**ETON RISE**		**TOTALS**
	S	**A**	**S**	**A**	**S**	**A**	
In favour	80	16	88	15	84	16	**299**
Against	9	0	4	1	2	1	**17**
Unreturned	8	2	13	4	18	3	**48**
Abstained	1	0	0	0	0	0	**1**
F **Totals:**	**98**	**18**	**105**	**20**	**104**	**20**	**365**

6. There were pre-trial reviews in both applications on 4 February 2010 at which the LVT consolidated the applications and ordered that any person wishing to oppose the variations must serve a statement in reply setting out their reasons.

7. Despite the number of those who balloted their opposition, only three tenants actively opposed the applications before the LVT: Jeremy Clyne ("Mr Clyne") lessee of flat 57 Eton Hall and Katherine Zographos ("Miss Zographos") and her brother Constantine joint lessees of 75 and 76 Eton Hall (all Shellpoint flats) and James Barnett ("Mr Barnett") lessee of flat 6 Eton Place (an Anston flat) each of whom served statements in reply setting out why they opposed the variations.

8. In broad terms, the applications sought variations to the leases G to enable the communal heating and hot water systems to be replaced with individual boilers in each flat with consequential provisions for recovery of those costs through the service charge and what have been described as "non-consequential" variations to the leases to enable the landlords to recover its costs of enforcing covenants contained in the leases through the service charge.

9. The only evidence served in support of the applications were two witness statements dealing with the heating and hot water issues and two witness statements explaining the background to the proposals. One was from Nicholas Goldreich ("Mr Goldreich") a solicitor and partner in Comptons who had been instructed by New Etons Residents H Association ("NERA"), a formally constituted and recognised residents association, and the other from Dr Seth Rankin ("Dr Rankin"), the tenant of flat 73 Eton Rise and a former chairman and committee member of NERA.

10. The hearing before the LVT took place on 19 April 2010 but was adjourned and resumed on 15 and 16 November 2010 by which time the applicants had adduced a further witness statement from Mr Goldreich. Refinements to the proposed variations appended to the applications were also provided, which were further refined after the conclusion of the hearing by email dated 22 November 2010. It is this last version of the variations to which we will refer in this decision. J

11. The LVT issued its decision on 13 January 2011 and the appellants applied to the LVT for permission to appeal. The LVT refused the application on 23 March 2011 (the "Refusal Decision"), dealing separately with each of the 20 grounds of appeal raised and to some extent adding to the reasons it had already given in the Decision.

12. On 15 August 2011, the President of the Upper Tribunal (Lands Chamber) granted the appellants permission to appeal because there was a "real prospect of success" and specifying that the appeal would be by way of review. It should be recorded that permission to apply for permission to appeal out of time was also granted.

Representation K

13. Before the LVT, the applicants, that is, the landlords and the 299 tenants were represented by Mr Stan Gallagher of counsel who was jointly instructed by the landlords' solicitors Teacher Stern LLP and the 299 tenants' solicitor Messrs Comptons ("Comptons"). The three active respondents represented themselves. Oral evidence was given by some of the witnesses.

14. Before this tribunal, the appellant-landlords were represented by Mr Howard Lederman of counsel, instructed by Teacher Stern LLP. The 299 applicant-tenants did not appeal the Decision, were not represented and have expressed no views about the merits of the appeal. The three active respondents before the LVT are the sole active respondents L to the appeal and, as below, represent themselves; Miss Zographos representing herself and her brother.

15. Each of the respondents was clearly well-acquainted and fully familiar with the issues relating to the non-consequential variations and well able to advance their arguments in opposition to the appeal. Indeed, as will become clear, they appeared at times to have a grasp of the issues superior to that of the appellants and at least some of their advisers who did not appear to understand the extent and scope of the new clauses they had asked the LVT for and were asking this tribunal for.

16. Whilst each of the respondents advanced different arguments, we shall treat their submissions in a composite manner, without M distinguishing between them save where necessary. We do so because the broad thrust of their positions was the same albeit that there were some differences of emphasis, and without intending any disrespect to them.

The legislation

17. Sections 37 and 38 of the 1987 Act permit applications to vary leases where the majority of parties concerned consent. This is in contrast to the separate and distinct section 35 jurisdiction which permits applications by any party to a long lease of a flat to vary a lease with unsatisfactory repairing, maintenance, insurance, service charge or similar provisions.

A 18. As Mr Lederman reminded us, the applications were made under section 37 and not under section 35. It is nonetheless necessary to set out the material parts of both provisions as reference has been made to them. Section 35 states so far as relevant:

> "35. *Application by party to lease for variation of lease.*
>
> (1) Any party to a long lease of a flat may make an application to a leasehold valuation tribunal for an order varying the lease in such manner as is specified in the application.
>
> (2) The grounds on which any such application may be made are that the lease fails to make satisfactory provision with respect to one or more of the following matters, namely—
>
> B (a) the repair or maintenance of—
>
> (i) the flat in question, or
>
> (ii) the building containing the flat, or
>
> (iii) any land or building which is let to the tenant under the lease or in respect of which rights are conferred on him under it;
>
> (b) the insurance of the building containing the flat or of any such land or building as is mentioned in para (a)(iii);
>
> (c) the repair or maintenance of any installations (whether they are in the same building as the flat or not) which are reasonably necessary to ensure that occupiers of the flat enjoy a reasonable standard of accommodation;
>
> (d) the provision or maintenance of any services which are reasonably necessary to ensure that occupiers of the flat enjoy a reasonable standard
> C of accommodation (whether they are services connected with any such installations or not, and whether they are services provided for the benefit of those occupiers or services provided for the benefit of the occupiers of a number of flats including that flat);
>
> (e) the recovery by one party to the lease from another party to it of expenditure incurred or to be incurred by him, or on his behalf, for the benefit of that other party or of a number of persons who include that other party;
>
> (f) the computation of a service charge payable under the lease…"

Sections 37 and 38 of the 1987 Act state:

> "37. *Application by majority of parties for variation of leases.*
>
> (1) Subject to the following provisions of this section, an application may
> D be made to a leasehold valuation tribunal in respect of two or more leases for an order varying each of those leases in such manner as is specified in the application.
>
> (2) Those leases must be long leases of flats under which the landlord is the same person, but they need not be leases of flats which are in the same building, nor leases which are drafted in identical terms.
>
> (3) The grounds on which an application may be made under this section are that the object to be achieved by the variation can not be satisfactorily achieved unless all the leases are varied to the same effect.
>
> (4) An application under this section in respect of any leases may be made by the landlord or any of the tenants under the leases.
>
> (5) Any such application shall only be made if—
>
> E (a) in a case where the application is in respect of less than nine leases, all, or all but one, of the parties concerned consent to it; or
>
> (b) in a case where the application is in respect of more than eight leases, it is not opposed for any reason by more than 10 per cent of the total number of the parties concerned and at least 75 per cent of that number consent to it.
>
> (6) For the purposes of subsection (5)—
>
> (a) in the case of each lease in respect of which the application is made, the tenant under the lease shall constitute one of the parties concerned (so that in determining the total number of the parties concerned a person who is the tenant under a number of such leases shall be regarded as constituting a corresponding number of the parties concerned); and
>
> (b) the landlord shall also constitute one of the parties concerned.
>
> F
>
> 38. *Orders ... varying leases.*
>
> (1) If, on an application under section 35, the grounds on which the application was made are established to the satisfaction of the tribunal, the tribunal may (subject to subsections (6) and (7)) make an order varying the lease specified in the application in such manner as is specified in the order.
>
> (2) If—
>
> (a) an application under section 36 was made in connection with that application, and
>
> (b) the grounds set out in subsection (3) of that section are established to the satisfaction of the tribunal with respect to the leases specified in the application under section 36,
>
> the tribunal may (subject to subsections (6) and (7)) also make an order varying each of those leases in such manner as is specified in the order.
>
> (3) If, on an application under section 37, the grounds set out in subsection G
> (3) of that section are established to the satisfaction of the tribunal with respect to the leases specified in the application, the court may (subject to subsections (6) and (7)) make an order varying each of those leases in such manner as is specified in the order.
>
> (4) The variation specified in an order under subsection (1) or (2) may be either the variation specified in the relevant application under section 35 or 36 or such other variation as the tribunal thinks fit.
>
> (5) If the grounds referred to in subsection (2) or (3) (as the case may be) are established to the satisfaction of the court with respect to some but not all of the leases specified in the application, the power to make an order under that subsection shall extend to those leases only.
>
> (6) A tribunal shall not make an order under this section effecting any H
> variation of a lease if it appears to the court—
>
> (a) that the variation would be likely substantially to prejudice—
>
> (i) any respondent to the application, or
>
> (ii) any person who is not a party to the application,
>
> and that an award under subsection (10) would not afford him adequate compensation, or
>
> (b) that for any other reason it would not be reasonable in the circumstances for the variation to be effected.
>
> (7) A tribunal shall not, on an application relating to the provision to be made by a lease with respect to insurance, make an order under this section effecting any variation of the lease—
>
> (a) which terminates any existing right of the landlord under its terms to J
> nominate an insurer for insurance purposes; or
>
> (b) which requires the landlord to nominate a number of insurers from which the tenant would be entitled to select an insurer for those purposes; or
>
> (c) which, in a case where the lease requires the tenant to effect insurance with a specified insurer, requires the tenant to effect insurance otherwise than with another specified insurer.
>
> (8) A tribunal may, instead of making an order varying a lease in such manner as is specified in the order, make an order directing the parties to the lease to vary it in such manner as is so specified; and accordingly any reference in this Part (however expressed) to an order which effects any variation of a lease or to any variation effected by an order shall include a reference to an order which directs the parties to a lease to effect a variation of it or (as the case may be) a reference to any variation effected in pursuance of such an order. K
>
> (9) A tribunal may by order direct that a memorandum of any variation of a lease effected by an order under this section shall be endorsed on such documents as are specified in the order.
>
> (10) Where a tribunal makes an order under this section varying a lease the tribunal may, if it thinks fit, make an order providing for any party to the lease to pay, to any other party to the lease or to any other person, compensation in respect of any loss or disadvantage that the tribunal considers he is likely to suffer as a result of the variation."

The case before the LVT

Introduction

19. The applicants were wholly successful in securing all of the L
variations they asked for to enable the communal boiler system to be replaced with a system providing for individual boilers in each flat and attendant changes to the service charge provisions and also to the insurance covenant, albeit that the LVT amended some of the wording in respect of which no complaint is made.

20. The applicants were wholly unsuccessful in securing the "non-consequential" variations they asked for (other than in respect of insurance) ie those variations which were not the consequence of, or required or necessitated by, replacement of the heating and hot water systems.

M

The three "non-consequential" variations sought

21. Three variations were sought. The first was to make additions to the usual proviso for re-entry contained in clause 5 of both the Shellpoint and Anston leases. Clause 5 of those leases ("original clause 5") provides as follows:

> "PROVIDED ALWAYS and it is hereby agreed and these presents are upon the express condition that if the said rents or any part thereof shall be unpaid for 21 days next after becoming payable (whether the same shall have been legally demanded or not) or if the lessee shall not duly perform or observe all the covenants and provisions hereby on the part of the lessee to be performed or observed then and in any of the said cases and thenceforth it shall be lawful for the lessor or any person or persons duly authorised by the lessor in that behalf

A into or upon the flat or any part thereof in the name of the whole to re-enter and the same to repossess and enjoy as if these presents had not been made but without prejudice to any right of action or remedy of the lessor in respect of any antecedent breach of any of the covenants by the lessee herein contained."

22. The LVT was asked to substitute the original clause 5 by a new clause 5 ("new clause 5") as follows (noting that new clause 5(i) is a mere repetition of original clause 5):

"5. Provided always it is hereby agreed and these presents are upon the express conditions that

(i) if the said rents or any part thereof shall be unpaid for 21 days next after becoming payable (whether the same shall have been legally demanded or not)
B or if the lessee shall not duly perform or observe all the covenants and provisions hereby on the part of the lessee to be performed or observed then and in any of the said cases and thenceforth it shall be lawful for the lessor or any person or persons duly authorised by the lessor in that behalf into or upon the flat or any part thereof in the name of the whole to re-enter and the same to repossess and enjoy as if these presents had not been made but without prejudice to any right of action or remedy of the lessor in respect of any antecedent breach of any of the covenants by the lessee herein contained

(ii) the lessor will use all reasonable and economically viable endeavours to recover the legal costs against a defaulting lessee above if any part of the legal costs shall not be recovered against a defaulting lessee under the provisions hereof the lessor may apply the balance of the unrecovered legal costs against
C the service charge in accordance with the provisions of para 13 of the Fourth Schedule and subject to the following provisions

(a) that no declaration has been made against the lessor pursuant to Section 20(c) the Landlord and Tenant Act 1985 or other statutory provision or otherwise prohibiting recovery of legal costs through the medium of the service charge

(b) The legal costs shall be subject to a detailed assessment on a full indemnity basis by the court if not agreed but capped as to the conducting solicitor's hourly charge rate to the maximum amount allowed by the court for solicitors of equivalent grade in central London area

(c) The lessor shall in the first instance be entitled to charge any reasonable legal and professional costs referred to above as a service charge and in
D particular shall be entitled to charge such costs on an interim basis during the conduct of any action prosecuted by it whether by way of litigation or otherwise before any court or tribunal or otherwise."

23. The second variation was to substitute clause 6(8)(b) of the Shellpoint leases with a new clause intended to enable the landlords to enforce covenants where a tenant initiated a complaint against a fellow tenant without requiring an indemnity and security from the tenant making the complaint. In respect of the Anston leases, the LVT was asked to insert a completely new covenant because they contain no equivalent to clause 6(8)(b).

24. By clause 6(8)(b) of the Shellpoint leases ("original clause 6(8)
E (b)"), the landlord covenants that:

"6(8)(b) If so required by the lessee to enforce the covenants and conditions similar to those contained herein on the part of the lessee entered into or to be entered into by the lessees of the other flats in the building so far as they affect the flat and the lessee indemnifying the lessor against all costs and expenses of such enforcement and giving reasonable security for such costs and expenses".

25. Under the substitute clause 6(8)(b) or, for the Anston leases, the new clause 6(8)(b) (together referred to as the "new clause 6(8)(b)") the landlord covenants that:

"6(8)(b) If so required by the lessee for the reasonable protection of the
F flat the lessor shall enforce the covenants entered into or to be entered into by a lessee of any one or more of the other flats in the property and the costs of such enforcement shall be recoverable by the lessor in accordance with the provisions of clause 5(ii) of the lease".

26. The third variation was to insert a new para into the Fourth Schedule of the Shellpoint and Anston leases which deals with what the landlords may recover from the tenants via the service charge. By clause 2(2) of the leases the tenant covenants to pay.

"a proportionate part of the expenses and outgoings incurred by the lessor in the repair maintenance renewal and insurance of the said building and the provisions of services therein and the other heads of expenditure as the same are set out in the Fourth Schedule hereto..."

27. The Fourth Schedule refers to clause 2(2) and is entitled G
"Lessor's expenses and outgoings and other heads of expenditure in respect of which the lessee is to pay a proportionate part by way of service charge". It then lists the various heads of expenditure. The LVT was asked to insert the following new para 12 or 13 (depending on the lease) at the end ("new para 12/13"):

"12/13. The reasonable cost of all proper steps deemed necessary by the lessor for (a) the recovery of arrears of outstanding service charges from lessees of the said property and (b) the enforcement of covenants entered into or to be entered into by a lessee of any one or more of the other flats in the property (such costs to include solicitors' and other professional fees incurred
in bringing claims to a court, tribunal or other competent body) providing that H
all reasonable and economically viable efforts shall be made by the lessor to recover such costs from defaulting lessees."

Nature, scope and effect of proposed new covenants

28. The leases contain the usual sort of covenants to be expected in a large block of flats granted on varying terms of between 99 and 999 years. So far as relevant, they contain the usual covenants by the tenant to pay the rent, service charge and general and water rates (clause 2(1), (2) and (3) and (4)), to pay the landlord's costs of abating any nuisance, the service of any section 146 notices, entering to carry out required
repairs to the demise which the tenant has failed to do and ensuring J
planning compliance and to indemnify the landlord for any damage caused by visitors and so on: clauses 2(5), (7), (12), (21) and (24). There are also covenants against alterations or the installation of machinery, using the demise for illegal or immoral purposes or as a trade or permitting a nuisance and assignment without the landlord's consent: clauses 2(14), (15), (16), (17) and (19) and 4. There then follow various covenants by the landlords to maintain, insure and so on the Etons.

29. The only provision in both of the leases dealing with recovery of the landlords' costs of enforcement of the covenants to which our attention was drawn was para 8 of the Fourth Schedule:

"8. The fees of the lessor's managing agents for the collection of the rents K
of the flats in the said building and for the general management thereof which shall be chargeable on the basis of all the flats being let at notional gross annual rentals equal to the current rack rental values thereof PROVIDED that such fees at not time exceed the maximum allowed in respect of property let at those rentals by the scales authorised for the time being by the Royal Institution of Chartered Surveyors and the said managing agents shall on the written request of the lessee furnish to him a statement showing the gross annual rentals assumed by them for this purpose and the calculation of their fees therefrom".

30. The proposed variations permit the landlords to recover the reasonable costs (legal or otherwise) of "all proper steps deemed
necessary by the lessor" in enforcing all of those covenants in the L
leases and any new ones entered into (para 12/13(b)). Singled out are the costs of recovering service charge arrears (para 12/13(a)), although that is probably contained within the words already referred to. Those are subject to the para 12/13 proviso that the landlord must have taken all reasonable and economically viable efforts to recover its costs from the defaulting tenant. The wording of para 12/13 is also sufficient to cover any costs of enforcing covenants requested by a tenant under new clause 6(8)(b).

31. Clause 5(ii) by contrast enables the landlords to recover as service charge their *legal costs* of re-entering, or forfeiting, the lease for
breach of covenant to the extent that they can not be recovered from the M
tenant in default subject to the various provisos set out in clause 5(ii)(a) to (c) whose meaning and effect, we confess, is somewhat opaque.

32. The new clauses therefore very considerably widen the nature, scope and effect of what costs the landlord may recoup from the tenants under the existing para 8 of the Fourth Schedule to the leases via the service charge covering as they do not just the normal costs to be expected in managing a block of flats but also the costs of enforcing covenants at the initiative of the landlords and/or the tenants as well as the landlords' own forfeiture proceedings and enforcement of covenants of particular interest to the landlords and their reversions. They go considerably further than the existing para 8, which itself is unusual in permitting recovery of the managing agents' fees for collecting rents

A which are usually, or often, the responsibility of the landlord as it is the landlord, not other tenants, who are interested in payment of rent. We should say for completeness that Mr Lederman did suggest that there was ambiguity as to whether para 8 was confined to the managing agents' fees for general management of the Etons, it being suggested that it might be wide enough to encompass legal fees. This argument was not developed further, no authorities were cited in support of it and it did not feature in his submissions so we do not consider it further save to say that it would appear to us to be a somewhat strained construction.

B *Evidence before the LVT*

33. The evidence before the LVT was that after protracted negotiations and considerable expense the landlords and the majority of the tenants (via NERA) had agreed that in return for the landlords agreeing to individualise the heating and hot water systems the tenants had agreed to the landlords' request for the non-consequential variations upon which the tenants had been balloted as already described. There was substantial evidence in support of individualisation.

34. There was no evidence at all as to why the non-consequential variations were required: no evidence of any past difficulties (financial or otherwise) of enforcing any covenants whether they be service C charge arrears or otherwise, whether enforcement be initiated by landlord or tenant; no evidence of landlords or tenants being inhibited from enforcing covenants due to the absence of the new covenants or, in the Shellpoint leases, the presence of the original clause 6(8)(b); no evidence of the likely financial impact of the proposed variations; no evidence of whether and if so how and why any so-called deficiencies in the leases had affected the management of the Etons. Neither was there any evidence about the financial stature of either landlord or their respective ownership, composition, management or otherwise. There was no evidence or suggestion that either landlord had been unable to properly manage the Etons or enforce any of the covenants due to being D unable to recover such costs via the service charge.

35. Such evidence as there was was confined to Mr Goldreich's explanation about the genesis of the non-consequential variations and some brief observations contained in the witness statement of Dr Rankin. In para 5 of his witness statement dated 30 March 2010, Mr Goldreich, having explained that he had been involved in protracted negotiations with the landlords' managing agents and solicitors with regard to the proposed changes to the heating and hot water systems, said this:

"5. The landlords' professional advisers were agreeable in principle to the E proposed substantive changes to the central heating and hot water systems but wished to address certain deficiencies in the flat leases which were affecting their management of the Etons. In particular, the main deficiencies included the existing provisions enabling the reciprocal enforcement of covenants between lessees and the lack of clauses enabling the effective recovery of service charge arrears from individual non-paying lessees. I was asked to draft the proposed lease variations in my capacity as the solicitor acting for the Residents Association in order to maintain even-handedness between the interests of the landlord and of the lessees and in order to safeguard the position of the lessees."

36. Mr Goldreich then explained that the new clauses as drafted by F him in conjunction with NERA and the landlords' managing agents and solicitors Teacher Stern were then submitted by ballot to the tenants, the overwhelming majority of whom accepted them. The front sheet of the ballot paper is before us, entitled "The Etons individual heating and hot water systems replacement project" with a space to vote on all of the amendments agreed between NERA and the landlords: all the variations were therefore presented as a single "take it or leave it" package upon which the tenants voted to accept all or none. We are told that it was accompanied by a bundle of documents explaining the nature and effect of the proposed amendments as well as a copy of them: these are not in the bundle and, so far as we can tell, were not before the LVT although the respondents thought they might have been although they were not provided with a copy of them. Mr Goldreich concluded that:

"10. …The impetus to change the individual heating and hot water systems G has therefore come from the lessees not from the landlords and the landlords have requested agreement of the non-consequential variations to address deficiencies in the existing leases and to facilitate their management of the estate as a precondition of their involvement in the central heating and hot water changes".

37. The LVT summarised the effect of Mr Goldreich's written and oral evidence in para 23 of the Decision (which is not challenged):

"23. Mr Goldreich gave some evidence concerning his involvement and that of [NERA]. He told us that the lessees wanted to change to the individual [heating and hot water] system. It was a condition of the landlord's agreement to this change that there be alterations to the terms of the lease, particularly H with regard to the recoverability of service charges and to the costs associated with recovery. He told us that he had been responsible for the drafting of the proposed amendments and that the clauses had been put to the body of lessees, but not as a separate matter from the question of the proposed new heating arrangements. He told us that if the landlord decided not to go ahead with the individual heating system because the terms could not be agreed, then there will be substantial cost to the lessees for upgrading the existing communal system. He was satisfied that the lessees were fully aware of what was being suggested, and that the clauses concerning the recoverability of service charges and the associated costs in the deed of variation were something of a quid pro quo".

38. Dr Rankin, in his witness statement dated 30 March 2010, J said that NERA had worked with the landlords and their solicitors and managing agents over the previous five to six years to permit the individualisation of the central heating and hot water systems "by amending the lease provisions in a way which is mutually acceptable to both the lessees and the landlords" (para 8). He confirmed that the non-consequential variations were at the instigation of the landlords who "have also sought to address a number of deficiencies in the existing leases in relation to the management of the blocks and administration of the service charges" concluding:

"23. …the proposed changes are to the benefit of the lessees as well as the landlords at the Etons, that they will assist both sides in the beneficial K management of the blocks and should be permitted as part and parcel of these applications.

"24. …I strongly believe and submit to [the LVT] that they are to the great benefit of all of the lessees, residents and tenants in occupation of the Etons…"

39. It was Dr Rankin's evidence that the landlords' managing agent had said that the applications to the LVT represented the last opportunity to individualise the heating and hot water systems and that if they did not succeed they "will have no alternative but to replace the communal systems before they fail completely". We comment here that practically speaking this is somewhat of a non-sequitur as by common L consent the non-consequential variations were not required to achieve individualisation of the communal heating and hot water system. The reasoning behind this statement appears to be that the landlords would only proceed if the non-consequential variations were also granted. It was in that sense that this represented a "last opportunity", there being no reason advanced as to why the landlords could not carry out the individualisation without the non-consequential variations.

Case as put to the LVT

40. In his written submissions to the LVT dated 19 April 2010, Mr Gallagher said that the applications were to enable replacement of the M communal heating and hot water systems "and also so as to up-date the covenants in the leases relating to the recovery of costs, insurance and reciprocal enforcement of covenants" (para 2). He later submitted that the non-consequential variations which "are not directly concerned with the proposed heating and hot water switch-over" are explained in Mr Goldreich's witness statement and "address various deficiencies in the leases which are affecting the management of the Etons". The latter submission was an assertion unsupported, as already noted, by any evidence.

41. There was then an adjournment, and revised variations were put before the LVT as well as supplemental written submissions dated 15 November 2010 in which Mr Gallagher submitted at para 2 that "the

objectives of the proposed variations (of dispensing with the central provision of heating and hot water in favour of individual units) can only be satisfactorily achieved if all of the leases demising flats in the Etons are varied to the same effect (section 37(3))". In other words, by this stage if not before, Mr Gallagher refined the applicants' case to one where all variations were to achieve that single object. That was consistent with the oral evidence of Mr Goldreich as recorded by the LVT and also the "single issue" nature of the ballot.

42. The LVT recorded in para 32 of the Decision that Mr Gallagher confirmed in final submissions "that the existing lease did not include the provision to recover costs and that the proposed amendments would not be retrospective and that accordingly the costs of these proceedings would not sought to be recovered from the respondents".

43. In its Refusal Decision, the LVT state that "it was never put to the [LVT] that the amendment of the service charge collection arrangements was necessary for the purpose of carrying out the works. In any event, the [LVT] dealt with this in para 48". The thrust of the Decision, and the Refusal Decision, is that the applicants put their case forward on the footing that there was a single "object" to be achieved by the proposed variations, viz replacement of the heating and hot water systems. This accorded with the respondents' understanding.

The Decision

44. As part of the section dealing with the "background" of the Decision, the LVT explained the reason (singular) for the applications:

"7. The reason for the request to vary was based on the assertion by the Applicants, or perhaps more particularly those lessees who supported the application, that the communal central heating system, now some 80 years old, was simply corroding away and not functioning satisfactorily. It was suggested that there were constant leaks which were expensive, time consuming and disruptive to remedy and that the wish therefore was to arrange for each flat to have its own individual central heating and hot water system, which is not possible under the present lease terms."

45. In the "findings" section of the Decision, the LVT, aware of course that it was exercising a discretion, sets out its understanding of the law thus:

"39. It seems to us however that the basis upon which any variation can be made is set out in Section 37 and 38 of the Act. We need to be satisfied as follows:

• Firstly, that the variation can only be achieved if all the leases are varied.
• That the applicant has garnered sufficient support to enable such an application to proceed.
• We shall not make an order varying the lease if it appears to us that the variation is likely to substantially prejudice any respondent, or indeed a person who is not a party to the application and that compensation would not compensate them.
• We should not make an order if for any reason it would not be reasonable in the circumstances for the variation to be effected.

"40. Further, section 38(8) gives us the power to order that the lease may be varied in such manner as we specify; and under section (10) of Section 38 there is the power for us to order compensation. These are the matters that we need to consider. These factors have been taken into account by us when we have determined the terms of the order and the proposed changes to the lease. In that regard we note in the draft order for the deed of variation the definition of the enabling works and of practical completion and also the provision in the body of the order providing for the deeds to be held in escrow until the cut-off date, and in the contents of the conditions precedent and transitional provisions in the order at paras 8 to 11 inclusive."

46. The LVT found that the replacement of the communal heating and hot water systems could only be effected by varying all the leases and that the requisite majority in support had been established. It considered and amended the applicants' draft order and determined that Miss Zographos and her brother might be substantially prejudiced by the installation of an individual heating system as they owned two flats albeit under a single lease but that they would be adequately compensated by an award under section 38(10): see paras 41 to 46.

47. The LVT then directed its attention to the wording of the variations to the leases in two paras, which form the kernel of this appeal, and made the following findings:

"47. In so far as the variations sought are concerned we think that we can probably take this quite shortly. The new covenants are contained in Schedule 2 and we are content with the wording put forward by the landlord, save and except that we are not persuaded that the changes in respect of the service charge provisions are reasonable. It does seem to us that the suggestions that the landlord should be able to substantially improve his position in respect of the recovery of costs against recalcitrant lessees to the detriment of other lessees, is inappropriate. In addition also, we do not see the need to expand the provisions at clauses 12/13 of the Fourth Schedule of the lease beyond that which is contained in the existing lease. We heard from Mr Goldreich that the reason for these clauses was a quid pro quo on the part of the landlord. Mr Gallagher in his submissions to us confirmed that it was open to the Tribunal to order some of the changes and not the others. However, he said that we should not [do] so because this was a quid pro quo arrangement. He suggested it would be objectionable for the tenants to accept part of the deed and walk away from the remainder. He reminded us that our discretion must be exercised judicially. Certainly, Mr Goldreich, as we have indicated, was of the view that these terms were included to sweeten the pill to the landlord for agreeing to the change from the communal to the individual heating system. For our part, we find it difficult to see what loss there is to the landlord in switching from a communal to an individual system. If anything, it ought to be of benefit to the landlord to avoid the responsibilities of a somewhat decrepit and potentially expensive communal arrangement. Furthermore, it seems to us that the original provisions of the lease, certainly insofar as the ability of the lessee to ask the landlord to enforce covenants, is conventional and perfectly adequate. The original clause provides that the landlord is entitled to an indemnity from the lessee in connection with any costs that may be incurred in seeking to enforce covenants, and we do not see that it is necessary for this power to be further extended.

"48. The original lease contains reasonable provisions for the recovery of service charge payments, although is deficient insofar as the recovery of costs are concerned. However, the changes proposed by the Applicants do seem to us to give to the landlord more than is justified and certainly in our view are likely to substantially prejudice the respondents and others and we find it is not reasonable in the circumstances for this particular variation to take effect. The payment of compensation would not be appropriate."

The grounds of appeal and arguments in support and against

48. The appellants seek an order that the Decision of the LVT in relation to the three non-consequential variations be set aside and that new clauses 5, 6(8)(b) and para 12/13 be inserted as requested.

49. The 20 grounds of appeal, and the very lengthy skeleton argument in support, are wide ranging and multi-faceted centring as they do on paras 47 and 48 of the Decision putting the same points in almost every different way conceivable. Without in any way wishing to detract from their complexity and sophistication, in our view there are three central lines of argument.

50. Firstly, the LVT failed to make any, or any adequate or readily understandable, finding as to what the "object" to be achieved by the proposed variations was. It should have found that there were dual objects to the variations; to improve management and administration of the Etons as well as replacing the heating and hot water systems.

51. Secondly, the LVT failed to proceed on the basis that as a matter of law the majority view should prevail unless either the variations would be likely to substantially prejudice the respondent or any non-party and compensation is inadequate or there is some other reason it would not be reasonable in the circumstances to make the variations pursuant to section 38(6). Great weight should have been (but was not) given to the facts that the agreement was supported by the overwhelming majority of tenants and as explained by Mr Goldreich and Dr Rankin was the product of protracted and complex negotiations involving a disparate and fluctuating body of lessees, some resident, some non-resident and amounted to a compromise of how to replace the communal heating and hot water systems in a manner mutually satisfactory to both sides which should not be scuppered by the very small minority of three active respondents.

52. Thirdly, the LVT made erroneous findings of law or approached the issues in ways which were wrong in law and did not give any or any adequate reasons for its findings. It substituted its own views of what was "needed" rather than accepting the views of the majority. It conflated what was "reasonable" with whether the object of the variations could only be satisfactorily achieved by making

A the variations. It found that the landlord was given more than was "justified". It said it was "inappropriate" for the landlords to improve their position in the negotiations when the reasons for and wisdom of the commercial compromise which had been reached were irrelevant to the exercise of its discretion.

53. The respondents submitted that the Decision is clear and intelligible. The LVT approached the applications on the basis that the "object" was replacement of the communal heating and hot water systems, that the non-consequential variations were required by the landlords as the "price" of replacing the heating and hot water systems which the LVT for the reasons stated found to be impermissible or
B in any event of substantial prejudice to the respondents or otherwise would not be reasonable in the circumstances to allow. This, they said, was all consistent with the evidence and how the case was presented before the LVT.

Adequacy of reasons given in the Decision

54. Regulation 18 of the leasehold Valuation Tribunals (Procedure) (England) Regulations 2003 does not give any guidance about the details of the reasons which must be given for a decision of the LVT. In *Lucie M*, Lawrence Collins J (as he then was) summarised the position in relation to all tribunals, even though he was dealing with a decision
C of the Special Educational Needs Tribunal:

> "10. The principles applicable to challenges of the tribunal's decision are not unique to this tribunal, but since several cases in relation to its decisions were cited I summarise their effect. First, proper and adequate reasons must be given, so that they are intelligible and deal with the substantial points that have been raised, and the reasons should deal, in short form, with the substantial issues raised in order that the parties can understand why the decision has been reached: *S* v *Special Educational Needs Tribunal and the City of Westminster* [1996] ELR 102 at 112; *R (on the application of B)* v *Vale of Glamorgan County Borough Council* [2001] ELR 529, 536: *Crean* v *Somerset County Council* [2002] ELR 152, 164 to 165. Secondly, and as a result of the first principle, the absence of reasons to explain why a case was rejected may make the decision
> D appear irrational: *Crean* at 167. Thirdly, where reasons are inadequate, it is not normally appropriate that the reasons should be amplified on the appeal to the High Court: *Oxfordshire County Council* v *GB* [2002] ELR 8, at 11 (CA).
>
> 11. Fourthly, a decision must be sufficiently specific and clear as to leave no room for doubt as to what has been has been decided: *London Borough of Bromley* v *Special Educational Needs Tribunal* [1999] ELR 260, 297 (CA)."

55. Those principles are broadly consistent with the three authorities considering the nature and extent of the duty of the judge to give reasons to which we were referred. In *Flannery* v *Halifax Estate Agencies Ltd* [2000] 1 WLR 377 the Court of Appeal said at pp381G-382C:

> E "We make the following general comments on the duty to give reasons.
>
> (1) The duty is a function of due process, and therefore of justice. Its rationale has two principal aspects. The first is that fairness surely requires that the parties especially the losing party should be left in no doubt why they have won or lost. This is especially so since without reasons the losing party will not know (as was said in *Ex parte Dave*) whether the court has misdirected itself, and thus whether he may have an available appeal on the substance of the case. The second is that a requirement to give reasons concentrates the mind; if it is fulfilled, the resulting decision is much more likely to be soundly based on the evidence than if it is not.
>
> (2) The first of these aspects implies that want of reasons may be a good self-standing ground of appeal. Where because no reasons are given it is
> F impossible to tell whether the judge has gone wrong on the law or the facts, the losing party would be altogether deprived of his chance of an appeal unless the court entertains an appeal based on the lack of reasons itself.
>
> (3) The extent of the duty, or rather the reach of what is required to fulfil it, depends on the subject matter. Where there is a straightforward factual dispute whose resolution depends simply on which witness is telling the truth about events which he claims to recall, it is likely to be enough for the judge (having no doubt summarised the evidence) to indicate simply that he believes X rather than Y; indeed there may be nothing else to say. But where the dispute involves something in the nature of an intellectual exchange, with reasons and analysis advanced on either side, the judge must enter into the issues canvassed before him and explain why he prefers one case over the other. This is likely to apply particularly in litigation where as here there is disputed expert evidence; but it is not necessarily limited to such cases.
>
> (4) This is not to suggest that there is one rule for cases concerning the G
> witnesses' truthfulness or recall of events, and another for cases where the issue depends on reasoning or analysis (with experts or otherwise). The rule is the same: the judge must explain *why* he has reached his decision. The question is always, what is required of the judge to do so; and that will differ from case to case. Transparency should be the watchword."

56. As was put by Lord Phillips MR (as he then was) in *English* v *Emery Reimbold & Strick Ltd* [2002] EWCA Civ 605; [2002] 1 WLR 2409, in [16]:

> "We would put the matter at its simplest by saying that justice will not be done if it is not apparent to the parties why one has won and the other has
> lost." H

In *South Bucks District Council* v *Porter (No 2)* [2004] UKHL 33; [2004] 1 WLR 1953 the House of Lords, considering the adequacy of reasons for a planning decision, said at [36]:

> "The reasons for a decision must be intelligible and they must be adequate. They must enable the reader to understand why the matter was decided as it was and what conclusions were reached on the "principal important controversial issues", disclosing how any issue of law or fact was resolved. Reasons can be briefly stated, the degree of particularity required depending entirely on the nature of the issues falling for decision. The reasoning must not give rise to a
> substantial doubt as to whether the decision-maker erred in law, for example J
> by misunderstanding some relevant policy or some other important matter or by failing to reach a rational decision on relevant grounds. But such adverse inference will not readily be drawn. The reasons need refer only to the main issues in the dispute, not to every material consideration. … Decision letters must be read in a straightforward manner, recognising that they are addressed to parties well aware of the issues involved and the arguments advanced."

Discussion and decision

Did the LVT err, and did it give adequate reasons?

57. We have carefully considered the various submissions and arguments made by the appellants, and the conclusion we have reached
is that when viewed in the light of the applications, the evidence K
adduced before the LVT and the case as put to it, the Decision properly, intelligibly and adequately addresses the central issues sufficient for any of the parties, including this tribunal, to understand what its decision is and why it has reached that decision consistent with the above-cited authorities.

58. In our judgment, it is clear from the Decision that the LVT found, on the basis of the evidence before it and the case put to it by the applicants' then counsel, Mr Gallagher, that there was a single object to be achieved by the proposed variations to the leases. Although
Mr Gallagher makes passing reference in paras 25 and 32 of his L
original submissions to what Mr Goldreich described as the "non-consequential" amendments, those submissions are almost wholly concerned with the issue of the heating and hot water system. By the time of Mr Gallagher's supplemental submissions in November 2010 this emphasis is even more pronounced. Indeed the parenthesis in para 2 of those submissions refers to a single object of the proposed variations, see [41] above. The LVT proceeded to assess the proposed variations, as it was entitled to do, by reference to this single object.

59. Mr Goldreich's evidence sought to "clarify and explain" the non-consequential variations which he had drafted. He said that the
landlords' professional advisers: M

> "wished to address certain deficiencies in the flat leases which were affecting their management of the Etons."

But Mr Goldreich's non-consequential variations, which, in our judgment mostly failed to achieve the landlords' intention (for reasons that we give below), were presented in the context of a single objective, namely to replace the communal heating and hot water system with individual boilers. This was the objective that Mr Gallagher spoke to in his supplemental submissions, and was clear to all concerned. That was the commercial deal or compromise reached by the parties. There was no, or no sufficient, evidence of any other object to be achieved by the non-consequential variations.

A 60. Having carefully considered all of the written evidence before the LVT and the submissions before us, we are not surprised that that was the common position of all parties by the end of the hearing before the LVT, however it might have been put or explained at the outset. Whilst the LVT could have expressly and deliberately considered each of the possible objects of the non-consequential variations as now submitted to this tribunal by Mr Lederman, it did not, and did not need to, because by the end of the hearing it was clear to all concerned that there was but one single object to be achieved.

61. That the LVT found a single object is in our judgment clear from para 7 of the Decision, which is carried through and consistent with the B rest of the Decision, especially paras 47 and 48 where the LVT tests the proposed variations against that "object" and gives its reasons as to why it will not exercise its discretion and grant the non-consequential variations sought.

62. When viewed in that light, the effect of paras 47 and 48 is that the object could be achieved without the non-consequential variations. That, Mr Lederman accepted, was correct as they were all wholly unrelated to replacement of the communal heating and hot water system save in one respect, namely, without new clauses 5(ii) and para 12/13 the communal heating and hot water systems could not be achieved or satisfactorily achieved. That, as the Refusal Decision notes, was not C an argument put forward before the LVT. Neither was it the subject of any evidence. Nor was it accurate as the leases all have adequate, standard-form covenants enabling the landlords to recover the service charges and there was no evidence of past difficulties in service charge collection.

63. Paras 47 and 48 have been subjected to lengthy critique and analysis suggesting that the LVT was impermissibly analysing, commenting on, disagreeing with and substituting its own views for the commercial wisdom or solution proposed by the compromise which the landlords and the majority of tenants had reached. At the same time, Mr Lederman submitted that while each variation should be tested D against its separate and distinct object, any substantial prejudice caused by the non-consequential variations was balanced out by getting the communal heating and hot water system replaced.

64. We do not accept these arguments. In our judgment, when properly understood, the LVT in para 47 goes through each of the three non-consequential variations and concludes that replacement of the communal heating and hot water system can be achieved without any of those proposed variations. Although it has not slavishly followed the statutory wording, that is the substance of its Decision: the landlords have used the replacement of the communal heating system to obtain the non-consequential variations which are not required to achieve that E object and, whilst it and the majority of tenants are fully entitled to reach such an agreement, if they can not point to any other object to be achieved by those variations they fail to satisfy the requirements of section 37(3) and the applications fail.

65. In the alternative and in any event, the LVT goes on to say in para 48 that the non-consequential variations are likely to substantially prejudice the respondents and others and are not reasonable in the circumstances. In stating that the landlord has got more than is justified the LVT is balancing the various aspects of the case against each other.

66. With no disrespect to Mr Lederman, what has happened in this F appeal is a complete re-working of the case without a full and proper analysis and understanding of the evidence and the issues. Indeed, as will become apparent, it was with some surprise that this tribunal was told that the appellants did not understand the full nature and effect of the non-consequential variations sought and that had they got what the applications in fact asked for it would have been "wrong". It may well be that had the appellants properly understood the true nature and effect of the non-consequential variations asked for they would have more readily understood why the LVT refused to grant them.

67. We will now embark upon a fully reasoned decision and exposition addressed to each of the issues now raised by Mr Lederman. We adopt his approach of considering each of the non-consequential variations by reference to the related evidence such as there was. It will become clear that in reality there was only one object. We will G also address, as appropriate, the various arguments aired before us, conscious that many were not advanced before the LVT because, as we have already said, it is clear to us on the evidence before us that there was but one object.

68. In embarking upon this lengthy exposition we are conscious that it might be said that that of itself proves that the Decision is insufficient. We do not accept that argument. We are doing so because we must address the central arguments presented to us and give as full an explanation as is necessary to explain why we dismiss the appeal. It was not necessary for the LVT to do so because most of these arguments were not addressed to them – since, as we have stated, this was treated H by all as a "single object" case.

69. We make clear that had we concluded that the LVT had erred such that the Decision must be set aside, we would have reached the same conclusion and our reasoning would have been as set out below. What follows should therefore be treated in the alternative as our reasons upon a review for refusing to allow the non-consequential variations. There are however two factors which have made it difficult to treat this as a straightforward appeal or review. Firstly, the appellants misunderstood, and did not want or intend, what they had in fact asked the LVT for. Secondly, it seems – but remains a little unclear – that the appellants have modified their position before the LVT, namely that J without the non-consequential variations they would not replace the communal heating and hot water system, to one where they have not reached a final decision.

Detailed reasons

70. The first question to address under section 37 is: what is or are the "object" or objects to be achieved by the non-consequential variations? As a matter of statutory construction, there may be single or multiple objects. In many respects, the object of a variation will be self-evident from the content of the variation itself. But it does not follow that that is necessarily the object, or purpose, of the variations. We K accept Mr Lederman's submission that it is for the applicants, not the tribunal, to identify the "object" or purpose which, broadly speaking, may be of infinite variety depending upon the facts and circumstances relating to the leases, buildings and flats in question.

71. What the object is is a question of evidence to be adduced by the applicants: what are they trying to achieve by the variations, and why? What problems or deficiencies are there or have there been in running the blocks and enforcing the leases? What is the purpose of the variation(s)? Without this information, or evidence, the tribunal can not make any findings as to the "object" to be achieved, nor can it properly exercise its discretion, which includes an evaluation of the L proposed variation. Whilst the section 37 jurisdiction is wider than and distinct from that of section 35, there is room for overlap. For example, where there are unsatisfactory provisions relating to recouping the costs of repairs or maintenance an application can be made to remedy those deficiencies where there is a sufficient supporting majority of tenants under section 37 and, if not, under section 35. We accept the submission of Mr Lederman that it is not for the tribunal to determine whether they approve of the object, but it is for the tribunal to make a finding, based upon the evidence, of what the object is.

72. The second question is: can the "object" be satisfactorily achieved by the proposed variation without varying all the leases (we summarise, M but do not lose sight of, the statutory wording)? There are two questions here: does the proposed variation achieve the object, and if so do all of the leases need to be varied to satisfactorily achieve that object? These of course presuppose that the leases do not already have sufficient or satisfactory provisions: if they do it is obvious that they need not be varied as there will be no object, or purpose, to the variation. Or, as Mr Barnett tersely put it, "if it ain't broke don't fix it".

73. These are questions of evidence to be adduced by the applicants: how do the proposed variations achieve that object or objects? Can that only be satisfactorily achieved if all the leases are varied? The nature and extent of the evidence will of course depend upon the variations sought. It is also a question for legal argument: as a matter of law do the

variations achieve the object or are they capable of doing so? We accept Mr Lederman's submission that it is for the applicants, not the tribunal, to select the solution or variation from what will frequently be one of a number of different options and if the majority of tenants are supportive then it is not for the tribunal to second guess them although, of course, the tribunal would be at liberty to make suggestions.

74. For convenience, we consider these questions and the related evidence together. We should say however that we do not accept Mr Lederman's submission that the majority view should prevail unless the section 38(6) grounds are made out. In our judgment, the purpose of section 37 is to enable the majority to apply to the tribunal for a variation to achieve a particular object: if they can not bring themselves within those requirements, then there is no jurisdiction to entertain the application or consider it further. The jurisdiction is relatively narrow, and is not intended to allow rewriting of leases merely because that is the will of the majority and in many cases may well seem sensible.

75. New clause 6(8)(b): Mr Lederman drew attention to the annexure to the applications where the stated object of new clause 6(8)(b) is lumped together with the heating-related variations which are described as being one of the "further changes relevant to Individual Heating Systems": that is plainly wrong as it has nothing to do with the replacement of the communal heating and hot water systems.

76. It was the evidence of Mr Goldreich that the non-consequential variations generally were to remedy deficiencies in the leases. New clause 6(8)(b) was, he said, to enable "reciprocal enforcement of covenants between lessees". This was echoed by Mr Gallagher in his opening skeleton argument. Mr Lederman quite properly accepted that new clause 6(8)(b) did nothing of the sort: it enabled a lessee to require the landlord to enforce a covenant against a fellow tenant at the ultimate cost of all tenants via the service charge.

77. Mr Goldreich went on to explain in para 8(3) of his first witness statement that the existing clause 6(8)(b):

> "placed an unreasonable and disproportionate burden on a complainant lessee and the change is therefore intended to be for the benefit of the body of the lessees as a whole. It also introduced the safeguard that enforcement action was to be for the reasonable protection of the flat in order to prevent abuse by either an unreasonable lessee or an unreasonable landlord".

78. In the absence of any evidence that the existing clause 6(8)(b) had inhibited Shellpoint tenants from requiring the landlord to enforce covenants or that it had in any way caused problems in the management of the Etons or that anyone regarded it as placing an unreasonable or disproportionate burden on a complainant tenant, Mr Goldreich's comments amount to no more than a theoretical observation of a possible opinion as to the practical effect of this clause. An alternative view advanced with some force by the respondents, particularly Miss Zographos, is that the existing clause contains within it an important brake on tenants making unnecessary and unreasonable complaints by requiring them to take the risk of enforcement especially where the complaint is flat-specific (such as a noisy upstairs neighbour) and does not affect anyone else. It also discourages landlords from initiating enforcement proceedings unless well founded as they can only recover their costs from the tenant in breach. The contractual intent of the original parties should not be altered without good reason or sound evidence of which there was none.

79. Notwithstanding that Mr Goldreich did not in his witness statement distinguish between the Shellpoint and Anston leases, those comments only apply to the Shellpoint leases. In relation to the Anston leases, there was no evidence that the absence of any covenant similar to clause 6(8)(b) had caused any problems in the enforcement of covenants or the management of the Etons. As the respondents argued, it does not follow from the absence of such a provision that there have been any problems or that the covenants have not been enforced by the landlords who have ample tools in their armoury to enforce covenants against tenants in breach and which, of course, one would expect a reasonable and responsible landlord to do. If successful the landlords' costs can be recovered from the tenant in breach. Again, the contractual intent of the original parties should not be altered without good reason or sound evidence.

80. Mr Lederman submitted that the absence of any such evidence did not prevent the LVT or this tribunal from finding that section 37(3) had been satisfied, and that great weight should be put on the fact that the overwhelming majority of tenants supported the new clause 6(8)(b) from which it could be inferred that they recognised there was a problem which needed to be resolved, and that this was the appropriate resolution of that problem. It was therefore not to address a purely theoretical problem but was something which the applicants, on advice, had agreed to and that it was unreal, impractical and unsatisfactory in an estate comprising 365 flats for all leases not to contain the same provision, at any rate, upon an application to vary.

81. That is a significant inference to be asked to make and it is not a suggestion that was put to the LVT. Had we been making our own decision by review, we would not have been prepared to so infer. In any event, if Mr Goldreich's evidence is taken at face value, there being none other, the object to be achieved by new clause 6(8)(b) was to enable reciprocal enforcement of covenants *between lessees*. If the tenant applicants have taken that at face value, they would (wrongly) be thinking that under the new clause one tenant could directly enforce a covenant against another tenant without resorting to the landlord. We do not think we are taking too pedantic an approach: sight must not be lost of the fact that a section 37 application is a serious matter affecting as it does the contractual rights of others for the remainder of long lease terms (in many cases more than 950 years) and so should be soundly based on evidence.

82. In our judgment, in the absence of evidence, the LVT was right to conclude as it did in the last two sentences of para 47 that the existing provisions of the leases are conventional and perfectly adequate and it was not "necessary" for the landlord (ie Shellpoint) to extend its powers of recovery of costs by substituting or inserting, as the case may be, new clause 6(8)(b). It was not necessary or required to achieve replacement of the communal heating and hot water systems.

83. Were we to be considering this aspect of the applications afresh by review, we would reach precisely the same conclusion but with slightly different reasoning. We would have found that *if* the object of new clause 6(8)(b) was to achieve reciprocal enforcement it failed to do so. There was insufficient evidence to support any finding other than the object of the clause being (part of the deal) to replace the communal heating and hot water system; but it did not in fact achieve that object as it was unrelated to it. There was no evidence or explanation that the object, whatever it may be, could not be "satisfactorily achieved" without modification or insertion of the new clause into all leases when, on the evidence, there was nothing to indicate that the existing arrangements in which the Anston leases had not clause 6(8)(b) had caused any problems or difficulties. We accept that great weight must be given to the views of the majority, but if they have failed to adduce even rudimentary evidence to satisfy the requirements of section 37(3) the applications must fail. Neither, for completeness sake, do we find the wording of existing clause 6(8)(b) (it not unusually puts the onus or risk upon the complainant tenant) or its absence from the Anston leases to be deficient or unsatisfactory.

84. New clause 5(ii) and para 12/13: Mr Lederman again drew attention to the annexure to the applications which states that the "object" of new para 12/13 was to "improve management of estate" and new clause 5(ii) is bundled in with all of the other variations described as being one of the "further changes relevant to individual heating systems".

85. Mr Goldreich said in his first witness statement that these variations were to resolve the other main deficiency of "the lack of clauses enabling effective recovery of service charge arrears from individual non-paying lessees". This was echoed by Mr Gallagher in his opening skeleton argument. Mr Lederman quite properly accepted that this was not right as all leases contain standard form covenants enabling the landlord to sue the tenant for non payment of service charge (clause 2(2)) failing which he must pay interest on arrears (clause 2(4)) and

ultimately forfeit his lease (clause 5). Nothing more is required to enable legally effective enforcement.

86. In para 8(1) of that witness statement, Mr Goldreich went on to elaborate:

> "This clause [5(ii)] imposes upon the landlord the requirement to make reasonable and economically viable efforts to recover legal costs in proceedings against the defaulting lessee. Thereafter, unrecovered costs may be applied against the service charges provided that no section 20C order is made against the landlord and that the costs are subject to a detailed assessment with a capped solicitors charging rate on the basis that those costs could be charged as service charges on an interim basis pending such recovery.
>
> "The purpose of this clause was both to permit and encourage landlords to seek recovery of legal costs against non-paying lessees rather than relying upon recovery through the service charge to the benefit of the general body of the lessees as a whole. This clause provides a safeguard in that it requires the landlords to "use all reasonable and economically viable endeavours" to recover their costs against individual defaulting lessees on the basis that only non-recoverable costs can be finally applied against the service charges and those will be subject to detailed assessment."

87. In para 8(4), he briefly refers to new para 12/13 as defining "the service charge elements necessary to give effect" to the other new clauses and that it contains safeguards such as "reasonable" and "proper" and so on.

88. As already noted, there was no evidence that the absence of any provisions enabling the landlords to recover their unrecovered costs of enforcing covenants via the service charge has caused any problems in the past or was even an issue, and no evidence as to the nature, ownership or financial standing of either landlord, that their financial position had made it difficult or impossible to enforce the covenants or otherwise manage the Etons. There was therefore no evidence of any practical difficulties of enforcing the covenants which is often the case, for example, where the landlord company is owned by the tenants and has one asset (the freehold), one source of income (rent from the leases), and one source of money to pay for managing and maintaining the property and enforcing the covenants (the service charge provisions) as was the case in *Thirlway* v *Troy* [2012] UKUT 302 (LC); LRA/57/2012.

89. Mr Lederman submitted that the object of these new provisions was principally to improve the management and administration of the Etons but was also linked to replacement of the communal heating and hot water systems by strengthening the provisions for recovery of the service charge inevitably due in respect of those works. The absence of any evidence was not an obstacle as great weight should be put on the fact that the overwhelming majority of tenants supported these new provisions and they had been drafted by NERA's lawyer, Mr Goldreich, who, in para 10 of his first witness statement, had explained that he was there "to draft and negotiate the wording of those new clauses as fairly as possible in order to safeguard the interests of all the lessees at the Etons". There was also the already recited evidence of Dr Rankin that the variations "are to the benefit of the lessees as well as the landlords". Again, appropriate inferences can be drawn.

90. When pressed, Mr Lederman was unable to advance any explanation of, or justification for, the width of these new provisions. Rather, and with commendable frankness, he said that neither he nor, on instructions, his client landlords had understood these new clauses to be quite so far reaching. They intended that only (a) the costs of tenant-initiated complaints under the new clause 6(8)(b) and (b) the legal costs of recovering sums due in respect of the service charge (ie enforcing clause 2(2)) should be recoverable from all tenants via the service charge. He accepted that, in consequence, certainly in so far as his clients were concerned, the applications had sought more than was wanted or, presumably, they thought had been agreed with the majority tenants. Had the LVT made the ordered variation of the leases by insertion of the new clauses it would have been, as Mr Lederman put it, "wrong".

91. This puts a quite different perspective on the applications and also upon the evidence before the LVT and therefore us. Quite how the landlords got so much more than they bargained for is unclear, especially as Mr Goldreich regarded himself as protecting the interests of the lessees rather than prejudicing them. Whether Mr Goldreich understood what he had drafted is unclear, although it seems reasonable to infer that he did as the wording of these new clauses is straightforward and was well understood by the respondents and, we have no reason to suppose otherwise, the LVT. Indeed, it appears to us that Mr Goldreich did understand the width of what he had drafted because in para 9 of his first witness statement he addresses the respondents' concerns thus:

> "9. In their written witness statements, the opponents to the changes [the respondents] have characterised the proposed non-consequential variations as effectively giving the landlords 'a limitless pot of money' and 'long-term open-ended costs' and 'no downside from taking legal action' and 'unnecessary legal and professional costs'. Contrary to those contentions I would ask [the LVT] to accept that the proposed new clauses do impose a system of checks and balances on the landlord by requiring them to act reasonably and properly and in an economically viable way. The recovery of costs through the service charge will, of course, remain subject to the provisions of section 20C and to the discretion accorded to the courts or the LVT by the requirement of for detailed assessment subject to capped charging rates. These clauses will also not affect the rights of the lessees to seek adjudication of the reasonableness of their service charges including legal costs in the LVT in accordance with the provisions of Section 27A nor have I seen any wish on the part of the landlords during my involvement in this process to interfere with those rights."

92. Whether the width of the new provisions had been accurately explained to the tenants in the ballot is also unclear as the supporting information was not in evidence. If the landlords misunderstood these new provisions, it may well be that the majority tenants did as well, in which case it may call into question the validity of the ballot (something challenged by the respondents before but rejected by the LVT which of course did not know about the fundamental misunderstanding made by landlords about the width of the non-consequential variations). If the width of these new provisions was understood by the tenants and their adviser (including Mr Gallagher) it would be relevant to the exercise of the jurisdiction to know why the majority tenants thought that such open-ended and far ranging variations would, as Dr Rankin put it, be to the benefit of the tenants. It may well be, we do not know, that the supporting tenants did not really want these variations at all but were prepared to swallow a bitter pill as the price of getting the communal systems replaced. If so, the parties have ended up in the position that the pill was far bitterer than expected. Or there may be some other reason.

93. Drawing those strands together, we are left in some doubt as to precisely what Mr Goldreich was saying the object of these variations was and how they achieved that object. If, as he says, it is to enable effective recovery of service charge arrears from individual non-paying lessees, it does not achieve that objective because there are already covenants enabling legally effective recovery by clauses 2(2) and 5 of the leases and also clause 2(4) enabling the landlord to be compensated by payment of interest to the extent that it is out of pocket for having to fund the non-paying tenant's service charge contribution. If he means it is to be viewed from a purely practical point of view, to relieve the landlords of the financial obligation or risk of enforcing the covenants, no evidence to justify or otherwise explain the same had been adduced.

94. In those circumstances, in our judgment, the LVT was right to conclude as it did in para 47 that the "original provisions of the lease, certainly insofar as the ability of the lessee to ask the landlord to enforce covenants, is conventional and perfectly adequate". This sentence refers not just to the existing clause 6(8)(b) but also to the normal, and standard-form enforcement provisions in the leases which we have already referred to. It must be read back to the fourth sentence, where the LVT does "not see the need to expand the provisions at [para 12/13] … beyond that which is contained in the existing lease". A counsel of perfection would have been for the LVT to specifically refer to and cite them, but in our judgment the provisions are so conventional that it is satisfactory for the LVT to have in effect "taken them as read". It is only had they been unusual that it would have been necessary for them to be recited verbatim. It was not necessary or required to change these

provisions to achieve the object of replacing the communal heating and hot water systems.

95. Were we to be considering this aspect of the applications afresh by review, we would reach precisely the same conclusion but with slightly different reasoning. We would have found that if the object was to enable "effective recovery of service charge arrears from individual non-paying lessees" it was not made out as all leases contained legally sufficient provisions and there was no evidence indicating any practical difficulties of recovery of service charge. Furthermore, new clause 5(ii) and para 12/13 covered far more than was required to achieve that limited object. There was insufficient evidence to support any finding other than the object being (part of the deal) to replace the communal heating and hot water system and, for the reasons already stated, that object could be "satisfactorily achieved" with making the proposed variations to any, let alone all, of the leases. Neither, for completeness sake, do we find anything deficient or unsatisfactory in the wording of the present provisions.

96. We would have reached the same conclusions even if we were to adopt Mr Lederman's submission that these variations were to achieve the wider object of improving the management and administration of the Etons. We would have gone further. We can not see how enabling the landlords to recover their costs of enforcement of all covenants could be sensibly regarded as an "improvement". Whilst it greatly improves or enhances the landlords' position, it can not properly be regarded as an improvement for the tenants. It is more likely, as Miss Zographos submitted, to worsen the tenants' position because it shifts all financial risk from landlord to tenant, the landlord (and its agents) now having an incentive (or no disincentive) to enforce, rather than reach a consensual agreement or otherwise, knowing full well, as they would, that all costs would be contractually recoverable from the tenants via the service charge whose flats would be at risk of forfeiture for non payment.

97. Whilst the reasonableness of the decision to enforce, the costs of enforcement and any settlement would be open to challenge under sections 19 and 20C of the Landlord and Tenant Act 1985 ("the 1985 Act") in the LVT, that is a recipe for litigation and for the landlord to run up further costs which would themselves be recoverable under the new provisions unless challenged under the said sections. As is well known, the section 20C jurisdiction is not intended to displace the contractual right of the landlord to recover his costs of enforcement and the discretion of the LVT under that section may only be exercised where it is "just and equitable in the circumstances" which is not a results-driven jurisdiction: see for example *Schilling* v *Canary Riverside Development PTE Ltd* LRX/26/2005 (LT). It frequently results in tenants having to pay the landlords' and their own costs because the landlord can not be shown to have acted unreasonably. Indeed, sections 19 and 20C were in part introduced to ameliorate the rigours of the contractual right of a landlord to recover its costs via the service charge.

98. The present contractual certainty of the landlords not being able to recover their costs from tenants through the service charge would be replaced by a certainty that they would be recoverable from all tenants subject to successful challenge under the 1985 Act. It follows, as again submitted by Ms Zographos and also Mr Barnett and Mr Clyne, that this would fundamentally alter the relationship between landlord and tenant, particularly where neither landlord is owned nor controlled by the tenants. As Mr Lederman acknowledged, the new provisions go beyond what the appellants intended or thought they had agreed with the majority tenants. The proposed variations may well be appropriate where the landlord is a single asset company owned and controlled by the tenants with no source of revenue other than the service charge and rent. But there was no evidence to justify this here.

99. In our view, the fact that the proposed variations go beyond what the appellants wanted or thought they had agreed with the majority tenants of itself indicates that the landlords do not regard these provisions as being required to achieve the object otherwise they would actually have asked for them. Weight should of course be given to the fact that the majority of tenants support these new provisions. However, for the reasons already stated, there is certainly a doubt in our minds as to whether or not they properly understood the full nature and extent of the new provisions when casting their ballots. Whilst there are only three active respondents, there were 17 opponents, 48 unreturned and one neutral. Even if they did understand what they were asking for, there is no explanation or evidence as to why they are requesting more than the landlords want.

100. We can not help but observe that the progress of these applications and the appeals offer a vivid illustration of why in this case the variations should not be permitted. In fact, the applications are the second attempt to vary the leases, the first attempt having been aborted after issuance. These applications however have been most unfortunate: whilst strongly based in respect of replacing the communal heating and hot water systems, in our judgment they have been poorly based and ill thought out in respect of the non-consequential variations. If the variations were already in place, the landlords would have a contractual right to recover all of its costs from the tenants. That right, we consider, would be of substantial value to the landlords. The tenants (or the minority tenant respondents) would then only be able to challenge those costs under sections 19 or 20C of the 1985 Act. The success of such a challenge may well be difficult and would trigger more litigation. It was also unfortunate that Mr Goldreich referred to the wrong clause numbers in the leases, and that the wrong clauses were reproduced in the annexures to the applications.

101. Conclusions on the first and second questions: In our judgment the appellants failed before the LVT and, had we been deciding the matter by review, would have failed before this tribunal to discharge their burden of proving the object or objects to be achieved by the non-consequential variations; that the variations achieve such object or objects; or that whatever the object or objects are they can only be satisfactorily achieved by varying all of the leases.

102. On analysis, and in reality, it was apparent to the LVT and is apparent to this tribunal that the only real object of the non-consequential variations was to get the communal heating and hot water systems replaced. In that respect, we respectfully agree with the observations of the LVT that the non-consequential variations went far beyond what was justified by, or necessary to achieve, replacement of the communal heating and hot water system. We also agree with the observations of the LVT that it is "inappropriate" for landlords to use such circumstances as an opportunity to substantially improve its contractual position – ie to demand new covenants unrelated to and not consequent upon achieving the object. In the rest of para 47, the LVT was properly considering the thinking or reasoning behind that state of affairs, and concluded that replacing the communal system was not just in the tenants' but also in the landlords' interests, a sentiment from which Mr Lederman did not dissent save to say that it was up to the parties what compromise deal was reached, and a sentiment which on the basis of the evidence before the LVT is well founded.

103. As we have said, Mr Lederman urged upon us that great weight should be placed on the will of the majority, and the view of the minority should not be able to thwart that will unless there were truly exceptional circumstances. Once, as here, the applicants have proved that they have the requisite section 37(5) majorities, it is for them to adduce evidence to prove that the variations satisfy section 37(3). Merely saying they agree to them is insufficient: the jurisdiction does not provide a statutory framework for implementing majority decisions or compromises with landlords but to vary leases in order to satisfactorily achieve a given object or objects. If the applicants fail to make a case out under section 37(3), their will has not been defeated by the minority: the majority (and landlords) has failed to bring itself within the statutory jurisdiction.

104. For the sake of completeness, we will briefly state what our decision would have been had we found that the applicants, or appellants, had satisfied the requirements of section 37(3) in relation to the non-consequential amendments. There would then be three further questions to determine.

105. The third question: would the proposed variations be likely substantially to prejudice the respondents to the applications (all non-supportive tenants) such that it can not be adequately compensated by an award under section 38(10) (section 38(6)(a))?

A 106. In our judgment, they would. Sight must not be lost of the fact that variations to leases interfere with the contractual bargain which the original parties reached and form the basis upon which their successors (landlords and tenants) acquired and will continue to acquire their respective interests. One of the singular aspects of these leases is that the landlords have covenanted to repair and maintain the Etons but at the cost of the tenants. How, when, by whom and at what cost the work is done is decided by the landlords over whom the tenants have no control as (so far as we are aware) the tenants do not own them. The landlords decide how to spend the tenants' money. At common law there is little control or basis for complaint by the tenants. As already B observed, statute has intervened to fill the gap to some extent by, principally, sections 19, 20 and 20C of the 1985 Act. We say "to some extent" because whilst the 1985 Act ameliorates the rigours of the law of contract, the source of the rigour is the landlord's contractual right to recover costs from the tenants and that base-line right is not altered by the Act.

107. It is in our judgment a quite exceptional, and substantially prejudicial, thing to enable the landlords to recover its costs not only of recovering the service charge but also of enforcement of all of its covenants from all tenants through the service charge, particularly where the landlords are not owned or controlled by the tenants and C there is no evidence that the landlords can not afford to do so or that the absence of such covenants has caused any difficulties in the past or will or is likely to in the future. It would enable the landlords to decide how, when, by whom and at what cost they should enforce covenants. That would shift all the financial risk and liability from the landlords to the tenants whose only control would be proceedings via the LVT and all the time, trouble, cost and uncertainty that involves. The appellants have put forward no justification for such a major risk and liability transfer and all that entails. In our view the proposed variation would significantly affect the way in which the landlords, given virtual financial impunity (subject to any 1985 Act challenge), make future D decisions about whether to enforce and pursue breaches of covenants.

108. If new clauses 5(ii) and para 12/13 were allowed, it is difficult to see why new clause 6(8)(b) would be required. In any event, enabling tenants to require the landlords to institute proceedings without the tenants taking personal financial liability as envisaged would, we agree with the respondents, serve to shift the risk from the complainant tenant to all other tenants even where the complaint had nothing to do with them and was specific to the complainant. The landlords would have little incentive to refuse.

109. If the object or objects of the non-consequential variations is or are separate and distinct from the replacement of the communal E heating and hot water system, they must be considered separately and the fact of substantial prejudice prevents these variations being granted, the language of section 38(6) being mandatory. If their object is the replacement of the communal heating and hot water systems, the result is the same as there is substantial prejudice. We doubt whether that prejudice should be weighed or balanced against the undoubted and substantial benefit of providing individual heating and hot water systems because section 37(3) is not directed at endorsing essentially commercial agreements or compromises between landlord and majority tenants, but even if it were to be so weighed or balanced the non-consequential variations should still not be allowed because they are F not required to achieve that object and their prejudice, long term and wide ranging as it is, outweighs the benefits of individualisation of the communal heating and hot water systems.

110. Either way, in our judgment an award of compensation would not afford adequate compensation. Whilst there was no evidence (valuation or otherwise) on the point, we are of the view that this sort of prejudice is not capable of financial compensation, affecting as it does the most important aspect of leasehold property and the attendant costs not just on a one off basis but for the remainder of the unexpired lease terms, many of which exceed 950 years.

111. We should say that we have been concerned by, and given thought to, the absolute and now somewhat modified position of the landlords. As we ventilated at the hearing in discussions with Mr Lederman, it seems to us that if the landlords were now to refuse to go G forward with the proposal to provide the tenants with individual boilers, which we understand would be cheaper and save the tenants a large amount of money in the future, that would afford the tenants the grounds to apply to the LVT for a pre-emptive order under sections 19 et seq of the 1985 Act that abandoning the proposals and simply replacing the communal system was unreasonable so that any costs over and above those incurred in providing individual boilers would be unrecoverable. Alternatively the tenants could make a similar application after the works had been completed. We simply mention this for completeness, without making any further observations.

112. The fourth question: is there any other reason it would not be H reasonable in the circumstances for the variations to be effected (section 38(6)(b)? On the evidence before the LVT, it was unnecessary to make any finding under this sub-section. Mr Lederman submitted that the sort of things this provision was addressed to was a situation where there were grounds for serious concern that say the ballot had been procured by fraud, undue pressure or the tenants had otherwise been misled.

113. Were we making our own decision by review, the fact that the evidence to the LVT and, initially, to this tribunal, described the nature and extent of the non-consequential variations inaccurately, whether wittingly or unwittingly, we would have found that it would not have been reasonable in the circumstances to allow these variations. We J would make this finding because such inaccuracy casts doubt upon the reliability of the ballot and precisely what it was the majority tenants in fact thought they were agreeing to. This conclusion is buttressed by the landlords' apparent shift in intention from not proceeding with the provision of individual boilers at all if they did not get the non-consequential variations they demanded to possibly doing so: had the majority tenants so realised and been balloted accordingly, perhaps they would not have voted for the non-consequential variations, remembering, of course, that there was no separate vote for them because all the variations were put to them as a "take it or leave it" package. We would therefore refuse to exercise the discretion on this K ground alone. In making these findings we bear in mind that any such decision would be by review only, which does not permit the adduction of new evidence. However, in our judgment that does not prevent us from taking cognisance of the landlords' misunderstanding and change of position.

114. The fifth question: in all the circumstances, should the tribunal exercise its discretion and make an order varying the leases? The only point we make here is that even if a tribunal is not prohibited from making an order by a section 38(6) finding, it must still exercise its discretion. If there are no section 38(6) factors, it may well conclude that the variations should be allowed. But not necessarily so. Given our L previous conclusions this question does not arise in this appeal.

Conclusion

115. For the reasons we have given above we are not satisfied that the object to be achieved by the non-consequential variations can not be satisfactorily achieved unless all the leases are varied to the same effect. If we are wrong in that conclusion we would not make an order varying the leases because it appears to us that the non-consequential variations would be likely to substantially prejudice the respondents and other persons who are not parties to the application and that an award under section 38(10) would not afford them adequate compensation M and in any event it would not be reasonable in the circumstances for the variations to be effected. In our opinion the LVT's Decision was sufficient and correct and we dismiss the appeal.

Reasons for refusal of application to amend the application

116. On the second day of the hearing, Mr Lederman applied for permission to amend the applications so that if the Decision was set aside this tribunal should reach its own decision but based upon the non-consequential variations limited to recovery of the appellant landlords' legal costs of recovering the service charge and of enforcing covenants pursuant to the new clause 6(8)(b) through the service charge.

117. We refused this application for the following reasons. Firstly, it would in our judgment be grossly unfair to the respondents to have

A to now deal with what in effect was a completely new application especially where they as, albeit sophisticated, lay persons had prepared evidence for and addressed their minds to the applications in their original form. Secondly, the respondents should have time to take legal advice on the nature and effect of the new non-consequential variations. Thirdly, the majority tenants would need to be informed of developments: they may now seek to adopt a different position now the landlords may not require the non-consequential variations as the price of replacing the communal systems; or they may wish to support the non-consequential variations in their current form in which case it would be important for their reasons to be before the tribunal. Fourthly,
B there comes a time in litigation where it really is too late for a party to amend its pleadings. And this is just such a case. It is in our judgment an **C**
unacceptable position for the landlords to have made applications to the LVT and then appealed its refusal to grant variations which the landlords neither understood nor wanted. All of the evidence and submissions had been addressed and prepared on the basis of the original variations. It is neither appropriate nor fair to allow the appeal to be heard on a new basis or indeed to adjourn to allow further evidence to be put forward. If the appellants wish to obtain the non-consequential variations in a reduced form, they will have to start again.

Appeal dismissed. **D**

NUISANCE

A
Stannard (t/a Wyvern Tyres) v Gore

COURT OF APPEAL

[2012] EWCA Civ 1248

4 October 2012

WARD, ETHERTON AND LEWISON LJJ

Nuisance – Rule in Rylands v Fletcher – Fire – Fire starting on business premises fuelled by tyres stored there for purposes of business of motor vehicle tyre supplier and fitter – Fire destroying
B ***adjoining property – Whether strict liability for damage to adjoining property arising under Rylands v Fletcher***

The appellant carried on business as a motor vehicle tyre supplier and fitter from premises on a trading estate. A fire broke out on the premises owing to an electrical fault; fuelled by the tyres stored there, the fire destroyed both the appellant's property and the respondent's adjoining premises. The respondent claimed damages from the appellant.

In the county court, the respondent's primary claim against the appellant in negligence was rejected; however, a secondary claim, in strict liability under the principle in *Rylands* v *Fletcher* (1868) LR 3 HL 330, was allowed. The Recorder found that the appellant had squeezed
C tyre stock into his premises, storing it haphazardly and untidily. He further found that although tyres did not ignite readily, once they caught fire from another source they were difficult to put out. He held that the requirements for liability under *Rylands* v *Fletcher* were met since: (i) the defendant had brought a dangerous thing, namely the tyres, onto his premises; (ii) the danger had "escaped" from the appellant's land onto the respondent's land; and (iii) the appellant's use of his land was "non-natural" and out of the ordinary because of the manner and numbers in which he stored tyres there.

On appeal, the appellant contended that the Recorder had erred in his application of the *Rylands* v *Fletcher* test for strict liability as it applied
D to fire cases.

Held: **The appeal was allowed. (1) In order for liability to arise under the rule in *Rylands* v *Fletcher*, it must be shown that the defendant, as the owner or occupier of land, has brought or kept or collected an exceptionally dangerous or mischievous thing on the land, in circumstances where he must have recognised, or should reasonably have recognised, judged by the standards appropriate to the relevant place and time, that there was an exceptionally high risk of danger or mischief if the thing escaped, however unlikely an escape seemed. The defendant's use of the land must be extraordinary and unusual, having regard to all the circumstances**
E **of time and place. Liability will then arise if the thing escapes from the defendant's property into or into the property of another, causing damage of a kind relevant to the rights and enjoyment of the claimant's land; damages for death and personal injury are not recoverable under the rule. It is not necessary to establish negligence on the part of the defendant, but an act of God, or the act of a stranger, will provide a defence. (2) There is no special rule for cases involving damage caused by the spread of fire. Damage caused by fire emanating from an adjoining property will rarely fall within the general *Rylands* v *Fletcher* rule since: (i) what must escape is the "thing" that has been brought onto the land, not the fire that is started or increased by the "thing"; (ii) while fire might be a dangerous thing, the occasions when fire, as such, is**
brought onto the land may be limited to cases where the fire is deliberately or negligently started by the occupier or one for whom he is responsible; and (iii) in any event, starting a fire on one's land might well be regarded as an ordinary use of the land. Per Lewison LJ: An occupier of land will not be liable to his neighbour for a fire that begins accidentally unless he is negligent in failing to prevent its spread. The general test of negligence might entail the taking of special precautions where the use in question involves the accumulation or storage of inflammable or readily combustible materials, but that will be a question of fact to be decided on a case
by case basis. (3) The appellant was not liable under the *Rylands* v *Fletcher* rule. The "thing" brought onto the appellant's premises was a large stock of tyres; tyres, as such, were not exceptionally dangerous or mischievous and there was no evidence that the appellant had recognised, or ought reasonably to have recognised, an exceptionally high risk of danger or mischief if the tyres escaped. The tyres had not in fact escaped; it was the fire that escaped, stoked by the tyres burning on the appellant's premises. The escape of the fire did not bring the case within the rule. In any event, keeping a stock of tyres on the premises of a tyre-fitting business, even a very large stock, was not, for the time and place, an extraordinary or unusual use of the land. Accordingly, since no negligence had been
proved, the claim failed.

Per curiam: **Etherton LJ and Lewison LJ took different approaches to the issue of whether liability for a fire under the rule in *Rylands* v *Fletcher* could coexist with the old statutory defence, in section 86 of the Fires Prevention (Metropolis) Act 1774, against the liability of an occupier of premises for a fire that started accidentally. Per Lewison LJ: The principle in *Rylands* v *Fletcher* could not defeat the section 86 defence for the consequences of a fire that started and spread accidentally, in the sense that it involved no negligence. Per Etherton LJ: Section 86 was intended to clarify**
the old customary law of *ignis suus*, regarding a person's liability for his own fire, and did not apply to liability under *Rylands* v *Fletcher*.

The following cases are referred to in this report.

Aldridge v *The Great Western Railway Co* (1841) 3 Man & G 515
Anderson v *Oppenheimer* (1880) 5 QBD 602
Attorney General v *Cory Bros & Co Ltd* [1921] 1 AC 521, 525, 530, 534, 536
Balfour v *Barty-King* [1957] 1 QB 496
Beaulieu v *Finglam* (1401) B&M 557
Becquet v *MacCarthy* [1831] 2 B&AD 951, 958
Black v *Christchurch Finance Co Ltd* [1894] AC 48
Boulcott Golf Club Inc v *Engelbrecht* [1945] NZLR 556 K
Burnie Port Authority v *General Jones Pty Ltd* (1994) 179 CLR 520
Bybrook Barn Centre v *Kent County Council* [2001] BLR 55
Cambridge Water Co v *Eastern Counties Leather plc* [1994] 2 AC 264
Carstairs v *Taylor* (1871) LR 6 Ex 217
Collingwood v Home & Colonial Stores [1936] 3 All ER 200, 205
Crogate v *Morris* (1617) 1 B&G 197
Donoghue v *Stevenson* [1932] AC 562
E Hobbs (Farms) Ltd v *The Baxenden Chemical Co Ltd* [1992] 1 Lloyd's Rep 54
Filliter v *Phippard* (1847) 11 QB 347, 357
Fletcher v *Rylands* (1866) LR 1 Ex 265, 279/280
Goldman v *Hargrave* [1967] 1 AC 645 (PC)
H&N Emanuel v *Greater London Council* [1971] 2 All ER 835

A *Harooni* v *Rustins Ltd* [2011] EWHC 1632 (TCC)
Heaven v *Pender* (1882-1883) 11 QBD 503
Job Edwards Ltd v *Birmingham Navigation Proprietors* [1924] 1 KB 341
Johnson v *BJW Property Developments Ltd* [2002] EWHC 1131 (TCC); [2002] 3 All ER 574
Jones v *Festiniog Railway Co* (1868) LR 3 QB
LMS International Ltd v *Styrene Packaging and Insulation Ltd* [2005] EWHC 2065; (TCC) 30 September 2005
Long Bee & Co v *Ling Nam Rubber Works* [1970] 2 Lloyd's Rep 247
Mason v *Levy Auto Parts of England Ltd* [1967] 2 QB 530, 542
McKillen v *Maybourne Finance Ltd* [2012] EWCA Civ 864
Mulholland & Tedd Ltd v *Baker* [1939] 3 All ER 253
Musgrove v *Pandelis* [1919] 2 KB 43
B *Piggot* v *The Eastern Counties Railway Co* (1846) 3 CB 229
Powell v *Fall* (1879-80) LR 5 QBD 597
Rainham Chemical Works Ltd v *Belvedere Fish Guano Company Ltd* [1921] 2 AC 465
Read v *J Lyons & Co Ltd* [1947] AC 156
Richards v *Easto* (1846) 15 M&W 246, 251
Rickards v *Lothian* [1913] AC 263, 280
Ross v *Fedden* (1872) 26 LT 966, 968
Rylands v *Fletcher* (1868) LR 3 HL 330
Sedleigh-Denfield v *O'Callaghan* [1940] AC 880
Smith v *The London and South Western Railway Company* (1869-70) LR 5 CP 98
Sochacki v *Sas* [1947] 1 All ER 344
C *Solomons* v *R Gertzenstein Ltd* [1954] 1 QB 565
Spicer v *Smee* [1946] 1 All ER 489
Starmark Enterprises Ltd v *CPL Holdings Ltd* [2001] EWCA Civ 1252; [2002] Ch 306
Stockport Metropolitan Borough Council v *British Gas plc* [2001] EWCA Civ 212
Sturges v *Hackett* [1962] 1 WLR 1257
Transco v *Stockport Metropolitan Borough Council* [2004] 2 AC 1
Turberville v *Stamp* (1702) 12 Mod 152; (1702) 1 Comyns 32
Vaughan v *Menlove* (1837) 3 Bing NC 468
Vaughan v *The Taff Vale Railway Company* (1858) 3 H&N 743
Viscount Canterbury v *Attorney-General* (1843) 1 Phillips 306
Williams v *Owen* [1955] 1 WLR 1293
Young v *Bristol Aeroplane Co Ltd* [1944] KB 718, 729

D This was an appeal by the appellant, Mark Stannard, trading as Wyvern Tyres, from a decision of Mr Recorder Potts, sitting in Worcester County Court, allowing a claim by the respondent, Robert Gore, for damages under the rule in *Rylands* v *Fletcher* in respect of property damage caused by fire.

Jonathan Waite QC and Michele de Gregorio (instructed by DAC Beachcroft LLP) appeared for the appellant; Philip Rainey QC and Nicholas Isaac (instructed by Beaumonts Solicitors LLP, of Hereford) represented the respondent.

Giving judgment, **Ward LJ** said:

E 1. Every law student is likely to remember how Mrs Donoghue suffered when a snail in a state of decomposition floated into her glass from the bottle of ginger beer manufactured by Mr Stevenson. Almost as memorable must be the case of Mr Ryland's reservoir bursting and flooding Mr Fletcher's colliery, which gave rise to the rule we now know as *Rylands* v *Fletcher*. Lest we cannot recall the iconic ruling by Blackburn J on behalf of the Exchequer Chamber, here it is:

> "We think that the true rule of law is, that the person who for his own purposes brings onto his lands and collects and keeps there anything likely to do mischief if it escapes, must keep it in at his peril and, if he does not do so, is prima facie answerable for all the damage which is the natural consequence F of its escape. He can excuse himself by shewing that the escape was owing to the plaintiff's default; or perhaps that the escape was the consequence of vis major, or the act of God; but as nothing of this sort exists here, it is unnecessary to enquire what excuse would be sufficient. The general rule, as above stated, seems on principle just. The person whose grass or corn is beaten down by the escaping cattle of his neighbour, or whose mine is flooded by the water from his neighbour's reservoir, or whose cellar is invaded by the filth of his neighbour's privy, or whose habitation is made unhealthy by the fumes and noisesome vapours of his neighbour's alkali works, is damnified without any default of his own; but it seems reasonable and just that the neighbour, who has brought something on his own property which was not naturally there, harmless to others so long as it is confined to his own property, but which he knows to be mischievous if it gets on his neighbour's, should be obliged to make good the damage which ensues if he does not succeed in confining it to his own property. But for his act in bringing it there no mischief could have accrued, and it seems G but just that he should at his peril keep it there so that no mischief may accrue, or answer for the natural and anticipated consequences. And upon authority, this we think is established to be the law whether the things so brought be beasts, or water, or filth, or stench." See *Fletcher* v *Rylands* (1866) LR 1 Ex 265, at pp279-280.

Whilst entirely concurring with that opinion, Lord Cairns, the Lord Chancellor, added the gloss that the land should not be used "for any purpose which I may term non-natural use" see *Rylands* v *Fletcher* (1868) LR 3 HL 330, 339.

2. Although *Donoghue* v *Stevenson* [1932] AC 562 has withstood the test of time, Lord Bingham of Cornhill was to say, in *Transco plc* v *Stockport Metropolitan Borough Council* [2003] UKHL 61; [2004] 2 AC 1 at [3]: H

> "Few cases in the law of tort or perhaps any other field are more familiar, or have attracted more academic and judicial discussion, than *Rylands* v *Fletcher*".

It was subjected to even more withering criticism by Mason CJ and the majority of the High Court of Australia in *Burnie Port Authority* v *General Jones Pty Ltd* (1994) 179 CLR 520 at p536 (though it is fair to say that the criticisms were not entirely accepted by the House of Lords in *Transco*): J

> "Notwithstanding the many accolades which have been, and continue to be, lavished on Blackburn J's judgment (see, eg, Wigmore, *Responsibility for Tortious Acts: Its History*, vol 7 (1894); *Harvard Law Review* 315, 383, 441 at 454: 'the master-mind of Mr Justice Blackburn'; Newark, *The Boundaries of Nuisance*, *Law Quarterly Review* vol 65 (1949) 480 at 487: "his great judgment"; *Salmond and Heuston on the Law of Torts*, 20th ed (1992), p314: 'always been recognised as one of the masterpieces of the Law Reports'), that brief exposition of 'the true rule of law' is largely bereft of current authority or validity if it be viewed, as it ordinarily is, as a statement of a comprehensive rule (see, eg, *Jones* v *Festiniog Railway Co* (1868) LR 3 QB, at p736, per Blackburn J: 'the general rule of common law'.) Indeed, it has been all but obliterated by subsequent judicial explanations and qualifications." K

In *Transco* itself Lord Hoffmann said, at [39]:

> "It is hard to escape the conclusion that the intellectual effort devoted to the rule by judges and writers over many years has brought forth a mouse."

3. Now we have another occasion to labour. How, if at all, does the rule apply where the damage to the claimant's land is caused by the "escape" of a fire which raged through the defendant's premises, its ferocity fed by the eventual ignition of the large stack of tyres which he had brought onto his land on which he carried out the business of a motor vehicle tyre supplier and fitter. On 22 August 2011, Mr Recorder Potts sitting in the Worcester County Court held that all the L requirements of *Rylands* v *Fletcher* were satisfied and accordingly judgment for the claimant was entered with damages to be assessed. The defendant now appeals.

The facts in a little more detail

4. The defendant carried on business as Wyvern Tyres supplying, fitting and balancing car and van tyres. The business was conducted at Unit 111 and 107a on the Holmer Trading Estate in Hereford. In the front part of the premises was the ramp and other pieces of equipment necessary for changing and fitting the tyres, vulcanising them and balancing them. To the side were two small offices. Along the length of the building to the rear were some six specially constructed racks for M the storage of tyres. The Recorder found that the defendant "squeezed stock" into the room, some tyres being "haphazardly and untidily above the racks as well as on the racks themselves" and others were "piled high in 'chimneys'". The claimant's premises were behind the defendant's. Between them was a space used as a further area for the storage of new and part-worn tyres. There were no racks in this area and the tyres were simply piled high vertically in these "chimneys", some of them being located along the wall which divided the claimant's premises from the defendant's. It was difficult to know how many tyres were held by the defendant but the Recorder was satisfied that at the material time about 3,000 tyres were stored in and outside the building.

A 5. Having heard the evidence of experienced former fire officers, the Recorder concluded that tyres have recognised particular characteristics so far as concerns fires and fire fighting as follows:

> "(a) They are not in themselves flammable, and in their normal state will not ignite unless there is sufficient flame or heat from another source.
>
> (b) Once however a primary fire has developed, and intensified it can produce sufficient heat or flame to ignite rubber composite tyres.
>
> (c) If tyres catch fire then combustion develops rapidly depending on the quantity of tyres present and how they are stored.
>
> (d) Once fire takes hold of tyres they are difficult to put out."

6. A fire did break out at about 6.15pm on 4 February 2008 some B time after the defendant had secured the premises, set the alarms at 4.50pm and left the site. It seems that the fire had started in the front workshop section and had quickly intensified. Such was its severity that 10 pumping fire appliances attended at the scene with three other special vehicles, a command unit, Ariel ladder platform and water carrier. Once the fire took hold its intensity grew rapidly. By 7.55pm, just over an hour after the original attendance, the entire Wyvern Tyres unit was fully ablaze; it was already unlikely that the claimant's premises could be saved and priority was given, in vain it seems, to saving the adjoining units. Alas they were also totally destroyed and for safety reasons were reduced to rubble the following day. The Recorder C made these findings:

> "4. I am... satisfied on the balance of probabilities that the fire originated in Wyvern's premises, and probably towards the front of the workshop area. From there it spread back into the workshop and tyre storage area, and to the side into the office and reception, and thence to the unit next door occupied by the double glazing business, and to the rear into Mr Gore's (the claimant's) premises which adjoined directly onto Wyvern's premises, the two being separated by what appears from Mr Gore's plan to be simply a single skin of brick."

The Recorder accepted the evidence of the fire officer who attended the fire that the cause of the fire was "something occurring in the wiring and appliances within Wyvern's premises". Thus he accepted that the D primary source was "electrical".

7. The Recorder asked whether the tyres made a difference. He held:

> "15. In my judgment, on the balance of probabilities they did. Something that stands out in this case in the descriptions of the fire by the witnesses, including fire officers, is its intensity and severity... This was a fire which spread with great rapidity and intensity. Indeed it was so aggressive that the Fire Service plainly had difficulty in bringing it under control, and could only do so some hours after they had first been called out and attended and after they had been reinforced with further equipment and personnel.
>
> 16. While it is clear that tyres per se are not readily combustible and will E not constitute a primary source of combustion, it is plain that they can catch fire from some other source as I have found to have existed in this case. Once the tyres were alight there was a huge problem for all concerned, including adjoining and neighbouring occupiers, because in the words of Mr Denton [the fire officer], 'Once a fire takes hold of tyres they are difficult to put out'. That is what, on the balance of probabilities, I find happened in the instant case with consequences that are of course only too apparent."

The judgment under appeal

8. The claim was brought both in negligence and in strict liability. The claim in negligence was rejected. The Recorder found, first, that F the primary cause of the fire lay in the wiring or electrical appliances in Wyvern's premises but there was nothing to show that such a state of affairs was the result of a failure to maintain or keep in good order the electrical system itself or all those electrical appliances that were located within the premises, as opposed to something that might have arisen entirely by accident. He was satisfied that the machinery was regularly checked and that the electrical work had been competently carried out and checked a reasonable time before this fire started. Secondly he found that it was not reasonable to expect the defendant to invest in an automatic alarm system or sprinkler system and in any event even if he should have done so, there was no evidence to suggest that such systems would have been likely to have prevented the fire from spreading to the claimant's premises. The third allegation of storing the tyres without any adequate measures in place to prevent G them catching fire or to minimise the chances of them catching fire was so vague and so lacking in specificity as to make it in practice impossible even to respond to it by way of defence other than by bare denial. There was no allegation and no evidence at all of what the defendant should or could have done to prevent the spread of fire that had undeniably broken out from reaching the tyres and thereby involving them in the conflagration. The claimant failed to prove that the defendant could have done something in terms of the storage of the tyres that would have prevented their igniting at all or would have reduced the chances of them doing so. Any complaint of a breach of the Regulatory Reform (Fire Safety) Order 2005 for the failure to take H such general fire precautions as would ensure so far as was reasonably practicable the safety of any of the defendant's employees could not take the claimant's case further because there were no employees and in any event nothing in that statutory duty conferred any right of action in civil proceedings. The Recorder accordingly held that:

> "...subject to the question of any *Rylands* v *Fletcher* liability, the failure of Mr Gore to establish to the satisfaction of the court any negligence on the part of Mr Stannard means that, apart from any *Rylands* v *Fletcher* liability, he has the benefit of a defence under section 86 of the Fires Prevention (Metropolis) Act 1774 on the basis that the fire was accidental."

J

Strict liability under Rylands v Fletcher

9. In a careful judgment, the Recorder analysed the authorities presented to him. Dealing specifically with fires in connection with the rule in *Rylands* v *Fletcher* he observed how the Fires Prevention (Metropolis) Act 1774 limited a defendant's liability to non-accidental fires, but he added:

> "Even then of course where a fire arose from something dangerous that the defendant had brought onto his land there could be nothing accidental about any fire that arose as a result, and what became the rule in *Rylands* v *Fletcher* continued to apply with full force, although of course it is the fire itself that is the dangerous thing that escapes the defendant's land, rather than whatever K caused the fire to arise in the first place."

10. As for *Rylands* v *Fletcher,* he noted:

> "The following well-known requirements to the rule in order that liability may be established:
>
> (i) the defendant must bring onto his land something that is dangerous;
>
> (ii) the danger must escape from the defendant's land to the claimant's land; and
>
> (iii) the use to which the defendant had put his land must be 'non-natural'."

L

He concluded that "there was plainly an escape within the meaning of the *Rylands* v *Fletcher* rule" and "liability therefore turns on whether or not Mr Stannard's activities on Wyvern's premises were dangerous and a non-natural use of his land". The only relevant activity was the storage of tyres. They were not in themselves flammable and would not ignite unless there was a sufficient flame or heat source. It is, however, not impossible for tyres to catch fire as obviously happened here and if they do ignite they have a special fire risk quality. This is that once alight they may burn rapidly and intensively such that they are difficult to put out. Given the haphazard way a large number of tyres were stored there was an exceptionally high risk of damage to the claimant's M premises if fire broke out. That was a foreseeable risk with the result that the defendant's activities in storing tyres in the numbers and ways that he did were dangerous within the meaning of the rule. As for non-natural use, the Recorder asked himself whether what the defendant did in storing and enlarging the tyre storage area for that purpose was out of the ordinary. Here the state of affairs created by the defendant was out of the ordinary. It was not normal and not routine, the standard of normality and routine being that of a tyre business storing tyres in an orderly fashion and in such numbers that are well within the capacity of its dedicated storage facilities. This was non-natural use. Consequently the requirements of the rule in *Rylands* v *Fletcher* were established and judgment was entered for the claimant accordingly.

A *Discussion: the only question in this appeal*

11. The appellant does not seek to challenge the findings of fact nor the dismissal of the claim in negligence. It is, however, submitted on the appellant's behalf that the Recorder erred in his application of the test for strict liability under the rule in *Rylands* v *Fletcher* as applied to fire cases. I propose to deal first with what seems to me to be the established formulation of the rule in classic *Rylands* v *Fletcher* cases, leaving for later consideration the question of whether there is a special or different rule of a *Rylands* v *Fletcher* kind to deal with damage caused by fire.

Classic Rylands v Fletcher

B 12. I need not repeat Blackburn J's formulation and the gloss added to it by Lord Cairns as set out at the beginning of this judgment. The first significant clarification was made nearly 40 years later in *Rickards* v *Lothian* [1913] AC 263, 280. There the Privy Council was concerned with property on the second floor of the premises, which were damaged by the continuous overflow of water from a lavatory basin on the top floor caused by the water tank having been turned on full and the waste-pipe plugged through the malicious act of some person. The Privy Council were of the opinion that the case did not come within the principle of *Rylands* v *Fletcher* because the matters complained of took place through no fault or breach of duty on the defendants' part but
C were caused by a stranger over whom and at a spot where they had no control. Lord Moulton added at pp279-280:

> "But there is another ground upon which their Lordships are of opinion that the present case does not come within the principle laid down in *Fletcher* v *Rylands*. It is not every use to which land is put that brings into play that principle. It must be some special use bringing with it increased danger to others, and must not merely be the ordinary use of the land or such a use as is proper for the general benefit of the community."

The Privy Council were of the opinion that the provision of a proper supply of water to the various parts of a house is not only reasonable, but has become, in accordance with prevailing sanitary views, an
D almost central feature of town life and it would be wholly unreasonable to hold an occupier responsible for the consequences of acts which he is powerless to prevent.

13. *Rainham Chemical Works Ltd* v *Belvedere Fish Guano Co Ltd* [1921] 2 AC 465 is often cited, but I confess that I find the jurisprudence on *Rylands* v *Fletcher* is hardly advanced by this case. At a time of war a process was invented whereby picric acid could be manufactured from dinitrophenol (DNP) and nitrate of soda. DNP had been used mainly for the manufacture of dyes, and it was a stable compound which did not explode easily. It was not in itself dangerous. Nitrate of soda was not an explosive but wood or bags impregnated with moist nitrate of soda will,
E when dry, burn fiercely if ignited. A hot flame is needed to ignite it and when ignited, large quantities of DNP become a dangerous explosive. While neither DNP nor nitrate of soda was, in itself, dangerous, they became a source of danger if stored in quantities and in close proximity to one another. It was proved that it was due to that cause that a massive explosion took place which caused damage to neighbouring property. On the evidence the manufacture of picric acid from DNP and nitrate of soda might or might not be dangerous in its character, but in that case it was being manufactured under dangerous conditions, and those dangerous conditions caused the accident. Accordingly the principle of *Rylands* v *Fletcher* became applicable. It was not, per Lord Carson,
F "seriously argued" that the defendant company was not liable for the damages caused by the explosion. Indeed it seems that before Scrutton LJ, the trial judge, it was admitted that the person in possession of the DNP was liable under the rule in *Rylands* v *Fletcher* for the consequences of the explosion. Thus the real issue in the case was stated by Lord Buckmaster at pp471-472 in these terms.

> "Now, the foundation of the action was a claim based upon the familiar doctrine established by the case of *Fletcher* v *Rylands*, which depends upon this – that even apart from negligence the use of land by one person in an exceptional manner that causes damage to another, and not necessarily an adjacent owner, is actionable:… In the present case the use complained of was that for the purpose of making munitions, which was certainly not the common and ordinary use of the land, two substances, namely, nitrate of soda and dinitrophenol, were stored in close proximity, with the result that on a fire breaking out they exploded G
> with terrific violence. It may be accepted that it was not known to either of the defendants that this danger existed, but that in itself affords no excuse, and the result is that the plaintiffs' cause of action is well founded and the only matter for determination is against whom the action should be brought."

The disputed question was whether responsibility lay at the door of the defendant company or the personal defendants who had a licence from the inventor to manufacture the required picric acid. Whilst, therefore, that case bears some similarly to the case before us (large quantities of materials which when once alight will cause damage to neighbouring property), there is little help on the principles to apply. H

14. Another explosion at a time of war gave rise to *Read* v *J Lyons & Co Ltd* [1947] AC 156. This was a claim by an inspector of munitions who was injured in the course of her duties while in a munitions factory operated by the respondents by the explosion of a high explosive shell during the process of manufacture. She failed because an essential condition of the application of the rule in *Rylands* v *Fletcher* was the "escape" from the land of something likely to do mischief if it escaped. Here the injuries were caused on the premises of the defendant. Doubts were expressed, *obiter*, as to whether the action lay where the claim was for damages for personal injury as distinguished from damage to property. Once again, this case gives us little help. J

15. *Cambridge Water Co* v *Eastern Counties Leather plc* [1994] 2 AC 264 is a much more important authority. There a chlorinated solvent used by the leather manufacturer seeped into the water company's borehole resulting in the water becoming unfit for human consumption. Although the water company had abandoned its claim in nuisance and could only succeed on the basis of the rule in *Rylands* v *Fletcher*, Lord Goff of Chieveley nonetheless considered it desirable to look at the nature of liability in both nuisance and *Rylands* v *Fletcher*.

16. Comparing nuisance and the rule in *Rylands* v *Fletcher,* he said at p299A-C: K

> "We are not concerned in the present case with the problem of personal injuries, but we are concerned with the scope of liability in nuisance and in *Rylands* v *Fletcher*. In my opinion, it is right to take as our starting point the fact that, as Professor Newark considered [in his seminal article on 'The boundaries of nuisance' (1949) 65 LQR 480], *Rylands* v *Fletcher* was indeed not regarded by Blackburn J. as a revolutionary decision: see, eg, his observations in *Ross* v *Fedden* (1872) 26 LT 966, 968. He believed himself not to be creating new law, but to be stating existing law, on the basis of existing authority; and, as is apparent from his judgment, he was concerned in particular with the situation where the defendant collects things upon his land which are likely to do mischief if they escape, in which event the defendant will be strictly liable for damage resulting from any such escape. It follows that the essential basis of
> liability was the collection by the defendant of such things upon his land; and L
> the consequence was a strict liability in the event of damage caused by their escape, even if the escape was an isolated event. Seen in its context, there is no reason to suppose that Blackburn J intended to create a liability any more strict than that created by the law of nuisance; but even so he must have intended that, in the circumstances specified by him, there should be liability for damage resulting from an isolated escape."

17. He questioned whether foreseeability of damage was a prerequisite for recovery of damages and said at p302B-D:

> "Blackburn J spoke of 'anything *likely* to do mischief if it escapes'; and
> later he spoke of something 'which he *knows* to be mischievous if it gets on M
> his neighbour's [property],' and the liability to 'answer for the natural *and anticipated* consequences.' Furthermore, time and time again he spoke of the strict liability imposed upon the defendant as being that he must keep the thing in at his peril; and, when referring to liability in actions for damage occasioned by animals, he referred, at p282, to the established principle that 'it is quite immaterial whether the escape is by negligence or not.' The general tenor of his statement of principle is therefore that knowledge, or at least the foreseeability of the risk, is a prerequisite of the recovery of damages under the principle; but that the principle is one of strict liability in the sense that the defendant may be held liable notwithstanding that he has exercised all due care to prevent the escape occurring."

And at p306A he concluded:

A "it appears to me to be appropriate now to take the view that foreseeability of damage of the relevant type should be regarded as a prerequisite of liability in damages under the rule."

In the result, since the plaintiffs were unable to establish that the pollution of their water supply by the solvent was foreseeable, the claim failed.

18. He considered that there was a striking similarity of function between the principle of reasonable user in nuisance (the principle of give and take as between neighbouring occupiers of land) and the principle of natural use in *Rylands* v *Fletcher*. As for the latter, he pointed out Lord Moulton's gloss in *Rickards* and added at pp308A-
B 309A:

"Blackburn J's statement of the law was limited to things which are brought by the defendant onto his land, and so did not apply to things that were naturally upon the land. Furthermore, it is doubtful whether in the House of Lords in the same case Lord Cairns, to whom we owe the expression 'non-natural use' of the land, was intending to expand the concept of natural use beyond that envisaged by Blackburn J. Even so, the law has long since departed from any such simple idea, redolent of a different age; and, at least since the advice of the Privy Council delivered by Lord Moulton in *Rickards*…, natural use has been extended to embrace the ordinary use of land.…

Rickards itself was concerned with a use of a domestic kind, viz the
C overflow of water from a basin whose runaway had become blocked. But over the years the concept of natural use, in the sense of ordinary use, has been extended to embrace a wide variety of uses, including not only domestic uses but also recreational uses and even some industrial uses.

It is obvious that the expression 'ordinary use of land' in Lord Moulton's statement of the law is one which is lacking in precision… A particular doubt is introduced by Lord Moulton's alternative criterion – 'or such a use as is proper for the general benefit of the community.' If these words are understood to refer to a local community, they can be given some content as intended to refer to such matters as, for example, the provision of services; indeed the same idea can, without too much difficulty, be extended to, for example, the provision of services to industrial premises, as in a business park or an industrial estate. But if the words are extended to embrace the wider interests of the local
D community or the general benefit of the community at large, it is difficult to see how the exception can be kept within reasonable bounds.… I myself, however, do not feel able to accept that the creation of employment as such, even in a small industrial complex, is sufficient of itself to establish a particular use as constituting a natural or ordinary use of land."

He did not, however, think it necessary to attempt any redefinition of the concept of natural or ordinary use because the storage and use of chemicals by the defendant was "an almost classic case of non-natural use", explaining at p309E that:

"It may well be that, now that it is recognised that foreseeability of harm
E of the relevant type is a prerequisite of liability in damages under the rule, the courts may feel less pressure to extend the concept of natural use to circumstances such as those in the present case; and in due course it may become easier to control this exception, and to ensure that it has a more recognisable basis of principle."

19. Of even greater importance for the understanding of the current state of the law is *Transco*. Here water pipes laid by the council from their tower block of flats to the water main fractured and the resulting flood caused the collapse of an embankment on part of their land through which the claimant had the right to run its gas main. The collapse of the embankment caused damage to the gas main. Transco's
F main claim was that the council was liable without proof of negligence under the rule in *Rylands* v *Fletcher*. It succeeded before the judge but was reversed by the Court of Appeal. Their Lordships were being asked to review and, if they thought it right to do so, to hold the rule in *Rylands* v *Fletcher* no longer to be good law. They were being invited "to follow the trail blazed by the majority of the High Court of Australia in *Burnie Port Authority* [supra] by treating the rule in *Rylands* v *Fletcher* as absorbed by the principles of ordinary negligence" (see [4]). Although Lord Bingham of Cornhill saw that there was "a theoretical attraction in bringing this somewhat anomalous ground of liability within the broad and familiar rules governing liability in negligence" he was unwilling "to suppress an instinctive resistance to treating a nuisance-based tort as if it were governed by the law of negligence." So the rule remains the law of the land. But the speech of Lord Bingham is the latest and last G word on the subject and bears close study.

20. Like Lord Goff, he recognised that Blackburn J did not conceive himself to be laying down any new principle of law and at [3] he said:

"It seems likely, as persuasively contended by Professor Newark (*The Boundaries of Nuisance* (1949) 65 LQR 480, 487-488), that those who decided the case regarded it as one of nuisance, novel only to the extent that it sanctioned recovery where the interference by one occupier of land with the right or enjoyment of another was isolated and not persistent."

Lord Bingham favoured retaining the rule but restating it so as to achieve as much certainty and clarity as is attainable, recognising that H new factual situations are bound to arise posing difficult questions on the boundary of the rule, wherever that is drawn. His restatement was set out in [9]-[11] of his speech. I extract these important observations:

"9. The rule in *Rylands* v *Fletcher* is a sub-species of nuisance, which is itself a tort based on the interference by one occupier of land with the right in or enjoyment of land by another occupier of land as such. From this simple proposition two consequences at once flow. First, as very clearly decided by the House in *Read*… no claim in nuisance or under the rule can arise if the events complained of take place wholly on the land of a single occupier. There must, in other words, be an escape from one tenement to another. Second, the claim cannot include a claim for death or personal injury, since such a claim does not J relate to any right in or enjoyment of land.…

10. It has from the beginning been a necessary condition of liability under the rule in *Rylands* v *Fletcher* that the thing which the defendant has brought on his land should be 'something which... will naturally do mischief if it escape out of his land" (LR 1 Ex 265, 279 per Blackburn J), 'something dangerous...', 'anything likely to do mischief if it escapes', 'something... harmless to others so long as it is confined to his own property, but which he knows to be mischievous if it gets on his neighbour's' (p280), 'anything which, if it should escape, may cause damage to his neighbour' (LR 3 HL 330, 340, per Lord Cranworth). The practical problem is of course to decide whether in any given case the thing which has escaped satisfies this mischief or danger test, a problem exacerbated by the fact that many things not ordinarily regarded as sources of mischief or danger may none the less be capable of proving to be such if they escape. K I do not think this condition can be viewed in complete isolation from the non-natural user condition to which I shall shortly turn, but I think the cases decided by the House give a valuable pointer.… Bearing in mind the historical origin of the rule, and also that its effect is to impose liability in the absence of negligence for an isolated occurrence, I do not think the mischief or danger test should be at all easily satisfied. It must be shown that the defendant has done something which he recognised, or judged by the standards appropriate at the relevant place and time, he ought reasonably to have recognised, as giving rise to an exceptionally high risk of danger or mischief if there should be an escape, however unlikely an escape may have been thought to be.

11. No ingredient of *Rylands* v *Fletcher* liability has provoked more discussion than the requirement of Blackburn J (LR 1 Ex 265, 280) that the L thing brought on to the defendant's land should be something "not naturally there", an expression elaborated by Lord Cairns (LR 3 HL 330, 339) when he referred to the putting of land to a 'non-natural use'… Read literally, the expressions used by Blackburn J and Lord Cairns might be thought to exclude nothing which has reached the land otherwise than through operation of the laws of nature. But such an interpretation has been fairly described as 'redolent of a different age' (*Cambridge Water* [1994] 2 AC 264, 308), and in *Read*… and *Cambridge Water*, at p308, the House gave its imprimatur to Lord Moulton's statement, giving the advice of the Privy Council in *Rickards*…

'… It must be some special use bringing with it increased danger to others, and must not merely be the ordinary use of land…'

I think it clear that ordinary user is a preferable test to natural user, making it M clear that the rule in *Rylands* v *Fletcher* is engaged only where the defendant's use is shown to be extraordinary and unusual. This is not a test to be inflexibly applied: a use may be extraordinary and unusual at one time or in one place but not so at another time or in another place (although I would question whether, even in wartime, the manufacture of explosives could ever be regarded as an ordinary user of land, as contemplated… in *Read*…). I also doubt whether a test of reasonable user is helpful, since a user may well be quite out of the ordinary but not unreasonable, as was that of *Rylands*, *Rainham Chemical Works* or the tannery in *Cambridge Water*. Again, as it seems to me, the question is whether the defendant has done something which he recognises, or ought to recognise, as being quite out of the ordinary in the place and at the time when he does it. In answering that question, I respectfully think that little help is gained (and unnecessary confusion perhaps caused) by considering whether the use is proper

A for the general benefit of the community.… An occupier of land who can show that another occupier of land has brought or kept on his land an exceptionally dangerous or mischievous thing in extraordinary or unusual circumstances is in my opinion entitled to recover compensation from that occupier for any damage caused to his property interest by the escape of that thing, subject to defences of act of God or of a stranger, without the need to prove negligence."

Lord Hoffmann agreed that the criterion of exceptional risk must be taken seriously and creates a high threshold for a claimant to surmount. Although giving reasons of their own, Lord Scott of Foscote and Lord Walker of Gestingthorpe agreed with Lord Bingham.

21. This is, therefore, the seminal authority for the test to be applied
B in a classic case of *Rylands* v *Fletcher*.

22. The proper approach I extract from those compelling authorities is this.

(1) The defendant must be the owner or occupier of land.

(2) He must bring or keep or collect an exceptionally dangerous or mischievous thing on his land.

(3) He must have recognised or ought reasonably to have recognised, judged by the standards appropriate at the relevant place and time, that there is an exceptionally high risk of danger or mischief if that thing should escape, however unlikely an escape may have been thought to be.

C (4) His use of his land must, having regard to all the circumstances of time and place, be extraordinary and unusual.

(5) The thing must escape from his property into or onto the property of another.

(6) The escape must cause damage of a relevant kind to the rights and enjoyment of the claimant's land.

(7) Damages for death or personal injury are not recoverable.

(8) It is not necessary to establish the defendant's negligence but an act of God or the act of a stranger will provide a defence.

23. If that is how the rule is to operate in any classic *Rylands* v
D *Fletcher* case, the next question is whether there is any other or some special rule for cases involving damage caused by the spread of fire.

Liability for damage caused by the spread of fire

24. Writing in the *Cambridge Law Journal* in 1931, that great master of the law of tort, Professor Percy Winfield, was driven to observe that there was a "somewhat confusing medley of remedies for the same wrong, which is traceable in the historical development of liability for the escape of fire," see PH Winfield: *Nuisance as a Tort* (1931) 4 CLJ 189. He concluded at p206:

> "The result seems to be that, at the present day, damage resulting from
> E the unintentional escape of fire may be redressed by any of the following remedies:
>
> "(i) The old action of trespass on the case.
>
> (ii) An action of the *Rylands* v *Fletcher* type: query whether this has totally absorbed (i).
>
> (iii) An action for nuisance.
>
> (iv) An action for negligence."

25. This appeal concerns head (ii) only and although any views I may express about the others will be treated as obiter, it is necessary to have some overview of the whole confusing picture. I propose, however, to do no more than take a quick canter over what is undoubtedly tricky
F terrain. The bye-line to AS Ogus' article "*Vagaries in Liability for the Escape of Fire*", (1969) CLJ 104 says it all:

> "How great a matter a little fire kindleth." (General epistle of James iii. 5)"

The old action of trespass on the case

26. The custom of the realm is that a person is liable for the damage caused by the escape of his fire – the *ignis suus* rule. By general custom of the realm the remedy for damage caused by the spread of fire was an action on the case *pur negligent garder son few*. The declaration was in the form "whereas by the custom of England a man is bound to keep his fire *ignem suum* safe and secure lest by default of custody of it loss should fall upon his neighbours in any way, the defendant *tam negligenter ac improvide apud custodivit, quod pro defectu debitae* G *custodiae* the fire spread and did damage…"

27. The essential features are that (1) fire spreading from a person's land is his fire; (2) that person is under a duty to contain the fire: (3) if it escapes, he is liable for the damage that results; (4) "negligenter" was a pleader's adverb: it did not mean negligently in the sense we now use the word. To quote Winfield again (Winfield: *The Myth of Absolute Liability*), (1926) 42 LQR 37, 49:

> "Exactly what it did mean must remain a matter of conjecture, but it excluded liability when the fire spread or occurred (i) by the act of a stranger – a man was not responsible for that, though he was for the act of his servant, his wife, his guest, or one entering his house with his leave or knowledge; (ii) H by misadventure which, to anticipate a modern term, seems to be equivalent to inevitable accident, or something which by no care reasonable in the circumstances could have been avoided."

28. The earliest case on the subject is *Beaulieu* v *Finglam* (1401) B&M 557 where, as reported by *Fifoot History and Sources of the Common Law* (1949) at pp166-167 Markham J. said:

> "A man is held to answer for the act of his servant or of his guest in such a case; for if my servant or my guest puts a candle by a wall and the candle falls into the straw and burns all my house and the house of my neighbour also, in this case I shall answer to my neighbour for his damage. J
>
> …
>
> I shall answer to my neighbour for each person who enters my house by my leave or my knowledge, or is my guest through me or through my servant, if he does any act, as with a candle or aught else, whereby my neighbour's house is burnt. But if a man from outside my house and against my will starts a fire in the thatch of my house or elsewhere, whereby my house is burned and my neighbours' houses are burned as well, for this I shall not be held bound to them; for this cannot be said to be done by wrong on my part, but is against my will."

29. Moving forward nearly three centuries to *Turberville* v *Stamp* (1702) 12 Mod 152, a fire started by the defendant in his field was held to be as much *ignis suus* as one lit in the house. Sir John Holt CJ said: K

> "Every man must so use his own as not to injure another. The law is general; the fire which a man makes in the fields is as much his fire as his fire in the house; it is made on his ground, with his materials, and by his order; and he must at his peril take care that it does not, through his neglect, injure his neighbour: if he kindle it at a proper time and place, and the violence of the wind carry it into his neighbour's ground, and prejudice him, this is fit to be given in evidence. But now here it is found to have been by his negligence; and it is the same as if it had been in his house."

30. These were times when fires were a scourge. The Great Fire of London in 1666 had had a devastating impact on crowded urban living. L Something had to be done. In 1707, the Fires Prevention (Metropolis) Act (6 and 31) was passed "for the better preventing the mischiefs that may happen by fire". Provision was made for improved safety, for the control of fires and for penal sanctions on servants carelessly setting fire to houses. The Act was repealed but re-enacted, eventually as the Fires Prevention (Metropolis) Act 1774, section 86 providing that:

> "No action, suit, or process whatsoever shall be had, maintained, or prosecuted against any person in whose house, chamber, stable, barn or other building, or on whose estate any fire shall accidentally begin, nor shall any recompense be made by such person for any damage suffered thereby, any law, usage or custom to the contrary notwithstanding." M

It may be that, as Sir William Holdsworth in his *History of England* vol XI p607 speculates, by this time lawyers were beginning to think it anomalous that a man should be liable for fire damage not caused by his negligence. It may have been that Parliament was simply resolving any doubt there was as to whether fire spreading from a person's house was *ignis suus*. Another explanation is implicit in Lord Tenterden's judgment in *Becquet* v *MacCarthy* [1831] 2 B&Ad 951, 958:

> "by the law of this country before it was altered by the statute of 6 Ann. c.31, s.6, if a fire began on a man's own premises, by which those of his neighbour were injured, the latter, in an action brought for such an injury would not be bound in the first instance to show how the fire began but the presumption

A would be (unless it were shewn to have originated from some external cause) that it arose from the neglect of some person in the house."

Thus one can surmise that the purpose of this statute was to remove the presumption of negligence on the defendant's part and cast the burden of proving negligence on the plaintiff.

31. Curiously, there seems to be little contemporaneous authority on what was meant by "accidentally". Lord Lyndhurst flirted with the idea that accidentally was used as the antithesis of wilfully: see his obiter remarks in *Canterbury* v *A-G* (1843) 1 Phil 306. The more informed view is expressed by Lord Denman CJ in *Filliter* v *Phippard* (1847)
B 11 QB 347, 357:

> "It is true that in strictness, the word *accidental* may be employed in contradistinction to wilful, and so the same fire might both begin accidentally and be the result of negligence. But it may equally mean that a fire produced by mere chance, or incapable of being traced to any cause, and so would stand opposed to negligence of either servants or masters. And, when we find it used in statutes which do not speak of wilful fires but make important provision with respect to such as are accidental, and consider how great a change in the law would be effected, and how great encouragement would be given to that carelessness of which masters may be guilty as well as servants, we must say that we think the plaintiff's construction [accidental as opposed to negligent] much the most reasonable of the two."

C

An action for negligence

32. By the 19th century a prototype of our modern claim in negligence was emerging. In *Vaughan* v *Menlove* (1837) 3 Bing NC 468 the defendant's haystack spontaneously combusted and it was alleged that he had "wrongfully negligently and improperly kept his haystack so that it became liable to ignite" and so be a danger to the claimant's property. The jury were left to consider the question of negligence and that direction was upheld. Tindal CJ said:

> "but there is a rule of law which says you must so enjoy your own property as not to injure that of another; and according to that rule the defendant is liable
> D for the consequences of his own neglect; and though the defendant did not himself light the fire, yet mediately, he is as much the cause of it as if he had himself put a candle to the rick; for it is well known that hay will ferment and take fire if it be not carefully stacked."

33. This was the age of the railways and much litigation followed. In *Aldridge* v *The Great Western Railway Co* (1841) 3 Man & G 515, Tindal CJ held that Mr Aldridge had to show some carelessness on the railway company's part for the emission of the sparks from their engine which set fire to his crop. *Piggot* v *The Eastern Counties Railway Co* (1846) 3 CB 229 was another case of sparks setting fire to the
E claimant's property. Tindal CJ held that:

> "The defendants are a company entrusted by the legislature with an agent of an extremely dangerous and unruly character, for their own private and particular advantage. And the law requires of them that they shall, in the exercise of the rights and powers so conferred upon them, adopt such precautions as may reasonably prevent damage to the property of third persons through or near which their railway passes. The evidence in this case was abundantly sufficient to show that the injury of which the plaintiff complains was caused by the emission of sparks, or particles of ignited coke, coming from one of the defendants' engines; and there was no proof of any precaution adopted by the company to avoid such a mischance. I therefore think the jury came to a right conclusion, in finding that the company were guilty of negligence, and that the
> F injury complained of was the result of such negligence."

We are seeing the emergence of a claim for negligence, the culmination of which in the 19th century was the well-known case in 1883 of *Heaven* v *Pender* (1882-1883) 11 QBD 503, which had nothing to do with fire but which is often thought to be the progenitor of the modern law.

34. Turning to more modern times, an example of the claim in negligence for fire damage is *Balfour* v *Barty-King* [1957] 1 QB 496, where an independent contractor negligently used his blow lamp to thaw frozen pipes and set fire to the combustible materials nearby. The fire spread to the claimant's property and destroyed it. The Court of Appeal held that the defendant was responsible for the negligence of the independent contractor. In *H&N Emanuel* v *Greater London* G
Council [1971] 2 All ER 835 the Court of Appeal held that an occupier was liable for the escape of fire caused by the negligence not only of his servant but also of his independent contractor and anyone else who was on his land with his leave and licence. The only occasion when the occupier would not be liable for negligence when the negligence is the negligence of a stranger.

An action in nuisance

35. The modern example of this is *Spicer* v *Smee* [1946] 1 All ER 489, where the plaintiff's bungalow was completely destroyed by a fire which originated in the defendant's bungalow owing to a defect in H
the electrical wiring. The wiring had been negligently installed by an independent contractor. Atkinson J held that the state of wiring in the defendant's bungalow constituted a nuisance on her property for which she was liable. The Fire Prevention (Metropolis) Act 1774, did not afford her a defence since the fire was due to the negligence or nuisance created by the defendant or those for whom he was responsible.

An action of a Rylands v Fletcher type

36. One simply cannot say that there is no liability for fire damage under the *Rylands* v *Fletcher* rule because the very man who established the rule said there was: see *Jones* v *The Festiniog Railway Co*. Lush J who was also a member of the Exchequer Chamber in *Rylands* v J
Fletcher, agreed with him. This was another case where the plaintiff's haystack had been fired by sparks from a railway engine. Blackburn J held at p736:

> "The general rule of common law is correctly given in *Fletcher* v *Rylands*, that when a man brings or uses a thing of a dangerous nature on his own land, he must keep it in at his own peril; and is liable for the consequences if it escapes and does injury to his neighbour. Here the defendants were using a locomotive engine with no express parliamentary powers making lawful that use, and they are therefore at common law bound to keep the engine from doing injury, and if the sparks escape and cause damage, the defendants are liable for the consequences, though no actual negligence be shewn on their part." K

I can readily understand this decision as an example of *Rylands* v *Fletcher*. The dangerous thing which the defendant railway company brought onto their land was a steam engine which depended for its locomotion on the burning of coal, particles of which would be belched forth from its maw onto the haystack adjoining the railway line. Although the engine itself remained on the defendant's land the sparks, which were an essential part of the machine, escaped and the danger posed by such an escape was high and it was foreseeable.

37. It is not as easy to rationalise the decision in *Musgrove* v *Pandelis* [1919] 2 KB 43, which remains for me, as for others, a troubling case. The defendant kept a car in a garage below the rooms L
occupied by the plaintiff. The defendant's servant, one Coumis, set about moving the car which necessitated applying the pressure pump to the carburettor to start the engine. For some unexplained reason there was an explosion in the carburettor and the petrol in the carburettor took fire. Instead of immediately turning the tap off which, if done, would have led to the fire burning itself out harmlessly, Coumis delayed and the continued supply of petrol extended the fire until it enveloped the car and ultimately the plaintiff's premises and furniture.

38. It is far from clear how the claim was brought. From the report of the proceedings at first instance – see [1919] 1 KB 314 – it would seem that the claim was brought to recover damages as a result of negligence. M
At pp44-45 of the Court of Appeal hearing, we are told that:

> "The action was tried before Lush J, who held that this enactment had no application to a case which fell, as in his view this case fell, within the principle of *Rylands* v *Fletcher*. Further he held that the fire which caused the damage did not begin accidentally, but as a result of the negligence of Coumis. He therefore gave judgment for the plaintiff."

Bankes LJ, after stating the facts, proceeded as follows at p46:

> "The plaintiff brought this action alleging that the fire was the result of Coumis's negligence. The negligence finally relied on was that he did not instantly turn off the petrol tap and so stop the further flow of petrol into the carburettor. … The defendant's main defence, apart from disputing the

negligence, was founded on section 86 of the Fires Prevention (Metropolis) Act 1774, and the argument has been chiefly directed to the construction of that enactment. Lush J took the view that the statute did not apply at all; and I agree. He also held that, if that view was not correct, the fire which caused the accident did not accidentally begin within the meaning of the Act. And there also I agree."

39. If the claim was indeed a claim in negligence, then the finding that the defendant was vicariously liable for the negligence of his servant would be the end of the matter. The defence under section 86 could not prevail in the light of that finding. There would be no need to consider *Rylands* v *Fletcher* and references to the rule could be treated as obiter. That would be one view of this case. Bankes LJ went on however to consider whether section 86 provided a defence where the case fell within *Rylands* v *Fletcher*. He said at p47:

> "*Rylands* v *Fletcher* is merely an illustration of that old principle [that a man must so use his own property as not to injure that of others], and in my opinion Lush J was right in saying that this case, if it falls within that principle, is not within the protection of the statute."

40. So Bankes LJ continued:

> "The question then, is whether this motor car, with its petrol tank full or partially filled with petrol, was a dangerous thing to bring into the garage within the principle of *Rylands* v *Fletcher*… I agree with Lush J that this motor car was dangerous within that principle. The defendant brought it, or caused it to be brought upon his premises, and he is responsible for the fire which resulted, and is not within the protection of the statute."

41. Warrington LJ was of the same opinion. He held, at p49:

> "If this motor car with the petrol in its tank was potentially dangerous, such as a man's own fire, then it was the defendant's duty to see that the potential danger did not become an actual danger causing damage to his neighbour. The Act of Geo 3 is no protection against that liability."

42. Duke LJ also agreed. He held, at p50:

> "I do not see how this case can be taken out of the principle of *Rylands* v *Fletcher*… In the present case there was petrol which was easily convertible into an inflammable vapour; there was the apparatus for producing a spark; and added to those there was a person supposed to control the combustion but inexperienced and unequal to the task. Taken together the presence of the petrol, and the production of the inflammable gas, all those combustibles together with the inexperience of the person placed in charge of them, it is impossible to say that this is not an instance of the principle laid down by Blackburn J."

He agreed that the case was outside any possible protection of the Act of Geo III.

43. *Musgrove* had not escaped judicial criticism. In *Collingwood* v *Home & Colonial Stores Ltd* [1936] 3 All ER 200, 205-206, Lord Wright MR said:

> "It was sufficient ground for the particular decision that the fire, that is to say the substantial fire, which actually caused the damage, was not caused without negligence. The fire which caused the damage was that which flowed from the original innocuous fire spreading, through the fault of the chauffeur, to the petrol in the tank, and that was clearly due to an act of negligence and, therefore, the protection of the statute did not apply. That was enough for the decision of the case… That having been laid down, the Court of Appeal proceeded to hold that the principle of *Rylands* v *Fletcher* would apply here… Well, I certainly have no desire to criticise in any way the actual decision in that case so far as it is based on the view that the real and substantial and destructive fire was the result of negligence. I confess, however, I find some difficulty about the other ground on which the decision was based, though, if it were necessary, I should follow the ruling of the Court of Appeal and apply it here if the case came within the scope of that ruling. But I do not think it does."

Romer LJ said at pp208-209:

> "I think at sometime it will be desirable if the House of Lords would consider the case of *Musgrove*, so far as the decision in the case was based upon *Rylands* v *Fletcher*. Of course the rule in *Rylands* v *Fletcher*, as it is well known, is a rule which relates to the escape from somebody's premises of a dangerous animal or thing brought by the owner upon those premises, and does not relate to a case like the present, or a case like the Court of Appeal had to deal with in *Musgrove,* where there had been an escape of nothing from the defendant's premises. … But what will have to be considered is whether *Musgrove*… can be supported seeing that the decision involves these two propositions, (i) that a motor car is – I am quoting from the judgment of Bankes LJ, a dangerous thing to bring into a garage, and (ii) that the use of one's land for the purpose of erecting a garage and keeping a motor car there, is not an ordinary and proper use of the land – two propositions which, but for that authority, I should myself respectfully have doubted."

And I respectfully agree with them.

44. In *Balfour*, the Court of Appeal did not think it necessary to consider the doctrine of *Rylands* v *Fletcher* as a separate head of liability though Lord Goddard CJ, giving the judgment of the court, did add at p505:

> "No doubt the doctrine of that case applies to fire, and is subject to the exception of the damage being caused by a stranger."

45. MacKenna J was clearly perplexed by *Musgrove*, pointing out in *Mason* v *Levy Auto Parts of England Ltd* [1967] 2 QB 530, 542 that the car had not escaped from the land, neither had the petrol in it, but he felt it his duty to follow the case unless it had been overruled or unless the principle did not apply to the facts of his case.

46. So we come back to *Transco*. This was not a fire case and the crucial question is whether, when their Lordships were re-stating the *Rylands* v *Fletcher* rule, they were leaving aside fire damage as some special category outside the general rule they were propounding. I do not see how that can be inferred. The House was being invited to follow the High Court of Australia in *Burnie Port Authority* and treat the rule in *Rylands* v *Fletcher* as absorbed by the principle of ordinary negligence. *Burnie Port Authority* was a fire case. The High Court of Australia necessarily had to consider the escape of fire as a case within *Rylands* v *Fletcher*. In the course of work being carried out to the Port Authority's premises by independent contractors employed by them, their welding activity resulted in sparks or molten metal falling on cartons containing highly inflammable polystyrene which caught fire. The ensuing conflagration damaged the claimant's premises and the adjoining premises. Their conclusion, as expressed by Mason CJ giving the majority judgment at p556:

> "In these circumstances, and subject only to the above-mentioned possible qualification in relation to liability in nuisance, the rule in *Rylands* v *Fletcher,* with all its difficulties, uncertainties, qualifications and exceptions, should now be seen, for the purposes of the common law of this country, as absorbed by the principles of ordinary negligence."

The House of Lords must have been aware that fire damage could be thought to be within the scope of *Rylands* v *Fletcher*. Moreover, *Musgrove* was cited to their Lordships, though only Lord Walker of Gestingthorpe referred to it in his speech, saying, at [107]:

> "The majority in *Burnie Port Authority* commented that the scope of the *Rylands* v *Fletcher* principle has been progressively restricted from within and without. Both those observations are correct up to a point, but the process has not been entirely one-way traffic. Since the middle of the 19th century many activities which were once regarded as unusually dangerous (such as running railways, which no longer use steam locomotives fuelled by coal manually shovelled into the firebox) have become commonplace. Other activities unknown in the 19th century (including all those connected with the internal combustion engine) have come on the scene, being regarded first as dangerous innovations (see *Musgrove*…) but now as basic necessities."

I am bound to conclude, therefore, that when their Lordships laid down their guidance for the application of *Rylands* v *Fletcher*, they did not exclude cases of the escape of fire and the principles they espoused should be applied in fire cases as well as in other more classic examples of escaping dangerous things.

My conclusions on fire and the rule in Rylands v Fletcher

47. I feel bound for the reasons I have given to hold that in an appropriate case damage caused by fire emanating from an adjoining property can fall within the *Rylands* v *Fletcher* rule. The appropriate case is likely to be very rare, as Lord Bingham said in *Transco*, at [5]:

"…Consideration of the reported English case law over the past 60 years suggests that few if any claimants have succeeded in reliance on the rule in *Rylands* v *Fletcher* alone."

Lord Hoffmann said, at [39]:

"…It is perhaps not surprising that counsel could not find a reported case since the second world war in which anyone had succeeded in a claim under the rule. It is hard to escape the conclusion that the intellectual effort devoted to the rule by judges and writers over many years has brought forth a mouse."

48. Cases of fire damage are likely to be very difficult to bring within the rule because (1) it is the "thing" which had been brought onto the land which must escape, not the fire which was started or increased by the "thing". (2) While fire may be a dangerous thing, the occasions when fire as such is brought onto the land may be limited to cases where the fire has been deliberately or negligently started by the occupier or one for whom he is responsible. Is this not a relic of the *ignis suus* rule? (3) In any event, starting a fire on one's land may well be an ordinary use of the land.

49. As for *Musgrove*, I reluctantly cannot conclude that it was decided *per incuriam*. I might be prepared to say, if forced to do so, that since the core finding was of negligence, then the findings based on *Rylands* v *Fletcher* were obiter. I prefer, however, simply to treat *Musgrove* as a fact-sensitive case. Liability must, as Lord Bingham has explained to us, always be judged in the light of prevailing conditions of time and place and *Musgrove* would most definitely not be decided today as it was nearly a century ago. *Musgrove* should, in my judgment, therefore simply be relegated to a footnote in the history of *Rylands* v *Fletcher*.

50. Fire cases and this appeal must be judged in accordance with the test to be derived from *Transco* which I have set out at [22] above. Applying those principles, I reach the following conclusions:

(1) The "thing" brought onto Wyvern's premises was a large stock of tyres.

(2) Tyres, as such, are not exceptionally dangerous or mischievous.

(3) There is no evidence that Mr Stannard recognised nor ought he reasonably to have recognised that there was an exceptionally high risk of danger or mischief if the tyres, as such, should escape.

(4) The tyres did not escape. What escaped was the fire, the ferocity of which was stoked by the tyres which were burning on, and remained burning on, Wyvern's premises. The Recorder was wrong to conclude it was the escape of fire that brought the case within *Rylands* v *Fletcher* principles.

(5) In any event, keeping a stock of tyres on the premises of a tyre-fitting business, even a very large stock, was not for the time and place an extraordinary or unusual use of the land. Here again the Recorder erred.

(6) Therefore *Rylands* v *Fletcher* liability is not established and, no negligence having been proved, the claim must fail.

(7) The moral of the story is taken from the speech of Lord Hoffman: make sure you have insurance cover for losses occasioned by fire on your premises.

51. I would therefore allow the appeal and dismiss the claimant's claim.

Etherton LJ said:

52. I pay tribute to the careful and comprehensive judgment of the Recorder. I nevertheless agree that this appeal should be allowed. It raises important points, but I can express my views quite briefly.

53. Firstly, in the light of the comprehensive review of the *Rylands* v *Fletcher* principle in *Transco*, I do not consider that the facts of the present case satisfy the basic requirement of the *Rylands* v *Fletcher* principle that there must have been an escape of something which the defendant has brought onto his or her land. Secondly, if there is a different requirement for so-called "fire" cases, where what has escaped from the defendant's land is fire generated from something that the defendant has brought onto his or her land, the principle still does not apply in the present case because tyres are not easily set alight and so do not pose any inherent danger of catching or causing fire. Thirdly, in any event, Mr Stannard's use of his property was not a non-natural use of his land for the purposes of the *Rylands* v *Fletcher* principle.

54. The classic statement of the strict liability principle by the Court of Exchequer Chamber in *Rylands* v *Fletcher* itself (at (1866) LR 1 Ex 265, 279) (approved by the House on Lords on appeal at (1868) LR 3 HL 330 with the addition of the requirement of non-natural user) requires an escape from the defendant's land of something which he or she has brought there and which is likely to cause harm to the claimant if it escapes:

"We think that the true rule of law is, that the person who for his own purposes brings on his lands and collects and keeps there anything likely to do mischief if it escapes, must keep it in at his peril, and, if he does not do so, is prima facie answerable for all the damage which is the natural consequence of its escape."

55. It is obvious that the principle of strict liability, as stated there, requires modification if it is to apply where the escape is not of something which the defendant has brought onto his or her land, but is fire which has spread from things which were brought onto the land, and the fire was not deliberately or negligently started by the defendant.

56. As the cases show, it is necessary to modify the classic statement of the *Rylands* v *Fletcher* principle in a radical way in order to encompass such a situation, but there is no clear consensus as to how the principle should be restated. In some of the fire cases it was regarded as sufficient that the defendant brought onto his land something which was (at the time) considered inherently dangerous, such as a train (*Festiniog Railway Company*) or a car (*Musgrove*). In other cases, such as *Mason*, it was the particular danger presented by the flammable nature of the materials brought onto the land by the defendant which was regarded as critical. The point was addressed by MacKenna J in *Mason* at p542 as follows:

"If, for the rule in *Musgrove*'s case to apply, there need be no escape of anything brought onto the defendant's land, what must be proved against him? There is, it seems to me, a choice of alternatives. The first would require the plaintiff to prove (1) that the defendant had brought something onto his land likely to do mischief if it escaped; (2) that he had done so in the course of a non-natural user of the land; and (3) that the thing had ignited and that the fire had spread. The second would be to hold the defendant liable if (1) he brought onto his land things likely to catch fire, and kept them there in such conditions that if they did ignite the fire would be likely to spread to the plaintiff's land; (2) he did so in the course of some non-natural use; and (3) the things ignited and the fire spread. The second test is, I think, the more reasonable one. To make the likelihood of damage if the thing escapes a criterion of liability, when the thing has not in fact escaped but has caught fire, would not be very sensible."

57. That was the approach taken by HH Judge Coulson QC, sitting as a High Court judge, in *LMS International Ltd* v *Styrene Packaging and Insulation Ltd* [2005] EWHC 2065 (TCC) 30 September 2005, in which he undertook an extensive analysis of the *Rylands* v *Fletcher* principle, with particular regard to the review by the House of Lords in *Transco* and the fire cases. He said as follows at [33]:

"In cases concerned with fire, the rule in *Rylands* v *Fletcher* requires two things. First, the defendant must have brought onto his land things which were likely to cause and/or catch fire, and kept them in such a condition that, if they ignited, the fire would be likely to spread to the claimant's land. To put it another way, those things must represent a recognisable risk to the owners of the adjoining land. Secondly, the actions on the part of the defendant must arise from a non-natural user of the defendant's land…"

58. A different formulation is put forward by Mr Philip Rainey QC and Mr Nicholas Isaac on behalf of Mr Gore, namely that the relevant danger which must be established is that the things collected on the defendant's land are likely to cause or facilitate the escape of the fire. They approve the formulation of Judge Coulson in *LMS International* if his expression "cause fire" is understood as meaning "cause the fire which escapes and damages the neighbouring land". In other words, they say the emphasis should not be on the danger posed by the flammability of the materials brought onto the defendant's land but on the danger posed by the role of the material in the escape of the fire from the defendant's land.

59. The lack of clarity and consensus as to the proper formulation of the *Rylands* v *Fletcher* principle in fire cases does not merely reflect poorly on a legal principle of strict liability 146 years after its first published judicial articulation. It calls into question whether the principle applies at all to fire cases. The *Rylands* v *Fletcher* principle was comprehensively reviewed by the House of Lords in *Transco*. That was not a fire case, but I consider that the reasoning and analysis of the House of Lords leave no scope for the application of the principle on the facts of the present case where there has been no escape of anything brought onto the defendant's land, but the escape is of fire, which was not created by the defendant.

60. The House of Lords in *Transco* had all the national, international and academic tools for a thorough review of the *Rylands* v *Fletcher* principle. They decided as a matter of policy to retain the principle, while restating it to achieve as much certainty and clarity as possible. As Lord Bingham said, at [8]:

> "There remains a third option, which I would myself favour: to retain the rule, while insisting upon its essential nature and purpose; and to restate it so as to achieve as much certainty and clarity as is attainable, recognising that new factual situations are bound to arise posing difficult questions on the boundary of the rule, wherever that is drawn."

61. Lord Bingham, with whom Lord Scott and Lord Walker agreed, described the core aspects of the principle in [10] and [11]. The latter para addressed the issue of non-natural user. The former identified, at the heart of the principle, the exceptional danger or mischief if there is an escape from the defendant's land. In describing that aspect, Lord Bingham said at [10]:

> "It has from the beginning been a necessary condition of liability under the rule in *Rylands* v *Fletcher* that the thing which the defendant has brought on his land should be 'something which... will naturally do mischief if it escape out of his land' (LR 1 Ex 265, 279 per Blackburn J), 'something dangerous...', 'anything likely to do mischief if it escapes', 'something... harmless to others so long as it is confined to his own property, but which he knows to be mischievous if it gets on his neighbour's' (p280), 'anything which, if it should escape, may cause damage to his neighbour' (LR 3 HL 330, 340, per Lord Cranworth). The practical problem is of course to decide whether in any given case the thing which has escaped satisfies this mischief or danger test, a problem exacerbated by the fact that many things not ordinarily regarded as sources of mischief or danger may none the less be capable of proving to be such if they escape. I do not think this condition can be viewed in complete isolation from the non-natural user condition to which I shall shortly turn, but I think the cases decided by the house give a valuable pointer. In *Rylands* v *Fletcher* itself the courts were dealing with what Lord Cranworth (LR 3 HL 330, 342) called 'a large accumulated mass of water' stored up in a reservoir, and I have touched on the historical context of the decision in para 3(3) above. *Rainham Chemical Works* involved the storage of chemicals, for the purpose of making munitions, which 'exploded with terrific violence'. In *Attorney General* v *Cory Bros & Co Ltd* [1921] 1 AC 521, 525, 530, 534, 536, the landslide in question was of what counsel described as an 'enormous mass of rubbish', some 500,000 tons of mineral waste tipped on a steep hillside. In *Cambridge Water*, the industrial solvents being used by the tannery were bound to cause mischief in the event, unforeseen on the facts, that they percolated down to the water table. These cases are in sharp contrast with those arising out of escape from a domestic water supply (such as *Carstairs* v *Taylor* (1871) LR 6 Ex 217, *Ross* or *Anderson* v *Oppenheimer* (1880) 5 QBD 602) which, although decided on other grounds, would seem to me to fail the mischief or danger test. Bearing in mind the historical origin of the rule, and also that its effect is to impose liability in the absence of negligence for an isolated occurrence, I do not think the mischief or danger test should be at all easily satisfied. It must be shown that the defendant has done something which he recognised, or judged by the standards appropriate at the relevant place and time, he ought reasonably to have recognised, as giving rise to an exceptionally high risk of danger or mischief if there should be an escape, however unlikely an escape may have been thought to be. "

62. Lord Bingham summarised the *Rylands* v *Fletcher* principle most succinctly in [11] as follows:

> "An occupier of land who can show that another occupier of land has brought or kept on his land an exceptionally dangerous or mischievous thing in extraordinary or unusual circumstances is in my opinion entitled to recover compensation from that occupier for any damage caused to his property interest by the escape of that thing, subject to defences of act of God or of a stranger, without the need to prove negligence."

63. Lord Hobhouse stated in [66] that "[t]he content of the rule has been clearly spelled out by Blackburn J [in *Rylands* v *Fletcher*]". Lord Scott similarly said in [75] that the classical exposition of the rule was to be found in Blackburn J's judgment.

64. The various analyses in *Transco* all endorse the original formulation of the principle in the Court of Exchequer Chamber and in the House of Lords in *Rylands* v *Fletcher*. They make clear that the relevant mischief or danger is not something inherent in what is brought onto the defendant's land, for the principle may apply where what is brought onto the land is in itself harmless (such as water). The speeches in *Transco* make clear that the relevant mischief or danger lies in what will happen if there is an escape from the defendant's land of what has been brought onto the land. Accordingly, they leave no scope for the formulations in the previous fire cases, which required, as a condition of liability, that there was some inherent danger (irrespective of escape) in what the defendant has brought onto his or her land, whether by virtue of flammability (eg *Mason*) or otherwise (eg *Festiniog Railway Company* and *Musgrove*).

65. I do not consider that *Transco* can properly be regarded as leaving entirely open the possibility of a radically different formulation of the principle in fire cases. While it is true that not all the fire cases were cited in the speeches or, apparently, in submissions in the House of Lords in *Transco*, at least one of the leading fire cases, *Musgrove*, was cited (and indeed mentioned by Lord Walker at [107]). In any event, it is clear that the House of Lords in *Transco* was undertaking a comprehensive and definitive review of the *Rylands* v *Fletcher* principle in order to provide legal certainty. Leaving out of account a significant area of activity, without even mentioning that it was the intention to do so, would have been entirely inconsistent with that objective.

66. In those circumstances, and in the light of the emphasis in *Transco* and elsewhere that the *Rylands* v *Fletcher* principle should not be expanded further, I can see no scope for the application of the principle in the present case where there was no escape of the tyres which caught fire and the fire itself was not created by Mr Stannard.

67. If I am wrong about that, both principle and the weight of authority support the conclusion that the *Rylands* v *Fletcher* principle only applies to fire cases, where the fire has not been created by the defendant, if the fire has spread from materials brought onto the defendant's land and those materials pose a particularly high risk because they are likely to cause or catch fire and, if they do, such fire will be likely to spread to the claimant's property. If I am wrong on my first point about *Transco*, that was a correct summary of the law by Judge Coulson in *LMS International*, following the approach of MacKenna J in *Mason*. I, therefore, reject the gloss put on Judge Coulson's summary on behalf of Mr Gore. In a standard non-fire *Rylands* v *Fletcher* case the culpability of the defendant is the act of bringing something onto his or her land which it is foreseeable poses an exceptional risk of harm to the claimant if it escapes. The foreseeable risk is causally related to (1) the defendant's act in bringing something onto his or her land, and (2) the likely consequences if that same thing escapes. In a fire case, where the fire has not been created by the defendant, the culpability must, by analogy, be the act of the defendant in bringing onto his or her land something which poses a foreseeable risk to the defendant involving the creation of fire and its escape. There must be a causal connection between the defendant's act in bringing something onto his or her land and the foreseeable risk of harm to the defendant in the event of an escape of fire. That causal connection can only be the inherent likelihood that what the defendant has brought onto his or her land will catch or cause fire and the fire will spread to the claimant's property. In the present case, the evidence was that tyres are not in themselves flammable and they are not easily set alight. The necessary link between the deliberate act of Mr Stannard (in bringing tyres onto his land) and the foreseeability of an exceptional risk of harm to Mr Gore in the event of an escape did not exist.

68. If that too is wrong, then I do not consider that it was open to the Recorder, in the light of the evidence, to conclude that Mr Stannard's use of his property was a non-natural use for the purposes of the *Rylands* v *Fletcher* principle. A non-natural use, in this context, is a use which is "extraordinary and unusual" and is a high threshold for a claimant to surmount: *Transco* at [11] (Lord Bingham) and [49] (Lord Hoffmann). It is a mixed question of fact and law. It would not be right to interfere with the Recorder's view unless it exceeded the generous ambit within which reasonable disagreement is possible. The Recorder reached his conclusion on non-natural use on the basis of the number of tyres relative to the size of Mr Stannard's premises and also the manner in which they were stored. That is, with respect, very far short of what would be required for a finding of extraordinary use such as to engage the *Rylands* v *Fletcher* principle. The commercial activity carried on by Mr Stannard as a motor vehicle tyre supplier was a perfectly ordinary and reasonable activity to be carried on in a light industrial estate. There was no evidence that the number of tyres and the method of their storage was out of the ordinary for similar premises carrying on that type of activity or that they posed a foreseeable or recognised danger. Indeed, as has been pointed out on behalf of Mr Stannard, any other conclusion would be inconsistent with the Recorder's rejection of Mr Gore's case in negligence. In those circumstances, it was not open to the Recorder to reach the conclusion that he did on this aspect.

69. For the sake of completeness, if I were wrong on all those points, I would uphold the Recorder's judgment in favour of Mr Gore. I would reject, both as a matter of precedent and principle, Mr Stannard's argument that section 86 of the Fires Prevention (Metropolis) Act 1774 Act excludes liability under the *Rylands* v *Fletcher* principle. That point was directly addressed and decided by the Court of Appeal in *Musgrove* and so is binding on this court. The alternative basis for the decision in that case was that the relevant fire, for the purposes of section 86 of the 1774 Act, was begun negligently and was outside the section for that reason. Nevertheless, the point that liability under *Rylands* v *Fletcher* is not exonerated by section 86 forms part of ratio of the Court of Appeal.

70. That also seems to have been the view of the Court of Appeal in *Collingwood*. In that case, Lord Wright MR and Romer LJ, with both of whom Macnaghten J agreed, questioned the correctness of the decision in *Musgrove* insofar as it held that there was liability under the *Rylands* v *Fletcher* principle on the facts of that case, but neither of them said that they also doubted the conclusion in *Musgrove* that *Rylands* v *Fletcher* liability falls outside section 86 of the 1774 Act.

71. In any event, that conclusion seems to me to be right in principle. It still remains a principle of statutory interpretation that, in cases of doubt, there is a presumption against interference with the common law: see generally *Bennion on Statutory Interpretation* (5th ed) section 269; *Craies on Legislation* (9th ed) 14.1.7 and 14.1.11. It is generally accepted that the word "accidentally" in section 86 cannot be read without some qualification and that it does not, for example, embrace a fire begun or continued negligently: *Filliter* at 357, *Balfour*, *Goldman* v *Hargrave* [1967] 1 AC 645 (PC). In this limited context, in view of the classical declaratory theory of the common law (viz judges merely expound and declare the existing law rather than create it), I do not consider that a particular problem to this approach is posed by the fact that *Rylands* v *Fletcher* was not decided until 1866, just as it has not been regarded as a problem for excluding negligently caused fires from section 86 of the 1774 Act that it was not until the 19th century that negligence began to emerge as the substantive allegation in an action on the case. In any event, there are perfectly sound and rational grounds to conclude that section 86 was enacted to clarify some limited areas of uncertainty in the customary law of *ignis suus*.

72. The factual and legal background to the enactment of section 86 was examined by AI Ogus in the excellent and well-known scholarly article *Vagaries in Liability for The Escape of Fire* [1969] CLJ 104. He there set out three hypothetical situations in relation to *ignis suus*. He concludes that section 86 was enacted to resolve the uncertainty in the second and third of those situations, namely: (II) where the defendant has lit a candle in his house, and an exceptionally strong wind, or the act of a stranger, knocks the candle over setting the house alight, and the fire then spreads to the plaintiff's land; and (III) where the defendant has lit a bonfire in his field, and, as a result of an exceptionally high wind or the act of a stranger, sparks are blown onto a pile of leaves, which ignite, the fire spreading to the neighbour's property. I see no reason to disagree with Mr Ogus' analysis that the only object of section 86 of the 1774 Act (and the equivalent provisions in its statutory predecessor (1707) 6 Anne c.31 x.6) was to resolve the doubts as to whether the defendant would be liable in those two particular situations.

Lewison LJ said:

73. I agree with Ward LJ that the appeal must be allowed. However, I would go further in limiting the scope of strict liability in relation to fire. Let me explain.

74. We have all heard of the Great Fire of London in 1666. But it was one in a long series of devastating fires which were common in mediaeval and Tudor England. London was ravaged by fire many times during the 12th and 13th centuries. Building codes dealing with fire prevention were promulgated in London as early as 1189. William Fitzstephen, writing in the 12th century, wrote that "the only plagues of London are the immoderate drinking of fools and the frequency of fires."

75. Fire was an equally serious hazard in other towns. The loss had to fall somewhere. These were the days before insurance. The custom of the common law of England laid the loss at the door of the person on whose property the fire had started. In *Beaulieu*, the custom was pleaded thus: "*quilibet de eodem regno ignem suum salvo et secure custodiat et custodire teneatur ne per ignem suum dampnum aliquod vicinis suis ullo modo eveniat*". The plea went on to allege that the defendant "*ignem suum improvide et negligenter custodivit*". The custom is translated in Baker & Milsom *Sources of English Legal History: Private Law to 1750* p610 as follows: "everyone in the same realm should keep and is bound to keep his fire safely and securely so that no damage befall any of his neighbours in any wise through his fire". The plea was that the defendant kept his fire "so negligently that for the want of due keeping of the aforesaid fire" the plaintiff's house was burned. Thirning CJ said "that a man should answer for his fire which by misfortune burns another's goods". Markham J said:

> "A man is bound to answer for his servant's act, as for his lodger's act, in such a case. For if my servant or lodger puts a candle on a wall and the candle falls into the straw and burns the whole house and also my neighbour's house, in this case I shall answer to my neighbour for the damage which he has suffered."

76. He added:

> "I shall answer to my neighbour for anyone who enters my house by my leave or with my knowledge, or is my guest or my servant's guest, if he does something with a candle (or whatever) by which my neighbour's house is burned. But if a man outside my house, against my will, sets fire to the thatch of my house or elsewhere, so that my house is burned and my neighbours' houses are burned as well, I shall not be held bound to answer to them for it; since it cannot be said to be ill-doing on my part when it is against my will."

77. In *Crogate* v *Morris* (1617) 1 B&G 197 the principle was pithily stated: "if my friend come and lie in my house, and set my neighbour's house on fire, the action lieth against me." Thus *Rolle's Abridgment* was able to state:

> "If fire (I know nothing of it) suddenly break out in my house and burn my goods, and also the house of my neighbour he shall have an action on the case against me."

78. It is noticeable that the custom is that a man should keep "his fire" (*ignis suus*) safe and secure; and that the judges' discussion is all about fires deliberately kindled (even in the case of a candle which is deliberately lit). The plea was that the fire had been kept "negligently". It is clear that "negligently" was not used in the way in which that adverb is understood in the modern law of tort. What it signified was that the defendant had failed to observe his legal duty to prevent his fire escaping and damaging others. Some of the subsequent cases are

A concerned with what counted as a man's fire or "*ignis suus*". That the custom was concerned with fires that had been deliberately kindled is borne out by the statement of the law in *Comyn's Digest: Action on the Case for Negligence* (A6):

> "An action lies, upon the general custom of the realm, against the master of a house if a fire be kindled there and consume the goods of another."

79. In a case of 1582 (Anon (1582) Cro Eliz 10), the defendant fired a gun at a fowl. In so doing he set fire to his own and his neighbour's house. The action was brought by way of action on the case. The court was of the view if the plaintiff "had counted on the custom of the realm
B as in [*Beaulieu*] the action had not been well brought". Although the court did not give reasons for this view, it was presumably because the fire was not deliberately kindled and was never within the defendant's control (ie it was not his fire or "*ignis suus*").

80. In *Turberville*, the real question was whether the defendant was liable for a fire deliberately kindled in a field; in other words, whether such a fire counted as his own ("*ignis suus*"). Sir John Holt justified the principle as follows:

> "Every man must so use his own as not to injure another. The law is general; the fire which a man makes in the fields is as much his fire as his fire
> C in his house; it is made on his ground, with his materials, and by his order; and he must at his peril take care that it does not, through his neglect, injure his neighbour: if he kindle it at a proper time and place, and the violence of the wind carry it into his neighbour's ground, and prejudice him, this is fit to be given in evidence. But now here it is found to have been by his negligence; and it is the same as if it had been in his house."

81. In a different report of the case in (1702) 1 Comyns 32 the judgment is summarised thus:

> "The case was afterwards adjudged in favour of the plaintiff by the whole court; for the action is as well for a fire kindled in the fields of the defendant as in his house, for it is the defendant's fire and kindled in his ground, and he ought
> D to have the same care of a fire which he kindles in his field as of that which is made in his house, for the duty to take care of both is founded upon this maxim, *sic utere tuo ut non lædas alienum*; but if the fire of the defendant by inevitable accident, by impetuous and sudden winds, and without the negligence of the defendant or his servants, (for whom he ought to be answerable) did set fire to the clothes of the plaintiff in his ground adjoining; the defendant shall have the advantage of this in evidence, and ought to be found not guilty. But here the verdict hath found negligence in the defendant. Therefore judgment for the plaintiff."

82. The report in 1 Salk 13 is even more abbreviated. There are five important points to be made about this case. First, liability is based on
E the general principle that a man "must so use his own as not to injure another" ("*sic utere tuo ut alienum non laedas*"). Second, on the facts liability was established because of the finding of negligence. Third, liability would have been avoided if the fire had been kindled at a proper time and place. Thus fourth, it is at least possible that even the common law required a finding of negligence. But if that is wrong, then it is clear that a defendant had a good defence if the fire was caused by inevitable accident or by act of a stranger. Fifth, the custom was directed towards a fire that a person had deliberately kindled, but which had subsequently got out of control. This, in my judgment, is implicit in the requirement to plead "*ignis suus*".

F 83. Even where liability is based on the principle *sic utere tuo ut alienum non laedas* a requirement of negligence is not necessarily excluded. In *Black* v *Christchurch Finance Co Ltd* [1894] AC 48 Lord Shand, delivering the advice of the Privy Council, said:

> "The lighting of a fire on open bush land, where it may readily spread to adjoining property and cause serious damage, is an operation necessarily attended with great danger, and a proprietor who executes such an operation is bound to use all reasonable precautions to prevent the fire extending to his neighbour's property (*sic utere tuo at alienum non laedas*)."

84. In other words, even where liability was founded on (or at least justified by) the maxim, liability was not absolute, but required the taking of all reasonable precautions.

85. In *Emanuel* (another case of a deliberately kindled fire) Lord G
Denning MR considered the scope of liability at common law. He said, at p838b-e:

> "After considering the cases, it is my opinion that the occupier of a house or land is liable for the escape of fire which is due to the negligence not only of his servants, but also of his independent contractors and of his guests, and of anyone who is there with his leave or licence. The only circumstances when the occupier is not liable for the negligence is when it is the negligence of a stranger. It was so held in a case in the Year Books 570 years ago, *Beaulieu*, which is well translated by Mr Fifoot in his book on the *History and Sources of the Common Law*. The occupier is, therefore, liable for the negligence of an independent contractor, such as the man who comes in to repair the pipes and H
> uses a blowlamp: see *Balfour*; and of a guest who negligently drops a lighted match: see *Boulcott Golf Club Inc* v *Engelbrecht* [1945] NZLR 556. The occupier is liable because he is the occupier and responsible in that capacity for those who come by his leave and licence: see *Sturges* v *Hackett* [1962] 1 WLR 1257.
>
> *But the occupier is not liable for the escape of fire which is not due to the negligence of anyone*. Sir John Holt himself said in *Tuberville* that if a man is properly burning up weeds or stubble and, owing to an unforeseen wind-storm, without negligence, the fire is carried into his neighbour's ground, he is not liable. Again, if a haystack is properly built at a safe distance, and yet bursts into flames by spontaneous combustion, without negligence, the occupier is not liable. That is to be inferred from *Vaughan*. So also if a fire starts without
> negligence owing to an unknown defect in the electric wiring: *Collingwood*; or J
> a spark leaps out of the fireplace without negligence: *Sochacki* v *Sas* [1947] 1 ALL ER 344." (Emphasis added)

86. Whatever the precise basis for liability analysed in modern terms, historically it was simply based on custom of the realm. As Lord Denning put it in *Emanuel*, at p839d-e:

> "There has been much discussion about the exact legal basis of liability for fire. The liability of the occupier can be said to be a strict liability in this sense that he is liable for the negligence not only of his servants but also of independent contractors and, indeed, of anyone except a 'stranger'. By the same token it can be said to be a 'vicarious liability', because he is liable for the K
> defaults of others as well as his own. It can also be said to be a liability under the principle of *Rylands* v *Fletcher*, because fire is undoubtedly a dangerous thing which is likely to do damage if it escapes. But I do not think it necessary to put it into any one of these three categories. *It goes back to the time when no such categories were thought of.* Suffice it to say that the extent of the liability is now well defined as I have stated it. The occupier is liable for the escape of fire which is due to the negligence of anyone other than a stranger." (Emphasis added)

87. This is an important point. Liability in mediaeval law (at least when it was concerned with liability by custom) did not depend on proving that the defendant was negligent in the modern sense. Nor did
it depend on proving that he was engaged in any particularly dangerous L
activity. As Sir William Holdsworth explains (*History of English Law* vol VIII p469):

> "The form which this stricter liability took was not the form taken by the rule in *Rylands* v *Fletcher*, for that rule was then the general rule of civil liability."

88. Apart from fire cases, others were also subject to strict liability by virtue of custom. One such category of people was the innkeeper who was liable by custom of the realm for the safe custody of the goods of his guests. This analogy was drawn on by counsel in *Turberville*, who
is recorded as having argued: M

> "that this action ought not to be grounded upon the common customs of the realm; for this fire in the field cannot be called iguis suus, for a man has no power over a fire in the field, as he has over a fire in his house. And therefore this resembles the case of an inn-keeper, who must answer for any ill that happens to the goods of his guest, so long as they are in his house; but he is not answerable, if a horse be stolen out of his close."

89. It was against this background that the first of the statutes restricting liability for fire damage was passed. This was the Act of 1707, section 6, which was originally temporary and applied to London only. According to Lord Denman CJ in *Filliter*, a decision of the Court of King's Bench, at pp354-355:

"The Act contemplates the probability of fires in cities and towns arising from three causes, the want of water, the imperfection of party walls, and the negligence of servants. The Act provided some means for supplying these material defects: but the third section was directed against the moral one, the carelessness or negligence of servants, which (it observes) often causes fires: and it imposes on *the servant* by whose negligence the fire may have been occasioned a fine of £100, to be distributed among the sufferers at the discretion of the churchwardens, or imprisonment for 18 months in case of nonpayment… The most usual cause of fires was assumed to be the negligence of servants: and the enactment might operate to induce habits of caution in that important class. The same statute, in the sixth section, enacts that, after a day named, no action shall be maintained against any person in whose house or chamber any fire shall accidentally begin, nor shall any recompence be made by such person for any damage suffered or occasioned thereby."

90. This was then re-enacted in 1772 and 1774, the last of the enactments remaining in force as section 86 of the Fires Prevention (Metropolis) Act 1774, which applies to the whole of England and Wales: *Richards* v *Easto* (1846) 15 M&W 246, 251. Section 86 provides:

"And… no action, suit or process whatever shall be had, maintained or prosecuted against any person in whose house, chamber, stable, barn or other building, or on whose estate any fire shall… accidentally begin, nor shall any recompence be made by such person for any damage suffered thereby, any law, usage or custom to the contrary notwithstanding… provided that no contract or agreement made between landlord and tenant shall be hereby defeated or made void."

91. It is important to emphasise the scope of the change effected by section 86. Not only does it apply notwithstanding any *custom* to the contrary (which would include liability under the *ignis suus* principle), it also applies notwithstanding any *law* to the contrary. This would, as a matter of ordinary statutory construction, include any other route at common law to liability: (compare *McKillen* v *Maybourne Finance Ltd* [2012] EWCA Civ 864).

92. Contemporaneous commentators (Blackstone *Commentaries on the Laws of England* (10th ed) 1787 p431 and Bacon *A New Abridgment of the Law* (5th ed) 1789 p85) thought that the Act had so altered the law that even where a fire had been caused or allowed to spread by negligence no liability arose. This treated the word "accidentally" as having been used in contrast to "wilfully". In *Viscount Canterbury* v *Attorney-General* (1843) 1 Phillips 306, Lord Lyndhurst LC at least flirted with this view, although it was not necessary for him to reach a decision.

93. *Vaughan* was a case of a haystack that spontaneously combusted. The allegation against Menlove was that he had "wrongfully, negligently and improperly" kept his haystack "so likely and liable to ignite and take fire, and in a state and condition dangerous" to Vaughan's cottages. It is important to note that this was not a case of a fire that had been deliberately kindled. At trial Patteson J directed the jury to consider whether the fire had been occasioned by "gross negligence" on the part of Menlove. The argument for Menlove on appeal was that the jury should have been asked to consider whether Menlove had "acted bona fide to the best of his judgment". The court rejected that argument and held that the case had been rightly left to the jury on the question of negligence. Liability thus depended on negligence.

94. *Viscount Canterbury* was a petition of right brought against the Crown by a former speaker of the House of Commons whose possessions had been destroyed in the fire that swept though the Houses of Parliament in 1834 (and to which we owe Barry and Pugin's building). The argument for the speaker was that the statute did not relieve against fires caused by negligence, contrary to the opinion that Blackstone had expressed in his commentaries. The attorney does not appear to have argued this point; because he succeeded on the ground that a petition of right did not lie for negligence on the part of the sovereign or his servants.

95. In *Filliter*, Lord Denman CJ confirmed that Blackstone's view was wrong. The question was: what did the word "accidentally" mean in the context of section 86? He said, at p357:

"It is true that, in strictness, the word *accidental* may be employed in contradistinction to wilful, and so the same fire might both begin accidentally and be the result of negligence. But it may equally mean a fire produced by mere chance, or incapable of being traced to any cause, and so would stand opposed to the negligence of either servants or masters. And, when we find it used in statutes which do not speak of wilful fires but make an important provision with respect to such as are accidental, and consider how great a change in the law would be effected, and how great encouragement would be given to that carelessness of which masters may be guilty as well as servants, we must say that we think the plaintiff's construction much the most reasonable of the two."

96. Thus the court held that a fire that had begun by negligence did not attract the protection of section 86. In *Burnie Port Authority*, Mason CJ, writing for the majority of the High Court of Australia, described this decision as "surprising". But it is too late to change that.

97. The railway age brought a new spate of litigation. Early locomotives were steam powered; and the steam was generated by means of fires in the fire box of the locomotive, deliberately kindled. Sometimes, however, the fire in the locomotive could not be kept fully under control; or sparks from the fire were carried up the locomotive's chimney together with the steam.

98. In *Aldrige*, the allegation was that sparks from the company's railway engine set light to the plaintiff's beans because the company carelessly managed their engine. Tindal CJ said, at p523:

"It is contended on the part of the defendants, that the plaintiff should be nonsuited; but I am not prepared to say that the fact of the engine emitting sparks may not amount to negligence. On the other hand, I cannot say that a verdict ought to be entered for the plaintiff. I think that the special case should be withdrawn, and that the parties should go on to trial. To entitle the plaintiff to recover, he must either shew some carelessness by the defendants, or lay facts before the jury from which it may be inferred."

99. Clearly liability was not strict but depended on negligence. In *Piggot*, sparks from the engine of a passing mail train set fire to the plaintiff's cart lodge. The claim against the railway company was that they "so carelessly, negligently, and unskilfully managed and conducted their said steam-carriage and steam-engine" that the plaintiff's cart house was set on fire. In other words the claim was framed in negligence. The point in the case was the admissibility of evidence; but Tindal CJ again described the legal principle as follows, at p240:

"The defendants are a company intrusted by the legislature with an agent of an extremely dangerous and unruly character, for their own private and particular advantage: and the law requires of them that they shall, in the exercise of the rights and powers so conferred upon them, adopt such precautions as may reasonably prevent damage to the property of third persons through or near which their railway passes. The evidence in this case was abundantly sufficient to shew that the injury of which the plaintiff complains was caused by the emission of sparks, or particles of ignited coke, coming from one of the defendants' engines; and there was no proof of any precaution adopted by the company to avoid such a mischance. I therefore think the jury came to a right conclusion, in finding that the company were guilty of negligence, and that the injury complained of was the result of such negligence."

100. Thus although the locomotive was regarded as "dangerous", liability still turned on negligence. The underlying reason may be that the use of steam engines was authorised by statute, and that therefore negligence was the only route to liability, although this reason is not explicitly stated. The ground began to shift in *Vaughan* v *The Taff Vale Railway Company* (1858) 3 H&N 743. This was another case of fire caused by sparks from the fire of a passing locomotive. Although the railway company had done everything that was practicable to the locomotives to make them safe, the railway embankment was covered with inflammable grass. It is probable that it was that which caught fire and from there the fire spread to the plaintiff's wood. The argument for the railway company was that they should not be liable if they had taken all reasonable care to prevent the fire. It was also argued that if the fire had started on the embankment and had spread from there to the plaintiff's wood, section 86 protected them in the absence of negligence. As often happened in the 19th century the ultimate result turned on the pleadings. There were two counts alleged against the

railway company. The first count alleged that they had negligently managed their locomotive and had failed to provide "the proper means for retaining the fire and igneous matter in the said locomotive whilst the same was being propelled along the railway". The second count alleged that the railway company knew that there was a danger that the combustible material on the bank would ignite; and that they therefore had a duty "to preserve and keep the bank in such a state and condition that fire should not be occasioned by reason of the ashes, etc, falling and settling thereon from and out of the locomotives, and to take all necessary precautions to prevent any fire which might be occasioned from extending to, and burning the wood of the plaintiff". The trial judge directed the jury, at p746, that:

> "if, to serve his own purposes, a man does a dangerous thing, whether he takes precautions or not, and mischief ensues, he must bear the consequences: that running engines which cast forth sparks is a thing intrinsically dangerous, and that if a railway engine is used, which in spite of the utmost care and skill on the part of the company and their servants is dangerous, the owners must pay for any damage occasioned thereby."

101. He also pointed out to the jury that the railway company could have kept the grass cut, or could have used gravel or sand to make a non-inflammable belt along the railway; and he asked them:

> "whether they did not think that there was inevitable negligence in the use of a dangerous thing calculated to do, and which did cause, mischief."

102. The jury found for the plaintiff; and the railway company appealed. The judgment of the Court of Exchequer was given by Bramwell B (who had also been the trial judge). He said, at pp751-752:

> "The first question then is: Was there evidence for the jury? And, as they may have found on either count, was there evidence in support of each? Next: Was the evidence such as to warrant the strong opinion of the learned judge?
>
> We are of opinion, on both these questions, in favour of the plaintiff. Here is confessedly the use of an instrument likely to produce damage, and producing it. This, according to general rules, would make the defendants liable. But two answers were suggested on their behalf. The first was, that if the fire originated on their own land they were protected by the 14 Geo 3, c78, section 86. But we are of opinion that the statute does not apply where the fire originates in the use of a dangerous instrument, knowingly used by the owner of the land in which the fire breaks out. It is impossible to suppose that the engine driver is liable to eighteen months imprisonment under section 84, and equally impossible to suppose there is no remedy against either master or servant, for what is a wrong by one or both. We are of opinion therefore that this answer fails."

103. One difficulty with this case is that the jury may have found that there was negligence on the part of the railway company (which went to the second count); in which case it would have been clear that section 86 would not have given the railway company a defence. But the reason that Bramwell B gave for the conclusion that the protection of the 1774 Act was excluded was that the fire was caused by the use of a dangerous instrument, *knowingly* used by the owner of the land. The dangerous instrument was the locomotive in which a fire was deliberately kindled and was burning in order to propel it. It is also noticeable that one of the ingredients of liability was that the instrument was "likely to produce damage". The judgement was reversed by the Court of Exchequer Chamber (1860) 5 H&N 679 on the ground that the use of the locomotive was authorised by Parliament and therefore that so far as the first count was concerned negligence had to be established. Since it was not clear whether the jury had found for the plaintiff on the first count or the second count, there would have to be a new trial. The point under section 86 did not arise for decision.

104. In *Smith* v *The London and South Western Railway Company* (1870) LR 5 CP 98, workmen employed by a railway company in cutting the grass and trimming the hedges bordering the railway, placed the trimmings in heaps near the line, and allowed them to remain there for 14 days, during very hot weather in the month of August. Fire from a passing engine ignited one of these heaps, and burned the hedge, and was carried by a high wind across a stubblefield and a public road, and burned the plaintiff's goods in a cottage about 200 yards away. The question was whether there was evidence of negligence to go before the jury. No one argued that the railway company was strictly liable. Bovill CJ said, at pp105-106:

> "I agree that the mere circumstance of the fire being caused by an engine of the company, is not enough to give a cause of action against them; but the plaintiff must shew some breach of duty on their part which occasioned the injury he complains of."

105. This decision was upheld by the Court of Exchequer Chamber (1870) LR 6 CP 14, among whom was Blackburn J. But as in the case of *Vaughan* v *The Taff Vale Railway Company*, the decision was based on the fact that the company had been authorised by statute to use steam locomotives on the railway. Blackburn J said, at pp21-22:

> "I take it that, since the case of *Vaughan* v *The Taff Vale Railway Company*, which was expressly affirmed in *Brand* v *Hammersmith Railway Co*, it is clear that when a railway company is authorised by their Act of parliament to run engines on their line, and that cannot be done without their emitting sparks, the company are not responsible for injuries arising therefrom, unless there is some evidence of negligence on their part."

106. So we come to the famous case of *Rylands* v *Fletcher* (sub nom *Fletcher* v *Rylands* (1866) LR 1 Ex 265). In the Court of Exchequer Chamber the judges were referred in argument to section 86 of the 1774 Act and also to *Filliter*. Neither found a place in the judgment. The well-known statement of principle formulated by Blackburn J was as follows, at pp279-280:

> "We think that the true rule of law is, that the person who for his own purposes brings on his lands and collects and keeps there anything likely to do mischief if it escapes, must keep it in at his peril, and, if he does not do so, is primâ facie answerable for all the damage which is the natural consequence of its escape. He can excuse himself by shewing that the escape was owing to the plaintiff's default; or perhaps that the escape was the consequence of *vis major*, or the act of God; but as nothing of this sort exists here, it is unnecessary to inquire what excuse would be sufficient. The general rule, as above stated, seems on principle just. The person whose grass or corn is eaten down by the escaping cattle of his neighbour, or whose mine is flooded by the water from his neighbour's reservoir, or whose cellar is invaded by the filth of his neighbour's privy, or whose habitation is made unhealthy by the fumes and noisome vapours of his neighbour's alkali works, is damnified without any fault of his own; and it seems but reasonable and just that the neighbour, who has brought something on his own property which was not naturally there, harmless to others so long as it is confined to his own property, but which he knows to be mischievous if it gets on his neighbour's, should be obliged to make good the damage which ensues if he does not succeed in confining it to his own property. But for his act in bringing it there no mischief could have accrued, and it seems but just that he should at his peril keep it there so that no mischief may accrue, or answer for the natural and anticipated consequences. And upon authority, this we think is established to be the law whether the things so brought be beasts, or water, or filth, or stenches."

107. It is, to my mind, striking that despite all the examples that Blackburn J gave, fire is not among them. He had already decided in *Smith* v *The London and South Western Railway Company* that strict liability at common law could be displaced by statute. In the light of the fact that the court had been referred in argument to section 86 of the 1774 Act and also to *Filliter*, he must have taken the view that section 86 had displaced strict liability; and accepted counsel's submission that:

> "…the liability for fire is restricted to mischief arising from negligence…"

108. Let it be supposed, however, that he thought that fire was a dangerous thing, and intended that the principle he formulated would apply to it. We can then restate the principle thus:

> "a person who for his own purposes brings on his lands and collects and keeps there [fire] must keep it in at his peril, and, if he does not do so, is primâ facie answerable for all the damage which is the natural consequence of its escape."

109. The principle thus formulated is consistent only with a fire deliberately kindled. A fire that starts accidentally and which damages the place where it begins can hardly be said to have brought on to the land by the occupier "for his own purposes". Nor in my judgment can he

be said without straining language to have brought the fire onto the land at all. In *Read*, Viscount Simon said of Blackburn J's formulation:

> "Blackburn J several times refers to the defendant's duty as being the duty of "keeping a thing in" at the defendant's peril and by "keeping in" he does not mean preventing an explosive substance from exploding but preventing a thing which may inflict mischief from escaping from the area which the defendant occupies or controls."

110. As is well-known the case went to the House of Lords (sub nom *Rylands* v *Fletcher* (1868) LR 3 HL 330). Neither section 86 nor any fire case was referred to in the course of argument. Lord Cairns formulated the principle thus, at p339:

> "On the other hand, if the defendants, not stopping at the natural use of their close, had desired to use it for any purpose which I may term a non-natural use, for the purpose of introducing into the close that which in its natural condition was not in or upon it, for the purpose of introducing water either above or below ground in quantities and in a manner not the result of any work or operation on or under the land, and if in consequence of their doing so, or in consequence of any imperfection in the mode of their doing so, the water came to escape and to pass off into the close of the plaintiff, then it appears to me that that which the defendants were doing they were doing at their own peril; and, if in the course of their doing it, the evil arose to which I have referred, the evil, namely, of the escape of the water and its passing away to the close of the plaintiff and injuring the plaintiff, then for the consequence of that, in my opinion, the defendants would be liable."

111. He then approved the formulation of Blackburn J in the court below. Lord Cranworth said, at p340:

> "If a person brings, or accumulates, on his land anything which, if it should escape, may cause damage to his neighbour, he does so at his peril. If it does escape, and cause damage, he is responsible, however careful he may have been, and whatever precautions he may have taken to prevent the damage."

112. Neither of their Lordships referred to fire. Nor does the way in which the principle is framed readily apply to fire unless the fire itself has deliberately been brought on to the defendant's land. Lord Cairns referred to the use of the land "for the purpose of" introducing something dangerous. Likewise, Lord Cranworth was clearly contemplating something dangerous deliberately brought onto or accumulated on the land. No consideration was given to how this principle might sit with section 86 of the 1774 Act.

113. *Festiniog Railway Co* was decided after *Fletcher* v *Rylands* in the Court of Exchequer Chamber and only a week or two before the appeal was heard in the House of Lords. The railway company ran steam locomotives on its railway. Although it had taken all reasonable precautions against the emission of sparks from the engine, nevertheless sparks from the engine set the plaintiff's haystack alight and burned down his barn. Blackburn and Lush JJ held that he was entitled to succeed. Blackburn J said, at p736:

> "The general rule of common law is correctly given in *Fletcher* v *Rylands*, that when a man brings or uses a thing of a dangerous nature on his own land, he must keep it in at his own peril; and is liable for the consequences if it escapes and does injury to his neighbour. Here the defendants were using a locomotive engine with no express parliamentary powers making lawful that use, and they are therefore at common law bound to keep the engines from doing injury, and if the sparks escape and cause damage, the defendants are liable for the consequences, though no actual negligence be shewn on their part."

114. The thing of the dangerous nature that the railway company had brought onto its land was the locomotive engine with the deliberately kindled fire. Lush J said, at p738:

> "I can see nothing in this statute to licence the company to use locomotive engines. In the absence of this licence the company are left to their liabilities at common law: that is, if they use a highly dangerous machine, they must do so at the peril of the consequences if it cause injury to others."

115. Again it is clear that the highly dangerous machine was the locomotive, and it was that dangerous thing that the railway company had deliberately brought onto its land. No argument was addressed to the court about the possible effect of section 86. *Jones* v *Festiniog Railway* was followed and applied in *Powell* v *Fall* (1880) 5 QBD 597, a case of a steam powered traction engine on the highway. Sparks from the engine set fire to the plaintiff's haystack. Since the fire started on the highway, section 86 could not have applied. The question for the court, at p598, was:

> "whether the owner of a locomotive engine propelled by steam along a public highway using a fire *for the purpose of generating the steam required* to propel such engine… and was managed and conducted with all reasonable care and without negligence, was liable to the plaintiffs for injury occasioning damage to a rick of hay standing on land adjoining the highway by sparks proceeding from such engine and firing the hay." (Emphasis added)

116. Mellor J answered that question in the affirmative and his decision was upheld by the Court of Appeal. In the Court of Appeal it appears to have been conceded that an action lay at common law, and the only question was whether there was a statutory exemption. As Bramwell LJ put it, at p601:

> "The plaintiffs are protected by the common law, and nothing adverse to their right to sue can be drawn from the statutes: the statutes do not make it lawful to damage property without paying for the injury."

117. Modern formulations of the principle in *Rylands* v *Fletcher* also emphasise the voluntary act of the defendant in bringing the dangerous thing onto his land. They also emphasise that it is that dangerous thing that must escape before strict liability is engaged. In *Transco*, Lord Bingham ended his discussion by saying, in [11]:

> "An occupier of land who can show that another occupier of land has brought or kept on his land an exceptionally dangerous or mischievous thing in extraordinary or unusual circumstances is in my opinion entitled to recover compensation from that occupier for any damage caused to his property interest by the escape of that thing, subject to defences of act of God or of a stranger, without the need to prove negligence."

118. Mr Rainey relied on the decision of the House of Lords in *Rainham Chemical Works*. The case concerned an explosion in a munitions factory. The facts are by no means clear, but what seems to have happened is that a fire broke out in the factory and ignited sodium nitrate stored there. The sodium nitrate in turn produced a hot flame which caused dinitrophenol to explode. The claim was tried by Scrutton LJ, who held that liability under the principle in *Rylands* v *Fletcher* had been established. His decision was upheld by both the Court of Appeal and the House of Lords. However, what was argued on appeal was not, in substance, the question of liability, but whether liability attached to the directors of the company personally. In *Read*, Viscount Simon (169), Lord Macmillan (174) and Lord Uthwatt (187) all pointed this out; and Viscount Simon also said that liability had been conceded before Scrutton LJ sitting at first instance as the trial judge. In *Cambridge Water Co*, for much the same reasons, Lord Goff also regarded *Rainham Chemical Works* as "fragile authority" on the scope of the principle. On the other hand, in *Transco*, *Rainham Chemical Works* was regarded by the Law Lords as being a case that demonstrated the utility of retaining the principle in *Rylands* v *Fletcher*, although the implications of that case were not discussed. Since there was no discussion of the principle either in the Court of Appeal or in the House of Lords in *Rainham Chemical Works*, I do not consider that it lays down any binding principle of law about the scope of the principle in *Rylands* v *Fletcher*, even though it may be regarded as a case which justifies retention of the principle. The manufacture of explosives may well be considered to be an inherently dangerous activity (even in wartime) sufficient to bring the principle into play. Thus the ingredients themselves are the dangerous things which have been deliberately brought onto the land. Section 86 cannot apply, because it applies only to fires; not to explosions.

119. I come now to *Musgrove*. Mr Musgrove rented rooms above a domestic garage, in which Mr Pandelis kept a car. Mr Pandelis sent his chauffeur, Mr Coumis, to clean the car. Mr Coumis had to move the car within the garage. For that purpose he went to the bonnet and turned on the petrol tap to allow the flow of petrol from the tank to the carburettor, and started the engine, when suddenly there was an explosion, and flames were seen to be coming from the carburettor.

A There was no woodwork within 18in of the carburettor, and if Mr Coumis had immediately turned off the tap of the pipe leading from the petrol tank the petrol in the carburettor would have soon burnt out, and the fire would have been prevented from spreading. But instead of doing so Mr Coumis wasted his time in looking for a cloth which he failed to find. He then went to the bonnet to turn off the tap, but was too late, for owing to the continued flow of the petrol into the carburettor the fire had spread to the body of the car. The garage itself then caught fire and the whole building was burnt, including Mr Musgrove's rooms overhead, together with a quantity of furniture belonging to him. Mr Pandelis relied, among other defences, on section 86. At first instance
B [1919] 1 KB 314, Lush J said, at p317:

> "But, nevertheless, I am of opinion that the statute affords the defendant no protection; for though the fire in the carburettor was accidental in a popular sense, I do not think it was accidental in the sense in which that term is used in the statute. If a man brings on to his premises a dangerous thing which is liable to cause fire, such as a motor car with petrol in it, the carburettor of which is not unlikely to get on fire when the engine is started, and a fire results, though without any negligence on his part, he must be held liable, the statute notwithstanding, for the rule is that he must keep such a thing under control at his peril."

120. The key point for him, therefore, was that the petrol was "liable
C to cause fire" and "not unlikely to get on fire". He also held that Mr Coumis was negligent in not immediately turning off the petrol tap. Lush J was upheld on appeal to this court ([1919] 2 KB 43). Each of the three Lords Justices gave judgments. Bankes LJ began by considering the state of the common law before liability for fire was restricted by statute. He said, at p46:

> "A man was liable at common law for damage done by fire originating on his own property (1) for the mere escape of the fire; (2) if the fire was caused by the negligence of himself or his servants, or by his own wilful act; (3) upon the principle of *Rylands* v *Fletcher*. This principle was not then known by that name, because *Rylands* v *Fletcher* was not then decided; but it was an existing
> D principle of the common law as I shall show presently."

121. Bankes LJ then referred to *Filliter*. He said (correctly) that that case decided that a fire negligently begun was not protected by the statute; and posed the question, at p47:

> "Why, if that is the law as to the second head of liability, should it be otherwise as to the third head, the liability on the principle of *Rylands* v *Fletcher*? If that liability existed, there is no reason why the statute should alter it and yet leave untouched the liability for fire caused by negligence or design. That the principle of *Rylands* v *Fletcher* existed long before that case was decided is plain. In *Vaughan* v *Menlove*, Tindal CJ says: 'There is a rule of law which says you must so enjoy your own property as not to injure that of
> E another.' Park J says: 'Although the facts in this case are new in specie, they fall within a principle long established, that a man must so use his own property as not to injure that of others.' *Rylands* v *Fletcher* is merely an illustration of that old principle, and in my opinion Lush J was right in saying that this case, if it falls within that principle, is not within the protection of the statute."

122. In my judgment, this reasoning is historically unsound. As MacKenna J pointed out in *Mason*, at common law there were not three separate routes to liability: there was only one. A householder was liable for the escape of his fire (*ignis suus*): no additional danger was needed to be proved. The liability was based on a custom of the realm and on no other principle. If the case was brought otherwise than on the custom
F of the realm (ie by action on the case) then negligence had to be proved. This view was shared by Holdsworth (whom I have already quoted) and by Lord Denning in *Emanuel* (whom I have also quoted). MacKenna J said, at p540E-541B:

> "There were not three heads of liability at common law but only one. A person from whose land a fire escaped was held liable to his neighbour unless he could prove that it had started or spread by the act of a stranger or of God. Filliter's case had given a special meaning to the words 'accidental fire' used in the statute, holding that they did not include fires due to negligence, but covered only cases of 'a fire produced by mere chance, or incapable of being traced to any cause'. But it does not follow, because that meaning may be given to 'accidental', that the statute does not cover cases of the *Rylands* v *Fletcher* kind where the occupier is held liable for the escape though no fault is proved against him. In such cases the fire may be 'produced by mere chance' or may G
> be 'incapable of being traced to any cause'. Bankes LJ was making a distinction unknown to the common law, between 'the mere escape of fire' (which was his first head) and its escape under *Rylands* v *Fletcher* conditions (which was his third), and was imputing an intention to the legislature of exempting from liability in the former case and not in the latter."

123. Likewise in his article on *Vagaries in Liability for the Escape of Fire* (1969) CLJ 104 AI Ogus says of Bankes LJ's analysis, at pp113-114:

> "This is, without doubt, a distortion of the position at common law. It has already been seen that the modern concept of negligence did not exist before H
> the 19th century. The argument on *Rylands* v *Fletcher* fares no better. Its origins are not to be found in the action on the case pur negligent garder son few. Though similar in result, their bases are substantially different. The old form of action took as its starting point the escape of fire from the defendant's land. The plaintiff need then only show that it was *ignis suus* which caused damage to his property. *Rylands* v *Fletcher*, on the other hand, requires the accumulation on the defendant's land of "something likely to do mischief if it escapes. The former is an example of tortious liability based on 'causal responsibility'. The latter is an example of strict liability attaching to the use of dangerous things. Nor is the quotation from *Vaughan* v *Menlove* of assistance. It was simply part of a discourse on the emerging concept of negligence, and in any case could apply equally well to the tort of nuisance." J

124. The Law Commission has also pointed out the fallacy in the reasoning. In their report on *Civil Liability for Dangerous Things and Activities* (1970) (Law Com No 32), Appendix I para 18, they said:

> "Indeed, it is difficult to escape a dilemma: if a liability of the *Rylands* v *Fletcher* type existed before its classic enunciation in the case from which it takes its name (as the Court of Appeal in fact argued in *Musgrove*) then the broad language of the statute, to which MacKenna J, drew attention can hardly be ignored; if on the other hand it were suggested that the doctrine of *Rylands* v *Fletcher* grew up after the 1774 Act then its development was necessarily limited by the mandate of Parliament."

125. Nor does the reference to the maxim that "a man must so use his K
own property as not to injure that of others" make good the reasoning. First, as shown by *Black* (and as Ogus says) even where the maxim does apply it may require proof of negligence. Indeed, that was the case in *Vaughan* v *Menlove*; where the debate was between an objective test of negligence on the one hand, and a more subjective test of acting in good faith. No question of strict liability arose in that case, despite reliance on the maxim. Thus in my judgment Bankes LJ misinterpreted what the court had decided in *Vaughan* v *Menlove*. Second, in so far as the custom was based on that maxim it was overridden by the statute. Even if it was not part of the *ignis suus* custom it was a "law" which the statute expressly disapplied. Third, later authority is inconsistent L
with Bankes LJ's proposition based on the maxim. Thus in *Emanuel*, Phillimore LJ said, at p842a-b:

> "The LCC were undoubtedly occupiers of this land. They were sued in that capacity and they did not call any evidence to suggest otherwise. As such they owed a duty to their neighbours which is best described in the old Latin maxim: *sic utere tuo ut alienum non laedas*. As Markham J put it in *Beaulieu*:
>
> 'I shall answer to my neighbour for him who enters my house by my leave or knowledge whether he is guest to me or my servant, if either of them acts in such a way with a candle or other things that my neighbour's house is burned." M

Since the Fire Prevention (Metropolis) Act 1774 it is I think necessary to insert the word "negligently" after the word "acts".

126. In my judgment, what *Filliter* decided was that "accidentally" meant "without negligence". The legislation must have been intended to change the common law, whether or not it was based on custom, otherwise Parliament would not have used the wide words that it did. On the footing that fires started by negligence did not attract the protection of section 86, what change must Parliament have intended? The answer, in my judgment, must be that Parliament intended to abolish liability based simply on the proposition that a man "must so use his own as not to injure another", without proof of negligence. Whether liability arose because of custom or because of some other principle of common law

does not matter. Both were overridden by section 86. This is confirmed by *Emanuel*.

127. It follows from this that I disagree with the view expressed by Judge Thornton QC in *Johnson* v *BJW Property Developments Ltd* [2002] EWHC 1131 (TCC); [2002] 3 All ER 574 and by Judge Coulson QC in *LMS International Ltd*, that MacKenna J was wrong in his appreciation of the common law, or in the effect of the 1774 Act.

128. As I have said, liability for fire was not the only form of strict liability imposed by custom of the realm. The liability of innkeepers was another such liability. In *Williams* v *Owen* [1955] 1 WLR 1293, Mr Williams left his car overnight in the hotel garage. A fire broke out and destroyed his car. Finnemore J held that the strict liability of an innkeeper was limited to loss of his guest's goods rather than to their destruction. But he went on to deal with section 86. He said, at pp1297-1298:

> "I think that this liability of the innkeeper was a custom of the realm. It is true it is embodied in common law rules, but then common law is the legal expression of custom, and it seems to me that that also would be an answer in this particular case. I suppose that by 1774 the legislature had appreciated what Lord Goddard CJ laid down in *Sochacki*: 'Everybody knows fires occur through accidents which happen without negligence on anybody's part.' Parliament in 1774 apparently thought it right that they should make it plain that whatever customs or usages there were to the contrary, in this country a man should not be held to be responsible for a fire which occurred accidentally – which I take to mean 'without negligence on his part'.
>
> Those are two points which I think would be enough to decide that there is in this case no absolute liability on the part of the innkeeper, first, because there was injury to the car and not theft or loss; and, secondly, because, in any event, as it was a fire the Act of 1774 would limit the liability of the innkeeper, so far as a fire is concerned."

129. In other words, section 86 overrode the custom. In my judgment this is plainly right because the Act expressly exempts a person from liability for accidental fires "any law, usage or custom to the contrary notwithstanding". This would include the custom of the realm upon which liability for "*ignem suum*" was based. The only question, then, is what Parliament meant by the word "accidentally"; and that had been decided by *Filliter*.

130. In *Musgrove*, Bankes LJ continued, at pp47-48:

> "The question then, is whether this motor car, with its petrol tank full or partially filled with petrol, was a dangerous thing to bring into the garage within the principle of *Rylands* v *Fletcher*? Mr. Hawke says a motor car is not a dangerous thing unless it is in such a condition that an accident is to be apprehended. But the expectation of danger is not the basis of the principle of *Rylands* v *Fletcher*. A thing may be dangerous although the danger is unexpected. I agree with Lush J that this motor car was dangerous within that principle. The defendant brought it, or caused it to be brought upon his premises, and he is responsible for the fire which resulted, and is not within the protection of the statute."

131. It will be recalled that the reason that Lush J thought the car was dangerous was that the petrol in its tank was "liable to cause a fire" and "not unlikely to get on fire". Warrington LJ agreed on that point; and said, at p49:

> "If this motor car with the petrol in its tank was potentially dangerous, such as a man's own fire, then it was the defendant's duty to see that the potential danger did not become an actual danger causing damage to his neighbour. The Act of Geo 3 is no protection against that liability."

132. However, both Bankes LJ and Warrington LJ went on to consider the question of negligence and both held that Lush J was right on that issue. That, as has subsequently been pointed out, is sufficient to support the decision itself. Duke LJ seems to me to have adopted a different line of reasoning. He said, at pp50-51:

> "I do not see how this case can be taken out of the principle of *Rylands* v *Fletcher*, which was thus stated by Lord Cairns LC in the very words of Blackburn J: "The true rule of law is, that the person who, for his own purposes, brings on his land and collects and keeps there anything likely to do mischief if it escapes, must keep it in at his peril." He can excuse himself by showing that the escape was owing to the plaintiff's default or perhaps that it was the consequence of *vis major* or the act of God. In the present case there was petrol which was easily convertible into an inflammable vapour; there was the apparatus for producing a spark; and added to those there was a person supposed to control the combustion but inexperienced and unequal to the task. Taking together the presence of the petrol, and the production of the inflammable gas, or those combustibles together with the inexperience of the person placed in charge of them, it is impossible to say that this is not an instance of the principle laid down by Blackburn J."

133. However, although he held that the principle of *Rylands* v *Fletcher* applied, he nevertheless went on to consider whether the fire was accidental for the purposes of section 86. If he had thought that the application of *Rylands* v *Fletcher* was a complete answer that consideration would have been unnecessary. On that question he said, at pp51-52:

> "That would dispose of this case but for the defendant's contention that he is excused by section 86 of the Fires Prevention (Metropolis) Act, 1774. In my opinion, the terms of that enactment fall far short of showing a definite intention to relieve a defendant in such a case as this. The actions against which the statute gives protection are in respect of fires which shall accidentally begin. I have the greatest doubt whether this fire began accidentally at any stage. If it was all one fire, it was begun not accidentally but intentionally. If progressive stages may be regarded it was not a fire which began accidentally without negligence at the stage when it became a conflagration involving goods and premises. The question may some day be discussed whether a fire, spreading from a domestic hearth, accidentally begins within the meaning of the Act, if such a fire should extend so as to involve the destruction of property or premises. I do not covet the task of the advocate who has to contend that it does. In the present case the fire, so far as it was a means of mischief, resulted from the negligent omission to turn off the petrol tap, an act which would have stopped the flow of petrol. All the witnesses who had any experience of such matters drew a distinction between fire in a carburettor, where the vapour can be instantly cut off, and such a fire as occurred in this case. The learned judge has found that this fire was due to negligence. I cannot disagree with him. Whatever may be the effect of the Act of Geo 3 upon the nice questions that have been discussed, this case is outside any possible protection of that statute."

134. This is, with all respect, a rather confused passage; but in my judgment the key finding here is that the spread of the fire was attributable to negligence.

135. *Musgrove* has been subsequently considered on a number of occasions. In *Job Edwards Ltd* v *Birmingham Navigation Proprietors* [1924] 1 KB 341, rubbish was tipped on land belonging to a canal company and on adjoining land belonging to mine owners. The rubbish on the mine owners' land was found to be on fire, and the canal company feared that the fire might spread to their own land. Having called on the mine owners to extinguish the fire, the canal company entered the mine owners' land (by agreement) and put out the fire. The question was whether the mine owners were liable to contribute to the cost. Bailhache J held that the mine owners had no duty to prevent the spread of the fire. He held, at p344, that:

> "where a fire occurs through no fault of the landowner, without his knowledge, and, as in this case, on matter brought on his land without his knowledge and against his will, he is not responsible for the spreading of such a fire to the adjoining land, but the neighbour is entitled to go upon his land and prevent the fire from spreading."

136. Thus the mine owners were not liable to contribute to the cost. The canal company appealed. Mr Vachell QC, for the canal company argued as follows, at pp344-345:

> "No doubt a person on whose estate any fire shall accidentally begin is exempted from liability by section 86 of the Fires Prevention (Metropolis) Act, 1774, for damage caused by that fire; and it may be admitted that a fire accidentally began upon land belonging to the respondents. But the fire which caused or threatened to cause damage to the appellants' property was not the fire which accidentally began on the respondents' land, when it might have been extinguished quickly and easily, but the fire which was from May to October, 1920, allowed to increase and become formidable: *Musgrove*."

137. Bankes LJ gave the leading judgment. He began by considering the distinction between a public nuisance (which he said a land owner had a duty to abate) and a private nuisance (which he said gave rise

A to no such duty). He then turned to the custom relating to fire, and explained it thus, at p352:

> "The case of fire has always been looked upon in our law as a somewhat exceptional case. It was no doubt the ancient law or custom of England that a person in whose house a fire originated which afterwards spread to his neighbour's property and destroyed it must make good the loss, but I do not consider that rule as opposed to the view I am putting forward in regard to liability for injury done by a private nuisance, as the ancient law no doubt considered a fire as a public nuisance owing to the danger of its spreading. The view of the law which I am taking does not touch a case where the private nuisance has been caused, or allowed to continue, by any act or default on the B part of the occupier of the land on which it exists."

138. This is a quite different explanation of the custom to that which he gave in *Musgrove*. The other point that he considered was the defence under section 86. As to that he said, at pp352-353:

> "Mr Vachell contended that whatever may have been the cause of the original fire it ceased to be an accidental fire within the meaning of the statute when the plaintiffs were informed of it, and that within the reasoning of the decision in *Musgrove* the fire as from that date must be treated as a second and independent fire. I cannot draw any such inference from the facts of the present case. In *Musgrove*, Lush J drew from the facts the inference that there were in substance either two fires, the first an accidental one which did no damage, and C the second which was due to negligence and did the damage; or alternatively that there was only one fire within the meaning of the statute, and that was the one due to negligence. This court agreed with the view of the learned judge, but the facts of that case are very special, and have in my opinion no bearing upon the case we are now dealing with."

139. It seems clear from this passage that Bankes LJ himself did not regard *Musgrove* as applying to what might be called an ordinary case of spread of fire. Bankes LJ (with whom Astbury J agreed) held that the mine owners owed no duty to the canal company and hence dismissed the appeal. Scrutton LJ dissented. On the question of common law duty, he said, at pp357-358:

D

> "There is a great deal to be said for the view that if a man finds a dangerous and artificial thing on his land, which he and those for whom he is responsible did not put there; if he knows that if left alone it will damage other persons; if by reasonable care he can render it harmless, as if by stamping on a fire just beginning from a trespasser's match he can extinguish it; that then if he does nothing, he has 'permitted it to continue', and become responsible for it. This would base the liability on negligence, and not on the duty of insuring damage from a dangerous thing under *Rylands* v *Fletcher*. I appreciate that to get negligence you must have a duty to be careful, but I think on principle that a landowner has a duty to take reasonable care not to allow his land to remain a receptacle for a thing which may, if not rendered harmless, cause damage to his neighbours."

E

140. This is a critical passage because it was subsequently approved by the House of Lords in *Sedleigh-Denfield* v *O'Callaghan* [1940] AC 880 who disapproved the reasoning of the majority. As far as section 86 was concerned, Scrutton LJ said, at pp360-361:

> "That statute (14 Geo 3, c78, section 86) provides that no action shall lie against any person in whose house or on whose estate "any fire shall accidentally begin." This fire undoubtedly began accidentally so far as the landowner and his agents were concerned. It has been decided that the statutory restriction of the previous common law liability does not exclude liability for fires caused by negligence of the owner or persons for whom he is responsible, F or by dangerous things for which the owner is responsible under the doctrine of *Rylands* v *Fletcher*. This leaves the difficult question – suppose the fire is caused by a trespasser, as if he throws down a match; and suppose the owner comes by immediately afterwards, sees the small fire, and could with no trouble extinguish it by stamping on it, but does not do so, so that the fire spreads and damages his neighbour, is he freed by the statute? He is then aware of a dangerous thing on his land which may damage his neighbour, and which by reasonable care he can prevent from damaging his neighbour, and he does nothing. I agree he is not an absolute insurer of that dangerous thing, for he did not himself create it, but I think on principle he is bound to take reasonable care of a dangerous thing which he knows to exist. Take the case of an ordinary house fire, where a coal leaps from the grate. If no one knows of the fire caused by the coal till it cannot be stopped, that fire may be within the protection of the statute, though Duke LJ doubted it in *Musgrove*. But suppose the owner sees it jump out, could extinguish it with a moment's trouble, and does not G trouble to do so, could he plead the statute to protect him? In *Musgrove*, where the real danger arose from the fact that the defendant's servant negligently did not turn a tap to stop a supply of petrol to a fire, the court treated the fire as two fires; I should respectfully have thought that it was safer to say that the fire was continued by negligence, and that the cause of action was not for a fire accidentally begun, but for negligence in increasing such a fire."

141. In *Collingwood*, Lord Wright MR also considered *Musgrove*. He said, at p206, that the ground for decision, in so far as it was not based on negligence was one about which he had "some difficulty"; and Romer LJ pointed out that in that case there had not been an escape of anything such as to bring the principle of *Rylands* v *Fletcher* into H play. Although he said that *Rylands* v *Fletcher* was based on a broader principle, he doubted the propositions on which the alternative ground rested. *Musgrove* was doubted again by Lord Porter in *Read*, referring to Romer LJ's observations in *Collingwood*. It was also regarded with scepticism by Lord Walker of Gestingthorpe in *Transco*, at [107].

142. In *Emanuel*, the fire in question had been deliberately started by workmen in order to burn rubbish on a demolition site. It spread to neighbouring land. The Court of Appeal based its decision entirely on the question of negligence. Bearing in mind that the fire had been deliberately started, *Rylands* v *Fletcher* would have provided a very short answer if the principle applied to a fire that spread accidentally. J But instead the Court of Appeal discussed negligence extensively in the context of the statutory defence under section 86. That approach is, in my judgment, quite inconsistent with *Musgrove*.

143. *Charlesworth & Percy on Negligence* say (para 13-112):

> "An alternative ground for the decision in *Musgrove* was that the 1774 Act was no defence when the fire originated from a dangerous thing. This proposition has been criticised, but it was adopted and followed in *Mulholland & Tedd Ltd* v *Baker* [1939] 3 All ER 253. Since it is the fire which is the dangerous thing, whether it is caused by petrol, paraffin or anything else, and the object of the statute is to give protection against accidental fires, it is difficult to understand why the statute should not protect as much in one case as K in the other. The presence of inflammable matter on premises is important when the question of negligence is being considered, so that if there is no negligence and it is found that the fire is accidental, it is submitted that the statute is a defence to the occupier of the land on which it begins, whatever may be the origin of the fire."

144. In my judgment, *Musgrove* is wrong in so far as it describes the basis of the common law before the earliest of the fire statutes. It invents an unhistorical justification for the basis of the rule. That justification has been criticised by judges, by scholars and by the Law Commission. It is inconsistent with a subsequent decision of this court L in *Emanuel*. For good measure, Bankes LJ himself said that it was decided on special facts; and, on the face of it, it is inconsistent with his own subsequent decision in *Job Edwards*. In *Collingwood*, Lord Wright MR said that whatever it decided *Musgrove* did not apply to the case of a fire caused by an unknown defect in electrical wiring. This was also the view of the Court of Appeal in *Stockport Metropolitan Borough Council* v *British Gas plc* [2001] EWCA Civ 212 (whose decision was affirmed sub nom *Transco*). It misinterprets both *Vaughan* v *Menlove* and *Filliter*, and wrongly distinguishes them. Moreover, it misstates the principle in *Rylands* v *Fletcher*, unless it is confined to fires deliberately kindled (or deliberately brought on to land). That, too, M was in effect decided by *Collingwood*, where Lord Wright considered whether *the electrical wiring* (which had been brought onto the land deliberately and where the fire started) rather than the fire fell within the principle in *Rylands* v *Fletcher*. *Musgrove* is clearly an extension of the principle in *Rylands* v *Fletcher* because the occupier did not bring the fire onto his land. This was also the view of the High Court of Australia in *Goldman* in which Windeyer J said that the principle in *Rylands* v *Fletcher* did not apply to the fire that started on Mr Goldman's land "simply because the respondent did not bring the fire upon his land, nor did he keep it there for any purpose of his own. It came there from the skies. And he did nothing to make its presence there more dangerous to his neighbours." Lastly, having regard to the findings of fact made by

Lush J at trial, the principle that Bankes LJ formulated was unnecessary to the decision, and wider than the facts of the case warranted.

145. In *Cambridge Water Co*, Lord Goff warned against developing the principle in *Rylands* v *Fletcher* into a rule of strict liability for damage cause by ultra-hazardous operations; and that warning was repeated by Lord Bingham in *Transco*. In *Transco*, the House of Lords were at pains to confine the scope of the principle to its proper ambit. Lord Bingham pointed out that it was a necessary condition of liability that the defendant had brought onto his land something which would be a source of mischief or danger if it escaped. That is not the case with a fire that is not deliberately kindled, because the defendant has not brought it onto his land. As I have already said he summarised his discussion by referring to a case in which an occupier of land "has brought or kept on his land an exceptionally dangerous or mischievous thing in extraordinary or unusual circumstances." A fire that accidentally ignites cannot be described as that; and nothing else escapes. Lord Scott also said, in [75], that the principle applied to:

> "cases where something or other, potentially dangerous, that the defendant has brought onto his land has escaped onto the plaintiff's land and there caused damage".

146. He described the element of escape as "essential". Lord Bingham also said that the use that the defendant made of his land would have to be "extraordinary and unusual" by reference to the time and place at which it was carried on before the principle would apply. This was plainly intended as a *narrowing* of the principle expounded in *Rylands* v *Fletcher*. In my judgment, no *extension* of the principle in *Rylands* v *Fletcher* can now be justified in the light of the decisions of the House of Lords in *Cambridge Water Co* and *Transco*. In my judgment, *Musgrove* cannot stand with the subsequent decisions of the House of Lords, and it is in conflict with the subsequent decision of this court in *Emanuel*. It wrongly distinguishes two previous decisions of courts of co-ordinate jurisdiction. In those circumstances, in my judgment, we are free to choose the correct path to follow: *Young* v *Bristol Aeroplane Co Ltd* [1944] KB 718, 729, *Starmark Enterprises Ltd* v *CPL Holdings Ltd* [2001] EWCA Civ 1252; [2002] Ch 306 ([64], [97]). In my judgment, *Musgrove* is unsound authority and should no longer be followed.

147. I turn then to the true scope of the defence under section 86 of the Act. In *Collingwood*, Lord Wright MR said of the Act, at p203:

> "That Act, as is well known, changed the law, because before that Act if a fire spread from a man's premises and did damage to adjoining premises, he was liable in damage on the broad ground that it was his duty at his own peril to keep any fire that originated on his premises from spreading to and damaging his neighbour's premises. The protection of the section is limited by the word "accidentally." The meaning of that is discussed in *Filliter*, where it was held that the section did not apply to a fire due to the negligence of the defendant or his servants."

148. The fire in that case was caused by an unknown defect in the electrical wiring system. There was no negligence established in relation either to the installation or the maintenance of the electrical system. The initial ignition was thus accidental. Nor, it seems, was there any negligence in controlling the fire once it had started, with the result that the statutory defence succeeded. Indeed the case was all the more striking because the damage was caused by water which had been deliberately applied in an attempt to extinguish the fire. *Spicer* was another case of a fire starting in the electrical system. However, Atkinson J held that the wiring had been negligently installed; and that in that state the wiring was a nuisance. It was for that reason that the statutory defence failed. The decision itself can be justified on the basis that the wiring had been negligently installed. The finding of nuisance is, in my judgment, more questionable.

149. In *Emanuel*, immediately after the passage I have quoted from Lord Denning's judgment in which he said that at common law a person was not liable for fire which escaped without negligence, he added, at p838f:

> "All those cases are covered, if not by the common law, at any rate by the Fire Prevention (Metropolis) Act 1774, which covers all cases where a fire began or spreads by accident without negligence. But that Act does not cover a fire which begins or is spread by negligence: see *Filliter*, *Musgrove* and *Goldman*."

150. In the same case, Phillimore LJ said, at p842:

> "Since the Fire Prevention (Metropolis) Act 1774 it is I think *necessary* to insert the word 'negligently' after the word 'acts'." (Emphasis added)

151. In *Sochacki*, the plaintiff lit a fire in the domestic grate in his room and went out one afternoon leaving the fire alight. A spark jumped out of the fire and set fire to the floorboards. The fire spread to an adjoining room where it caused damage. Although there was neither a fireguard nor a fender, Lord Goddard said that no negligence was involved. He held that the principle in *Rylands* v *Fletcher* did not apply, because although the fire had been deliberately kindled, the kindling of the fire was an ordinary use of land. He continued, at p345A:

> "It is not the case of a fire starting on one owner's premises and spreading to the premises of an adjoining owner. If a fire is negligently or improperly started by a person on his land, as for instance, lighting a bonfire which spreads, he may be liable, not merely to an adjoining owner who suffers damage, but to any other person who suffers damage. If I happen to be on somebody else's land at a time when a fire spreads to that land and my motor car or property is destroyed, I have just as much right against the person who improperly allows the fire to escape from his land as the owner of the land on which I happen to be. I do not doubt that for a moment, but here the fire was being used by a man in a fireplace in his own room. There was an ordinary, natural, proper, everyday use of a fireplace in a room."

152. There was no express discussion of section 86, but it is plain that Lord Goddard decided the case on the basis of negligence alone.

153. In *Solomons* v *R Gertzenstein Ltd* [1954] 1 QB 565, a fire started as a result of an electrical short circuit. That set fire to some wood and in due course to a stack of paper. Lord Goddard held that section 86 applied. He said, at pp573-574:

> "In my opinion, it was a short circuit that set fire to the wood in the neighbourhood of the ventilator and that in turn set fire to the stack of paper. Pausing here, it does not appear that this paper ever burnt freely; no doubt it caused a lot of smoke, and I accept the evidence that there was some flame seen, but it was not that stack apparently that caused the sudden sheet of flame which caused the real damage here. The cause of that is obscure; the only explanation was that offered by the fire officers, that the heating of the paint and varnish caused an accumulation of gas which suddenly ignited and rushed upwards. However, I do not propose to deal further with this because I cannot hold that placing packing paper and cardboard cartons in this recess behind the balustrade was negligent. Business of the sort carried on by the first defendants necessitates having a stack of packing and wrapping material at hand. This material is not highly inflammable like loose tissue paper or shavings would be. It is common knowledge that it takes a good deal to get closely packed thick paper well alight, though it will smoulder. But in any case I cannot see how it can be negligent to store this paper in what was a convenient recess any more than it would be to store it in one of the rooms occupied by the first defendants. They had no reason to suppose that there was likely to be a short circuit which would fire the panelling in the immediate neighbourhood of the stack, which I may mention was never burnt through. On the evidence before me I am not prepared to find that the fire was caused by the negligence of any of the defendants, and I hold that it was accidental and need only refer on this matter to *Collingwood*."

154. Thus Lord Goddard held that section 86 applied where a fire accidentally began and set fire to material that was not readily inflammable; and that to store that material was not negligent. His process of reasoning was the same as that in *Sochacki*. The case went to the Court of Appeal, where Lord Goddard's decision was reversed; but on a completely different point.

155. However, where there is negligence liability attaches, even if the negligence in question is that of an independent contractor. This is shown by another decision of Lord Goddard, this time sitting in the Court of Appeal, in *Balfour*. In that case the fire started as the result of the negligent use of a blow torch by an independent contractor. The use of fire had, therefore, been deliberate. The argument for the successful

plaintiff, at p498, was that "If negligence be shown, it matters not against whom, the fire is not accidental." Lord Goddard said, at p504:

> "The precise meaning to be attached to 'accidentally' has not been determined, but it is clear from these last two cited cases that where the fire is caused by negligence it is not to be regarded as accidental. Although there is a difference of opinion among eminent text writers whether at common law the liability was absolute or depended on negligence, at the present day it can safely be said that a person in whose house a fire is caused by negligence is liable if it spreads to that of his neighbour, and this is true whether the negligence is his own or that of his servant or his guest, but he is not liable if the fire is caused by a stranger.
>
> Who, then, is a stranger? Clearly a trespasser would be in that category, but if a man is liable for the negligent act of his guest, it is, indeed, difficult to see why he is not liable for the act of a contractor whom he has invited to his house to do work on it, and who does the work in a negligent manner."

156. I come now to the important case of *Goldman*. A tree on Mr Goldman's land was struck by lightning and caught fire. He cleared a space round the tree of combustible material, and arranged for the tree to be cut down. However, once it had been cut down instead of putting out the fire, he left it to burn itself out. Unfortunately, it did not, and it spread to Mr Hargrave's land. The question was whether Mr Goldman was liable. The Privy Council held that an occupier of land was under a general duty of care, in relation to hazards, whether natural or man-made, occurring on his land, to remove or reduce such hazards to his neighbour. The existence of the duty must be based on knowledge of the hazard, ability to foresee the consequences of not checking or removing it, and the ability to abate it; and the standard of care required of the occupier is founded on what it was reasonable to expect of him in his circumstances. There are a number of important points that arise out of this case. First, the Privy Council rejected, at p656B, the argument that Mr Goldman had adopted the fire as his own "as *suus ignis* – and had made use of it for his own purpose or advantage." In my judgment this encapsulates the principle of *ignis suus* that I have already discussed. It reinforces the proposition which I believe to be correct, namely that a fire that was never deliberately kindled fell outside the principle of *ignis suus*. In order to count as "*ignis suus*" the fire must have been made for the purpose or advantage of the occupier. Second, the case was not one in which the occupier had brought a source of danger (ie the fire) onto his land. This is equally consistent with the rationale of *Rylands* v *Fletcher*, viz. that an occupier of land has brought something dangerous onto his land *for his own purposes*. Third, the Privy Council held that in the case of fire there was no difference between a fire that started from natural causes and one that had been started by human agency. Lord Wilberforce put it thus, at p661B-G:

> "Their Lordships would first observe, with regard to the suggested distinction, that it is well designed to introduce confusion into the law. As regards many hazardous conditions arising on land, it is impossible to determine how they arose – particularly is this the case as regards fires. If they are caused by human agency, the agent, unless detected *in flagrante delicto*, is hardly likely to confess his fault. And is the occupier, when faced with the initial stages of a fire, to ask himself whether the fire is accidental or man-made before he can decide upon his duty? Is the neighbour whose property is damaged bound to prove the human origin of the fire? The proposition involves that if he cannot do so, however irresponsibly the occupier has acted, he must fail. But the distinction is not only inconvenient, it lacks, in their Lordships' view, any logical foundation.
>
> Within the class of situations in which the occupier is himself without responsibility for the origin of the fire, one may ask in vain what relevant difference there is between a fire caused by a human agency, such as a trespasser, and one caused by act of God or nature. A difference in degree – as to the potency of the agency – one can see but none that is in principle relevant to the occupier's duty to act. It was suggested as a logical basis for the distinction that in the case of a hazard originating in an act of man, an occupier who fails to deal with it can be said to be using his land in a manner detrimental to his neighbour and so to be within the classical field of responsibility in nuisance, whereas this cannot be said when the hazard originates without human action so long at least as the occupier merely abstains. The fallacy of this argument is that, as already explained, the basis of the occupier's liability lies not in the use of his land: in the absence of 'adoption' there is no such use; but in the neglect of action in the face of something which may damage his neighbour. To this, the suggested distinction is irrelevant."

157. It is important to note that this passage is directed to a case in which the occupier is without responsibility for the *origin* of the fire: that is to say the initial ignition. This is clear from the subsequent structure of the Privy Council's advice because Lord Wilberforce went on to consider the separate question of liability for the *spread* of the fire. Thus the Privy Council went on to consider whether, once the fire had started, there was a duty; and held that there was a duty to do what was reasonable in the circumstances to prevent the spread of the fire. This was precisely the approach of Scrutton LJ in the *Job Edwards* case. On the facts, Mr Goldman was negligent in having left the fire to burn out, rather than extinguishing it. The last point in the case was whether Mr Goldman was entitled to rely on the statutory defence under section 86 (which applies in some Australian states). On that question Lord Wilberforce said, at pp664D-665D:

> "The words 'shall accidentally begin' are simple enough, but the simplicity is deceptive. Read literally they suggest that account need be taken of nothing except the origin of the fire and that given an accidental beginning, no supervening negligence or even deliberate act can deprive a defendant of the benefit of the statute. But further reflection suggests a doubt both because such a result seems capable of producing absurdity and injustice, and because of the inherent difficulty of saying what the expression 'any fire' is intended to mean. A fire is an elusive entity; it is not a substance, but a changing state. The words 'any fire' may refer to the whole continuous process of combustion from birth to death, in an Olympic sense, or reference may be to a particular stage in that process – when it passes from controlled combustion to uncontrolled conflagration. Fortunately, the Act has been considered judicially and, as one would expect, the process of interpretation has taken account of these considerations. In *Filliter*, Lord Denman explained the purpose of the earlier Act (6 Anne, c31, s6) as being to remove the supposed common law liability of a person 'in whose house a fire originated which afterwards spread to his neighbour's property' and held that it did not apply to a fire caused deliberately or negligently. This was carried further in *Musgrove*, where a fire started accidentally in the carburettor of a car, but spread because the chauffeur negligently failed to turn off the petrol tap. The Court of Appeal held that the Act did not apply. Bankes LJ put it that the Act relieved an owner for a mere escape of fire from his premises but did not relieve him against a claim for damages for negligence. The fire which caused the damage was, he thought, not the spark which caused the initial ignition, but the raging fire which arose from the act of negligence. Their Lordships accept this interpretation: it makes sense of the statute, it accords with its antecedents, and it makes possible a reasonable application of it to the facts of the present case, that is to say, that the fire which damaged the respondents' property was that which arose on 1 March as the result of the negligence of the appellant. The statutory defence therefore fails."

158. It will be noted that Lord Wilberforce treated *Musgrove* as having turned on negligence and refrained from endorsing any wider basis for defeating the statutory defence. The decision in *Goldman* can, in my judgment, be summarised thus:

(i) An occupier of land is not liable for the initial outbreak of fire, whether due to natural causes or human agency, unless he himself has brought the fire onto the land;

(ii) He has a duty to do what is reasonable to prevent the spread of the fire. If he fails to do what is reasonable to prevent the spread of fire he is negligent;

(iii) If he is negligent in preventing the spread of the fire the statutory defence under section 86 will fail. If not, it will succeed.

159. The law as expounded in *Goldman* has since been accepted as being the law of England and Wales too: *Leakey* v *National Trust* [1980] QB 485; *Bybrook Barn Centre* v *Kent County Council* [2001] BLR 55*.

160. I turn now to consider a short line of cases about fire that have developed since *Musgrove*. *E Hobbs (Farms) Ltd* v *The Baxenden Chemical Co Ltd* [1992] 1 Lloyd's Rep 54 was a case in which the use of a grinding machine in a barn emitted sparks that set combustible material alight. The fire spread to adjoining property. Sir Michael Ogden QC found that the contractor who operated the grinding machine

---

* Editor's note: Also reported at [2000] 2 EGLR 63; [2000] 20 EG 158

had been negligent, with the result that the statutory defence under section 86 failed. Since negligence was established, the decision is plainly correct. What he said about *Rylands* v *Fletcher* was therefore *obiter*. In *Johnson* a fire was deliberately kindled in a domestic grate. It spread to adjoining property because a fire brick lining to the chimney had been negligently removed. As Judge Thornton QC put it, in [44]:

> "It follows that the defendant in this case is to be held vicariously liable for the damage caused by the escape of fire into and onto the claimant's premises since the fire only escaped and caused damage because of the negligent workmanship of the defendant's independent contractor for whom the defendant is to be held separately liable."

161. I agree that on the facts found in that case, section 86 was no defence. I do not, however, agree with Judge Thornton's more general observations on liability for the spread of fire.

162. I have already mentioned *Mason*. MacKenna J followed *Musgrove* with obvious reluctance. He recognised correctly that the principle of *Rylands* v *Fletcher* could not be directly applied to a case where the occupier had not brought the fire onto the land. Nothing that the occupier had brought on to the land had escaped. He therefore had to find a modified and extended version of the principle. The principle that he formulated, at p542C-D was that an occupier of land was liable if:

> "(1) he brought onto his land things likely to catch fire, and kept them there in such conditions that if they did ignite the fire would be likely to spread to the plaintiff's land; (2) he did so in the course of some non-natural use; and (3) the things ignited and the fire spread."

163. This formulation has been followed in subsequent cases at first instance: see, for example *LMS International*; *Harooni* v *Rustins Ltd* [2011] EWHC 1632 (TCC).

164. The unfortunate result of *Musgrove* compelled MacKenna J reluctantly to extend the principle of *Rylands* v *Fletcher*. It is equally unfortunate that in neither *Cambridge Water Co* nor *Transco* did the House of Lords discuss the modified principle that had been applied to fire cases. But in the light of *Transco* the extension of the principle in *Mason* was, in my judgment, a wrong turning in the law. No extension of the principle in *Rylands* v *Fletcher* can be justified. I would therefore overrule *Mason*, a result that MacKenna J would no doubt have welcomed.

165. Where a person brings combustible materials onto his land, he may well owe his neighbour a duty to take reasonable precautions to prevent their combustion or to take reasonable precautions to prevent the spread of fire. The principle was well put by Lord Macmillan in *Read*, at pp172-173:

> "Every activity in which man engages is fraught with some possible element of danger to others. Experience shows that even from acts apparently innocuous injury to others may result. The more dangerous the act the greater is the care that must be taken in performing it. This relates itself to the principle in the modern law of torts that liability exists only for consequences which a reasonable man would have foreseen. One who engages in obviously dangerous operations must be taken to know that if he does not take special precautions injury to others may very well result. … The sound view, in my opinion, is that the law in all cases exacts a degree of care commensurate with the risk created."

166. In my judgment, the law is as stated in *Goldman* at least as regards fires that have not been deliberately kindled. An occupier of land will not be liable to his neighbour for a fire that begins accidentally unless he is negligent in failing to prevent its spread. The general test of negligence may entail the taking of special precautions where the use in question involves the accumulation or storage of inflammable or readily combustible materials. But that is a question of fact to be decided on a case by case basis. This conclusion is, I believe, also consistent with the subsequent decision of the Privy Council in *Long Bee & Co* v *Ling Nam Rubber Works* [1970] 2 Lloyd's Rep 247.

167. It may be said that the retention of a special rule of liability in the case of fire is justified as a matter of policy. But, in my judgment, that route is barred by *Cambridge Water Co* and *Transco*.

168. In the present case, the Recorder found that the primary source of the fire lay in the wiring of electrical appliances in Mr Stannard's unit. However, he also found that:

> "… there is nothing to show that such a state of affairs was the result of failing to maintain or keep in good order the electrical system itself, of those electrical appliances that were located within Wyvern's premises, as opposed to something that may have arisen entirely by accident."

169. He also found that there was no allegation or evidence about what (if anything) Mr Stannard should have done to prevent the spread of the fire from reaching the tyres, and thereby involving them in the conflagration. He thus concluded that, subject to the principle in *Rylands* v *Fletcher*, Mr Stannard made good his defence under section 86. Having referred to a number of case both on the principle in *Rylands* v *Fletcher* and the modified principle applied to fire cases, he concluded, at [64]:

> "The only relevant activity was the storage of tyres. Was this dangerous within the *Rylands* v *Fletcher* rule, ie did it constitute a foreseeable and exceptional high risk of damage to Mr Gore's premises? Tyres are not in themselves flammable, and will not ignite unless there is a sufficient flame or heat source. It is, however, not impossible for tyres to catch fire (as obviously happened here), and if they do ignite, they have a special fire risk quality. This is that once alight they may burn rapidly and intensively, such that they are difficult to put out… If fire broke out there was an exceptionally high risk of damage to Mr Gore's premises because of the rapidity and intensity of the fire that would be created by the tyres that Mr Stannard had stored on Wyvern's premises."

170. Although the Recorder said that he drew support from *Mason*, in fact he went much further than that case. One of MacKenna J's conditions for liability was that the occupier brought onto his land things that were likely to catch fire. This was also the view of Judge Coulson QC in *LMS International* and of Akenhead J in *Harooni*. On the Recorder's findings the tyres were not likely to catch fire. Thus although the scope of *Rylands* v *Fletcher* has been narrowed each time the highest courts have considered it, the Recorder in fact extended it beyond any previous expression of the principle. He imposed strict liability where it had not existed before. Accordingly, even if I am wrong in thinking that the extended principle in *Rylands* v *Fletcher* does not defeat a defence under section 86 against liability for the consequences of a fire that starts and spreads without anyone's negligence, in my judgment, the Recorder was wrong to hold Mr Stannard liable on the facts that he found.

Appeal allowed.

RESTRICTIVE COVENANTS

Trustees of the Coventry School A Foundation v Whitehouse and others

CHANCERY DIVISION

[2012] EWHC 2351 (Ch)

10 August 2012

HH JUDGE SIMON BARKER QC, SITTING AS A HIGH COURT JUDGE

Restrictive covenants – Benefit – Successors in title – Restrictive covenant prohibiting use of land for any noisy pursuit or occupation* B *or for any purpose that might grow to be a nuisance or annoyance – Covenant benefiting retained land of vendor – Planning permission granted for development of school on land – Section 84 of Law of Property Act 1925 – Application for declarations concerning nature and extent of restriction – Whether covenant continuing to benefit retained land in hands of successors in title of original vendor – Whether proposed development likely to breach covenant

The claimant school trustees owned 44 acres of unregistered freehold land in Coventry, which they had laid out and used as playing fields. The land was subject to a restrictive covenant, contained in a 1931 conveyance of that land from a charity and benefiting the adjacent C retained land of the vendor; the covenant prohibited the erection of buildings "for any noisy... pursuit or occupation or for any purpose which shall or may be or grow to be in any way a nuisance damage annoyance or disturbance to the vendors and their successors in title or... which may tend to depreciate or lessen the value of the vendors adjoining or adjacent property". The vendor's retained land had later been developed for housing and the dwellings sold on; the defendants were the owners of four of those dwellings. The vendor charity continued to exist but had ceased to hold any interest in the retained land and its records relating to the land had been passed to the City Archive Department.

D In 2010, the claimants obtained planning permission to construct a two-storey school building on their land, together with car parking, access road, vehicle drop-off point and landscaping. The development was intended to accommodate the claimants' two existing junior schools under one roof and cater for expansion in numbers, from approximately 280 pupils to 400 pupils. The defendants contended that the development was prohibited by the restrictive covenant since a school would be result in traffic noise and other loud noise as pupils arrived and left; they submitted that the noise would cause nuisance, damage, annoyance or disturbance, or depreciate the value of their property, within the meaning of the restriction.

E The claimants applied to the court for declarations, under section 84(2) of the Law of Property Act 1925, that: (i) the restrictive covenant had ceased to benefit the retained land; or (ii) in any event, the development would not constitute a breach of that covenant. The defendants were joined as being representative of numerous others who, as owners of properties in the development, shared the same interest.

Held: **The application was dismissed. A declaration under section 84 of the 1925 Act operates in rem. Where the claimants sought a negative declaration, to the effect that the proposed development would not breach any of the restrictions imposed by the covenant, the burden of proof lay on them to satisfy the court, by reference to the balance of probabilities, that the declaration should be made. (1) The claimants were not entitled to the first declaration sought** F **since the restrictive covenant continued to benefit the retained land. The benefit of the covenant ran with the retained land by annexation. The requirements of annexation were met, in that: when the covenant was entered into, the vendor owned other land; the covenant related to or "touched and concerned" at least part of that other, retained land; and some of the defendants now owned part of the retained land to which the covenant related. The covenant also met the further requirement that the land to be benefited should be so defined as to be easily ascertainable. It is not sufficient, for the purposes of annexation, to show by evidence wholly outside the document in question that the covenant does in fact touch and concern identifiable land of the covenantee.** G **The land to be benefited must be described in the document itself by express words or necessary implication, although it may be necessary to have regard to evidence outside the document fully to identify that land. Whether the benefited land is easily ascertainable falls to be determined a matter of fact and degree in each case. The covenant in the 1931 conveyance met the test for annexation. It was easily ascertainable from the instrument itself that the vendor's retained land was intended to benefit from the covenant. Any ambiguity as to the extent of the retained land was easily resolved first by an enquiry to the vendor charity and then by recourse to its records held by the City Archive Department.** H **The fact that two stages of enquiry were involved did not mean that the benefited land was not easily ascertainable. The trail was short, clear and well marked and what it revealed was clear and free from ambiguity. The covenant was enforceable by three of the defendants, whose properties had been built on the retained land; the property of the remaining defendant lay outside that land and therefore did not benefit from the covenant. (2) The claimants were not entitled to the second declaration that they sought since they were unable to establish, on the balance of probabilities, that it was unlikely that the development of the school might grow to be in any way a nuisance, damage or annoyance. The word "nuisance", as it** J **appeared in the conveyance, was not to be construed as meeting the criteria necessary to establish the tort of nuisance. "Nuisance" referred to something that materially interfered with the ordinary comfort of existence, while "annoyance" had a wider meaning of something that reasonably troubled the mind of an ordinary sensible inhabitant of a house; the former affected the senses while the latter affected the intellect, although, in both cases, the impact was to be judged by reference to the sensible person, whose qualities included a reasonable measure of robustness and common sense. On the evidence, the impact of additional traffic from the operation** K **of the school, by way of noise, parking and obstruction, was such that it might be or grow to be a nuisance or annoyance within the meaning of the covenant. The second declaration should be refused on that ground. As to depreciation or lessening of property values, the protection of the covenant in that regard was intended to benefit only the original vendor; moreover, there was no evidence that the operation of the school would cause any such depreciation.**

The following cases are referred to in this report.

Crest Nicholson Residential (South) Ltd v *McAllister* [2004] EWCA Civ 40; [2004] 1 WLR 2409

A *Dennis and others* v *Davies* [2008] EWHC 2961 (Ch)
Federated Homes Ltd v *Mill Lodge Properties Ltd* [1980] 1 WLR 594
Hampstead & Suburban Properties Ltd v *Diomedous* [1969] 1 Ch 248
In Re B [2009] AC 11
In Re S-B [2010] 1 AC 678
Investors Compensation Scheme Ltd v *West Bromwich Building Society* [1988] 1 WLR 896
Marquess of Zetland v *Driver* [1939] Ch 1
Tod-Heatly v *Benham* (1889) LR 40 Ch D 80
Walter v *Selfe* (1851) 4 De G & Sm 315

This was the hearing of a claim by the claimants, the trustees of the Coventry School Foundation, against the defendants, Steven
B Whitehouse, Mervyn Lewis, Stephen Tate and Keith Griffiths, for declarations, under section 84 of the Law of Property Act 1925, as to the nature and extent of the restriction imposed by a covenant affecting land.

Edward Denehan (instructed by Band Hatton LLP, of Coventry) appeared for the claimants; Mike O'Brien QC (instructed by the Law Partnership Solicitors LLP, of Coventry) represented the defendants.

Giving judgment, **HH Judge Simon Barker QC** said:

Background to the proceedings

C 1. This action was commenced under CPR Part 8. The claimants are the Trustees of the Coventry School Foundation (TCSF). They seek two declarations under section 84(2) of the Law of Property Act 1925 (respectively section 84 (2) and LPA 1925) for the benefit of some 44 acres of unregistered freehold land in the Coundon district of Coventry, which they own and which is currently laid out and used as playing fields (TCSF's land).

2. By a conveyance dated 29 October 1931 (the "1931 conveyance"), TCSF (then the Bablake School & Foundation) acquired TCSF's land as part of a plot of some 51 acres from the estates trustees of the Sir Thomas White's Charity (STWC). At that time, STWC also owned and
D retained adjoining or adjacent land, broadly to the north, which was let to tenant farmers and included land and properties referred to as Manor House Farm and as Moat House Farm.

3. On 30 March 2010, TCSF was granted planning permission to develop part of TCSF's land (the site) by the construction of a one and two storey school building (the school), car parking, access road, vehicle drop off point and landscaping (collectively the development). The development is intended to be a junior school to accommodate TCSF's two existing junior schools in Coventry under one roof and to cater for expansion in numbers from about 280 pupils to 400 pupils. The site is at the northern boundary of TCSF's land where it adjoins
E Norman Place Road and Duncroft Avenue.

4. By the declarations, TCSF seeks to establish (1) that a restrictive covenant burdening TCSF's land (the covenant) does not benefit the adjoining or adjacent land which was retained by STWC as the vendor imposing the covenant or (2) if it does or may do so, that the development will not constitute a breach of the covenant.

5. Although it appears from endorsements on the 1931 conveyance that TCSF was aware of the covenant until 1972 at least and it is the case that TCSF has retained the services of the same firm of solicitors throughout, it seems that the existence of the covenant was subsequently forgotten or overlooked until it was referred to in a letter from a local
F resident to the planning authority during the consultation process in December 2009.

6. This is said to have prompted research by TCSF which established that the covenant was registered against the land conveyed in 1931 as a Class D(ii) land charge. TCSF was aware that the original covenantee (STWC) still exists, not least because STWC nominates one of TCSF's governors. Assisted by its solicitors, TCSF then sought to identify the land of the original covenantee (STWC) intended to, or which might, benefit from the covenant.

7. The covenant itself refers to STWC's "adjoining or adjacent property" and the plan to the 1931 conveyance indicates that STWC owned and retained land to the north of TCSF's land, either including or the other side of Hangman's Lane (which is now Norman Place Road and Duncroft Avenue). The plan also identifies a farm to the north east G
(Manor House Farm). STWC no longer owns any land adjoining or adjacent to TCSF's land.

8. TCSF's solicitors were informed that STWC had passed its records to Coventry City's Archives Department. From STWC's estates book of maps and particulars dating back to 1808 and its cropping book for 1934, TCSF was able to ascertain that the land to the north of TCSF's land and TCSF's land had been known as Moat House Farm and that the retained part of Moat House Farm (the retained land) was sold in 1935 to a local solicitor, Mr Thomas Cooksey Jnr, for housing development. This is readily apparent from the closing entry in the cropping book: "The whole farm was sold 1935 to Mr Thomas H
Cooksey Jnr for building purposes". TCSF also produced in evidence a photocopy of Coventry City Council's general works committee's plan dated 10 September 1935 which shows the layout and names of roads for the development of the retained land; this plan also shows the redesignation of Hangman's Lane as Norman Place Road and Duncroft Avenue.

9. Having obtained this information, TCSF then took legal advice and as a result decided to send a circular letter (the letter) addressed to the owners of each of the 1,200 or so properties built on the retained land and other nearby land. The letter, which is dated 18 March 2011, was drafted and sent by TCSF's solicitors and was hand delivered by a J
process server over several days shortly thereafter.

10. The letter refers to the 1931 conveyance, quotes the potentially relevant parts of the covenant, refers to the 30 March 2010 planning permission, refers to counsel's advice to TCSF that the covenant is not enforceable "because no person or persons has the benefit of the covenant and/or because the development of the site will not constitute a breach of the restriction", and states that as a matter of prudence proceedings would be issued for declarations pursuant to section 84 (2). The letter was accompanied by several enclosures.

11. The letter also invites the addressee's consent and continues that "if you decide that you are unable to consent, you will be made a K
defendant to the court proceedings which [TCSF] intends to issue. If you contest [TCSF]'s claim you may be at risk as to costs if the court decides that the advice we have given our client is correct". The letter invites responses by hard copy (for which a reply form and pre-paid envelope was enclosed) or e-mail.

12. A number of recipients of the letter perceived it to be a threat, perhaps because even silent abstention would lead to joinder as a defendant to proceedings. Whilst that reaction is understandable, the other side of the coin is that, once aware of the covenant and assuming that TCSF wished to press on with the development, the only safe course for it to take was to seek the court's sanction. L

13. Over the course of March, April and May, TCSF's solicitors received 129 individual replies, 124 expressing consent and five objecting. In addition, Mr Steven Whitehouse, the first defenant, (SW) replied by letter dated 31 March 2011, supported by more that 170 signatories as owners or co-owners who had received the letter, requesting an extension of time for their response. This was granted, but apparently no further response was received. Evidently, SW organised a residents' meeting and an ad hoc group, the Bablake Residents Action Group (BRAG), was formed.

14. As there was not unanimous consent to TCSF's assertion that the covenant was not enforceable, TCSF proceeded to issue proceedings on M
26 May 2011.

15. Given the number of potential defendants and the commonality of their interest in the proceedings, TCSF selected SW, who is the owner of 146 Norman Place Road, and Mr and Mrs Devenport, who are the owners of 110 Duncroft Avenue, as representative defendants pursuant to CPR Part 19.6. Mr and Mrs Devenport were reluctant to be involved in the proceedings, and, at a CMC on 6 July 2011, they were removed as parties and directions were given for an acknowledgment of service and evidence to be served and filed by SW and for a further CMC.

16. SW served and filed an acknowledgment of service on 15 July 2011 indicating an intention to seek an injunction and/or damages. The acknowledgment purports to be filed by SW and Mr Mervyn Lewis

A (ML) as representatives of BRAG. Although SW was represented by solicitors on the record, this self-styling was entirely informal and does not constitute BRAG (which is not a legal person) a party or affect the capacity in which the defendants are parties to the proceedings. However, it is not entirely a matter of semantics because, so submits Mr Edward Denehan, counsel for TCSF, the defendants (collectively Ds) have misled themselves into thinking that they are spokespersons for others whereas in fact they are representatives of a potentially large number of people having the same interest; I shall return to this point when considering the evidence of Ds and their witnesses.

17. At the following CMC, on 14 October 2011, ML, who owns 26
B Coundon Green, was formally joined as a defendant together with Mr Stephen Tate (ST), who owns 122 Norman Place Road, and Mr Keith Griffiths (KG), who owns 236 Hollyfast Road. All three were joined on the basis that they have the same interest, namely that each of them and SW own properties built on the retained land; and that, if the covenant is enforceable, they are beneficiaries of it, and, further, they object to the development being built on the site.

18. So that there is no misunderstanding about what Ds seek to achieve by their opposition, none of Ds objects to the development being built further to the south on TCSF's land or to an access road with a dropping off point provided the access road joins the public highway
C on Scotts Lane, which runs along the south eastern boundary of TCSF's land and is undeveloped (the land opposite being a reservoir). Indeed, none of Ds or the persons they represent could object because the land to the south and east (and, for that matter to the west) of TCSF's land was not part of STWC's adjoining or adjacent land at the time of the 1931 conveyance.

19. It is the case that some of Ds and a number of other witnesses object to the fact that the development will bring an independent school to the area on the grounds that their property values will or may be adversely affected, but that alleged "blight" would not occur if the development takes place on the southern part of TCSF's land, so on that
D ground there is no objection to an independent school per se.

20. Mr Denehan identifies two issues for the court to decide in this case: (1) Can Ds enforce the covenant (or perhaps more broadly, is there anyone who can enforce the covenant)? (2) If so, will the development, or the operation of the school on the site once the development is complete constitute a breach of the covenant?

21. The Rt Hon Mike O'Brien QC, who is counsel for Ds, contends on their behalf that the answer to both questions is in the affirmative, but that, as the claim is brought by TCSF, it is for TCSF to prove on the balance of probabilities that the answer to both questions is in the negative.

E 22. Mr Denehan disagrees and submits that on issue (2) the evidential burden is on Ds to establish a breach of the covenant.

23. Section 84 para 2 provides:

> "The court shall have power on the application of any person interested–
>
> (a) to declare whether or not in any particular case any freehold land is, or would in any given event be, affected by a restriction imposed by any instrument; or
>
> (b) to declare what, upon the true construction of any instrument purporting to impose a restriction, is the nature and extent of the restriction thereby imposed and whether the same is, or would in any given event be, enforceable and if so by whom.
>
> F Neither subsections (7) and (11) of this section nor, unless the contrary is expressed, any later enactment providing for this section not to apply to any restrictions shall affect the operation of this subsection or the operation for purposes of this subsection of any other provisions of this section."

And section 84 para 5 provides:

> "Any order made under this section shall be binding on all persons, whether ascertained or of full age or capacity or not, then entitled or thereafter capable of becoming entitled to the benefit of any restriction, which is thereby discharged, modified or dealt with, and whether such persons are parties to the proceedings or have been served with notice or not."

24. Thus, it is open to the owner of burdened land and the owner of benefiting land to seek a declaration. In the instant case, the owner of the burdened land seeks declarations. If made, the declarations will G
operate in rem.

25. The answer to where the burden of proof lies is provided by the terms of the declaration sought, which is that "the development and use of that part of the Foundation's land shaded yellow on the plan annexed ('the site') by the erection thereon of a single and two storey school building, car parking, access road, vehicle drop off point and landscaping does not constitute a breach of any of the restrictions imposed by the covenant contained in the said conveyance".

26. It seems to me that as TCSF has chosen to come to court and seek a declaration in the negative, it is for TCSF to satisfy the court that the declaration should be made, not for Ds to prove that it should not be H
made. I therefore agree with Mr O'Brien QC on this point.

27. I have not been specifically referred by either counsel to authority on the balance of probabilities. I take is as now well settled that there is a single standard of proof, that is the straightforward balance of probabilities, ie more likely than not, and that the standard of proof does not vary with the gravity of the consequences for the party affected by the decision. Further, if and to the extent that the court is called upon to draw conclusions as to inherent probabilities or inherent improbabilities or to draw inferences, when so doing the standard of proof remains the same, but better evidence may be required to persuade a judge to draw a conclusion or inference which is out of the ordinary; thus, at all J
times the court must have regard to the quality of the evidence before it when making a finding or reaching a factual conclusion on the balance of probabilities (see *In Re B* [2009] AC 11, in particular the speech of Lord Hoffmann at pp11-15, *In Re S-B* [2010] 1 AC 678, in particular the speech of Baroness Hale at pp11-13).

The covenant

28. There are five restrictions under the covenant. Only a part of one restriction is relevant, although reference may be made to other restrictions when addressing one aspect of Ds' case. The covenant and the first restriction provide as follows: K

> "And the purchasers do hereby for themselves and their successors in title and to the intent and so as to bind not only themselves personally but also all further owners and tenants of the property hereby conveyed or any part thereof and to bind such property into whosesoever hands the same may come covenant with the vendors and their successors in title in manner following that is to say:
>
> (1) That no house or other building erection or structure already erected or hereafter to be erected upon the land hereby conveyed or any part thereof shall at any time be erected, used or occupied for the purposes of or as a club, inn, hotel, public house, beerhouse or place for the sale of ale, beer, wines, spirits or other intoxicating liquors to be consumed on or off the premises nor shall the said land or any building erection or structure already erected or hereafter to L
> be erected thereon be used for the sale of fried fish or for any noisy noxious or offensive trade business pursuit or occupation or for any purpose which shall or may be or grow to be in any way a nuisance damage annoyance or disturbance to the vendors and their successors in title or their lessees or tenants or which may tend to depreciate or lessen the value of the vendors adjoining or adjacent property and the purchasers and their successors in title shall not sell or permit to be sold on any part of the land hereby conveyed any intoxicating liquors wines ale beer or spirits."

Is the covenant enforceable?

29. STWC no longer owns any of the retained land or any nearby M
land which might benefit from the covenant. TCSF accepts that successors in title of STWC may have the benefit of the covenant if the benefit passed to them as a matter of law. Thus, SW, ST and KG are accepted as potential beneficiaries, but not ML because his property is beyond, to the north-west, the retained land. ML's case is that the land nevertheless was owned by STWC and was sold separately and subsequently, which appears to be borne out by the documents exhibited to ML's witness statement.

30. It is common ground that in this case that the benefit of the covenant can only run, if at all, by annexation.

31. As the covenant arises under the 1931 conveyance, section 78(1) of the LPA 1925 applies. Section 78(1) provides:

A "(1) A covenant relating to any land of the covenantee shall be deemed to be made with the covenantee and his successors in title and the persons deriving title under him or them, and shall have effect as if such successors and other persons were expressed.

For the purposes of this subsection in connexion with covenants restrictive of the user of land, "successors in title" shall be deemed to include the owners and occupiers for the time being of the land of covenantee intended to be benefited."

32. As to the criteria to be met for there to be annexation, it is accepted by Mr Denehan on behalf of TCSF that (1) in 1931 when the covenant was entered into STWC had other land, (2) the covenant B relates to, or "touches and concerns", at least part of STWC's other land, and (3) three of Ds (SW, ST and KG) now own part of the retained land retained by STWC to which the covenant relates (ie part of the retained land). What TCSF does not accept is that the land owned in 1931 by STWC and having the benefit of the covenant is easily ascertainable.

33. On this point in issue, both Mr Denehan and Mr O'Brien QC referred in some detail to *Marquess of Zetland* v *Driver* [1939] Ch 1, *Federated Homes Ltd* v *Mill Lodge Properties Ltd* [1980] 1 WLR 594*, and *Crest Nicholson Residential (South) Ltd* v *McAllister* [2004] EWCA Civ 40; [2004] 1 WLR 2409.

34. In *Marquess of Zetland*, giving the judgment of the Court of C Appeal, Farwell J said at p8:

"the land which is intended to be benefited must be so defined as to be easily ascertainable, and the fact that the covenant is imposed for the benefit of that particular land should be stated in the conveyance and the persons or the class or persons entitled to enforce it. The fact that the benefit of the covenant is not intended to pass to all persons into whose hands the unsold land may come is not objectionable so long as the class of persons intended to have the benefit of the covenant is clearly defined."

35. The *Marquess of Zetland* case was not cited in *Federated Homes Ltd*. In the latter case, the Court of Appeal left open, as Chadwick LJ noted in Crest Nicholson at p30, the question whether section 78 only D effects annexation when the land intended to be benefited is described in the instrument itself (by express words or necessary implication, albeit that it may be necessary to have regard to evidence outside the document fully to identify that land) or whether it is enough that it can be shown, from evidence wholly outside the document, that the covenant does in fact touch and concern land of the covenantee which can be identified.

36. Chadwick LJ, with whose judgment Auld LJ and Arden LJ agreed, concluded at p33 that the question had already been answered in *Marquess of Zetland*. Thus, although the provisions of section 78 have rendered it unnecessary for the conveyance to state that the covenant E is to be enforceable by persons deriving title under the covenantee or under his successors in title and the owner or occupier of the land intended to be benefited, or that the covenant is to run with the land intended to be benefited, there is nothing in *Federated Homes* which suggests that it is no longer necessary that the land intended to be benefited should be so defined that it is easily ascertainable.

37. Chadwick LJ identified, at p34, policy reasons as good reasons for that requirement. It is desirable that a purchaser of burdened land should be able to identify from entries on the relevant register not only that the land is so burdened but also the land for which the benefit of the covenant was taken so that he may identify who can enforce the F covenant. It would be oppressive to require a purchaser of burdened land where the benefiting land was not described in the instrument to make inquiries as to what, if any, land the original covenantee retained and what, if any, of that land the covenant did, or might have, "touched and concerned"; and, it should be kept in mind that the question of enforceability may arise long after the date of the instrument by which it is imposed.

38. Applying these principles to the facts of this case: (1) The land remains unregistered and TCSF's title is derived from the 1931 conveyance; (2) that instrument identifies STWC (strictly STWC's then present estates trustees) as "the vendors" and refers in the body of the restriction to "the vendors" having "adjoining or adjacent property"; and G (3) the plan to the 1931 conveyance shows that STWC has adjoining or adjacent land to the north and north-east of TCSF's land. All of that is immediately ascertainable from the instrument itself.

39. Thus, it is easily ascertainable from the 1931 conveyance that STWC's land to the north and north-east of the land sold to TCSF was intended to benefit from the covenant.

40. Mr Denehan submits that that is not enough, and that the 1931 conveyance raises as many uncertainties and ambiguities as it solves or appears to solve, including in particular the extent of the land intended to benefit; that one such uncertainty or ambiguity is the boundaries of the benefiting land; and, therefore, that the "easily ascertainable" test is H demonstrably shown to have been not met.

41. Mr O'Brien QC submits that because it is permissible to look outside the instrument itself for evidence from which to identify the benefiting land, it is a matter of fact and degree in each case whether the benefiting land is easily ascertainable.

42. I accept Mr O'Brien QC's submission.

43. In so far as the particular instrument leaves the benefiting land undefined, the other evidence to which resort may be had must be readily available or accessible and of such evidential quality as to eliminate uncertainty or ambiguity. J

44. In this case, BC explains in his witness statement that inquiry was made of STWC and that that inquiry produced the information that (1) STWC no longer owned any land; and, (2) STWC's records had been passed to the City Archives Department. BC explains that inquiry there produced an answer: (1) the southern boundary of the land retained by STWC corresponds with the southern boundary of the land sold to Mr Cooksey Jnr; (2) the land retained by STWC and the land sold to TCSF was mapped in STWC's estates records, which appear to date back to 1808, as a single estate; and, (3) STWC's cropping book for 1934 contains a plan of the retained land and is closed with the final entry: "The whole farm was sold 1935 to Mr Thomas Cooksey Jnr for K building purposes."

45. There is no evidence on behalf of TCSF that the ascertainment of this information was onerous in any way. Certainly, it involved two stages of inquiry, but that of itself would not necessarily render ascertainment other than easy. STWC was the obvious first point of inquiry. If the charity was no longer in existence the trail might well have become difficult or impossible to follow. If STWC's records had been lost or destroyed, the trail might also have become difficult or impossible to follow. However, STWC was known to be in existence; it nominated one of TCSF's governors.

46. In this case, the trail was, as it must be, short, clear and well L marked. What was revealed at the end of the trail is also clear and free from uncertainty or ambiguity.

47. On this finding, ML's land is located outside the benefiting land. ML's land was not part of the retained land (ie the land sold to Mr Cooksey Jnr). Mr O'Brien QC draws attention to the fact that TCSF's solicitors caused the letter to be delivered to ML's property and submits that, although not adjoining TCSF's land, ML's property may be described as adjacent thereto. On that basis, Mr O'Brien QC submits that it is easily ascertainable that ML has the benefit of the covenant.

48. If the land on which ML's home now stands was intended to benefit from the covenant, that fact is not easily ascertainable; at least, M not on the available evidence. Indeed, reference to ML's title provides a good illustration of a trail which is not short, clear and well marked.

49. As a final point, I observe that the fact that it is not easy to ascertain whether there is also other land retained by STWC, in addition to that sold to Mr Cooksey Jnr, intended to have the benefit of the covenant is not a reason for concluding that land which can easily be ascertained as intended to benefit is somehow tainted by the uncertainty surrounding other adjoining or adjacent land and, in consequence, is to be deprived of that benefit.

50. I therefore decline to make the first declaration sought by TCSF.

* Editor's note: Also reported at [1980] 1 EGLR 113; [1979] 254 EG 39

A *Will the development and the use of the site for the school with car parking, access road, vehicle drop off point and landscaping not constitute a breach of the restrictions in the 1931 conveyance?*

51. Referring to a well known passage in the speech of Lord Hoffmann in *Investors Compensation Scheme Ltd* v *West Bromwich Building Society* [1988] 1 WLR 896 at p912, Mr Denehan submits that the determination of this question turns upon the proper construction of the covenant, and in particular the restrictions, as understood by a reasonable person having regard to the covenant in the context of the 1931 conveyance as a whole and having the background knowledge which would reasonably have been available to the parties in the
B situation in which they were at the time of the 1931 conveyance.

52. The search or inquiry is not for the particular subjective intentions of the parties but for the objective meaning of the language of their instrument. The parties have chosen and agreed upon specific language and this is to be construed objectively so as to ascertain what the law regards as and terms: the intention of the parties.

53. The restrictions in the covenant are set out in five sub-clauses. Although only part of the first sub-clause is in issue, it is necessary to refer to other sub-clauses. This is because Ds contend, by their evidence and submissions, that the process of erecting the school or indeed any other substantial building on the site would be noisy and/or a nuisance,
C damage, annoyance, or disturbance and therefore prohibited by the first restriction.

54. Ds say that in addition to the noise caused by the erection of the school and preparation and laying of the car park, access road and drop off point and the landscaping over a period of many months, there would also be noise, nuisance, damage, annoyance, and/or disturbance caused by lorries and other vehicles delivering building materials and removing tarmac, trees and other materials, and by the building work itself, as the initial ground works are undertaken and as the works progress over most, if not all, of that period.

55. In response, Mr Denehan points to the various restrictions and observes and submits that (1) only certain types of building (those
D associated with the sale of alcohol and/or fried fish) are prohibited by sub-clause (1), and even then it is the use to which the building is to be put rather than the process of erection which is objectionable; (2) the inescapable implication is that the erection of other buildings is not prohibited; (3) if that is not a complete answer, sub-clause (3) expressly envisages that TCSF's land may be levelled and that foundations and drainage may be laid for "the building of any house or other building thereon"; accordingly, (4) Ds' contention is the result of a misreading or misunderstanding of the plain words of the covenant and is misconceived.

E 56. I agree with Mr Denehan. Ds' case on this point is hopeless.

57. Before addressing the second declaration sought further, it is important to have clearly in mind the particular restrictions in issue. In the language of the 1931 conveyance they are as follows:

> "…nor shall the said land or any building erection or structure already erected or hereafter to be erected thereon be used for
>
> any noisy… trade business pursuit or occupation or
>
> for any purpose which shall or may be or grow to be in any way a nuisance damage annoyance or disturbance to the vendors and their successors in title or their lessees or tenants or
>
> which may tend to depreciate or lessen the value of the vendors adjoining or adjacent property" (layout altered).

F

The above extract of text from sub-clause (1) of the restrictions is not intended to limit or detract from the correct approach to understanding the meaning of the covenant, but to identify the particular matters to which the evidence of the parties was, or was intended to be, directed.

58. Mr Denehan observes that the covenant prescribes only what cannot be done by TCSF. Where the covenant is silent, TCSF is not susceptible to any objection by Ds or those whom they represent.

59. Thus, for example, Mr Denehan submits that, contrary to the apparent belief of one or more of Ds' witnesses, the covenant does not require that TCSF's land be used only as playing fields; indeed, there is nothing in the 1931 conveyance that would prohibit TCSF from laying concrete over the entire 44+ acres.

60. Further, Mr Denehan submits that the terms of the covenant do G not exclude noise altogether; rather, the restriction prohibits the use of TCSF's land for a trade, business, pursuit or occupation which is noisy. Mr Denehan submits that the operation of a school is a pursuit, and he accepts that there are times in the school day when noise is created (for example if and when children are outside during break times). However, submits Mr Denehan, the development of the site by the erection of the school with its attendant parking and access road will not result in TCSF's land being "used for any noisy … pursuit or occupation".

61. Ds' case is that noise is relevant in three ways. First, there is the express prohibition against the use of TCSF's land for any noisy pursuit or occupation. Secondly, there is the prohibition against the use H of TCSF's land for "any purpose which shall or may be or grow to be in any way a … damage annoyance or disturbance to the vendors['] successors in title". Thirdly, Ds submit that noisy use of TCSF's land "may tend to depreciate or lessen the value of" their land which is part of what is described in the covenant as "the vendors adjoining or adjacent property". The second and third points raise noise as an integral or contributory factor, and I shall return to those later.

62. As to the first way in which noisy use is referred to by Ds, Mr O'Brien QC submits that there will be traffic noise and other loud noise as up to 400 school children arrive and leave during term time. The evidence to support this submission includes the oral testimony of J witnesses who are local residents and who experience the term time comings and goings of pupils to and from Hollyfast Primary School and Coundon Court School and College nearby. There is also DVD evidence of arrivals and departures at these schools and at the existing Bablake School in Coventry. In both cases, the numbers of children arriving and departing (1) far exceed the maximum capacity for which the development has planning permission and (2) include large numbers of children over the age of 11 years who make their own way to school and contribute significantly to the noise levels at arrival and departure times.

63. The prohibition in the covenant is of any noisy trade, business, K pursuit or occupation. Thus, the focus of the restriction is on noise as a feature of the use to which the burdened land is put.

64. Circumstanced as the parties were at the time of the 1931 conveyance, I think it unlikely that their objective intentions would have included the prohibition of the operation of an independent junior school for up to 400 pupils on the grounds of it being a business, noisy pursuit or occupation. In 2012, the position is the same. Schools for hundreds rather than tens of children are a feature of everyday urban life, just as delivery and collection of young children by motor vehicles is a feature of everyday life.

65. On the evidence before me, that is the evidence of Mr Peter L Leonard, TCSF's acoustics expert, at its highest from Ds' point of view during cross-examination was that the increase in traffic noise will be noticeable and on occasions will be at a level which, if sustained continuously would cause harm or injury; but there is no evidence that over the course of 125 or 30 minutes twice a day the noise will be more than noticeable. However, there is already noise of children arriving at and leaving nearby schools, both on foot and in parents' vehicles. Taken as a whole, the evidence is insufficient to support a finding that it is as likely as not that the operation of the school will result in TCSF's land being used for a noisy pursuit or occupation.

66. On this issue, I accept Mr Denehan's submission that the M covenant was not intended to prohibit the development or the operation of the school.

67. Mr O'Brien QC submits that the primary basis on which Ds contend that TCSF fails to establish its entitlement to a declaration in respect of breach is "annoyance" and "tendency to diminish property values"; and that to a lesser extent Ds rely upon "nuisance" and "damage".

68. The restriction prohibiting a nuisance, damage, annoyance or disturbance is to be measured (1) not only in the present but also in the future and (2) as a possibility not a probability. Thus, it is for TCSF to establish on the balance of probabilities that it is unlikely that

A the development may grow to be in any way a nuisance, damage or annoyance.

69. Nuisance is a term to be construed according to "plain and sober and simple notions among the English people", *Walter* v *Selfe* (1851) 4 De G & Sm 315; in more modern language, nuisance or annoyance are to be "determined by the courts according to robust and common-sense standards", *Hampstead & Suburban Properties Ltd* v *Diomedous* [1969] 1 Ch 248, Megarry J at p258C.

70. In his submissions, Mr O'Brien QC did not identify any particular features distinguishing "nuisance" or "damage" from "annoyance". Similarly, the evidence adduced by Ds and the witnesses B on their behalf took the line that the development would or may grow to be a nuisance or annoyance without distinguishing nuisance from annoyance. There is nothing in the language of the covenant or the 1931 conveyance indicating that "nuisance" is to be construed as meeting the criteria necessary to establish the tort of nuisance, and indeed no such suggestion is made by either party.

71. In the *Shorter Oxford English Dictionary*, the definition of annoyance refers to nuisance and vice versa. However, the use of the phrase "nuisance or annoyance" in a restrictive covenant is not to be understood as the repetition of synonyms.

72. Both Mr Denehan and Mr O'Brien QC refer in their submissions C to *Tod-Heatly* v *Benham* (1889) LR 40 Ch D 80. In that case, the Court of Appeal held that the establishment of a hospital for the treatment of out-patients suffering from ear, nose, throat, skin, eye and other diseases was a breach of a covenant against doing any act "which shall or may be or grow to the annoyance, nuisance, grievance, or damage" of the covenantees.

73. Cotton LJ referred to annoyance as something to be determined by judges having regard to the evidence before the court and to whether reasonable, sensible people would be annoyed; thus, an act which is an interference with the pleasurable enjoyment of a house is an annoyance (p94). In particular:

D

> "It is not sufficient... to shew that a particular man objects to what is done, but [the court] must be satisfied by argument and by evidence, that reasonable people, having regard to the ordinary use of a house for pleasurable enjoyment, would be annoyed or aggrieved by what is being done." (Cotton LJ p94)

In that case, the circumstances which led Cotton LJ to make a finding of possible annoyance and therefore breach of covenant were that there was credible evidence of a reasonable apprehension that the servant of a family might run against someone going to or from the hospital who has a contagious disease and possibly take it into the house of his or her employer in the neighbourhood; that was a reasonable apprehension of E risk from other people meeting the patients coming to the hospital.

74. Damage is not confined to pecuniary damage (Lindley LJ, p96). Finding that the establishment of the hospital was a breach of covenant, Lindley LJ confessed to having thought at one time that the covenantor might succeed in showing that there was really nothing to complain of except fanciful annoyance, but "upon the evidence" the Lord Justice was "satisfied that what has been done really does annoy a great many people on this estate" (p96).

75. Bowen LJ was clear that "annoyance" and "nuisance" must have different meanings or one would not have been put in the covenant. Bowen LJ identified the characteristics of a nuisance as being directed F to material interference with the ordinary comfort of existence and those of "annoyance" as having a wider meaning and being directed to something which reasonably troubles the mind and pleasure of an ordinary sensible inhabitant of a house, not a fanciful person on one side of the line or a skilled person who knows the truth on the other side of the line (p98). Thus, the distinction appears to be between something which affects the senses (nuisance) and something which affects the intellect (annoyance).

76. Expressing his conclusion as a test, Bowen LJ said, at p98: "And if it is not an unreasonable thing for an ordinary person who lives in the neighbourhood to be troubled in his mind by the apprehension of such risk, it seems to me there is a danger of annoyance, though there may not be a nuisance" and again, at p99, as: "The measure is what is reasonable enjoyment of a man's premises... according to what G ordinary people... may think it."

77. The evidence which weighed with Bowen LJ as the "strong part" was that because people suffering from diseases were collected in the vicinity of the hospital that people in the neighbourhood should be apprehensive as to the consequences was not unreasonable (p99).

78. Over time the circumstances and risks which cause or may cause apprehension in the minds of sensible people and concern that they may be adversely affected and, in consequence, annoyed will change. Similarly, what constitutes a "material interference with the ordinary comfort of existence", or a nuisance, will also change over time. Conceptually, the distinction in the case of an annoyance is between a H real apprehension or concern about being adversely affected on the one hand and experiencing a minor irritation on the other; and, in the case of a nuisance the distinction is between a material interference with the ordinary way of life on the one hand and a trivial inconvenience on the other. In either case, the impact, whether mental or sensory, is to be judged by reference to the sensible person, whose qualities include a reasonable measure of robustness and common sense.

79. What then are the matters raised by Ds, as representatives of the owners of the properties now standing on the retained land, in relation to the use of the development as being for any purpose which shall or may be or grow to be in any way a nuisance or annoyance? J

80. Mr Denehan identifies a number of matters raised in the evidence of Ds and the other witnesses giving evidence on behalf of Ds in their representative capacity which he classifies as "red herrings". These are: (1) that the Former Pupils Association (FPA) has permitted the sale of alcohol; (2) that the campaign by BRAG demonstrates that the development is a nuisance and/or an annoyance; (3) that the tone of the letter is threatening; and, (4) that TCSF is a wealthy foundation that will bring a private school to an area where residents send and will continue to send their children to local state schools. In my judgment, Mr Denehan is correct in that (1) the sale of alcohol and other occurrences at past summer galas organised by the FPA are not in issue in these K proceedings, neither is any action to be taken by TCSF in relation to future FPA galas; (2) as already explained, BRAG is not a party to these proceedings and the members of BRAG are not exclusively covenantees, thus the views of BRAG, as distinct from the views of the persons represented by Ds, are not relevant; (3) I have already addressed the letter, and I do not consider its content to constitute a threat or, for that matter, to constitute a breach of the covenant; and, (4) the financial position of TCSF is irrelevant and, save possibly in respect of traffic matters (the point being that there is no catchment area and, being young, the children are unlikely to walk or travel by public transport to and from the school, and in consequence most children with L be transported to and from the school by their parents), the status of the school is irrelevant.

81. As to particular matters constituting possible breaches of covenant, these were listed by Mr Denehan as: (1) that the school will be a blot on the landscape; (2) covenantees will be overlooked; (3) covenantees will suffer the loss of a view and of accompanying amenity; (4) the development will cause or result in harm to wildlife; (5) the development will or may increase the likelihood of flooding or subsidence; and, (6) traffic associated with the operation of the school will or may be or grow to be a nuisance or annoyance.

82. As to (1), the evidence of Mrs Julia McNaney (JMcN), the M recently appointed vice chairman of the governors and trustees of TCSF, is that the design of the development is anything but a blot on the landscape and she referred to the improvements to be made to the landscaping and fencing along the boundary of TCSF's land with Norman Place Road and Duncroft Avenue. Architect's plans and CGI impressions of the school, the surrounding areas and the landscaping are in evidence. Mr Denehan adds that the covenant could have but did not make building on the site subject to the covenantee's approval, and that, if relevant, the existing structures on the site (portacabins and a large shed) are less appealing to the eye than the school.

83. I accept that the design of the school may not be to everyone's liking, and that Ds and persons having the same interest, or some of

A them, do or might consider the development to be an eyesore. However, reasonable people may reasonably disagree over what constitutes good architectural design and what does not, that reasonable people may reasonably disagree over what constitutes good architectural design and what does not, and there is no evidential basis on which I can make a finding that the development will or may be or grow to be in any way a nuisance or damage on this basis; nor is there a legal basis for a finding that Ds or other covenantees, as sensible people, would or might be or might grow to be annoyed or find their senses offended in a way constituting a nuisance by the design of the school.

84. As to (2), JMcN's evidence as to the design of the school is that
B the windows at 1st floor level will face Duncroft Avenue, and in this respect BC's evidence is incorrect. However, JMcN's evidence is also that the windows will be in the dining hall and the indoor sports hall and that these windows will be set high in the walls. Thus, JMcN's evidence is that local residents will not be overlooked. That evidence is not entirely consistent with the architect's CGI impressions of the façade of the school. The evidence of SW and Mr Mark Eaves (ME), of 139 Norman Place Road, is that being overlooked and losing privacy will be or grow to be a nuisance and annoyance. As to SW's concerns, the architect's plans (especially the drawing at [1/6] in the trial bundle) indicate that SW's property is some distance from the school (the width
C of Norman Place Road , the gardens of 147 Norman Place Road and 112 Duncroft Avenue, and the width of Duncroft Avenue) and that 147 Norman Place Road is likely to have windows facing SW's property, albeit at an oblique angle, in much closer proximity than the school. As to ME's property, the rear bedrooms, which are the primary source of ME's concern, will not be overlooked by the school but will look down through trees and landscaping onto the staff car park and the cycle park.

85. Mr Denehan fairly observed that in both cases there is a greater likelihood of being observed by a passer by in the street and that the usual action of a neighbour not wanting to be overlooked, or looked
D in upon, is to draw the curtains or pull the blinds. When giving their evidence, both SW and ME appeared to be reasonable and sensible men, and, in fairness, Mr O'Brien QC characterised the concerns they expressed on behalf of themselves and their respective families about being overlooked as a factor to be taken into account, implicitly acknowledging that such concerns are insufficient in their own right to constitute or possibly grow to constitute a nuisance or annoyance in breach of the covenant. In my judgment, even as an additional factor, the incremental weight would be insufficient to tip the scales against TCSF to any perceptible degree.

86. As to loss of a view and accompanying amenity, SW gave
E evidence about enjoying the view from his property of hockey being played on artificial pitches; this aspect is of particular interest to SW because he was, for some 14 years until recently, a regular hockey player on these pitches.

87. In this context, Mr O'Brien QC referred to *Dennis and others* v *Davies* [2008] EWHC 2961 (Ch); [2009] EWCA Civ 1081**, which concerned objections by other residents of homes on an estate at Heron Island in the Thames near Caversham. Those proceedings concerned two covenants, a permission covenant (by which a proposed alteration or addition had first to be approved by the estate's management company) and a nuisance and annoyance covenant (prohibiting anything being
F done which may be or become a nuisance or annoyance to others on the estate). On the facts, the views of the Thames from each property on the estate were a design feature and a material factor adding to the owner's or occupier's enjoyment. Having conducted a view and considered the evidence, the trial judge was in no doubt that what Mr Davies proposed to do would be a breach of the nuisance and annoyance covenant.

88. In fairness, Mr O'Brien QC again conceded that applying a robust common sense approach, SW's loss of a view would not justify refusal of a declaration.

89. ST gave evidence of also enjoying the view, albeit as an incident to being at an open window for the purpose of smoking, and of enjoying the view looking into TCSF's land when walking his dogs around the G
perimeter. The loss of a view from ST's property could not realistically feature as a basis for refusing to make the declaration sought, and much of the perimeter to TCSF's land is not benefiting land. To the extent that ST might enjoy the view as he walks along benefiting land, first, once outside his own boundaries, he is neither an owner nor an occupier of such land and the covenant does not promise or protect a view as such and, secondly, any such view would be a glimpse of playing fields through trees and other well established undergrowth all encased by what JMcN described as "stalag-like fencing".

90. Mrs Kay Johnson (KJ) of 44 Duncroft Avenue, also gave
evidence of enjoying a view of TCSF's land as playing fields from H
her property (which is at or near the corner of Duncroft Avenue and Drummond Close). KJ's view requires her to look over or around the bungalows at Fullers Close and across the open space of Christ The King Catholic Junior School before sighting TCSF's land. In any event, the development will not be in KJ's line of sight. KJ has lived in Duncroft Avenue for more than 50 years; she has experienced the development of bungalows at what is now Fullers Close and the sale of land by TCSF for the building of the Christ The King School. It transpired from KJ's oral evidence that her real concern in relation to aspect is that once the school has been built it will be expanded by the
addition of further buildings. However, any breach of covenant that J
may arise in that eventuality is a future question and is outside the scope of whether or not the development may in the future grow to be a nuisance damage annoyance or disturbance.

91. As to (4), notwithstanding KJ's concerns about the wildlife on TCSF's land, there is no evidence that the development or the operation of the school will, or may possibly in the future, cause harm to wildlife, and, if so, how that will, or may be or grow to be, a nuisance or annoyance. Thus, the question of whether there might be a breach of the covenant simply does not arise.

92. As to (5), the only evidence of flooding or risk of subsidence is
in relation to other burdened land at Fullers Close. K

93. The real area of controversy in this case is about traffic and the operation of the school.

94. The question for me is whether the sensible person would consider that the traffic attending the school, that is the noise, parking and obstruction, and congestion caused by such traffic, is a use of TCSF's land for a purpose which shall or may be or may grow to be in any way a nuisance or annoyance?

95. I must approach this question having regard to the written and oral evidence of TCSF's and Ds factual witnesses, the DVD evidence, and the expert evidence of Mr Leonard, Mr Aulak and Mr Silcock.

96. In so doing, I must also have some regard to the fact that L
planning permission has been granted and a section 106 agreement entered into for traffic measures. In consequence, the main entrance to TCSF's land will be relocated to a point on Norman Place Road approximately midway between the junctions with HR and Duncroft Avenue; a right turn filter lane will be introduced for vehicles travelling east along Norman Place Road onto the access road; safety measures will include the siting of one or more pedestrian refuges on Norman Place Road ; and, TCSF has a proposal for marshals to keep traffic flowing on TCSF's land.

97. I should also have some regard to Ds' and other factual witnesses' M
evidence about the possibility of parking restrictions being introduced, but in so doing I must take into account that any such measures would be preceded by consultation.

98. However, the issue concerning the covenant is about the private rights of the covenantees and the corresponding obligations of the covenantor. I begin by returning to Mr Denehan's submission that Ds have misled themselves into thinking that they are spokespersons for others whereas in fact they are representatives of a potentially large number of people having the same interest. Mr Denehan further submits that Ds and their other witnesses of fact are not entitled to speak to the concerns expressed by members of BRAG and that I should, in addition, disregard both letters from residents exhibited to SW's

** Editor's note: Also reported at [2010] 1 EGLR 81; [2010] 03 EG 104

second statement and newspaper articles, such as those exhibited to the statement of Mrs Marie Whitehouse (MW), who is SW's wife.

99. To a considerable extent, I agree with Mr Denehan, but this submission and approach can be carried too far in a case such as this where Ds are representative of a large number of people with the same interest. The views of persons who are members of BRAG but do not have the benefit of the covenant are obviously irrelevant. So too, newspaper articles and the views of journalists and commentators are also irrelevant. Correspondence from owners of properties on the benefiting land (specifically the 35 letters exhibited to SW's second statement) is not evidence in the sense that the individuals have not made witness statements and either been tendered for cross-examination or had their statements adduced in accordance with rules permitting their statements to be adduced notwithstanding their absence. However, such correspondence is not necessarily of no evidential weight whatsoever.

100. To take one example, Mr Denehan submits that, even if (which he does not accept) the court can infer that there are residents in Norman Place Road or Duncroft Avenue who have had their front gardens converted to accommodate vehicle parking and have a dropped kerb onto the highway, there is no evidence that any parking across the vehicular access will cause annoyance to any owner and, as Mr Denehan puts it, "it may be a matter of indifference" to all such persons.

101. In evaluating that submission, I observe (1) having viewed the site and walked around the area with counsel, I am aware that a number of properties do use their front gardens for off-street parking; (2) there is absolutely no evidence underpinning the submission that parking across the access way to an off street parking space may be "a matter of indifference"; (3) from a walk along Norman Place Road, particularly the section opposite the site, it is impossible not to notice the banners and posters adorning properties and in windows proclaiming opposition to the development; (4) BC refers to having TCSF or its solicitors having received 124 letters in response to the letter consenting to the development, but there is not a single property owner giving evidence in support of TCSF; and (5) common sense suggests that sensible people will not be indifferent to temporary obstruction of their access way to parking, and, depending on the frequency of such an occurrence, may, at one end of the scale, reasonably consider it to be an irritation and, at the other end of the scale, reasonably consider it to be an annoyance. In such circumstances, it is legitimate for the trial judge to test the water by reviewing evidence, whether hearsay or formal, such as the correspondence exhibited by SW; a review reveals correspondence complaining of obstruction to access ways to rear garages and an expression of concern that there will be difficulty in leaving and accessing the road from a property (200 Norman Place Road). In addition, KG has given evidence about his driveway being obstructed by parents dropping off and collecting children at Hollyfast Primary and Coundon Court Schools. Returning to Mr Denehan's submission as to covenantees being indifferent to the risk or fact of their driveways being obstructed, I reject that as at odds with the evidence, inherently improbable and unrealistic.

102. The conclusion I draw from SW's evidence on this point, which makes clear that all letters exhibited to his statement were from persons willing to give formal evidence, including orally, is that there is very strong opposition to the development and that the impact of additional traffic at drop off and collection times (ie the combined effect of additional noise, parking and obstruction, traffic congestion) are widely held fears and concerns amongst the class of persons represented by Ds, ie those with the benefit of the covenant.

103. However, the question is whether those fears and concerns relate to the use of TCSF's land for a purpose which shall or may be or grow to be in any way a nuisance or annoyance?

104. There is cogent evidence that Norman Place Road, Duncroft Avenue and Hollyfast Road (HR) are already busy roads. Mr Leonard, who gave expert evidence in the form of a noise impact assessment, measured the existing traffic flows at the junction of Norman Place Road and Duncroft Avenue between 07.30 and 15.30 on a weekday in January 2012 to be on average 714 vehicles per hour (vph). He worked on the assumption that traffic flow would increase to 1192 vph, calculated that that would represent a 16% increase in noise, and concluded that such an increase would be barely perceptible. In cross-examination, Mr Leonard accepted that the reason the noise increase would be barely perceptible was because the additional vehicles were assumed to pass evenly over the space of an hour. He acknowledged that he had assumed that all vehicles would be moving and that he had not factored in opening and shutting of doors or engines starting. More importantly, when asked to recalculate the noise increase if 400 additional vehicles were to pass within 15 minutes, Mr Leonard calculated that the increase would be in the order of 50% and opined that such an increase would be noticeable; that is noticeable above an already noisy threshold.

105. Mr Harminder Aulak gave oral expert evidence on behalf of TCSF, speaking to his own report and the written reports of his colleague Mr Philip Cook, about traffic and parking management. Mr Paul Silcock gave expert evidence on behalf of Ds. Mr Aulak is a disarmingly sensible and responsible person; he gave evidence in the same manner, and that is clearly his default setting. He factors his own approach into the assumptions and predictions he makes about the behaviour of others, and it is this which leads him to his views that the access road and drop off point with parking spaces for 19 vehicles, additional parking spaces for up to 26 vehicles if the children's designated play area is used for parking, and the further possibility of access to parking in the overflow staff parking area for a possible 12 vehicles will all be used in an orderly manner by parents who will not be troubled by the inconvenience of have to queue to enter the access road from the right on Norman Place Road and/or to exit the access road to the left or the right onto Norman Place Road because any inconvenience will be outweighed by being able to drop off the child or children within the school's perimeter rather than on the public highway. Mr Aulak also appeared to assume that the school's traffic marshals might assist in ensuring that any on Site roads to be crossed by children could be safely crossed. This was an expectation on JMcN's part as she naturally put safety of the children as a very high priority. However, TCSF's plan relating to traffic marshals is that they are to concentrate on keeping traffic flowing.

106. Mr Silcock's evidence was to the effect that in the real world traffic does not flow at a uniform rate, rather it tends to bunch. As to utilisation of parking spaces along the access road or in the play area, Mr Silcock's view is that parents will need to be "incentivised" to opt into such an arrangement for drop off and collection. Mr Silcock readily accepted that TCSF would do what it could to ensure that traffic on the site and that the movement of pupils was safe. Mr Silcock's view is that what happens in reality is that parents are on the whole in a hurry, and they tend to park where there is a space convenient to the route they wish to take and proximate to the school in question. Mr Silcock agreed with Mr Denehan that as a general proposition most people want to behave responsibly; but, Mr Denehan did not ask the follow up question of whether, in Mr Silcock's professional opinion borne out of his work, such intentions are a reliable guide to how people do or are likely to behave. Mr Silcock's view of additional parking for more than 19 vehicles was that while a number of parents might be prepared to use the play area for drop off and collection, the majority would not. For these reasons, Mr Silcock is of the opinion that the access road and drop off scheme will not work as TCSF intends. Mr Silcock also expressed the view, in response to cross-examination about some parents wanting a quick drop off and others wanting a chat, that the effect will be to block people (that is covenantees) in quite frequently.

107. Overall, Mr Silcock's view is that the proposed access point and arrangements for drop off and collection will cause traffic congestion in the locality, particularly along Norman Place Road and HR. Such congestion will obviously affect other covenantees wishing to access those roads at or around school drop off and collection times.

108. Mr Silcock was asked to view a DVD of parents dropping off children at TCSF's current site of Bablake School (which is a combined senior and junior school) and state whether what he observed was consistent with his evidence as to parents' intentions; as the DVD

A clearly showed, his answer was that many parents were not acting lawfully or reasonably.

109. In my judgment, Mr Silcock's evidence provides the more reliable guidance for the court on traffic and parking management.

110. In relation to difficulties being caused to parking on the public highway or the extended verges, I accept SW's evidence that there is a long established practice; that, by convention, local residents accommodate one another; and, that this practice will be disrupted at drop off and collection times. However, I also accept Mr Denehan's submission that this is an opportunity which local residents, such as SW and MW, enjoy but it is not a legal right and it is not protected by
B the covenant.

111. However, the position is different for those covenantees who have off street parking on their own property and whose access is at risk of obstruction by inconsiderate parking. The risk of such obstruction will be a twice daily occurrence, five days a week, for 30 or more weeks of every year. There is no precise evidence as to the number of covenantees who would be affected, but the correspondence exhibited to SW's second statement includes a statement that the conversion of front gardens to parking areas has increased since planning permission was given for the school. At one level that may be viewed as hearsay evidence of which some notice may be taken; at another level, it may
C be of assistance when considering inherent probabilities. Either way, on all the evidence, including my view of the locality, I am satisfied and find that the prospect of such obstruction occurring to a number of covenantees is not fanciful but is a realistic likelihood.

112. If fortification from other circumstantial evidence is needed, I bear in mind KG's evidence and DVD, and the photographic and DVD evidence of drop off at the current Bablake School. This evidence is more than sufficient to tackle and bring down Mr Denehan's submission that the incidence of delinquent parking by parents is much exaggerated as a risk. Looked at from the perspective of TCSF proving its case, there is no evidence of parents having a propensity to avoid
D parking obstructively or illegally in practice.

113. Mr Denehan's submission that covenantees with rights to access their property for parking will be protected from obstruction by the law and by its enforcement is shown to be unrealistic by the evidence of current parking at Bablake School and by the evidence of KG and ST as to what actually happens on a day-to-day basis.

114. Mr Denehan also submits that annoyance at obstructive parking reveals a sensibility which is not protected by the covenant. Adding particulars, Mr Denehan submits that the risk of obstruction on say 400 occasions for up to half an hour over the course of the year is a trivial fear. In my judgment, it is not. It seems to me to be at least comparable to the fear regarded as "strong evidence" by Bowen LJ in *Tod-Heatly* and regarded as a breach of a nuisance and annoyance covenant by E
Cotton LJ and Lindley LJ.

115. Taken in the round, the traffic issues (noise, parking and obstruction, and congestion) caused by the operation of the school are such that they shall or may be or grow to be a nuisance or annoyance. Although it is not necessary for me so to do, I make this finding on the balance of balance of probabilities having regard to all the evidence in this case.

116. Finally, I turn to the provision in the covenant that "...the said land or any building or structure... hereafter to be erected thereon [shall not] be used... for any purpose... which may tend to depreciate or lessen the value of the vendors adjoining or adjacent property". F

117. Mr Denehan submits that the specified purpose of this part of the covenant (protection of the value of the vendors' adjoining or adjacent property) is to be contrasted with the specified purpose of all the preceding restrictions in the covenant (protection of the vendors and their successors in title or their lessees or tenants). Mr Denehan submits that the significance is that the protection of value for the benefit of only the vendors, ie STWC, and that it relates to the whole of the retained land and not to any or various parts or parcels thereof. Mr Denehan submits that the reason for this is clear, that STWC wished to protect the value of the retained land as a whole because STWC intended to sell that land in one lot for development, which is what happened in 1935. G
Thus, Mr Denehan submits, the protection of value of the whole of the retained land would matter to STWC but not to a developer who would divide it up into building plots.

118. In my judgment, having regard to the language and structure of the first restriction and the covenant as a whole, there is considerable force in Mr Denehan's submissions, and I accept them.

119. Even if those submissions are wrong, the evidence as to the effect of the development on value does not enable me to reach a conclusion that it may cause a tendency to depreciate or lessen the value of any covenantee's property because the acceptable margin for difference of opinion as to valuation set by the RICS is 10% and there H
is no suggestion that any property value will be affected to any such degree.

120. Pausing here, Mr Denehan's submission as to the purpose of this part of the restriction is also significant in that it serves to confirm that the area of land intended to have the benefit of the covenant is that sold to Mr Cooksey Jnr (ie the retained land).

121. In conclusion, I decline to make the second declaration sought by TCSF.

122. That is my judgment.

Application dismissed.

Estates Gazette Law Reports

Cumulative Index 2005-2012

Supplement to
Estates Gazette Law Reports 2012, Vol 3

TABLE OF CASES

Titles of cases shown in bold type.
Names of parties reversed shown in ordinary type.

References are to year, volume number and page,
e.g. [**2012**] 1 **153** refers to volume 1 of 2012 at p 153

INDEX OF SUBJECT MATTER

References are to year, volume number and page,
e.g. **[2012]** 1 **9** refers to volume 1 of 2012 at p 9

COMPULSORY PURCHASE — *continued*

Lawfulness

Local Government Act 1972

Local authority making compulsory purchase order in respect of village green — CPO purportedly made under power in s121 of 1972 Act to acquire land for purpose authorised by any public general Act — Authority relying on purpose of promoting social and environmental wellbeing of their area within s2 of Local Government Act 2000 — Secretary of state refusing to confirm CPO — Whether permissible to acquire land compulsorily purely for "wellbeing" purpose — Held, compulsory acquisition of land not permitted purely for wellbeing purpose under s2 of 2000 Act — This precluded by prohibition in s121(2)(a) of 1972 Act on compulsory acquisition for analogous purpose of benefit, improvement or development of area under s120(1)(b) — Decision not to confirm CPO upheld

***R (on the application of Barnsley Metropolitan Borough Council)* v *Secretary of State for Communities and Local Government (Admin)*, [2012] 2 1**

Material considerations

Town and Country Planning Act 1990

Local planning authority granting planning permission to rival parties for development on same site — Resolution to exercise compulsory purchase powers under s226(1)(a) of 1990 Act in favour of one party — That party promising to develop another site funded by CPO site development — Whether local planning authority entitled to take into account prospect that development of CPO site would cross-fund other development either as s226(1A) well-being benefit or other material consideration — Held, impermissible under s226(1A) to take into account commitment by developer of CPO site to secure development of unconnected site to achieve further well-being benefits for the area — Such matters not otherwise material considerations relevant to exercise of s226(1)(a) power

***R (on the application of Sainsbury's Supermarkets Ltd)* v *Wolverhampton City Council and another (SC)*, [2010]2 103**

Notice to treat

Withdrawal

Ss4 and 31 of Land Compensation Act 1961 — Whether notice to treat can be withdrawn at any time up to and within six weeks of valid notice of claim — Whether notice to treat can be withdrawn only within six-week period following notice of claim — Whether notice of claim served — Whether notice of claim must contain details of calculations of compensation — Decision, notice to treat not withdrawn and tribunal having jurisdiction — Notice to treat can be withdrawn only under s31 of 1961 Act within six-week period following notice of claim or, where no claim served, of determination of compensation by Lands Tribunal — Notice of claim must show how compensation is calculated

***Acton* v *Trustees of Birmingham South West Circuit Methodist Church Manses Trust (LT)*, [2006] 3 101**

COMPULSORY PURCHASE ACT 1965 — See **Compulsory purchase**

CONDITIONS PRECEDENT – See **Town and country planning** (Planning permission)

CONSENT — See **Restrictive covenants**

CONSENT ORDER — See **Boundaries**

CONSERVATION (NATURAL HABITATS ETC) REGULATIONS 1994 — See **Sale of land (Environment)**

CONSTRUCTION

Adjudication

Appointment of adjudicator

Housing Grants, Construction and Regeneration Act 1996 — Written framework agreement for contractor to provide building consultancy services on various projects of local authority employer — Individual projects to be subject of further agreement — Adjudication clause naming local authority as specified nominating body for purposes of choosing adjudicator — Adjudicator appointed by RICS at request of contractor — Enforcement of adjudicator's award — Whether adjudicator having jurisdiction — Whether relevant construction contract in writing for purpose of s107 of 1996 Act — Whether contractual adjudication provisions contrary to s108 — Held, contract "in writing" where letter setting out proposed terms for project accepted by conduct of parties — Such agreement "otherwise than in writing by reference to terms which are in writing" within s107(3) — Non-compliance of contractual adjudication provisions with s108 resulting in incorporation of statutory Scheme for Construction Contracts in entirety — Provision for one party to act as specified nominating body contrary to policy of impartiality under Act — Adjudicator validly appointed by RICS — Award enforced

***Sprunt Ltd* v *London Borough Council (TCC)*, [2012] 2 19**

Jurisdiction — Building contract — Dispute — Contract providing for either party to request nomination of adjudicator by RIBA — Contractor wanting adjudicator with legal qualifications — Contractor contacting particular adjudicator to check availability and suggesting name in application to RIBA — Adjudicator appointed and giving decision — Whether adjudicator's appointment and decision valid — Whether implied term that party not seeking unilaterally to influence nominator's decision — Whether apparent bias — Held, appointment and decision valid — Unnecessary to imply term — No apparent bias

***Makers UK Ltd* v *Camden London Borough Council (TCC)*, [2008] 3 1**

Enforcement

Construction — Adjudication — Enforcement — S107 of Housing Grants, Construction and Regeneration Act 1996 — Clause 9A.7.1 of JCT Management Contract — Basis for adjudicator's jurisdiction where no written contract existing between parties — Whether parties reaching final determination by agreement on disputed item so as to render adjudicator's decision non-binding — Whether enforceability of decision affected by mistakes — Whether appropriate to adjourn enforcement proceedings pending final account — Held, adjudicator's jurisdiction in absence of clear written contract founded on exchange of written submissions in which parties not disputing jurisdiction — Final determination by agreement possible only by agreement reached after adjudicator giving decision and not by agreement during adjudication process — Mistakes not affecting enforceability of decision provided adjudicator answering questions referred to him and not proceeding in obviously unfair manner — Inappropriate to adjourn enforcement proceedings pending outcome of final account

***SG South Ltd* v *Swan Yard (Cirencester) Ltd (TCC)*, [2010] 2 7**

JCT prime cost building contract — Adjudicator giving decision on dispute referred under contract — Further dispute referred to second adjudication — Contractor seeking enforcement of second adjudication — Whether second adjudicator deciding issue already determined in first adjudication — Whether amendment to second adjudicator's decision constituting amendment beyond correction of clerical error — Held, decision of adjudicator binding only so far as deciding dispute as referred or as expanded within course of adjudication process — First adjudicator purporting to decide issue as to contract fee rate not referred to him and over which parties conferring no jurisdiction such that decision on that issue not binding — That issue therefore validly raised and adjudicated on in second adjudication — Second adjudicator entitled to apply slip rule to correct obvious arithmetical error — Second adjudicator's decision enforceable

***Redwing Construction Ltd* v *Wishart (TCC)*, [2011] 1 7**

Set-off — Stay of execution — First adjudicator deciding contractor entitled to sums under terms of contract — Second adjudicator deciding employer entitled to claim liquidated damages — Contractor seeking to enforce first adjudicator's decision — Whether employer entitled to set off entitlement to liquidated damages — Whether stay to be granted by reason of contractor's parlous financial position and contractual obligation to repay sums due — Held, sums found due to contractor in adjudication decision subject to right offset-off where clear and unequivocal decision in further adjudication as to employer's entitlement to recover specific sum by way of liquidated and ascertained damages — Stay of execution granted where contractor not in financial position to repay sums awarded and this representing a change from position when relevant contract made and not due to employer's failure to pay sums awarded by the adjudicator — Further grounds for stay where terms of contract requiring imminent repayment of sum claimed

***JPA Design & Build Ltd* v *Sentosa (UK) Ltd (TCC)*, [2009] 3 15**

Jurisdiction

Building contract providing for referral of disputes to arbitration in accordance with Construction Industry Model Arbitration Rules — Arbitrator appointed in contract dispute between employer and contractor — Parties reaching settlement agreement — Employer seeking appointment of new arbitrator in new dispute over alleged defects in work — Whether original arbitrator retaining jurisdiction over matters raised in new dispute — Whether arbitrator *functus officio* having discharged duties in respect of matters referred — Whether new dispute settled by earlier agreement — Held, arbitrator retaining jurisdiction and not *functus officio* — Extent of jurisdiction determined by contract under which appointed — Jurisdiction remaining save to extent precluded or limited by settlement — Arbitrator entitled to determine dispute as to whether defects encompassed by settlement agreement – Arbitrator correctly so finding

***Dawes* v *Treasure & Sons Ltd (TCC)*, [2011] 1 1**

Housing Grants, Construction and Regeneration Act 1996 — Building contract containing adjudication clause — Party serving notice of adjudication and requesting appointment of adjudicator — Party then failing to serve referral documents and allowing adjudication to lapse — Same party commencing new adjudication of same dispute — Whether losing right to have dispute adjudicated by failure to pursue first adjudication — Whether adjudicator having jurisdiction — Whether decision vitiated by apparent bias where expressing preliminary view before receiving submissions of one party — Held, failure to proceed with adjudication not resulting in loss of

CONSTRUCTION — *continued*
Adjudication — *continued*
Jurisdiction — continued

right to have dispute adjudicated — No implied term that claimant losing right to adjudicate if deliberately and without good reason failing to serve referral documents by the due date — No scope for concept of abuse of process of adjudication — Adjudicator having jurisdiction — Expression of preliminary view for assistance of parties not resulting in apparent bias — Award enforceable

***Lanes Group plc* v *Galliford Try Insurance Ltd (t/a Galliford Try Rail) (CA),* [2012]** 1 **27**

Liquidated and ascertained damages

Adjudicator making award to contractor in building dispute — Summary judgment sought on claim to enforce award — Whether principle in *Bramall & Ogden Ltd* v *Sheffield City Council* (1983) 29 BLR 73 applying — Whether employer entitled to set off claim for liquidated and ascertained damages (LADs) against sum due — Whether entitled to stay pending second adjudication in which employer referring party — Held, contractor entitled to rely upon *Bramall* principle in respect of LADs claim where partial possession granted and contract not providing for sectional completion — No entitlement to set off LADs claim against adjudicator's award — No stay to await outcome of second adjudication

***Avoncroft Construction Ltd* v *Sharba Homes (CN) Ltd (TCC),* [2008]** 2 **7**

Set-off

Contractor repaying sums to employer pursuant to adjudicator's decision — Subsequent decision of adjudicator regarding interim certificate finding overpayment by contractor on balance of account — Interim order for costs made against contractor in concurrent arbitration — Whether adjudicator having jurisdiction to order repayment by employer — Whether contractor entitled to set off amount of overpayment against costs — Held, adjudicator finding repayment due from employer either expressly or by inference as logical and inevitable consequence of valuation — Adjudicator having jurisdiction to order repayment — Arbitration costs award not trumping adjudicator's decision binding on parties for time being — Mutual set-off of debts arising from same transaction and same underlying dispute

***Workspace Management Ltd* v *YJL London Ltd (TCC),* [2009]** 3 **11**

Statutory demand

Cross-claim — Insolvency Rules 1986 — Adjudicator's award enforced by Technology and Construction Court — Statutory demand in respect of sum awarded by adjudicator — Application to set aside statutory demand on grounds of cross-claim equalling or exceeding debt specified in statutory demand — R6.5(4)(a) of 1986 Rules — Whether discretion to set aside demand exercisable where demand founded on judgment enforcing adjudicator's award and debtor having ability to pay — Held, discretion under r6.5(4)(a) to be exercised with regard to all relevant circumstances — No rule preventing creditor from pursuing insolvency proceedings where debtor having genuine cross-claim — No presumption against allowing debtor to rely on substantial and genuine cross-claim save where oppressive not to do so — Ability of debtor to pay sums specified in statutory demand not decisive reason for refusing to set aside demand

***Shaw* v *MPF Foundations & Piling Ltd (Ch),* [2010]** 1 **1**

Building contract
JCT standard form

Clauses 24 and 25 — Delays to works — Liquidated and ascertained damages (LADs) payable if works not completed by completion date — Certificate of non-completion issued — Employer serving notice to withhold LADs from payment due under interim certificate — Time extension granted to contractor after payment of reduced sum — Whether affecting entitlement to deduct LADs — Held, employer entitled to withhold payment in reliance upon non-completion certificate where preconditions met and proper notices served — Subsequent grant of time extension not depriving employer of right to rely upon certificate

***Reinwood Ltd* v *L Brown & Sons Ltd (HL),* [2008]** 2 **1**

Notice of default — VAT — Employer querying amount of contractor's VAT assessment and refusing to pay — Contractor serving notice of default — Whether VAT assessment required to set out basis of calculation — Whether employer raising valid objection to assessment — Whether valid notice of default served — Whether contractor entitled to determine contract — Held, "assessment" of value need only set out a figure and not the basis upon which it is reached — Part of supply chargeable at standard rate on ground that commercial rather than residential accommodation — Valid assessment given — No valid objection raised where no grounds of objection given but amount simply queried — Valid notice of default served— Consequently, valid notice of determination served

***Reinwood Ltd* v *L Brown & Sons Ltd (No 2) (CA),* [2008]** 3 **21**

Liquidated and ascertained damages

Penalty — Delay — Contract providing for sectional completion of works— First section of work delayed — Employer granting time extension for part of delay but deducting liquidated and ascertained damages in respect of remainder — Knock-on effect on later sections — Whether contractor entitled to time extension for whole of delay to later sections regardless of how delay arising under first section — Whether "cascading" liability for LADs for later sections constituting penalty — Held, contractual provisions workable — No entitlement to time extension for delays to later sections flowing from culpable delay to first section — Liability for LADs fair and not constituting penalty

***Liberty Mercian Ltd* v *Dean & Dyball Construction Ltd (TCC),* [2009]** 1 **1**

NEC Engineering and Construction Contract
Arbitration

Notice — Validity of service — Housing Grants, Construction and Regeneration Act 1996 — Parties entering into contract in terms of NEC Engineering and Construction Contract (2nd ed) 1995 — Contract specifying address for service of notice to refer adjudicator's decision to arbitration — Employer faxing notice to contractor's solicitor — Solicitor confirming receipt of notice and forwarding it to relevant individuals at employer company — Whether arbitration provisions in clause 93.1 contrary to Pt II of 1996 Act — Whether notice an effective communication under cause 13 of the contract — Whether appropriate to extend time for service — Held, arbitration provisions in clause 93.1 not fettering right to refer dispute to adjudication at any time and therefore compatible with Pt II of 1996 Act — Notice effective only if served at address prescribed by clause 13 of contract regardless of whether received in time by relevant personnel — Employer's notice effectively served in accordance with clause 13.2 since address of contractor's solicitor being that last notified for receipt of communications relevant to adjudication — Arbitration notice relevant to adjudication since preventing adjudicator's decision from becoming final and binding — Extension of time therefore not required

***Anglian Water Services Ltd* v *Laing O'Rourke Utilities Ltd (TCC),* [2010]** 3 **103**

Subcontract
"Pay when paid" clause

Insolvency — Clause in subcontract entitling main contractor to withhold payment in event of insolvency of employer — Insolvency defined in terms of s113(1) of Housing Grants, Construction and Regeneration Act 1996 by reference to four routes under Insolvency Act 1986 prior to amendment by Enterprise Act 2002 — Employer going into administration by self-certifying route under 2002 Act — Whether main contractor entitled to withhold payment to subcontractor — Whether pay-when-paid clause to be construed including all routes to administration under legislation as amended — Held, contractor not entitled to withhold payment — Pay-when-paid clause seeking to relieve main contractor of liability otherwise existing — Clear words needed to achieve that effect — Lack of clarity to be resolved against main contractor — Pay-when-paid clause not to be rewritten to take account of2002 Act amendments

***William Hare Ltd* v *Shepherd Construction Ltd (CA),* [2010]** 2 **10**

CONSTRUCTION CONTRACT

Adjudication
Final determination

Limitation — Subcontractor on building project sub-subcontracting road surfacing works — Subcontractor deducting sums from final account of sub-subcontractor on grounds of defective work — Deduction not challenged until after expiry of limitation period for subcontractor's contractual claim in respect of alleged defective work — Adjudicator ordering subcontractor to pay outstanding sum — Whether subcontractor entitled to bring court proceedings for final determination of dispute and return of sum paid — Whether such claim time-barred — Implied term that party who pays moneys in compliance with decision of adjudicator entitled to final determination of dispute in legal proceedings and repayment of moneys so far as final determination in its favour — Limitation period for such claim six years from date of payment — Subcontractor's claim not statute-barred

***Jim Ennis Construction Ltd* v *Premier Asphalt Ltd (TCC),* [2009]** 3 **7**

CONSTRUCTION INDUSTRY MODEL ARBITRATION RULES – See **Construction** (Arbitration)

CONSTRUCTION OF CLAUSE – See **Option to purchase**

CONSTRUCTIVE TRUST – See **Title to land**

Pallant v Morgan equity
Lease

Company having lease of building — Company operating amusement arcade on ground floor — Demerger of property and leisure sides of group of which company a part — Freehold of building transferred to property side — New owner operating break clause in lease and seeking possession on ground of intention to redevelop — Whether constructive trust arising such that company entitled to remain in occupation of ground floor for full term of lease — Held, Pallant v Morgan constructive trust resting on

reservation in lease amounting to "consent or agreement" preventing prescriptive right to light from arising — Whether such consent required to refer expressly to light — Held, reservation making clear that right to light not absolute and indefeasible — reservation amounting to agreement or consent — Express reference to light unnecessary

***RHJ Ltd* v *FT Patten (Holdings) Ltd (Ch),* [2007] 3 1; (CA), [2008] 2 11**

Remedies

Mandatory injunction or damages — Extension to commercial building interfering with right of light enjoyed by neighbouring listed building — Whether small award of damages sufficient to compensate for infringement of right — Whether oppressive to grant mandatory injunction requiring removal of offending parts of extension — Held, damages in lieu of mandatory injunction appropriate only where injury small, capable of being estimated in money and adequately compensateable by small money award and where injunction would be oppressive to servient owner — Injury not small when nature of listed building and owner's investment in its restoration taken into account

***HXRUK II (CHC) Ltd* v *Heaney (Ch),* [2010] 3 15**

Rights of way

Derogation from grant

Right of way for lorries to access spring water bottling plant in accordance with relevant planning permission — Temporary planning permission imposing condition as to provision of visibility splay — Owner of servient tenement interfering with visibility splay — Subsequent refusal of planning permission for bottling plant — Whether interference with splay amounting to derogation from grant — Held, right of way construed in light of knowledge of parties at time of grant — Provision of splay contemplated by parties to be essential to right in light of planning condition in force at time of grant — Blocking of splay amounting to derogation from grant — Mandatory injunction granted — Damages awarded

***Carter* v *Cole (CA),* [2009] 2 15**

Estoppel

Compulsory purchase of land for new highway — Highway running through vehicular access route of business premises — Right of way granted from premises to new highway — Position of access later moved pursuant to planning permission — Whether council retaining ransom strip such that owner of premises having no highway frontage — Whether council estopped from demanding payment for licence to cross ransom strip at location of new access — Held, ransom strip not forming part of highway — Council estopped from denying permanent right of access across ransom strip to owner of premises

***Bexley London Borough Council* v *Maison Maurice Ltd (Ch),* [2007] 1 19**

Extent of right

Track — Claim that gates erected on servient tenement interfering with right of way over track — Gateposts set wider than track itself — Whether right of way extending to "swing space" allowing tolerance for wide loads provided wheel base of vehicle remaining within track — Whether including "verge space" on either side of physical track allowing vehicles to swing onto verge where necessary for manoeuvring — Held, right of way confined to width of physical track — No right to swing space in absence of evidence that occasional swinging onto verge inevitable at time of grant and as to points where that occurring

***Oliver* v *Symons (CA),* [2012] 2 9**

Vehicular access to certain leasehold land gained across adjoining freehold land in same ownership — Whether right of way existing across freehold land for that purpose — Effect of express right of way in favour of freehold and contained in 1946 conveyance of both freehold and leasehold land — Whether right of way in favour of leasehold land to be assumed when determining price payable on leasehold enfranchisement claim — Held, no right of way in favour of leasehold land to be assumed — Right of way in favour of freehold land not benefiting leasehold land either expressly or impliedly — Conveyance disclosing no sufficient intention to grant right in favour of leasehold land — Right incapable of arising by prescription where purported dominant and servient tenement in same ownership — No right to be implied by necessity where adequate access afforded by virtue of ownership of both parcels of land — Leasehold land to be valued on assumption that landlocked

***Re Clarke's appeal (UT:LC),* [2012] 3 11**

Interference

Obstruction — Right of deviation — Construction on right of way of wall and pillars forming part of gateway — Substantial interference with right of way — Declaration of right to deviate from route of right of way to use gateway — Whether right of deviation precluding existence of actionable interference with right — Held, creation of right of way over new route not extinguishing right over old route — Existing right remaining unless released by holder of right — Obstruction of right of way continuing to be actionable interference although injunction may be refused

***Heslop* v *Bishton (Ch),* [2009] 2 11**

Pathway

Leasehold Reform Act 1967 — Boundary on mid-line of path between properties — Extent of rights of way — Whether rights of way over moiety of pathway outside boundary — Extent of rights passing with conveyance on leasehold enfranchisement — Whether rule in *Wheeldon* v *Burrows* applying on enfranchisement — Whether rights passing on enfranchisement under s62 of Law of Property Act 1925 — Held, rule in *Wheeldon* v *Burrows* not applying to conveyance on enfranchisement — Rights passing on enfranchisement under s62 of 1925 Act and s10(2)(i) of 1967 Act — Conveyance on enfranchisement encumbered by any rights in favour of adjoining property

***Kent* v *Kavanagh (CA),* [2006] 2 127**

ECONOMIC LOSS – See **Negligence**

EFFECTIVE CAUSE — See **Estate agents** (Buyer's agent)

EFFECTIVE CAUSE OF SALE — See **Estate agents**

EIA DIRECTIVE — See **Civil practice and procedure** (Protective costs order)

ELECTRICITY ACT 1989 — See **Compensation** (Electricity wayleaves)

ENFORCEMENT — See **Charging order; Construction; Mortgage; Sale of land** (Positive covenants)

ENFORCEMENT NOTICE — See **Town and country planning**

ENFRANCHISEMENT — See **Leasehold enfranchisement** (Leasehold Reform, Housing and Urban Development Act 1993)

ENFRANCHISEMENT OF HOUSE — See **Leasehold enfranchisement**

ENTERPRISE ACT 2002 — See **Construction** (Subcontract)

ENVIRONMENT — See **Town and country planning**

ENVIRONMENTAL ASSESSMENT

Environmental impact assessment

Intensive agricultural purposes

Application of manure to grassland farm — Whether increase in productivity — Whether "intensive agricultural purposes" — Whether intervention to natural surroundings amounting to "intensive agricultural purposes" — Held, increase in productivity to normal state not amounting to "intensive agricultural purposes" — Intervention not for purposes of "intensive agricultural purposes".

***Secretary of State for Environment, Food and Rural Affairs* v *Alford (Admin),* [2005] 3 11**

EQUITABLE MISTAKE — See **Mortgage**

EQUITY OF REDEMPTION — See **Mortgage**

ESCROW — See **Landlord and tenant** (Lease); **Sale of land** (Completion)

ESTATE AGENTS

Enforceability

Estate Agents Act 1979 — Estate Agents (Provision of Information) Regulations 1991 — Agent engaged by client to sell property under contract described as sole agency agreement — Client selling property to buyer not introduced by agent — Another agent involved in that sale — Whether first agent entitled to damages for loss of commission — Whether contract satisfying requirements of 1979 Act and 1991 Regulations as to provision of information on client's liability to pay remuneration — Whether contract enforceable — Held, possible to construe contract in light of statutory scheme such that no liability arising on part of client — Contrary interpretation resulting in noncompliance with statutory requirements as to provision of information — Non-compliance causing substantial prejudice to client such that contract not enforced in exercise of discretion under s18(6) of 1979 Act

***Great Estates Group Ltd v Digby (CA),* [2011] 3 101**

LAND REGISTRATION — *continued*
Alteration of registered title — *continued*
Rectification for mistake — *continued*

grounds of mistake — Held, charge valid where executed in proper form by registered proprietor — Registration of charge not a mistake unless mortgagee having notice of defect in mortgagor's title — Lodging of caution appropriate means of putting prospective mortgagee on notice
***Barclays Bank plc* v *Guy (CA),* [2008]** 2 **74**

Land Registration Act 2002 — Land used as access way to reach yard from highway — Application to remove access way from title to yard and include it in title to cottage abutting both highway and access way — Deputy adjudicator finding that access way belonging to neither party — Whether legal presumptions as to ownership of soil of road applying — Para 6(3) of Sched 4 to 2002 Act — Whether exceptional circumstances justifying refusal to rectify title — Held, two rebuttable presumptions applying to ownership of soil of road — First presumption that owner of land adjoining highway also owning soil of half of highway up to middle line — Second presumption that conveyance of land adjoining highway including soil of highway up to halfway line if transferor owning that soil — Neither presumption applying so as to confer ownership of access way on proprietor of cottage — Further hearing ordered to determine whether exceptional circumstances justifying refusal to rectify
***Paton* v *Todd (Ch),* [2012]** 2 **19**

Standing
Possessory title to land registered — Application by parish council for rectification or closure of part of title — Para 5 of Sched 4 to Land Registration Act 2002 — R24 of Adjudicator to Her Majesty's Land Registry Practice and Procedure Rules 2003 — Council having no standing at common law for private law proceedings — Whether application raising public or private law issues — Whether incumbent on adjudicator to strike out application — Held, application raising only private law questions — Application to be struck out owing to applicant's lack of standing
***Wells* v *Pilling Parish Council (Ch),* [2008]** 2 **29**

Practice and procedure
Cancellation order
Land Registration Rules 2003 — Application to register title to land cancelled by registrar on order of adjudicator and removed from day list — Stay of execution of order not granted by adjudicator — Cancellation order subsequently overturned by High Court on appeal — Whether possible to restore application in day list with original priority date — Whether applicants for registration obliged to make fresh application — Held, registration application capable of reinstatement on day list with original date of priority following successful appeal against cancellation order without need for fresh application — Preservation of integrity of day list as real-time record insufficient grounds to depart from general ability of appeal court to restore struck-out claim with retrospective effect — Restoration of application to be made on terms respecting rights of intermediate incumbrancers
***Franks* v *Bedward (Ch),* [2010]** 3 **29**

Reference to adjudicator
Land Registration Act 2002 — Unilateral notice of right of way against title to land — Objection to landowner's application to cancel notice — Dispute referred to adjudicator — Objector seeking to withdraw objection — Adjudicator refusing request — S73(7) of 2002 Act — Whether objector entitled to withdraw objection — Whether adjudicator having discretion to permit withdrawal — Held, objector not entitled to withdraw objection once dispute referred to adjudicator – Adjudicator lacking jurisdiction to permit withdrawal — Reference once made capable of disposal only by determination of underlying rights
***Silkstone* v *Tatnall (Ch),* [2010]** 3 **25**

Land Registration Act 2002 — Unilateral notice of right of way against title to land — Objection to landowner's application to cancel notice — Dispute referred to adjudicator — Objector seeking to withdraw objection — S73(7) of 2002 Act — Whether objector able to terminate reference by withdrawing objection — Whether adjudicator having discretion to terminate reference on withdrawal — Held, neither party able to terminate reference by unilateral act so as to avoid decision on the merits — Adjudicator having discretion in dealing with withdrawal of objection — Just course depending on stage at which withdrawal occurring — Adjudicator properly refusing to terminate reference and proceeding to substantive decision
***Silkstone* v *Tatnall (CA),* [2011]** 3 **47**

Restrictions
Application to register restriction against property on ground of beneficial interest — Objection to registration received — Land Registry adjudicator conducting trial and finding no beneficial interest established — Application cancelled — S73(3) of Land Registration Act 2002 – Whether adjudicator entitled to conduct trial — Whether obliged to register restriction where arguable claim to relevant interest disclosed pending determination of claim by courts — Held, adjudicator entitled to resolve disputes concerning substantive rights where necessary rather than merely conducting summary process to determine whether arguable claim exists — Adjudicator having jurisdiction to conduct trial for purpose of determining reference under s73(3) — Appropriate exercise of discretion on facts of case
***Jayasinghe* v *Liyanage (Ch),* [2010]** 1 **61**

Procedure
Cancellation order
Land Registration Act 2002 — Application to register title to land cancelled by registrar on order of adjudicator and removed from day list— Adjudicator not granting stay of execution — Cancellation order overturned by High Court on appeal by applicants — Subsequent order for re-entry of application in day list with original priority date — Whether jurisdiction to make such order — Held, court having jurisdiction under CPR 52.10(2)(a) to make order restoring registration application with original priority date — Such order appropriate where no third party interests affected — Jurisdiction exercisable at discretion of court in case where third-party interests affected — Applicants not obliged to make new registration application
***Franks* v *Bedward (CA),* [2011]** 3 **39**

Rectification
Fraud
Signature of original registered proprietor forged on transfer — Transferee under forged transfer registered as proprietor — Transferee transferring property to purchaser — Bank advancing loan to purchaser on security of charge on property — Application by purchaser and bank to register titles — Application by original registered proprietor to be restored as proprietor by rectification of register — Exercise of discretion — Held, application for rectification refused — Statutory indemnities not relevant to exercise of discretion
***Nouri* v *Marvi (Ch),* [2006]** 1 **71**

Registration of title
Lease of roof space
Leasehold Reform, Housing and Urban Development Act 1993 — Collective enfranchisement — Landlord granting lease of roof space after initial notice of enfranchisement — Whether lease of roof space liable to acquisition under s2(1)(b) of 1993 Act before initial notice registered — Whether lease void under s19 of 1993 Act — Whether Land Registry should accept application to register title to roof-space lease — Held, roof space required for management purposes — Lease would have been liable to acquisition under s2(1)(b) of 1993 Act — Lease void under s19 — No entitlement to have lease registered
***Kintyre Ltd* v *Romeomarch Property Management Ltd (Adjud to HMLR),* [2006]** 1 **67**

LAND REGISTRATION ACT 1925 — See **Adverse possession** (Human rights); **Landlord and tenant** (Rent)

LAND REGISTRATION ACT 2002 — See **Land registration**; **Mortgage**; **Sale of land** (Contract — Recession) **Title to land**

LAND REGISTRATION RULES 2003 — See **Land registration**

LANDLORD AND TENANT — See also **Defective premises** (Duty of landlord); **Right to manage**
Appurtenant rights
Reservation
S62 of Law of Property Act 1925 — Rights of common of pasture — Rights registered as attaching to farm — Whether rights passing to tenant of farm under s62 — Whether reserved to landlords under tenancy agreement as rights "affecting" farm — Whether rights waived prior to grant of tenancy— Held, reservation of rights in favour of vendor's retained land requiring express words — Rights of common not rights "affecting" farm — Rights of common not among those reserved to landlords — Rights passing to tenant under s62 — No waiver of rights prior to grant of tenancy
***Hall* v *Moore (CA),* [2009]** 2 **31**

Assignment
Breach of contract
Lease assigned without landlords' consent — Equitable assignee — Whether wrongful inducement of breach of contract — Whether breach of restrictive covenant — Whether grant of mandatory injunction to reverse unlawful assignment — Whether landlords' entitlement to injunction prior right to forfeiture — Held, assignment of lease without landlord's consent constituting breach of covenant — Tort of inducing breach of contract — Restrictive covenant binding equitable assignees — Breach of restrictive covenant — Grant of mandatory injunction priority remedy
***Test Valley Borough Council* v *Minilec Engineering Ltd (in liquidation) (Ch),* [2005]** 2 **113**

Consent to assignment
Headlease of residential property comprising flats let on subleases — Assignee incorporated for purpose of acquiring headlease — Whether unreasonable for freeholders to require personal guarantee from director

LANDLORD AND TENANT — *continued*
Assignment — *continued*
Consent to assignment — continued

of assignee as condition of consent to assignment — Whether income from sublessees adequate security for covenants in headlease — Whether three-year guarantee sufficient — Held, reasonable to require guarantee in circumstances of case — Income from subleases not adequate security — Guarantee to be for residue of term until further assignment
***Landlord Protect Ltd* v *Dolman (CLCC),* [2007]** 2 **21**

Lease providing that landlord's consent to assignment not to be unreasonably withheld subject to requirement for provision of surety — Whether requirement ineffective as attempt to limit unconditional proviso in s19(1) of Landlord and Tenant Act 1927 that consent not to be unreasonably withheld — Whether rent on review to be calculated on basis of one letting or two – Whether parties bound by determination of independent expert on review — Held, requirement for surety effective as precondition to tenant's ability to apply for consent — Rent review on basis of two lettings if producing a higher figure — Parties not bound by independent expert's determination in event of erroneous interpretation of lease provisions
***Level Properties Ltd* v *Balls Brothers Ltd (Ch),* [2007]** 2 **26**

Underlettings required to be at greater of passing rent or open market rental value — Proposed underletting at passing rent subject to reverse premium — Application for consent to underletting — Application refused on grounds of unsatisfactory financial arrangements and insufficient covenant strength of proposed subtenant — Whether decision made within reasonable time — Whether decision withholding consent reasonable — Whether landlord's expert evidence concerning effect upon value of reversionary interest relevant — Held, decision made within reasonable time — Expert evidence concerning reversionary values relevant — Consent reasonably withheld
***NCR Ltd* v *Riverland Portfolio No 1 Ltd (No 2) (CA),* [2005]** 2 **42**

Guarantee

Assignment of lease — Assignor company and its owner guaranteeing assignee's rent and service charge liability — Landlord undertaking to notify owners of rent arrears — Whether landlord entitled to claim under guarantee where no such notification given — Whether notification a condition precedent to liability on guarantee — Whether breach of notification requirement going to root of contract so as to discharge further such liability — Held, notification requirement not amounting to condition precedent to operation of guarantee but instead existing as separate obligation on landlord — Notification requirement not intended to be condition going to root of contract — Breach not intended to result in release of guarantee liability — Landlord entitled to claim on guerentee
***Greene King plc* v *Quisine Restaurants Ltd (CA),* [2012]** 2 **64**

Contracts (Rights of Third Parties) Act 1999 — Former tenant guaranteeing liabilities of assignee — Supplemental deed between assignee and landlord limiting recovery against tenant or any previous tenant to assets of assignee — Whether deed intended to limit liability of former tenant as guarantor — Whether former tenant entitled to rely upon provisions of deed as third party upon whom benefit conferred under s1(1)(b) of 1999 Act — Held, deed leaving rent reserved by underlease unchanged but limiting recovery against either tenant or previous tenant — Previous tenant entitled to rely upon deed as third party upon whom benefit conferred
***The Prudential Assurance Co Ltd* v *Ayres (Ch),* [2007]** 2 **31**

Contracts (Rights of Third Parties) Act 1999 — Former tenant guaranteeing liabilities of assignee — Supplemental deed between assignee and landlord stated to limit recovery against tenant or any previous tenant to assets of assignee — Whether deed intended to limit liability of previous tenant as guarantor — Whether previous tenant entitled to rely on provisions of deed as third party upon whom benefit conferred under s1(1)(b) of 1999 Act — Held, construction of deed according only to ordinary meaning of words used not appropriate — Deed intended to restrict recovery by landlord or previous tenant against individual partners of assignee — Deed not intended to affect liability of previous tenant as guarantor — Deed not conferring benefit on previous tenant within meaning of1999 Act
***The Prudential Assurance Co Ltd v Ayres (CA),* [2008]** 1 **5**

Sale at auction — Contract conditional on obtaining landlords' consent to assignment — Landlords seeking guarantee by sole director of purchaser to be released on subsequent assignment provided that reasonable alternative security provided — Whether landlords' requirements unreasonable under terms of lease — Whether proviso conferring on landlords greater protection than that to which entitled under headlease — Whether purchaser entitled to rescind – Held, landlord not generally entitled to guarantee of liabilities of assignee extending beyond period during which assignee liable to landlord by privity of estate — Landlord adequately protected by entitlement to refuse consent to subsequent assignment — Requirement of reasonable alternative security unreasonable
***Landlord Protect Ltd* v *St Anselm Development Co Ltd (CA),* [2009]** 2 **23**

Law of Property Act 1925, s63

Assignment of lease — Break clause personal to original tenant but capable of assignment to specified assignees — Assignment of lease without express assignment of break clause — Exercise of break clause — Whether break clause validly exercised — Whether clause "touching and concerning" land — S142 of 1925 Act — Whether clause having reference to subject matter of lease — Whether clause passing under s63 with assignment of lease — Held, break clause validly exercised — Hybrid break clause "touching and concerning land" and having reference to subject matter of lease under s142 — Break clause constituting a right for purposes of s63
***Harbour Estates Ltd* v *HSBC Bank plc (Ch),* [2005]** 1 **107**

Liability for rent

Sublease of commercial premises — Sublease assigned without consent of landlord — Transfer of sublease not registered — Arrears of rent accruing — Whether consent to assignment required on proper construction of sublease — Landlord and Tenant (Covenants) Act 1995 — Land Registration Act 2002 — Whether assignor having continuing liability for rent under tenant's covenants — Whether recovery from assignor limited to amounts specified in notices served under s17 of 1995 Act indicating intention to recover rent from former tenant — Held, sublease requiring consent of landlord to assignment — No statutory requirement to give consent in reasonable time where request for consent not "served" in manner required by sublease — Assignment therefore ineffective to relieve assignor of liability on tenant's covenants — Same result flowing from failure to register transfer — Non-registration resulting in transfer becoming void and sublease reverting to assignor — Assignor therefore current not former tenant — Landlord not limited to recovering sums specified in s17 notices
***E.ON UK plc* v *Gilesports Ltd (Ch),* [2012]** 3 **23**

Virtual assignment

Lease of commercial premises — Premises occupied by subtenant — Tenant transferring all economic benefits and burdens and management responsibilities to third party — Consent of landlord not obtained for assignment — Whether tenant sharing or parting with possession of demised premises — Whether breach of covenant against alienation without landlord's consent — Held, possession given ordinary meaning of right to exclude all others from property — Subtenant in possession of demised premises pursuant to underlease — Tenant not in possession such that virtual assignment not amounting to sharing or parting with possession — Assignee not in possession in sense of receiving or having right to receive rents — Rents collected as agent for tenant with assignee having purely contractual right to retain them thereafter under terms of virtual assignment
***Clarence House Ltd* v *National Westminster Bank plc (CA),* [2010]** 1 **43**

Lease of commercial premises — Premises occupied by underlessee — Tenant transferring all economic benefits and burdens and management responsibilities to third party — Consent of landlord not obtained for assignment — Whether breach of covenant against alienation without landlord's consent — Held, covenant against assignment applying only to legal assignment unless context showing otherwise — Virtual assignment not amounting to underletting or declaration of trust within meaning of covenant — Virtual assignment amounting to parting with possession of demised premises contrary to covenant where demised premises underlet and tenant relinquishing power to deal with property to assignee
***Clarence House Ltd* v *National Westminster Bank plc (Ch),* [2009]** 2 **19**

Assured shorthold tenancy
Insured tenancy deposit scheme

Landlord and tenant — Assured shorthold tenancy — Insured tenancy deposit scheme — Prescribed information — Housing Act 2004 – Tenant paying deposit – Deposit lodged in insured tenancy deposit scheme under s212 of2004 Act — Landlord failing to provide tenant with prescribed information regarding scheme within required period – Information subsequently provided — Ss 213(6), 214(3) and (4) of 2004 Act — Whether tenant entitled to refund of deposit — Whether landlord liable to make penalty payment — Held, deposit refundable and penalty payable only where prescribed information not given — Landlord not so liable where information given in writing albeit outside 14-day time period
***Harvey* v *Bamforth (CC),* [2008]** 3 **66**

Termination

Fixed-term tenancy for one year less one day — Tenancy providing for rent as a yearly sum payable quarterly — Statutory periodic tenancy — Termination — S21 notice served by reference to quarterly statutory periodic tenancy — Whether statutory periodic tenancy quarterly or yearly— Proper construction of s5 of Housing Act 1988 — Held, statutory periodic tenancy quarterly — S5(3)(d) requiring ascertainment of period for which rent last payable
***Church Commissioners for England* v *Meya (CA),* [2006]** 2 **39**

Assured tenancy
Possession

Adjournment — Ground 8 claim for possession — Mandatory ground of possession — S7 of Housing Act 1988 — Claim for possession— Application for possession to enable tenant to reduce rent arrears below ground 8 threshold — Whether county court having jurisdiction to grant

LANDLORD AND TENANT — *continued*
Assured tenancy — *continued*
Possession — continued

adjournment — Held, court cannot normally grant adjournment for purpose of enabling tenant to reduce rent arrears below ground 8 threshold limits
***North British Housing Association Ltd* v *Matthews; London & Quadrant Housing Trust* v *Morgan (CA),* [2005]** 1 **31**

Variation
Registered social landlord — Standard form of tenancy agreement — Unfair Terms in Consumer Contracts Regulations 1999 — Tenancy agreement containing contradictory provisions for variation of terms — Whether conferring unilateral power on landlord to vary terms without approval of tenants — Whether such provision "not binding" on tenants as unfair term — Held, no right of unilateral variation of terms — Doubt raised by contradictory terms to be resolved in favour of tenant as consumer — Unilateral variation clause unfair and not binding unless Office of Fair Trading guidelines taken into account
***Governors of Peabody Trust* v *Reeve (Ch),* [2008]** 3 **61**

Breach of covenant
Damages
Damages assessed on negotiating basis — Date of breach — Lease liable to forfeiture on date of breach — Whether hypothetical negotiations should assume that lease liable to forfeiture at breach date — Held, presumption that damages under negotiating basis assessed at date of breach — Lord Cairns' Act raising quasi-equity considerations — Court entitled to depart from usual presumption in appropriate cases
***Lunn Poly Ltd* v *Liverpool & Lancashire Properties Ltd (CA),* [2006]** 2 **29**

Break clause
Agency
Commercial underlease containing break clause permitting tenant to end lease by giving six months' notice — Notice given in name of other company in tenant's group — Landlord asserting notice invalid and underlease continuing — Whether notice valid — Whether notice given by tenant as required by underlease — Whether given by third party acting as agent for tenant — Held, notice not given by tenant invalid even if reasonable recipient would think it was so given — Subsequent ratification by tenant of notice served by third party not effective to save notice — Notice given by third party as principal not as agent for tenant
***Hexstone Holdings Ltd* v *AHC Westlink Ltd (Ch),* [2010]** 2 **13**

Assignment
Commercial lease — Lease assigned with licence of landlord — Licence to assign containing break clause — Right to break personal to defendant as assignee — Defendant assigning to associated company with consent of landlord – Subsequent reassignment without landlord's consent — Landlord refusing consent on grounds relating to risk of defendant exercising break clause — Whether defendant able to exercise break clause while not the tenant — Whether right to break reviving on reassignment — Whether consent to reassignment unreasonably withheld — Held, break clause a right incident to lease passing to assignee on assignment — No commercial sense in construing licence to assign so as to permit termination of lease by party after ceasing to be tenant — Right not reviving on reassignment — Claimant not acting unreasonably in refusing consent to avoid perceived possibility of loss of significant rental income
***Norwich Union Life & Pensions Ltd* v *Linpac Mouldings Ltd (Ch),* [2009]** 3 **147**

Conditions
Lease of commercial premises — Tenant purporting to break lease — Whether complying with conditions of break clause — Whether break invalidated by failure to make all payments due to landlord under lease by break date — Whether payment by cheque on day before break date sufficing — Whether outstanding sums of interest remaining due under lease — Held, implied agreement by landlord to accept payment by cheque to be inferred from course of conduct — Tender of cheque on day before break date sufficient to make payment by break date — Break nonetheless ineffective by reason of outstanding sums of default interest due under lease — Such sums accruing and falling due day by day regardless of whether demanded by landlord
***Avocet Industrial Estates LLP* v *Merol Ltd (Ch),* [2012]** 1 **65**

Material compliance
Lease containing break clause subject to material compliance by tenant of all obligations — Tenant carrying out large programme of repairs — Landlord declining to inspect or comment on specification during repair works — Whether material compliance requiring strict compliance — Whether tenant having materially complied with repair obligations — Whether lease terminating on termination date — Held, material compliance not requiring strict compliance — Tenant having materially complied — Lease terminating on termination date
***Fitzroy House Epworth Street (No 1) Ltd* v *The Financial Times Ltd (TCC),* [2006]** 1 **19**

Lease containing break clause subject to material compliance by tenant of all obligations — Tenant undertaking large programme of repairs — Landlords declining to inspect or comment upon specification during repair works — Whether material compliance requiring strict compliance — Whether tenant having materially complied with repair obligations — Whether lease terminating on termination date — Held, material compliance not requiring strict compliance — Tenant having materially complied — Lease terminating on termination date
***Fitzroy House Epworth Street (No 1) Ltd* v *The Financial Times Ltd (CA),* [2006]** 2 **13**

Notice
Acknowledgment of receipt — Commercial lease containing tenant's break clause exercisable on six months' prior notice — Notice to be valid only where given by hand or posted to appropriate address unless receipt by other means acknowledged by landlord — Tenant giving notice by fax — Landlord bringing claim for rent for period after break date — Acknowledgment of receipt of break notices — Whether acknowledgment retrospectively validating break notices — Whether lease continuing — Held, faxed notices invalid unless and until receipt acknowledged by landlord — Notice invalid at critical moment six months prior to break date incapable of validation by later acknowledgment — Lease continuing — Tenant liable for rent
***Orchard (Developments) Holdings plc* v *Reuters Ltd (CA),* [2009]** 1 **13**

Vacant possession
Commercial lease — Tenant exercising option to break lease after first year of term — Valid exercise of option depending on vacant possession being given by break date — Tenant's workmen remaining on premises for several days after break date to complete works necessary to comply with repairing obligations — Whether vacant possession given — Whether landlord waiving requirement to give vacant possession on break date — Held, test of vacant possession same as in domestic or commercial sale — Property to be empty of people and also of chattels that would substantially prevent or interfere with enjoyment of the right of possession of a substantial part of the premises — Tenant not giving vacant possession where remaining in occupation and control of premises without agreement of landlord — Lease not validly broken
***Ibrend Estates BV* v *NYK Logistics (UK) Ltd (CA),* [2011]** 3 **1**

Certainty of term
Possession
Law of Property Act 1925 — Residential premises occupied under agreement with housing association — Agreement expressed to take effect from month to month and to be terminable by association only on occurrence of specified events — Whether tenancy on terms of agreement invalid for uncertainty of term — Whether association entitled to possession on service of notice to quit — S146(6) of 1925 Act — Held, agreement not capable of taking effect as tenancy in accordance with its terms — Purported fetter on right to serve notice determining periodic tenancy effective only if fetter applying for specified period only — Agreement none the less taking effect under s146(6) as tenancy for 90 years subject to right to determine on specified events — Housing association not entitled to possession
***Mexfield Housing Co-operative Ltd* v *Berrisford (SC),* [2011]** 3 **11**

Collateral contract
Dilapidations
Agreement for lease — Whether collateral contract that no terminal schedule of dilapidations would be served upon expiry of lease — Whether position affected by conduct of negotiations "subject to formal deed" — Held, labelling of negotiations as "subject to contract" not precluding formation of collateral contract at time of main contract though not before — No contract concluded between parties save that resulting from grant of lease — Collateral contract not providing defence to claim by assignee of landlord for dilapidations
***Business Environment Bow Lane Ltd* v *Deanwater Estates Ltd (CA),* [2007]** 2 **51**

Dilapidations — Whether tenant entering into lease in return for assurance that no terminal schedule of dilapidations would be served at its end — Whether proviso in lease agreed as substitute for assurance — Whether position affected by conduct of negotiations "subject to formal deed" — Held, landlord bound by assurance — Proviso in lease no substitute for assurance — "Subject to contract" principle not affecting binding nature of assurance where lease entered into in reliance upon it
***Business Environment Bow Lane Ltd* v *Deanwater Estates Ltd (Ch),* [2007]** 2 **37**

Commercial lease
User covenant
Construction — Law of Property Act 1925 — Underlease containing covenant not to use demised premises for parking motor vehicles for sale on forecourt — Subtenant of underlessee displaying parked cars for sale around perimeter of premises — Whether underlessee thereby in breach of covenant — Whether covenant making underlessee liable for acts of subtenant — Whether underlessee to be deemed to covenant for itself and third parties deriving title under it by reason of s79 of 1925 Act — Held, user covenant

LANDLORD AND TENANT — *continued*
Commercial lease — *continued*
User covenant — continued

in active voice indicating that lessee covenanting on own behalf not to use premises in a certain way — Underlessee not thereby covenanting on behalf of third parties — S79 of 1925 Act not applying since displaced by parties' expression of contrary intention

***Roadside Group Ltd* v *Zara Commercial Ltd (Ch),* [2010]** 3 **41**

Covenants
Access by landlord

Commonhold and Leasehold Reform Act 2002 — Tenant of flat complaining of leaking pipes — Lease covenant requiring tenant to permit access by landlord and its agents for purpose of examining state and condition of flat — Tenant refusing entry to company secretary of landlord to inspect pipes but willing to allow access by another agent of landlord — Whether landlord entitled to insist on access by company secretary — s168(4) of 2002 Act — Whether tenant in breach of terms of lease — Held, covenant requiring grant of access to any person falling within specified category for purpose of examining state and condition — Company secretary as agent of landlord entitled to access for that purpose — Refusal of access would be breach of lease terms — No determination under s168(4) of 2002 Act made in light of draconian nature of such relief

***Beaufort Park Residents Management Ltd* v *Sabahipour (UT:LC),* [2012]** 1 **53**

Exoneration clause

Lessee claiming specific performance and damages against landlord in respect of breeches of lease covenants — Exoneration clause in lease relieving landlord of liability for damage suffered through fault of agents employed by it save to extent that such liability insured against — Whether such clause exonerating landlord from liability in respect of lease covenants — Held, exoneration clause applying only to claims in tort brought on basis of vicarious liability — Clause not applying to breaches of landlord's repairing covenants — Clause applying only to physical damage and not covering loss of amenity

***Newman* v *Framewood Manor Management Co Ltd (CA),* [2012]** 2 **45**

Landlord and Tenant (Covenants) Act 1995

Covenant limiting liability of landlord until disposal of reversion — Whether agreement within s25(4) of 1995 Act — Whether agreement frustrating purpose of Act — Whether liability of landlord capable of release only unders8 procedure — Whether liability of landlord capable of being released by agreement in lease or otherwise — Held, agreement to release landlord on parting with reversion not voided by s25(1) of 1995 Act — Parties to lease not restricted to s8 procedure to release liabilities — Parties able to release liability by agreement in lease or otherwise — Act providing additional release route

***Avonridge Property Co Ltd* v *Mashru (HL),* [2006]** 1 **15**

Landlord and Tenant (Covenants) Act 1995

Covenant limiting liability of landlord until disposal of reversion — Whether agreement within s25(4) of 1995 Act — Whether agreement frustrating purpose of Act — Whether liability of landlord capable of release only unders8 procedure — Whether liability of landlord capable of being released by agreement in lease or otherwise — Held, agreement to release landlord on parting with reversion not voided by s25(1) of 1995 Act — Parties to lease not restricted to s8 procedure to release liabilities — Parties able to release liability by agreement in lease or otherwise — Act providing additional release route

***Avonridge Property Co Ltd* v *Mashru (HL),* [2006]** 1 **15**

Parting with or sharing possession

Company in demised premises — Whether tenant sharing occupation — Whether sharing of occupation parting with or sharing possession — Whether breach of covenant — Whether breach of covenant against sharing possession capable of remedy — S146 notice — Whether notice specifying breach — Whether s146 notice effective — Forfeiture — Relief — Whether grant of relief from forfeiture — Held, sharing occupation not sharing possession – No breach of covenant — Breach of covenant against sharing possession capable of remedy — S146 notice failing to specify breach — Notice ineffective — If re-entry effective, no relief from forfeiture

***Akici* v *LR Butlin Ltd (CA),* [2006]** 1 **34**

Service charges

Lease providing for certification of service charges by borough treasurer of council lessors — Whether finance director of new landlord entitled to certify service charges — Liability for refuse collection — Liability for management fees — Held, finance director not entitled to certify service charges — Contractual mechanism for recovering service charges no longer having effect — Defendant liable for proportion of cost of refuse collection— No justification for figure of 10% in management fees

***St Modwen Developments (Edmonton) Ltd* v *Tesco Stores Ltd (Ch),* [2007]** 1 **63**

Set-off

Assignment — Agreement between landlord A and tenant — Landlord A agreeing to erect building and grant lease to tenant — Tenant taking lease — Landlord A assigning reversion — Landlord B acquiring reversion — Tenant admitting liability for post-assignment arrears of rent to landlord B — Tenant claiming damages for breaches of agreement with landlord A — Whether tenant entitled to set off claim for damages against arrears of post-assignment rent — Held, tenant not entitled to set off claim for damages for breaches of agreement by landlord A against post-assignment arrears of rent due to landlord B

***Edlington Properties Ltd* v *JH Fenner & Co Ltd (QB),* [2006]** 1 **29**

User clause

Prohibition against business of "victualler" or "coffee house keeper" — Proposed underletting to Pret A Manger — Whether use for sale of sandwiches and drinks for consumption off or with ancillary consumption on premises prohibited — Whether "victualler" meaning "licensed victualler" — Whether "coffee house keeper" including use of premises for selling drinks — Held, "victualler" having ordinary general meaning of person supplying food and drink and not limited to "licensed victualler" — "Coffee house keeper" including use of premises for selling drinks — Proposed user breaching user clauses

***Joint London Holdings Ltd* v *Mount Cook Land Ltd (CA),* [2005]** 3 **119**

Defective Premises Act 1972
Relevant defect

Landlord's repairing covenant — Tenant injured by putting hand through glass panel in door — Glass not in disrepair but of unsafe type — Whether unsafe glass amounting to relevant defect under s4 of 1972 Act even though no disrepair — Whether landlord liable for failure to make safe — Held, s4 duty extending no further than landlord's covenant to maintain and repair — Duty to maintain not encompassing duty to make safe

***Alker* v *Collingwood Housing Association (CA),* [2007]** 2 **43**

Demise
Roof

Sublease — Extent of demise — Whether references in provisions relating to decorations and insurance relevant to extent of demise — Meaning of "buildings and structures" — Whether roof part of demise under sublease — Whether sublessee liable to repair roof — Held, references to "buildings and structures" in unrelated provisions no guide to extent of demise — Contribution provision in respect of common areas best guide — Roof not demised — Sublessee not liable for repair

***Delgable Ltd* v *Perinpanathan (CA),* [2006]** 1 **78**

Dilapidations
Damages

Covenant to deliver up in good and substantial condition — Roof — Repair or replacement — Whether measure of damages cost of patch repairs or more extensive overcladding demanded by incoming tenant — Second limb of s18(1) of Landlord and Tenant Act 1927 — Whether some repairs superseded by structural alterations carried out after termination of tenancy — Held, patch repairs sufficient to comply with covenants in tenancy — Stance of incoming tenant relevant but not determinative — Works to roof not superseding patch repairs — Former tenant liable for full cost of patch repairs

***Carmel Southend Ltd* v *Strachan & Henshaw Ltd (TCC),* [2007]** 3 **15**

Headlease governed by Part II of Landlord and Tenant Act 1954 — Grant of licence to sublet — Sublease subject to same covenants as headlease — Undertaking by head lessee's parent company that sublessee's repairing obligations confined to keeping property wind- and watertight with any greater obligation to be performed at expense of parent — Expiry of headlease — Whether landlord entitled to substantial damages for dilapidations — Whether any diminution in value of reversion – Whether landlord's position affected by undertaking — Held, no entitlement to substantial damages— No diminution in value of reversion where landlord able to enforce repairing covenants against sublessee — Undertaking not adversely affecting reversion

***Lyndendown Ltd* v *Vitamol Ltd (CA),* [2007]** 3 **11**

Landlord and Tenant Act 1927 — Liquidators of tenant company surrendering leases of commercial premises — Landlord submitting proofs of debt to liquidators in respect of disrepair and alleged liability of tenant to reinstate demised premises — Whether debts provable — Whether liability on reinstatement obligation arising prior to end of lease term — Whether damages claimed for breach of covenant to keep in good repair properly taking into account rent-free period and reduced rent granted to new tenant — S18(1) of 1927 Act — Held, debts not provable — No liability on reinstatement obligation as at date of surrender since tenant having full period of term created by leases in which to carry out requisite reinstatement — Impermissible to take account of rent-free period and reduced rent on re-letting when assessing damages for breach of covenant to keep demised premises in good repair — Such matters inconsistent with requirement under

LANDLORD AND TENANT — *continued*
Housing Act 1985 — *continued*
Succession — *continued*

to tenancy — Tenancy ceasing to be secure where surviving tenant not residing in property— Landlords entitled to possession
***Solihull Metropolitan Borough Council* v *Hickin (CA),* [2010]** 2 **147**

Housing Act 1988
Assured tenancy

Determination of rent — Local Government and Housing Act 1989 — Former long residential tenancy coming to an end — Tenant becoming assured tenant by virtue of Sched 10 to 1989 Act — Rent determined by rent assessment committee disregarding tenant's improvements — Improvements not disregarded on subsequent reference under s13(4) of 1988 Act — Whether words to be read into statutory provisions so as to permit disregard of improvements — Held, tenant not meeting conditions for disregard of improvements where no assured tenancy at date of improvements since such tenancies not yet existing — No provision in 1989 Act for deeming long residential tenancy to have been assured tenancy — Legislation not unintelligible, absurd or totally unreasonable such as to justify reading in words so as to permit application of disregard
***Hughes* v *Borodex Ltd (Admin),* [2009]** 2 **47**

Determination of rent — Local Government and Housing Act 1989— Former long residential tenancy ending — Tenant becoming assured tenant by virtue of Sch 10 to 1989 Act — Initial rent determined by rent assessment committee disregarding tenant's improvements — Improvements not disregarded on subsequent reference under s13(4) of Housing Act 1988 — Whether words to be read into statutory provisions so as to permit disregard of improvements – Held, paras 9 and 10 of Sch 10 applying only to fixing of initial rent at outset of assured tenancy — No disregard of improvements permitted on subsequent determinations under ss 13 and 14 — No justification for reading words to contrary into statute
***Hughes* v *Borodex Ltd (CA),* [2010]** 2 **25**

Rent review clause — Assured tenancy providing for annual rent increase both during term and any subsequent statutory periodic tenancy — Tenants holding over after end of contractual term as statutory periodic tenants — Managing agents proposing substantial rent increase – S13(1) of 1988 Act — Whether rent review clause in original assured tenancy continuing in effect — Whether rent assessment panel having jurisdiction to determine rent — Held, statutory periodic tenancy governed by s13(1)(a) not (b) — Statutory scheme governing increase of rent — Rent review clause in original assured tenancy ceasing to have effect — Rent assessment panel having jurisdiction to determine rent
***London District Properties Management Ltd* v *Goolamy (Admin),* [2009]** 3 **67**

Possession order

Assured tenancy — Position of assured tenant after making of possession order with enforcement suspended on terms — S9(2) of 1988 Act — Standard form N28 — Whether assured tenancy continuing after date for possession specified in order — Whether tenant becoming tolerated trespasser — Held, result depending upon correct construction of county court order — Order on form N28 having effect of ending assured tenancy on date for possession specified in order — Former tenant thereafter having status of tolerated trespasser
***White* v *Knowsley Housing Trust (CA),* [2007]** 2 **61**

Protected and statutory tenants Transitional protection — Tenant originally protected by Rent Act 1977 — Joint tenancies — S34(1)(b) of 1988 Act — Series of joint tenancies with fluctuating membership — Original tenant protected by 1977 Act ceasing to be a joint tenant — Whether remaining joint tenants under successor joint tenancies entitled to protection of s34(1)(b) — Held, protection continuing for joint tenancies only where joint tenant originally protected prior to enactment of1988 Act
***Secretarial & Nominee Co Ltd* v *Thomas (CA),* [2005]** 3 **37**

Rent assessment

Practice and procedure — Rent assessment committee taking into account supplemental submission from tenant — Submission not copied to landlord — Whether breach of natural justice — Correct approach to valuation of security of tenure — Held, fairness requiring that all parties see all relevant evidence — Further material submitted to committee after conclusion of oral proceedings to be sent *inter partes* not just to committee — Security of tenure advantageous to tenant such that committee not entitled to treat it as valueless — Committee's assessment not to be subjected to artificial threshold of £25,000 representing limit for security of tenure
***Earl Cadogan* v *Chehab (Admin),* [2010]** 1 **35**

Housing Act 1996
Demoted tenancy

Possession order — Secure tenancy demoted on ground of anti-social behaviour by tenant's family members — Possession notice served by local authority landlords — Review panel upholding notice — S143D of Housing Act 1996 — Application to court for possession order – County court conducting judicial review of panel's decision and granting possession order – Extent of county court's review jurisdiction – Art 8 of European Convention on Human Rights – Whether s143D(2) compatible with Art 8 — Held, Art 8 requiring that court asked by local authority to make possession order in respect of a person's home to have power to assess proportionality of making order and to resolve any relevant dispute of fact — S143D(2) of 1996 Act to be construed under s3(1) of the Human Rights Act 1998 as permitting such exercise on application for possession order against demoted tenant — Tenant therefore able to raise proportionality issues in county court possession proceedings and county court able to exercise powers necessary for that purpose
***Manchester City Council* v *Pinnock (SC),* [2010]** 3 **113**

Forfeiture

Possession — Landlords bringing claim for unpaid service charges — Tenant filing no acknowledgement of service or defence — Default judgement entered for landlords — Whether landlords entitled to possession on ground of forfeiture of lease — Whether such claim precluded by s81 of 1996 Act — Whether default judgement a "final determination" that service charge payable within meaning of s81 — Held, default judgement constituting final determination for purposes of s81 — Such judgement final for that purpose notwithstanding absence of specific time limit for application to set aside — Landlords entitled to possession
***Church Commissioners for England* v *Koyale Enterprises and another (CC),* [2012]** 2 **42**

Housing Act 2004
Assured shorthold tenancy

Tenancy deposit scheme — Dwelling — Landlords failing to protect tenants' deposits within 14 days of receipt by complying with initial requirements of authorised tenancy deposit scheme or to notify tenants of such compliance as required by s213 of 2004 Act — Whether landlords liable to sanctions prescribed by s214 for failure to comply with those requirements — Whether compliance within 14-day period necessary to avert sanctions — Whether compliance by date of hearing sufficient – Held, 14-day time limit not forming part of initial requirements or notification requirement under s213 — Initial requirements not including any time limit imposed by particular scheme — Tenant's counterclaim invoking s214 sanctions provided landlord complying with requirements by date of hearing
***Vision Enterprises Ltd (t/a Universal Estates)* v *Tiensia; Honeysuckle Properties* v *Fletcher (CA),* [2010]** 3 **53**

Tenancy deposit scheme

Assured shorthold tenancy — Landlords not complying with statutory requirement to deal with deposit in accordance with authorised scheme and to provide prescribed information in that regard — Claim by tenant for three times deposit as sanction under s214(4) of 2004 Act — Deposit paid into scheme and then repaid to tenant prior to hearing — Whether sanction still applying in respect of continuing failure to provide prescribed information — Whether that requirement ceasing on repayment of deposit — Held, landlord's obligation to provide prescribed information continuing regardless of return of deposit — Sanction under s214(4) applying where prescribed information not provided by hearing date — Information supplied by Deposit Protection Service insufficient to comply with landlord's obligation to provide prescribed information personally
***Suurpere* v *Nice (QB),* [2011]** 3 **19**

Housing association
Possession

Property occupied under occupancy agreement granted by fully mutual housing co-operative association — Agreement expressed to take effect from month to month and be terminable by association only in event of occupant's breach of agreement — Association giving notice to quit — Whether occupancy agreement creating periodic tenancy terminable by four weeks' notice regardless of whether breach occurring — Whether creating contractual licence with benefit of restriction on termination — Held, lease terminable only in event of breach void for uncertainty of term — Agreement purporting to grant such lease ineffective to create lease either in law or equity — Such agreement not to be construed as contractual licence where both parties intending lease — Exclusive possession and payment of rent on monthly basis giving rise to common law periodic tenancy terminable by notice regardless of whether breach occurring
***Mexfield Housing Co-operative Ltd* v *Berrisford (CA),* [2010]** 2 **137**

Human rights
Possession

Joint tenancy — Notice to quit served by one joint tenant bringing tenancy to an end — Local authority landlords bringing possession proceedings against other former joint tenant — Whether local authority landlords acting unlawfully in procuring notice to quit — Whether breach of right to respect for home under Article 8 of the European Convention on Human Rights — Held, interference with right to respect for home — Interference disproportionate owing to lack of procedural safeguards
***McCann* v *United Kingdom (ECtHR),* [2008]** 2 **45**

LANDLORD AND TENANT — *continued*
Service charges — *continued*

such lease not falling within exception from qualifying long-term agreements in s20ZA(3)(1)(*d*) — Contract for initial 12-month period then from year to year not a qualifying long-term contract — Contract for window cleaning not a contract for "qualifying works" — Cap on recovery where consultation requirements not met to be applied to service charge for each individual dwelling
***Paddington Walk Management Ltd* v *Governors of the Peabody Trust (CC),* [2009]** 2 **123**

Recoverability — Time limit — S20B(1) of 1985 Act — Landlord invoicing tenant for advance payments of estimated service charge followed by balancing charge based on actual expenditure over year — Whether certain items of service charge to be disallowed on ground that incurred more than 18 months before demand for payment — Whether advance payment invoices constituting demands within meaning of s20B(1) — Held, invoices for advance payments and for balancing payments both constituting demands for payment of service charges within s20B(1) — Advance payment invoices relating to prospective costs only such that none of costs in question incurred more than 18 months before relevant demand— Balancing payment invoices relating only to costs incurred once amount of advance payments exhausted — All such costs incurred within the 18-month period – No items to be disallowed under s20B(1)
***Holding & Management (Solitaire) Ltd* v *Sherwin (UT:LC),* [2011]** 1 **29**

Costs

Proceedings in leasehold valuation tribunal — Landlord bringing LVT proceedings to determine liability of tenant for unpaid service charges — Whether legal and surveyor's costs of those proceedings recoverable from tenant under terms of lease — Whether those costs incurred in provision of services for better use and enjoyment of the property — Held, relevant enforcement provision in lease placing costs of enforcement on non-defaulting tenant not tenant enforced against — Relevant paragraph in lease unambiguous and no grounds for rectification — Enforcement costs not incurred in provision of services for better use and enjoyment of property — Collection of service charge not a service enabling tenants better to enjoy property but only a by-product of such services
***Twenty Two Clifton Gardens Ltd* v *Thayer Investments SA (UT:LC),* [2012]** 2 **56**

Landlord and Tenant Act 1987

Management order

Whether permissible for order to extend to land not within curtilage of buildings containing leasehold flats — Held, works to servient tenement ancillary to easement amounting to function in connection with management of premises — Management order not confined to curtilage of buildings containing leasehold properties — Order able to include servient tenement over which leaseholders having easement
***Cawsand Fort Management Co Ltd* v *Stafford (LT),* [2007]** 1 **85**

Notices

Relevant disposal — Purchase notice by requisite majority of tenants — Time limits — Tenants receiving documents concerning relevant disposal — Documents obtained by tenants' own solicitor — Whether requisite majority of tenants "served" with document giving notice of relevant disposal — Held, notice not "served" — Service requiring physical delivery making recipient aware time limit for service of purchase notice has been triggered
***Savva* v *Galway-Cooper (CA),* [2005]** 3 **40**

Variation of lease

Landlord applying to vary leases of flats to permit replacement of communal heating and hot water system with individual boilers in each flat — Landlord seeking further "non-consequential" variations to permit recovery through service charge of costs of enforcing all lease covenants — Application made with approval of requisite majority of tenants — Ss37 and 38 of 1987 Act — Identification of object of application — Whether object satisfactorily achieved without non-consequential variations — Whether such variations unreasonable and prejudicial to tenants — Held, application identifying single object of achieving replacement of heating and hot water system — That object satisfactorily achieved without non-consequential variations — Accordingly no power to make non-consequential variations even though requisite majority of tenants in favour — Non-consequential variations also unreasonable and prejudicial to tenants contrary to s38(6) where having effect of shifting all financial risk of enforcement from landlord to tenant thereby fundamentally altering relationship between the two
***Shellpoint Trustees Ltd* v *Barnett (UT:LC),* [2012]** 3 **115**

Landlord and Tenant (Covenants) Act 1995

Authorised guarantee agreement

Assignment — Administrative receivership — Landlord requiring administrative receivers to offer authorised guarantee agreement — Personal liability of receivers under s44 of Insolvency Act 1986 — Whether reasonable requirement of landlord — Held, question of construction of lease — Lease not requiring AGA from third parties — Whether reasonable requirement of landlord amounting to triable issue
***Legends Surf Shops plc (in administrative receivership)* v *Sun Life Assurance Society plc (Ch),* [2005]** 3 **43**

Guarantee

Lease of commercial premises — Assignment — Landlord consenting to assignment on terms that guarantor of existing tenant also guaranteeing obligations of assignee — Landlord seeking to recover unpaid rent from guarantor — Ss24 and 25 of 1995 Act — Whether guarantee void and unenforceable as frustrating operation of 1995 Act — Held, s24 intended to ensure obligations undertaken as guarantor for tenant ending on assignment of lease — Requirement for guarantor to enter into further guarantee of assignee's obligations frustrating operation of s24 — Such guarantee void and unenforceable against guarantor under s25(1)
***Good Harvest Partnership LLP* v *Centaur Services Ltd (Ch),* [2010]** 1 **29**

Landlord covenant

Release on assignment — Lease excluding named landlord's liability following assignment of reversion — Covenant otherwise expressed as binding successors in title — Whether personal covenant — Whether agreement to release in advance of assignment — Whether agreement excluded by s25(1) — Whether landlord remaining liable following assignment — Held, covenant remaining landlord covenant — Agreement an attempt to release liability — Prohibited by s25(1) — Landlord remaining liable
***Avonridge Property Co Ltd* v *Mashru (CA),* [2005]** 1 **15**

Liability of former tenants

Notices served on former tenant demanding payment of rent while rent under review — Further notices demanding additional rent once review completed — Whether notices valid — Whether s17(2) or (4) appropriate provision for recovery of additional rent — Whether former tenant entitled to indemnity from assignee — Held, s17(4) applying where rent liable to be raised on rent review — Landlord obliged to serve s17(2) notices for unreviewed rent even where no default in order to preserve possibility of recovering additional rent under s17(4) once review completed — Former tenant entitled to indemnity from assignee where payments to landlord fairly and reasonably incurred even though no legal liability to pay
***Scottish & Newcastle plc* v *Raguz (No 2) (CA),* [2007]** 1 **47**

Notices served on former tenant demanding payment of rent while rent under review — Further notices demanding additional rent once review completed — Whether notices valid — Whether s17(2) or (4) appropriate provision for recovery of additional rent — Whether former tenant entitled to indemnity from assignee — Held, landlord serving appropriate notices under s17(2) — Revised rent an additional and separate fixed charge becoming "due" upon determination of rent review and recoverable by s17(2) notice
***Scottish & Newcastle plc* v *Raguz (No 2) (HL),* [2008]** 3 **115**

Lease

Escrow

Escrow conditions — Doctrine of relation back — Whether tenant bound by lease from date of delivery of lease in escrow and escrow conditions fulfilled at later date — Held, tenant not bound from date upon which lease delivered by landlord in escrow
***Dyment* v *Boyden (CA),* [2005]** 1 **19**

Yield up and reinstatement obligations

Obligation to restore premises to original condition and to reinstate to specification — Whether obligations mutually exclusive — Whether tenant having choice of obligation — Whether notice to comply with obligations void for uncertainty — Held, obligations not alternatives — Tenant obliged to comply with both save as to inconsistent works — Notice not void
***Fairgate International Ltd* v *Citibank International plc (CA),* [2005]** 2 **48**

Lease of commercial premises

Right of inspection

Former garage with underground fuel tanks — Landlord having right of entry to inspect and examine state of repair and condition of premises — Landlord wanting to drill boreholes to test for environmental contamination from tanks — Whether right of inspection extending to intrusive works for purpose of inspection — Held, power to inspect and examine not confined to superficial visual inspection — Implied power to undertake intrusive work for purposes of inspection or examination
***Possfund Custodian Trustee Ltd* v *Kwik-Fit Properties Ltd (CS:OH),* [2008]** 3 **113**

Lease or licence

Advertising hoarding sites

Grant of right to use advertising hoarding site — Whether grant of licence or lease — Whether occupation satisfying s23 of Landlord and Tenant Act 1954 — Held, agreement granting permission to erect and maintain advertising display constituting licence — Further agreement granting right to define site as tenancy — Occupation under tenancy satisfying s23
***Clear Channel UK Ltd* v *Manchester City Council (Ch),* [2005]** 1 **128**

LANDLORD AND TENANT — *continued*
Lease or licence — *continued*
Advertising hoarding sites — *continued*

Grant of right to use sites for advertising hoardings — Whether grant of licence or lease — Whether occupation satisfying s23 of Landlord and Tenant Act 1954 — Held, agreement relating to undefined area and not granting exclusive possession — Permission to erect and maintain advertising display a licence
***Clear Channel UK Ltd* v *Manchester City Council (CA),* [2006]** 1 **27**

Leasehold Reform, Housing and Urban Development Act 1993
Lease extension
Determination of premium — Premium including element representing rent payable under existing leases — Construction of rent review provisions — Whether provision for payment of market ground rental value requiring payment of nominal ground rent or market rent for lease of land with right to construct the demised premises — Held, nominal ground rent payable — Concept of nominal ground rent not inconsistent with that of market ground rent
***Elmbirch Properties plc* v *Schaefer-Tsoropatzadis (LT),* [2007]** 2 **167**

S42 notice Validity of service of notice claiming new lease — Landlord and Tenant Act 1987 — Whether landlord notifying tenant of address for service of notices under s48 of 1987 Act — Whether tenant entitled to serve notice at address last furnished as landlord's address under s47 — Held, communication from landlord's agent not reasonably understood as notification of address for s48 purposes — Tenant entitled to serve notice at address given under s47
***Glen International Ltd* v *Triplerose Ltd (CA),* [2007]** 2 **81**

Mesne profits
Assessment
Property divided into flats — Agreement to let property to company for subletting to individual tenants — Company retaining possession after expiry of tenancy — Keys to two of seven flats subsequently returned to landlord — Landlord claiming mesne profits — Whether payable by reference to entire property or date on which individual flats returned — Correct method of assessment — Whether occupants of flats individually liable in addition to company — Held, mesne profits assessed by reference to individual flats in property where these independently commercially exploitable units — Measure of mesne profits calculated by reference to open market letting value less any deductions necessary to reflect actual loss of landlord or actual benefit to occupant — Award to represent greater of those two figures — Occupants of flats individually liable for period of occupation notwithstanding liability of company for whole building
***Compatriot Holdings Ltd* v *City Partnership (Ch),* [2009]** 3 **153**

New tenancy
Requisite intention
Lease of business premises — Application for new tenancy — Landlord opposing application on ground of intention to redevelop within s30(1)(*f*) of Landlord and Tenant 1954 Act — Tenant applying for summary judgment dismissing ground of opposition — Whether landlord required to show requisite intention as at date of anticipated trial or date of summary judgment hearing — Held, date of hearing at which necessary intention must be shown to exist always date of substantive trial of landlord's ground of objection — Summary judgment hearing determining whether landlord having real prospect of proving necessary intention by anticipated future trial date
***Somerfield Stores Ltd* v *Spring (Sutton Coldfield) Ltd (in administration) (No 2) (Ch),* [2010]** 3 **37**

Non-payment of rent
Mitigation
Tenant ceasing use of premises and ceasing to pay rent — Claim by landlord for rent arrears — Whether landlord obliged to mitigate loss — Whether landlord acting wholly unreasonably in failing to relet premises — Whether damages adequate remedy — Held, ordinarily not unreasonable for landlord to leave it to tenant to propose assignee or substitute — Damages not adequate remedy
***Reichman* v *Beveridge (CA),* [2007]** 1 **37**

Option to determine
Tenant's break clause
Assignment — S63 of Law of Property Act 1925 — Break clause personal to original tenant but capable of assignment to specified assignees — Assignment of lease without express assignment of break clause — Exercise of break clause — Whether break clause validly exercised — Whether clause "touching and concerning" land — S142 of 1925 Act — Whether clause having reference to subject matter of lease — Whether clause passing under s63 with assignment of lease — Held, break clause validly exercised — Hybrid break clause "touching and concerning land" and having reference to subject matter of a lease under s142 — Break clause accordingly a right for purposes of s63
***Harbour Estates Ltd* v *HSBC Bank plc (Ch),* [2005]** 1 **107**

Tenant's break option
Option notice — Letter purporting to exercise option to break lease containing incorrect date — Whether lease provisions requiring option notice to contain date of termination — Whether letter effective exercise of option to break lease — Held, option clause not requiring notice to contain date of termination — Break clause effectively exercised
***Peer Freeholds Ltd* v *Clean Wash International Ltd (Ch),* [2005]** 1 **47**

Nuisance
Caveat lessee principle
Lease of flat — Water ingress caused by defect in light wells in part of building outside tenant's demise — Defect existing prior to grant of lease — Whether landlord's repairing covenant imposing obligations beyond mere obligation to repair — Whether landlord liable in nuisance — Whether *caveat* lessee principle applying to complaints regarding state and condition of parts of building not included in demise — Held, no liability of landlord beyond mere obligation to repair — No liability in nuisance — *Caveat* lessee principle applying where complaint relating to state and condition of other parts of building in which demised premises located
***Jackson* v *JH Watson Property Investment Ltd (CC),* [2008]** 1 **33**

Possession order
Housing Act 1980
Stay of execution — Housing Act 1980 — Possession order made in county court — High Court granting stay of execution outside time limits in s89 of 1980 Act pending application for permission to appeal — Stay set side in absence of occupants — Stay reinstated on appeal — Whether s89 applying to High Court — Whether appellate court having inherent jurisdiction to stay execution beyond s89 time limits — Held, s89 applying to High Court as well as county court — S89 none the less not displacing inherent jurisdiction of court exercising appellate jurisdiction to grant stay pending appeal
***Admiral Taverns (Cygnet) Ltd* v *Daniel (CA),* [2009]** 1 **35**

Stay of execution
Housing Act 1980 — County court making possession order in favour of landlord — High Court granting stay of execution outside time limits in s89 pending application for permission to appeal — Court setting aside order for stay in absence of defendants — Whether s89 applying to appellate court — Whether appellate court having power to stay execution beyond s89 time limits — Held, s89 not applying to court exercising appellate jurisdiction — Appellate court having power to order stay of execution pending determination of appeal
***Admiral Taverns (Cygnet) Ltd* v *Daniel (QB),* [2008]** 3 **59**

Suspension
Housing Acts 1985 and 1988 — Possession orders suspended on terms — Effect of suspended possession order upon rights of assured and secure tenants — Whether breach of suspended possession order ending assured tenancy — Whether revival of secure tenancy by discharge of possession order reviving right to buy previously established — Whether former secure tenant still able to apply for discharge of suspended possession order where order ceasing to be enforceable upon payment of arrears and costs — Held, assured tenancy not ended by breach of terms of suspension but continuing until possession delivered up — Revival of secure tenancy by discharge of possession order reviving right to buy pursuant to notice previously served — Court having power to make proleptic order for discharge of possession order in event of tenant paying off all arrears and costs — Terms of discharge may be different from terms of suspension — Tenant in breach of terms of suspension still able to apply to court for relief in cases where no proleptic discharge provision or where provision that possession order not to be enforced once all arrears and costs paid
***White* v *Knowsley Housing Trust; Islington London Borough Council* v *Honeygan-Green; Porter* v *Shepherds Bush Housing Association (HL),* [2009]** 1 **131**

Rent
Liability of original tenant
Indemnity by assignee — S24 of Land Registration Act 1925 — Notice served on original tenant by landlord — S17 of Landlord and Tenant (Covenants) Act 1995 — Unresolved rent reviews — Form of s17 notice Whether original tenant liable for rent arrears — Whether indemnifier liable — Whether form 1 with para 4 required to be served in all cases where rent review outstanding — Whether forms 1 and 2 both required to be served — Whether wider arguments precluding indemnity — Held, where rent subject to rent review, form 1 with para 4 must be served whether or not rent in arrears — When review determined, form 2 must be served — Arrears not payable by original tenant where incorrect form 1 and no form 2 served — Wider arguments not precluding recovery of arrears under indemnity
***Scottish & Newcastle plc* v *Raguz (No 2) (Ch),* [2006]** 3 **119**

Set-off
Building agreement — Landlord building factory and granting lease — Assignment of reversion — Claim for arrears of rent — Tenant counterclaiming damages for breach of building agreement — Whether tenant entitled to set off claim for damages against rent — Whether lease excluding equitable set-off — Held, tenant not entitled to set off claim for

LANDLORD AND TENANT — *continued*
Rent — *continued*
Set-off — *continued*

damages against original landlord against rent due after transfer of landlord's interest — Set-off not excluded by terms of instant lease
***Edlington Properties Ltd* v *JH Fenner & Co Ltd (CA),* [2006]** 2 **18**

Turnover rent
VAT — Whether VAT included in turnover for purpose of calculating turnover rent — Whether "gross amount of the total sales" including VAT– Held, turnover rent not including VAT
***Debenhams Retail plc* v *Sun Alliance & London Assurance Co Ltd (Ch),* [2005]** 1 **26**

VAT — Whether VAT included in turnover for purpose of calculating turnover rent — Whether "gross amount of the total sales" including VAT — Held, turnover rent including VAT — Gross amount of total sales including VAT
***Debenhams Retail plc* v *Sun Alliance & London Assurance Co Ltd (CA),* [2005]** 3 **34**

Rent Act 1977
Possession order
Protected tenancy — Tenants occupying flat for 45 years — Landlord seeking possession — S98(1)(*a*) of Rent Act 1977 — Provision of suitable alternative accommodation by landlord— Landlord wanting to maximise realisation of investment by selling with vacant possession — Whether reasonable to make possession order — Correct approach to reasonableness — Held, test whether reasonable to make order not whether reasonable for landlords to seek one — Reasonableness to be determined by considering all angles including effect on both parties if order made and if not made — Emotional and social consequences of move for elderly tenants capable of outweighing inability of landlords to realise vacant possession value
***Whitehouse* v *Lee (CA),* [2009]** 2 **75**

Statutory tenancy
Estoppel — *Tomlin* order — Claim for possession on ground that statutory tenancy terminating on non-occupation as residence — Claim compromised by *Tomlin* order and agreement by landlord to repair premises for tenant's return — Second claim for possession — Whether legal status of tenant with statutory tenancy determined by agreement in *Tomlin* order — Whether essential background facts to the agreement estopping landlord from relying upon evidence that tenant had given up occupation and did not intend to return to premises — Held, landlord estopped by *Tomlin* order agreement and background facts — Tenant's status determined by agreement — Tenant having statutory tenancy and intention to occupy
***The Carphone Warehouse UK Ltd* v *Malekout (CA),* [2006]** 2 **35**

Rent review
Assumptions
Lease of commercial premises — Tenant using site as drive-through hamburger restaurant — Valuation to be on assumption that modern single-storey warehouse built on site — Whether describing type of building only — Whether implying a particular use — Whether planning permission for retail use to be assumed — Held, word "warehouse" describing type of building only — Second-generation standard warehouse unit intended — Planning permission for Class B8 storage and distribution use to be assumed — Prospect of obtaining planning permission for Class A1 retail use to be taken into account
***McDonalds Real Estate LLP* v *Arundel Corporation (Ch),* [2008]** 2 **53**

Delay
Parties acquiring respective interests of tenant and landlord in premises —Previous landlord not carrying out rent review due under terms of lease — New landlord seeking review — Arbitrator determining rent due — Whether landlord entitled to possession for non-payment of rent arrears by tenant — Whether available defence of unreasonable delay in seeking rent review — Whether claim for rent arrears time-barred under s19 of Limitation Act 1980
***Idealview Ltd* v *Bello (QB),* [2010]** 1 **39**

Letting for mixed uses
Premises consisting of shop with residential accommodation above — Tenant ceasing business use of shop — Landlord seeking possession — Whether tenancy attracting protection of 1977 Act on cessation of business use — Whether premises let as a dwelling — Held, premises let for mixed uses not let as a dwelling for purposes of 1977 Act — Tenant unable unilaterally to switch from protection of Landlord and Tenant Act 1954 to that of 1977 Act by cessation of business use — Landlord not assenting to change of user
***Tan* v *Sitkowski (CA),* [2007]** 1 **113**

Review date
Assured tenancy of residential property — Clause providing for rent increase with effect from first Monday in June — Whether time of the essence— Whether on proper construction lease permitting increase from later date — Held, lease not specifying single date from which rent increase must start — Rent increase permitted once in any year on four weeks' notice to take effect on any date on or after first Monday in June
***Riverside Housing Association Ltd* v *White (HL),* [2007]** 2 **69**

Time of essence
Agreement provided for rent increase on specified annual date — Rent increased by notice on non-specified date — Whether time not of the essence of rent increase date — Whether tenants bound by rent increase by variation of agreement, estoppel or waiver — Held, presumption of time not of the essence not applicable to contractual date for rent increase — No variation of agreement — Estoppel not applying to further cause of action — No waiver
***Riverside Housing Association Ltd* v *White (CA),* [2006]** 1 **45**

Trigger notice
Time of the essence — Whether time of the essence for service of trigger notice — Whether trigger notice valid — Held, time not of the essence for service of trigger notice — Trigger notice valid – Counternotice valid
***Lancecrest Ltd* v *Asiwaju (CA),* [2005]** 1 **40**

Triggering review
Lessor having right to rent review — Review upwards/downwards — Whether presumption that tenant can require rent review where review upwards/downwards — Whether landlord required to have review — Tenant's alternative contention in absence of 2003 rent review — Whether rent payable 60% of rent payable in 1982 — Held, no presumption with upwards/downwards reviews that tenant entitled to review — Landlord not obliged to have a review — Previous rent remained payable
***Hemingway Realty Ltd* v *Master Wardens and Commonalty of Freemen of the Art or Mystery of Clothworkers of the City of London (Ch)* [2005]** 2 **36**

Repairs
Breach
Remedies — Damp in basement — Failure of damp-proof membrane — Whether disrepair — Whether landlord having obligation to carry out works to prevent damp from entering basement — Whether breach of express term to use reasonable endeavours to repair — Whether implied term to effect that landlord will remedy any defective part of "structure" causing damage to demised premises that tenant is obliged to maintain — Where landlord having right but no obligation to repair whether landlord must inform tenant whether and if so when and how it intends to carry out works — Whether tenant entitled to withhold any part of rent — Whether tenant's decision to vacate basement reasonable — Whether tenant entitled to specific performance and damages — Held, damp caused by defectively installed damp-proof membrane — No breach of express covenant by landlord to use reasonable endeavours to repair — No implied term imposing repairing liability on landlord — No obligation on landlord to inform tenant of repairs — Tenant not entitled to withhold rent — Tenant's decision to vacate basement reasonable — No remedies for tenant
***Janet Reger International Ltd* v *Tiree Ltd (Ch),* [2006]** 3 **131**

Breach of covenant
Lease of luxury residential property — Landlord's repairing covenants — Tenant withholding rent and claiming damages for breaches of covenant — Whether standard of service and repair to be expected higher owing to nature of property and level of rent charged — Whether landlord obliged to employ professional managing agent — Whether reasonable time for carrying out repairs to be judged stringently in case of high-class property at high rent— Held, standard of service and repair determined by terms of covenants not level of rent — No duty to employ professional managing agent —— Duty of landlord only to remedy disrepair in reasonable time — Reasonable time judged objectively by reference to time an owner-occupier would take
***Langham Estate Management Ltd* v *Hardy (CC),* [2008]** 3 **125**

Dilapidations
Roof — Landlord's claim for replacement of roof — Tenant repairing roof immediately prior to termination of lease — Whether roof in disrepair— Whether replacement of damaged sheets or complete replacement of roof required to put roof in covenanted condition — Fees — Whether landlord entitled to fees — Held, roof capable of repair by replacement of sheets — Roof in covenanted condition at term date — Claim for fees dismissed
***Riverside Property Investments* v *Blackhawk Automotive (TCC),* [2005]** 1 **114**

Main structure
Floor joists — Liability for repair — Whether floor joists "main structures" — Whether landlord or tenant liable to repair floor joists — Meaning of "in particular" — Whether meaning given to "in particular" in *Toff* v *MacDowell* (1995) 69 P&CR 535 correct — Held, floor joists "main structures" — Landlord responsible for repair — Repair costs recoverable in service charge — Construction of "in particular" in *Toff* incorrect
***Marlborough Park Services Ltd* v *Rowe (CA),* [2006]** 2 **27**

Right of inspection
Former garage premises with underground fuel tanks — Clause permitting entry by landlord to inspect and examine condition of premises — Landlord

LANDLORD AND TENANT — *continued*
Repairs — *continued*
Right of inspection — continued

wanting to enter and drill boreholes for purpose of assessing soil contamination — Whether permitted under terms of lease — Whether clause confined to inspections of non-intrusive nature — Held, intrusive investigation of kind proposed by landlord interfering with tenant's possession of premises — Such investigation not permitted in absence of clear words in lease — Inspection not to cause material disturbance to tenant or material damage to premises
***Possfund Custodial Trustee Ltd* v *Kwik-Fit Properties Ltd (CS: IH)*, [2009]** 1 **39**

Roof

Demise — Sublease — Extent of demise — Whether references in provisions relating to decorations and insurance relevant to extent of demise — Meaning of "buildings and structures" — Whether roof part of demise under sublease — Whether sublessee liable to repair roof — Held, references to "buildings and structures" in unrelated provisions no guide to extent of demise — Contribution provision in respect of common areas best guide — Roof not demised — Sublessee not liable for repair
***Delgable Ltd* v *Perinpanathan (CA)*, [2006]** 1 **78**

Residential tenancy
Disrepair

Economic loss — Tenant agreeing to move to different property on terms that landlord carrying out certain works to that property — Long-running dispute over defects in the works — Landlord seeking possession for rent arrears — Whether tenant able to advance counterclaim for damages for economic loss caused by abandonment of business venture owing to time taken up by dispute — Whether landlord under relevant duty of care — Held, works not undertaken for purpose of furthering or assisting defendant's business — Unreasonable for defendant to rely on them for that purpose — Not fair, just or reasonable to impose duty of care on landlord in respect of pure economic loss unrelated to landlord and tenant relationship — Tenant's counterclaim unarguable in law
***Eaton Square Properties Ltd* v *Shaw (QB)*, [2011]** 3 **13**

Right to buy
Housing Act 1985

Grant of long lease — S125 notice — Reference period for estimate of contribution to improvement works — Successive notices giving different reference periods — Whether applicable period in original notice — Held, scheme of Act providing for single s125 notice — Parties free to agree terms differing from notice — Reference period to be determined by construction of lease
***Hyams* v *Wilfred East Housing Co-operative Ltd (LT)*, [2007]** 1 **89**

Relevant valuation date — Claims to exercise right to buy in 1991 and 2001 — Relevance of concepts of abandonment, waiver and estoppel — Whether 1991 or 2001 claim established — Whether relevant valuation date in 2001 — Held, relevant valuation date in 2001
***Copping* v *Surrey County Council (CA)*, [2006]** 1 **42**

Relevant valuation date — Claims to exercise right to buy in 1991 and 2001 — Whether claim to exercise right to buy can be abandoned or impliedly withdrawn — Whether 1991 claim abandoned — Whether relevant valuation date in 2001 — Held, claim to exercise right to buy capable of abandonment or implied withdrawal — 1991 claim abandoned — Relevant valuation date in 2001
***Copping* v *Surrey County Council (QB)*, [2005]** 2 **56**

Tenant making right-to-buy application — Suspended possession order made against tenant — Right to buy ceasing to be exercisable under s121 — Secure tenancy revived upon discharge of rent arrears — Whether right-to-buy application revived — Whether tenant entitled to mandatory injunction under s138(1) requiring landlord to convey long lease — Held, right-to-buy application ceasing to be exercisable where s121 applying — Application not revived along with tenancy — Conditions for grant of injunction unders138 not fulfilled
***Islington London Borough Council* v *Honeygan-Green (QB)*, [2007]** 3 **23**

Tenant making right-to-buy application — Suspended possession order made against tenant — Terms of order breached — Right to buy ceasing to be exercisable under s121 — Secure tenancy revived upon discharge of rent arrears — Whether right-to-buy application revived — Whether tenant entitled to mandatory injunction under s138(1) requiring landlord to convey long lease — Held, right-to-buy application revived when tenancy revived — No outstanding matters relating to grant — Mandatory injunction appropriate
***Islington London Borough Council* v *Honeygan-Green (CA)*, [2008]** 2 **133**

Right to light
Rooftop development

Extent of demise — Landlord obtaining planning permission to build additional storey on roof of terrace of maisonettes — Development to involve blocking skylights in roof thereby depriving parts of maisonettes of natural light — Tenant of one maisonette claiming development a trespass — Whether roof and air space above it forming part of demise of maisonettes — Whether derogation from grant — Whether breach of right to light — Held, demise including roof and reasonable air space above skylights — Deprivation of light from skylights amounting to derogation from grant and breach of right to light acquired by user or implied grant
***Dorrington Belgravia Ltd* v *McGlasham (CC)*, [2009]** 1 **27**

Right to manage
Exclusions

Commonhold and Leasehold Reform Act 2002 — Exclusion of non-residential parts — Lease of part for business and ancillary residential accommodation — Whether part occupied or intended to be occupied for residential purposes — Held, part of premises occupied and intended to be occupied for residential purposes — Right to manage not excluded
***Gaingold Ltd* v *WHRA RTM Co Ltd (LT)*, [2006]** 1 **81**

Secure tenancy
Possession order

S85 of Housing Act 1985 — CPR 3.1(2)(a) — Arrears of rent — Possession order suspended on conditions as to regular payment of arrears — Failure to comply with terms — Landlords obtaining warrant for possession— Subsequent payment of entire arrears — Whether tenant entitled to have order discharged or rescinded notwithstanding breach of conditions — Whether CPR 3.1(2)(a) permitting retrospective amendment of conditions — Held, no entitlement to discharge or rescission of order where conditions breached — CPR 3.1(2)(a) not to be used to rewrite valid order where no change of circumstances
***Porter* v *Shepherds Bush Housing Association (CA)*, [2008]** 2 **125**

Succession

Joint tenant becoming sole tenant by right of survivorship — Tenancy subsequently becoming secure tenancy after Housing Act 1980 coming into force — Ss87 and 88(1)(*b*) of Housing Act 1985 — Whether son of tenant entitled to succeed to secure tenancy — Whether tenant herself a successor so as to exclude that right — Held, s88(1) not referring to events occurring before 1980 coming into force — "Successor" referring to successor under secure tenancy — Son entitled to succeed to tenancy
***Birmingham City Council* v *Walker (HL)*, [2007]** 2 **58**

Tolerated trespasser

Suspended possession order — Rent arrears cleared — No offer of new tenancy — No application under s85 of Housing Act 1985 — Whether secure tenancy revived — Whether grant of new tenancy — Whether members of family entitled to secure tenancy by succession — Held, tolerated trespasser status continuing — No grant of new secure tenancy — Agreement that tolerated trespasser status only continuing — No rights of succession
***Newham London Borough Council* v *Hawkins (CA)*, [2005]** 2 **51**

Service charges
Building contracts

Lease permitting recovery of service charges in respect of actual cost incurred by landlord in relevant year — Works done under building contracts providing for payment of contractor by landlord after certification of value of works by architect — Whether landlord entitled to recover costs certified or paid after end of relevant year — Whether errors in calculation invalidating service charge demands — Held, no entitlement to sums certified outside relevant year — "Actual cost" including sums certified and falling due for payment within relevant year even where not yet paid — Errors not invalidating service charge demands — Lands Tribunal empowered to amend demands under s27A of Landlord and Tenant Act 1985
***Barrington* v *Sloane Properties Ltd (LT)*, [2007]** 3 **91**

Caretaker's flat

Notional rent — Lease entitling landlord to recover annual sum equivalent to market rent of accommodation provided rent-free for caretaker — Whether landlord entitled to recover notional rent for caretaker's flat — Held, landlord entitled to recover notional rent for caretaker's flat
***Earl Cadogan* v *27/29 Sloane Gardens Ltd (LT)*, [2006]** 2 **89**

Demand

Landlord and Tenant Act 1985 — Service Charges (Summary of Rights and Obligations, and Transitional Provision) (England)Regulations 2007 — Landlord bringing county court proceedings against tenants for unpaid service charges — Section 21B of 1985 Act — Whether service charge demands complying with section 21B served before county court proceedings commenced — Whether demands "accompanied" by summary of tenant's right and obligations prescribed by regulation 7 of 2007 Regulations — Held, summary sent after relevant demand not "accompanying" demand and insufficient to validate it — Enclosure of copy of 2007 Regulations insufficient to cure absence of required heading from summary — No valid demand served before county court proceedings begun — Those proceedings premature
***Tingdene Holiday Parks Ltd* v *Cox (UT:LC)*, [2011]** 3 **30**

void as disposal of acquirable common parts under s19(1)(ii)and s2(1)(b) and (3) — Whether interest under lease liable to acquisition on behalf of participating tenants under s19(2) and (3) — Held, no severance of reversion where long lease granted rather than freehold interest — Insurance rent and service charges under lease amounting to "rent" such that lease not capable of enlargement into freehold under s153 of Law of Property Act 1925 — Lease void under s19(1)(a)(ii) since caretaker's flat a "common part" acquirable on behalf of participating tenants — Light well a common part but not necessary to acquire it on behalf of participating tenants — Grant of lease not a disposal of freeholder's interest within s19(2) and (3) so as to render interest acquirable on behalf of participating tenants
***Panagopoulos* v *Earl Cadogan (Ch),* [2010]** 1 & 2 **151**

Application to acquire part of building containing 20 flats — That part divisible into two sets of 10 flats — Whether property self-contained part of a building within s3 — Whether that term denoting a part incapable of further subdivision into smaller self-contained parts — Held, property qualifying for enfranchisement where comprising self-contained part of building and none of statutory exclusions in s4 applying — S3 imposing no further requirement that property indivisible into smaller self-contained parts
***41-60 Albert Palace Mansions (Freehold) Ltd* v *Craftrule Ltd (Ch),* [2010]** 2 **45**

Charitable Housing Trust — Whether flats on long leases held for charitable purposes — Whether tenants of charitable housing trust entitled to exercise rights of collective enfranchisement — Held, only flats let as social housing held for charitable purposes — Tenants holding long leases entitled to seek collective enfranchisement
***Brick Farm Management Ltd* v *Richmond Housing Partnership Ltd (QB),* [2005]** 3 **57**

Counternotice — Counternotice held invalid in earlier proceedings — Landlord ordered to serve further counternotice admitting claim within time limit — Second counternotice not admitting entitlement to enfranchise— Whether landlord able to serve second counternotice denying entitlement — Whether nominee purchaser entitled to order under s25 that premises be acquired on terms of initial notice — Whether landlord entitled to order varying time limit to serve third counternotice admitting entitlement — Held, where court determining issue as to entitlement of tenant to collective enfranchisement, landlord cannot serve further counternotice denying entitlement — Nominee purchaser entitled to acquire on terms of initial notice — No jurisdiction to vary time limit to permit landlord to serve third counternotice admitting entitlement
***Brick Farm Management Ltd* v *Richmond Housing Partnership Ltd (No 2) (Ch),* [2006]** 2 **46**

Deferment rates — Extended lease — Whether appropriate to apply same percentage for capitalisation rate as for deferment rate — Whether deferment rate for central London properties applicable to properties elsewhere — Held, no valuation rationale for applying same capitalisation rate — Same deferment rate applied as for central London properties
***Nicholson* v *Goff (LT),* [2007]** 1 **83**

Determination of premium — Non-participating tenants — Whether hope value to be attributed to flats of non-participating tenants — Relativity in relation to flats of participating tenants — Hope value of caretaker's flat — Marriage value — Section 42 notice served by participating tenant prior to collective enfranchisement claim — Whether notice to be taken into account for purpose of calculating marriage value — Held, small allowance for hope value of one participating flat where existing lease having short unexpired term — Relativity of participating tenants' existing leases calculated by percentage deduction of 25% from market value to allow for benefit of 1993 Act rights — No hope value attributed to caretaker's flat where no evidence of willingness of all relevant leaseholders to release covenant requiring provision of residential caretaker — Flat in respect of which section 42 notice served excluded from marriage value calculation since no participating tenant immediately before binding contract entered into in pursuance of section 13 notice
***Earl Cadogan v Cadogan Square Ltd; Cadogan Square Ltd v Earl Cadogan (UT:LC),* [2011]** 3 **127**

Flats — Calculation of premium — Deferment rate — Whether uplift from generic *Sportelli* rate appropriate where unexpired lease terms less than 20 years and property market near top of cycle at valuation date — Whether deferment rate for houses not flats to be applied where prospect of reconverting to single dwelling on expiry of leases — Held, adjustment to *Sportelli* deferment rates may be appropriate having regard to property cycle in cases where lease having unexpired term of less than 20 years — Where property market at or near top of cycle real growth rate may be reduced to reflect lower growth rate in short-term compared with long-term trend — This resulting in overall uplift to deferment rate — Uplift determined by reference to result that would be arrived at in negotiations between hypothetical parties — Prospect of reconversion to single dwelling-house at term not justifying application of deferment rate for houses
***Cadogan Square Properties Ltd* v *Earl Cadogan (UT:LC),* [2011]** 1 **155**

Initial notice — Four out of seven qualifying tenants seeking to enfranchise — S13 initial notice signed by three individual tenants and director of fourth corporate tenant — Whether notice validly signed "by" corporate tenant within s99(5) of 1993 Act — Whether signature of two directors or one director and company secretary required under s36A of Companies Act 1985 — Whether notice invalid on ground that not given by majority of qualifying tenants — Company signing notice under s13 of 1993 Act required to provide signature of two directors or director and company secretary or affix common seal under s36 of 1985 Act — Notice not validly signed by corporate tenant where signature of only one director provided — Notice invalid since not signed by majority of qualifying tenants
***Hilmi & Associates Ltd* v *20 Pembridge Villas Freehold Ltd (CA),* [2010]** 2 **41**

Initial notice served and registered on behalf of qualifying tenants under s13 of 1993 Act — Landlord thereafter granting long lease of basement flat to nominee — Whether lease void under anti-avoidance provisions as disposal of acquirable common parts within ss19(1)(ii) and 2(1)(b) — Held, common parts comprising those areas of building available for shared use or benefit — No requirement that area be accessible by tenants to constitute common part — Caretaker's flat amounting to common part where tenants sharing services of caretaker for purpose of which flat provided— Caretaker's flat a common part which necessary to acquire on enfranchisement for its proper management or maintenance
***Panagopoulos* v *Earl Cadogan (CA),* [2011]** 1 **33**

Internal floor area — Common parts — Building containing both residential and commercial units — Residential lessees serving notice to acquire freehold — S4(1) of 1993 Act — Landlord disputing right to acquire freehold on ground that more than 25% of the internal floor area neither residential nor comprised in the common parts — Whether common parts including those for use of commercial occupants only — Held, all common parts qualifying as such regardless of whether used by residential lessees — Internal floor area including balconies let with flats
***Marine Court (St Leonards on Sea) Freeholders Ltd* v *Rother District Investments Ltd (CC),* [2008]** 1 **39**

Landlord's counternotice — Validity — Failure to state if premises within area of estate-management scheme — Requirements of landlord's counternotice — Requirements of s21 — Whether additional requirements of reg 4 of Leasehold Reform (Collective Enfranchisement)(Counternotices) (England)Regulations 2002 — Whether reg 4 requirements mandatory or directory — Whether landlord's counternotice valid — Held, requirement in reg 4 to serve negative statement specifying premises not within EMS area directory — Landlord's counternotice valid
***7 Strathray Gardens Ltd* v *Pointstar Shipping & Finance Ltd (CA),* [2005]** 1 **53**

Premium payable by nominee purchaser to reversioners — Deferment rate — Upper Tribunal (UT) giving guidance that deferment rate for flats with reversions of less than five years to be appropriate net rental yield for the property with end allowance of 5% — Whether guidance vitiated by failure to consider possibility of tenants having right to remain at end of lease term by virtue of Sched 10 to Local Government and Housing Act 1989 — Whether hope value properly awarded in respect of non-participating flats with reversions of less than 80 years — Held, decision of UT not vitiated by failure to consider 1989 Act where parties accepting UT's approach at the time — UT entitled to give guidance on deferment rate for reversions of less than five years — Guidance open to qualification in case where party raising issue of possible applications of 1989 Act at end of lease term — Hope value of non-participating flats properly taken into account so far as attributable to possibility of tenants seeking new leases by negotiation rather than as of right under 1993 Act
***Carey-Morgan* v *Trustees of Sloane Stanley Estate (CA),* [2012]** 3 **38**

Purchase price — Building containing flats — Deferment rate— Marriage value — Paras 4 and 3(1) of Sched 6 to 1993 Act — Whether participating tenants' existing interests to be valued on assumption that no 1993 Act rights applying for purpose of calculating marriage value — Held, package comprising participating tenants' interests and freehold interest to be valued both before and after enfranchisement to obtain marriage value — Each of two elements in package to be valued on consistent basis — This requiring that existing interests of participating tenants to be valued on same "no Act rights" assumption as required to be applied to existing freehold interest under paras 4 and 3(1)
***McHale* v *Earl Cadogan (CA),* [2011]** 1 **36**

Purchase price — Deferment rate — Flats — Enfranchisement of block of flats in Hove — Whether uplift to generic 5% *Sportelli* rate for flats justified in light of greater obsolescence and management risks and lower capital growth rates than in prime central London (PCL) area — Held, 0.25% addition justified for obsolescence to reflect fact that property values lower

than in PCL while building costs comparable — Addition of 0.5% justified to reflect lower capital growth rates in Brighton and Hove area than in PCL — Further addition of 0.25% justified to reflect greater management problems of flats over houses — Deferment rate of 6% applied
***Ashdown Hove Ltd* v *Remstar Properties Ltd (LVT),* [2010]** 3 **61**

Purchase price — Deferment rate — Hope value — Enfranchisement of 1930s block of flats by 50% of qualifying tenants — Leasehold valuation tribunal applying 5% *Sportelli* deferment rate and making no allowance for hope value — Whether higher deferment rate justified for increased risk of obsolescence and variation in growth rates in non-prime areas — Held, appropriate percentage of hope value to be awarded for non-participating flats — No evidence that non-prime property prices more volatile than in prime areas so as to justify different deferment rate — Obsolescence concerned with risk of deterioration to building not increase in value as site for redevelopment — Property having no features increasing risk of obsolescence or differentiating from generality of properties in that regard — Hope value for non-participating flats assessed at 10% of marriage value where 50% of flats not participating and leases having 65 years to run
***Culley* v *Daejan Properties Ltd (Ch),* [2009]** 3 **165**

Qualifying tenants — Lessee converting property into four flats and assigning two flats each to two separate transferees with joint transfer of common parts — S5(5) of 1993 Act — Whether each transferee a qualifying tenant for purposes of Pt I of Ch1 — Whether transferees entitled to exercise right of collective enfranchisement — Held, assignment of part only of property comprised in demise resulting in severance of leasehold estate and apportionment of obligations of tenant accordingly — Each transferee becoming tenant only of two flats assigned to him — Neither excluded from being qualifying tenant by s5(5) — Both qualifying tenants entitled in principle to exercise right of collective enfranchisement
***Smith* v *Jafton Properties Ltd (CA),* [2012]** 1 **77**

Right to amend initial notice — Whether claim including property that no right to acquire — Whether self-contained part of a building — Provision of "relevant services" — Held, purposive approach to amendment provisions in para 15(2) of Sched 3 to Act not permissible — Amendment allowed since claim including property that claimant not entitled to acquire — Specified premises not self-contained part of building where not possible to provide relevant services independently by reconfiguring existing services
***Oakwood Court (Holland Park) Ltd* v *Daejan Properties Ltd (CLCC),* [2007]** 1 **121**

Tenant's notice proposing purchase price of £210 — Landlord's counternotice proposing £130,000 — Whether counternotice valid — Whether *Cadogan* v *Morris* test applying to landlord's counternotice — Whether *Cadogan* v *Morris* test of unrealistic price invoking objective or subjective condition — Held, counternotice valid — *Cadogan* v *Morris* test not applying to landlord's counternotice — *Cadogan* v *Morris* test invoking subjective condition — Landlord entitled to reply upon price proposed in good faith
***9 Cornwall Crescent London Ltd* v *Kensington and Chelsea Royal London Borough Council (CA),* [2005]** 2 **131**

Terms of acquisition — S1(4)(a) of 1993 Act — Land proposed to be retained by freeholder — Whether rights offered over retained land sufficient to comply with s1(4)(a) — Test to be applied — Enfranchisement premium — Deferment rate — Held, rights offered to be considered in light of freeholder's proposals as a whole — Offered rights insufficient where effectively taken away by rights retained under s21(3)(d) — Deferment rate of 7% applied
***Ulterra Ltd* v *Glenbarr (RTE) Co Ltd (LT),* [2008]** 1 **103**

Vesting order — S24 of 1993 Act — LVT determining price payable for freehold — Terms of transfer subsequently agreed — Nominee purchaser applying for vesting order — Whether application made out of time — Whether time running from date of LVT decision or subsequent agreement of terms of transfer — Held, LVT having power to issue more than one decision dealing with issues in dispute in stages — Terms of transfer not taken to be determined at time of determination of price — Time running from date of agreement of terms of transfer — Application for vesting order made in time
***Goldeagle Properties Ltd* v *Thornbury Court Ltd (CA),* [2008]** 3 **69**

Vesting order — Time limit — Terms of acquisition agreed between tenants' nominee purchaser and freeholder save for nature of title guarantee — Nominee purchaser finally agreeing to Sched 7 limited title guarantee offered by freeholder — Application for vesting order under s24 — Whether application made in time — Whether all terms of acquisition agreed for that purpose on date positively agreed or date nominee purchaser having no further prospect of resisting term proposed by freeholder — Held, Act requiring terms to be positively agreed — Implicit agreement by silence insufficient— Dispute capable of arising even if result a foregone conclusion— Fact that outcome of dispute inevitable not justifying conclusion that no dispute existing and term agreed — Application for vesting order made in time where made within relevant period after date all terms positively agreed
***Pledream Properties Ltd* v *5 Felix Avenue London Ltd (Ch),* [2011]** 1 **42**

Counternotice

Initial notice of collective enfranchisement served on then freeholder — Purchaser of freehold reversion serving counternotice before purchase registered — Tenants applying to acquire freehold on terms of initial notice — Effect of s19 of 1993 Act — Whether valid counternotice served — Whether validated retrospectively upon registration of purchaser's title — Whether expert evidence relevant to bona fides of price proposal in initial notice — Held, no valid counternotice — S19 not retrospectively validating acts of purchaser done without legal right — Expert valuation evidence not probative of issue of good faith in relation to tenants' proposals
***Renshaw and others* v *Magnet Properties South East LLP (CC),* [2008]** 1 **42**

Determination of price

Applications for enfranchisement or extension of leases — Appropriate deferment rate — Whether hope value to be excluded from valuation — Method of valuation for hope value where included — Held, setting of general deferment rate appropriate — Deferment rate to be calculated as risk-free rate minus housing market growth rate plus risk premium — Deferment rate 4.5% for houses and 4.75% for flats in prime central London area — Departure from generic deferment rate justified only if compelling evidence indicating necessary owing to particular features of individual case — Hope value to be excluded from valuation on collective enfranchisement claim
***Earl Cadogan v Sportelli (LT),* [2007]** 1 **153**

Collective enfranchisement — Block of flats — Purchase price — Deferment rate — Whether departure from general rate of 5% for flats justified — Location — Obsolescence — Management risks — Valuation date — Whether date of landlord's counternotice or date of decision of leasehold valuation tribunal — Held, general deferment rate for flats applicable — No adjustment for location — No adjustment for obsolescence where fully reflected in vacant possession value and risk premium — No difference between flats managed by head lessee and those managed directly by freeholder — Valuation date that of counternotice
***Hildron Finance Ltd* v *Greenhill Hampstead Ltd (LT),] 2008]** 1 **179**

Deferment rate — Evidence — Whether freeholder should be excluded from adducing evidence challenging generic deferment rate laid down in *Earl Cadogan* v *Sportelli* — Held, leave to amend required for freeholder to contend for deferment rate outside range contended for in pleadings on appeal — Evidence directed to challenging guidance of Lands Tribunal of kind given in *Sportelli* to be excluded save in exceptional circumstances
***Earl Cadogan* v *Erkman (LT),* [2009]** 1 **87**

Deferment rates — Absence of market evidence on deferment rates — Conventional 6% — Settlement evidence — Financial markets — Property-specific factors — RICS *Red Book* valuation manual — Relevance of *Red Book* guidance to statutory valuations — Held, deferment rates determined from evidence — Settlement evidence unreliable without agreed analysis — In absence of market sales evidence, financial market evidence of index-linked gilts acceptable — 4½% appropriate deferment rate on Cadogan Estate — No adjustments between individual flats and collective enfranchisement — One-quarter per cent adjustment between houses and flats on Cadogan Estate — Regard to be had to statutory valuation definitions as interpreted by case law and not to *Red Book*
***Arbib* v *Earl Cadogan; Earl Cadogan* v *55/57 Cadogan Square Freehold Ltd; Earl Cadogan* v *Moussaieff; Day* v *32 Rosary Gardens (Freehold) Ltd; In re Earl Cadogan's Appeal (LT),* [2005]** 3 **139**

Hope value — Deferment rate — Whether hope value to be excluded from purchase price — Whether general deferment rate laid down by Lands Tribunal to be applied — Held, hope value to be excluded from lease extension and collective enfranchisement claims under 1993 Act — No error in Lands Tribunal's approach to deferment rate
***Earl Cadogan* v *Sportelli; Earl Cadogan* v *27-29 Sloane Gardens Ltd; Earl Cadogan* v *Grandeden Property Management Ltd; Howard de Walden Estates Ltd* v *Maybury Court Freehold Co Ltd (CA),* [2008]** 1 **137**

Head lessees

Collective enfranchisement — Whether landlord's leaseback proposals required to be specified in s21 counternotice —Whether possible to serve leaseback notice at any time up to acquisition by nominee purchaser — Effect of definition of "appropriate time" in Sched 9 — Held, not possible to require leaseback unless specified in counternotice — Definition of "appropriate time" imposing condition subsequent preventing leaseback specified in counternotice if flat having qualifying tenant by date of acquisition by nominee purchaser
***Cawthorne* v *Hamdan (CA),* [2007]** 1 **67**

Head lessee — Chapter II of 1993 Act — Buildings incorporating flats and common parts — Whether head lessee entitled to individual lease extension

LEASEHOLD ENFRANCHISEMENT — *continued*
Leasehold Reform, Housing and Urban Development Act 1993 — *continued*
Head lessees — continued

in respect of flat within the building — Held, lessee under lease including flat together with other premises qualifying as "tenant of a flat" for purposes of Chapter II — Head lessee of such premises entitled to exercise right of individual lease extension
***Howard de Walden Estates Ltd* v *Aggio; Earl Cadogan* v *26 Cadogan Square Ltd (HL),* [2008]** 2 **57**

Lease extension

Deceased landlord — Landlord leaving property to son in will and naming him as one of her executors — No application for grant of probate — Tenant of property seeking to exercise right to new lease under Chapter II of Leasehold Reform, Housing and Urban Development Act 1993 — Whether identity of landlord unknown in absence of grant of probate — Whether tenant entitled to grant of new lease by vesting order under s50 — Whether required to serve s42 notice — Whether method of service under s18(1) of Law of Property (Miscellaneous Provisions) Act 1994 available to tenant of deceased landlord for service of s42 notice — Held, death of landlord not rendering successor in title unidentifiable or unable to be found within s5(1)(b) of 1993 Act if personal representatives capable of being identified and found — Real property left by will vesting in executor on death and not dependent on grant of probate — Deceased's will sufficient evidence of son's title as landlord — Tenant to serve s42 notice — Service on executors in usual way or on personal representatives under s18(1) of 1994 Act
***Power* v *Stanton (CC),* [2010]** 3 **71**

Landlord opposing grant of new lease on ground of intention to redevelop premises — Proposal to combine appellant's flat and flat below to create duplex — S47(2) of 1993 Act — Meaning of "premises" — Whether apt to describe part of building consisting of two vertically adjacent flats — Whether "premises" required to constitute recognisable unit within building — Held, "premises" referring to objectively recognisable physical space rather than any part of building identified by landlord — Two vertically adjacent flats in larger building not "premises"
***Majorstake Ltd* v *Curtis (HL),* [2008]** 1 **44**

Lease of building — Lessee giving notice to freeholder to acquire new lease of one flat in building under s42 of 1993 Act — Contractual term of lease expiring — Subsequent claim to acquire freehold of entire building under s8 of Leasehold Reform Act 1967 — Whether lessee tenant of entire property so as to qualify under 1967 Act — Whether para 5(1) of 1993 Act continuing lease of entire building or only of flat to which s42 claim relating — Held, para 5(1) continuing lease only in respect of flat to which s42 claim relating — Lessee no longer tenant of entire property
***Ackerman* v *Lay (CA),* [2009]** 1 **50**

Leasehold enfranchisement — Acquisition of freehold — Premium — Deferment rate — Whether departure from generic rate in *Earl Cadogan* v *Sportelli* justified for 88-year reversion outside prime central London area — Whether additional value to be attributed in respect of potential for development of undeveloped land — Held, departure from generic deferment rate appropriate where adduced evidence showing increase in risk factors of volatility, illiquidity, deterioration or obsolescence — Decisions of other LVTs of little evidential value on deferment rate — Insufficient evidence for higher deferment rate on facts — Evidence not supporting additional development value for undeveloped land
***Sherwood Hall (East End Road) Management Co Ltd* v *Magnolia Tree Ltd (LC),* [2010]** 1 **181**

Lessee serving s42 notice on freeholder claiming lease extension — Notice served more than two years after lessee completing purchase of flat but less than two years after registration of title — S39(2) of 1993 Act — Whether a qualifying tenant for required period of two years prior to service of notice — Whether s39(2) concerning only period during which lessee having registered legal title — Held, only legal owners of leasehold entitled to enfranchise — Requirement to be qualifying tenant for two years before relevant date requiring registration with legal title throughout that period — No right to enfranchise where only equitable title held for part of that period
***The Wellcome Trust* v *Baulackey (CC),* [2010]** 1 **125**

Premium — Deferment rate — Obsolescence — Growth rates in capital value of residential properties — Additional management burden associated with flats — Whether evidence supporting departure from 5% rate in *Earl Cadogan* v *Sportelli* — Held, increase in risk premium justified to reflect greater risk of deterioration in lower-value properties outside PCL area — Further increase appropriate to reflect slower long-term growth rates for residential properties in West Midlands — Additional management allowance for flats owing to greater burden imposed on landlords by Service Charge (Consultation Requirements)(England) Regulations 2003 — Overall deferment rate increased to 6%
***Zuckerman* v *Trustees of the Calthorpe Estate (LC),* [2010]** 1 **187**

Valuation — Relativity — Property in Midlands — No sufficient evidence of comparable transactions — Leasehold valuation tribunal adopting relativity percentage derived from graph of previous Midlands LVT decisions — Whether alternative graph providing better evidence of relativity — Held, inappropriate to rely on graph based on small sample of Midlands decisions — No evidence of particular regional variations in relativity percentages other than in prime central London — Graph based solely on settlement evidence rejected — Best evidence derived from composite "LEASE graph" of relativity compiled by Leasehold Advisory Service from LVT decisions between 1994 and 2007
***Re Coolrace Ltd's appeal (UT:LC),* [2012]** 2 **69**

Leaseback

Lease extension — Buildings incorporating flats and common parts — S56(1) — Whether head lessees entitled to individual lease extension in respect of flat within building — Paras 2 and 3 of Sched 3 to Leasehold (Collective Enfranchisement and Lease Renewal) Regulations 1993 — Whether courts should order payment of deposit where landlord disputing tenant's right to lease — Held, right of individual lease extension not exercisable by head lessees — *Maurice* v *Hollow-Ware Products Ltd* [2005] 2 EGLR 71 overruled — Until tenant's notice held invalid courts should exercise discretion to order payment of deposit unless mitigating circumstances — Decision of High Court still binding upon county courts where former exercising same first instance jurisdiction as conferred on latter by Leasehold Reform Acts
***Howard de Walden Estates Ltd v Aggio; Earl Cadogan v 26 Cadogan Square Ltd (CA),* [2007]** 3 **141**

Leaseback notice

Vesting order — Appeal to Lands Tribunal — Landlord requesting lease-back of flat — Request made after determination of price by LVT and before hearing of appeal by Lands Tribunal — Whether landlord entitled to make request for lease-back — Whether Lands Tribunal having jurisdiction to determine landlord's request — Held, Lands Tribunal having jurisdiction so to do — Held, Lands Tribunal having jurisdiction to decide validity of lease-back notice — Lease-back notice too late and invalid
***Cawthorne* v *Hamdan (LT),* [2006]** 3 **183**

Leasehold valuation tribunal

Application to LVT to determine premium and other terms — Hearing adjourned on terms — LVT dismissing application as abuse of process — Whether LVT having jurisdiction to strike out applications as abuse — Whether LVT properly exercising striking-out power — Whether appellant entitled to appeal — Whether appellant "appearing" before LVT — Held, LVT having jurisdiction to strike out applications as abuse of process — LVT improperly using power by failing to give opportunity of being heard — Appellant "appearing" before LVT by making application for adjournment in writing and entitled to appeal
***De Campomar* v *Trustees of the Pettiward Estate (LT),* [2005]** 1 **83**

Notices — Tenants' original notice not complying with s13(3) of 1993 Act — Whether tenants prohibited by s13(8) and (9) from serving further notice with 12 months of original notice — Whether estopped from challenging efficacy of own notice — Held, original notice of no legal effect — No withdrawal or deemed withdrawal of notice so prohibition on further notices not applying — Tenants not estopped from relying upon invalidity of original notice
***Sinclair Gardens Investments (Kensington) Ltd v Poets Chase Freehold Co Ltd (Ch),* [2007]** 3 **29**

Qualifying tenant of flat

Tenant of block of 28 flats requesting new extended lease of each flat — Effect of request resulting in new leases of each flat but old lease remaining in respect of other parts of building — Whether tenant qualifying tenant of each flat — Whether 1993 Act containing mechanism to address apportionement — Whether common law apportionment rules applying — Held, s56(1)(a) having clear application — Tenant entitled to leases of individual flats — Common apportionment rules applying to address apportionment problems
***Maurice* v *Hollow-Ware Products Ltd (Ch),* [2005]** 2 **71**

S42 notice

New extended lease claim — S42 notice — Landlord intending to redevelop flat with further flat on floor below — Objection to new extended lease — Whether two contiguous flats on different floors "any premises" within meaning of s47(2)(b) of 1993 Act — Held, "any premises" can include two contiguous flats on different floors — "Premises" will also include flat and adjoining box-room
***Majorstake Ltd* v *Curtis (CA),* [2006]** 3 **50**

Jurisdiction of leasehold valuation tribunal
Extent of land to be acquired

Initial notice claiming additional communal garden land — Counternotice admitting claim but disputing right to acquire all additional land — Whether county court having exclusive jurisdiction to determine dispute as to extent of land — Whether LVT having jurisdiction — Held, s21 of 1993Act bifurcating disputes between county court and LVT — County court

NEGLIGENCE — *continued*
Solicitor — *continued*
Assessment of damages — continued

— Whether chance that under correct s25 notice no statutory compensation payable — Assessment of chance — Whether trial judge failing properly to assess evidence relating to speculative chance — Whether conclusions on hypothetical facts conclusions on facts — Held, trial judge failing to reach proper conclusions on evidence of speculative chance — Damages increased
***Talisman Property Co (UK) Ltd* v *Norton Rose (CA),* [2006]** 3 **59**

Notice under s25 of Landlord and Tenant Act 1954 wrongly relying upon ground (*f*) — Admitted liability for incorrect notice — Liability to pay statutory compensation — Identity of tenant — Whether party in occupation found to be tenant — Whether chance that under correct s25 notice no statutory compensation would have been payable — Assessment of chance — Held, by serving incorrect notice on occupier, landlord missing opportunity to argue that occupying party was the tenant and no compensation would then have been payable — Assessment of that chance at 30% — Damages assessed at 30% of statutory disturbance compensation
***Talisman Property Co (UK) Ltd* v *Norton Rose (Ch),* [2006]** 1 **145**

Limitation period

Cause of action — Negligent advice alleged in respect of entry into building agreement and subsequent loss of rights by waiver of certain contractual provisions — Date upon which cause of action accruing — Whether damage resulting at later date than entry into agreement or waiver of provisions — Whether loss contingent — Whether claims time-barred — Held, damage consisting of loss of chance of renegotiating agreement occurring at date agreement entered into — Damage immediate not contingent where benefit of contractual provision lost — Claims time-barred
***Watkins* v *Jones Maidment Wilson (CA),* [2008]** 1 **149**

Limitation Act 1980 — Party purporting to be owner of property instructing solicitor to act on transfer of it — Transfer fraudulent — Actual owner alleging breach of duty of care by solicitor in failing to verify identity of client — Whether claim statute-barred under section 2 of 1980 Act — Whether cause of action accruing on completion of fraudulent transfer or on later registration of title by fraudulent transferee — Whether solicitor owing continuing duty of care after completion — Held, cause of action accruing on completion of fraudulent transfer — Forged transfer exposing any purchaser of property to risk of litigation and therefore reducing price willing to pay — Consequent diminution in open market value of property at date of completion of forger transfer — This constituting actual loss to owner — No special facts creating continuing duty of care after completion of transfer for which solicitor retained
***Nouri* v *Marvi (CA),* [2010]** 3 **79**

S14A(5) of Limitation Act 1980 — Three-year limitation period — Knowledge required for bringing action — Leases containing options to determine personal to original tenant — Assignments to subsidiary company — Negligent failure to advise consequence of assignments — Options ceasing to have effect — Whether claimant entitled to rely upon s14A(5) three-year limitation period — Whether appellants having requisite knowledge for bringing action — Whether appellants having knowledge of material facts — Whether claimant knowing enough to investigate further — Held, claimant having sufficient knowledge to investigate further with view to instigating proceedings before writ — S14A(5)three-year period having expired
***3M United Kingdom plc* v *Linklaters & Paines (CA),* [2006]** 2 **53**

Surveyors
Damages

Supervision of building contract — Extent of duties of supervising building surveyors/architects — Defects — Alternative remedial schemes — Measure of damages — Defective Premises Act 1972 — Whether building surveyors negligent — Whether any claim under 1972 Act — Whether alternative measure of damages to reinstatement costs — Determination of alternative remedial schemes — Measure of damages — Held, building surveyors/ architects negligent — No claim under 1972 Act — Cost of repair appropriate measure of damages — Less expensive remediation scheme appropriate — Damages awarded
***Catlin Estates Ltd* v *Carter Jonas (a firm) (TCC),* [2006]** 2 **139**

Surveyor negotiating grant of leases on behalf of landowners — Surveyor failing to negotiate turnover rent — Whether breach of duty by surveyor — Whether damages to be assessed on valuation basis or loss of earnings basis — Date at which loss to be assessed — Held, surveyor in breach of duty — Chance lost to obtain turnover rent — Damages to be assessed on valuation basis by reference to comparative value of landlord's interest under lease actually granted and that which should have been obtained — Valuation by reference to values as at date of transaction
***Earl of Malmesbury* v *Strutt & Parker (QB),* [2007]** 3 **153**

Valuation

Causation — Appropriation of debt payment — Surveyor valuing property for purpose of further advance by lender to borrower — Borrower defaulting and property sold — Proceeds of sale insufficient to pay both original loan and further advance — Whether lender appropriating sale proceeds to repay further advance — Whether lender having cause of action against surveyor for negligence — Held, proceeds of sale appropriated to further advance by automatic payment into that account by lender's computer system — Such appropriation communicated to surveyor in schedules served in connection with claim — Loss on main account resulting from appropriation by lender and not from any duty of care by surveyor in relation to further advance for which valuation provided — Surveyor not liable for loss on main loan account
***Capital Home Loans Ltd* v *Countrywide Surveyors Ltd (CC),* [2011]** 3 **153**

Duty of care — Surveyor valuing property for mortgagee in relation to buy-to-let transaction — Purchaser buying property in reliance on capital valuation and estimated rental figure in surveyor's report — Whether surveyor instructed by mortgagee owing duty of care to buy-to-let purchaser in relation to valuation — Whether liable in damages for negligent capital and rental valuation — Held, valuer instructed by mortgagor owing no duty of care to purchaser of residential property as buy-to-let investment — Not foreseeable that such purchaser will rely on valuation produced for mortgagee rather than obtaining own advice on capital and rental values — No liability for damages arising
***Scullion* v *Bank of Scotland plc (t/a Colleys) (CA),* [2011]** 3 **69**

Identification of property — Surveyor agreeing to value property — Surveyor misled by owner into valuing different property on nearby plot — Mortgage lender advancing loan in reliance upon valuation — Whether appellant liable to lender for breach of duty — No allegation of failure to exercise reasonable care and skill — Whether unqualified obligation on surveyor to value property to which instructions relate — Held, normal retainer of surveyor to inspect and value property containing absolute obligation to inspect and value correct property — Valuation of different property amounting to breach of duty notwithstanding exercise of reasonable care and skill
***Platform Funding Ltd* v *Bank of Scotland plc (CA),* [2008]** 3 **83**

NET INTERNAL AREA — See **Agreement for lease** (Specific performance)

NHBC AGGREEMENT – See **Negligence**

NOISE — See **Nuisance**

NON-COMPLETION CERTIFICATE — See **Construction** (Building contract)

NON-COMPLIANCE — See **Town and country planning**

NO-SCHEME WORLD — See **Compulsory purchase**

NOTICE TO COMPLETE — See **Sale of land**

NOTICE TO QUIT — See Agricultural holdings

NOTICE TO TREAT — See **Compulsory acquisition**; **Compulsory Purchase**

NOTICES — See also **Agricultural holdings** (Notice to quit) (Notice to remedy repairs — Service); **Compulsory acquisition**; **Landlord and tenant** (Assignment) (Landlord and Tenant Act 1987) (Landlord and Tenant (Covenants) Act 1995 — Lease; Yield up and reinstatement obligations) (Option to determine — Tenant's break option) (Rent review— Trigger notice) (Service charge); **Leasehold enfranchisement** (Leasehold Reform, Housing and Urban Development Act 1993 — Claim for new lease; Collective enfranchisement; Notice); **Licence**; **Option agreement**
Option
Validity

Option in lease to acquire freehold — Exercise of option requiring at least three months' notice expiring within specified period — Date of expiry of notice relevant to valuation and completion — Notice expressed to expire before three months — *Mannai Investment Co Ltd* v *Eagle Star Life Assurance Co Ltd* [1997] 1 EGLR 57 —Whether notice valid — Held, notice invalid
***Peaceform Ltd* v *Cussens (Ch),* [2006]** 3 **67**

Service
Time limit

Share sale agreement specifying that advance written notice be given of any claim for breach of warranty with any proceedings to be issued and served

NOTICES — *continued*
Service — *continued*
Time limit — *continued*

within 12 months thereafter — Agreement specifying certain methods of service including personal delivery to recipient's address — Whether notice served at time when delivered to recipient's home or later on deemed receipt by recorded delivery — Whether subsequent claim served in time — Held, service by personal delivery effected only where documents handed over to recipient personally — No service by personal delivery where documents simply left at recipient's address — Methods of service specified in agreement permissive not exclusive — Service of notice therefore occurring when actually seen by recipient not later date of service by recorded delivery — Claim documents served outside time limit
***Ener-G Holdings plc* v *Hornell (CA),* [2012] 3 61**

NUISANCE AND ANNOYANCE — See also **Compensation; Compensation for injurious affection; Landlord and tenant; Restrictive covenants; Rights to light; Tree preservation orders**

Flooding
Liability of sewerage undertaker

Flooding from culvert causing damage to property — Whether sewerage undertaker having statutory duty to maintain — Culvert forming part of channel originally consisting of natural stream — Whether character of culvert altered to that of sewer — Relevance of changes to other parts of channel — Held, status of other parts of channel not dictating status of culvert — Increase in receipt of surface water not affecting status where cessation of natural ground-water flow not proved — Character of culvert not changed to that of sewer
***Raglan Housing Association Ltd* v *Southampton City Council (CA),* [2008] 1 155**

Surface water

Measured duty of care — Local authority selling lower part of school playing field for new housing development — Developer negligently blocking culvert so that water from field flooding gardens of pre-existing houses on adjoining road — Whether local authority owing measured duty of care to abate nuisance as owners of higher land from which water flowing — Extent of duty — Held, scope of measured duty of care depending on all circumstances — Duty confined to steps that reasonable to expect occupier of land on which hazard arising to take to abate nuisance — Duty of local authority extending to co-operation with solution involving remedial works on their retained land — Authority not obliged to meet entire cost of works in circumstances where problem caused by negligence of another party whose liability established
***Lambert* v *Barratt Homes Ltd (Manchester Division) (CA),* [2010] 2 59**

Human rights
Remedy

Just satisfaction — Human Rights Act 1998 — Occupants of properties near sewage treatment works claiming damages for nuisance and breach of human rights — Whether damages awarded to holder of proprietary interest in nuisance affording just satisfaction to other persons in same household without legal right to occupy — Whether possible to "top up" damages awarded to holder of proprietary interest under s8 of 1998 Act — Whether alternative remedies relevant — Damages for nuisance intending to compensate holder of proprietary interest only — Effect of nuisance upon other occupants in practice taken into account in valuing loss of amenity claim — That factor relevant in deciding whether damages under s8 of 1998 Act necessary to afford just satisfaction to family members without proprietary interest — Damages under s8 unlikely to be awarded to holder of proprietary interest to top up damages in nuisance
***Dobson* v *Thames Water Utilities Ltd (CA),* [2009] 1 167**

Noise
Injunction

Motor racing circuit — Local residents claiming noise from circuit excessive and seeking to restrict number of race days — Whether planning permission for circuit effecting change in character of neighbourhood — Whether residents acquiescing in infringement of rights — Whether damages to be awarded in lieu of injunction — Held, planning authority unable to authorise nuisance — Permission not changing character of neighbourhood where no strategic decision affected by considerations of public interest — Planning agreement not more appropriate form of control — No acquiescence in infringement of rights — Damages to be awarded in lieu of injunction
***Watson* v *Croft Promo-Sport Ltd (QB),* [2008] 2 149**

Odour
Landfill

Householders seeking damages in nuisance for odour from waste disposal site —Site operator having permit for activites on sit — Whether permitted activities capable of giving rise to claim in nuisance in absence of brach of permit or negligence — Held, claims governed by conventional, well-settled principles of law of nuisance — Use not automatically reasonale by reason of compliance with terms of specific waste permit for the site — No requirement to adapt common law to "march in step with" system of regulatory controls under EU and domestic environmental legislation — No need to prove breach of permit or negligence in order to establish claim
***Barr* v *Biffa Waste Services Ltd (CA),* [2012] 2 157**

Private nuisance
Planning permission

Claim in private nuisance for noise emanating from motor sport stadium and track — Planning permission existing for motor sports uses — Whether planning permission relevant to issue of whether noise continuing nuisance — Whether such permissions changing character of locality — Held, long-established noisy activities with benefit of planning permission capable of changing nature of locality so that noise not constituting nuisance — Whether planning permission resulting in change in nature of locality a question of fact in each case — Permissions for stadium and track changing character of locality so that noise part of that character and not actionable nuisance
***Lawrence* v *Coventry (t/a RDC Promotions)(CA),* [2012] 1 165**

Rule in *Rylands* v *Fletcher*
Fire

Fire starting on business premises fuelled by tyres stored there for purposes of business of motor vehicle tyre supplier and fitter — Fire destroying adjoining property — Whether strict liability arising under *Rylands* v *Fletcher* for damage to adjoining property — Held, liability under *Rylands* v *Fletcher* arising only where dangerous thing brought onto defendant's property and subsequently escaping — No special rule for cases involving damage caused by spread of fire — No liability where escape of fire started or increased by dangerous thing rather than of thing itself — Escape of fire fuelled by tyres not giving rise to strict liability since tyres themselves not dangerous and not escaping
***Stannard (t/a Wyvern Tyres)* v *Gore (CA),* [2012] 3 129**

OBLIGATION TO REPAIR — See **Landlord and tenant** (Nuisance)

OCCUPIERS' LIABILITY — See **Defective premises** (Duty of landlord)

OPTION AGREEMENT

Assignment of lease
Put option to take assignment

Put option held by three parties — Lease registered in names of two parties — Whether option validly exercised by two of three parties — Lease re-registered in names of three parties — Whether option validly exercised by three parties where lease not vested in such parties throughout — Held, two of three parties unable to exercise put option — Lease not required to be held by three parties throughout put option period
***BP Oil UK Refining Ltd* v *Lloyds TSB Bank plc (CA),* [2005] 1 61**

Notice
Validity

Provision for extension of option period on service of notice in writing and making of further payment — Whether notice served in time — Whether letter amounting to valid notice — Whether requirement to make payment within original option period — Held, notice served in time — Letter clearly requiring option to be extended — Payment within reasonable time sufficient
***Rennie* v *Westbury Homes (Holdings) Ltd (Ch),* [2007] 2 95**

OPTION TO PURCHASE

Valuation
Construction of clause

Lease of land for development of golf course — Lease containing option for lessee to purchase land — Lessee seeking to exercise option — Contradictory valuation provisions in lease – Whether land to be valued for existing golf course use or at full market value including potential for residential development — Held, clause containing contradictory provisions to be given sensible meaning as a whole taking into account factual background known to parties at time lease entered into — Lease to be construed as commercial agreement — Commercial parties unlikely to have intended lessee to obtain windfall by purchasing land for sum taking no account of substantial hope value — Such value to be taken into account as part of "full market value"
***Multi-Link Leisure Developments Ltd* v *North Lanarkshire Council (SC),* [2011] 1 67**

OPTIONS — See also **Agreement for lease**; **Landlord and tenant** (Option to determine); **Notices**; **Perpetuities**

Specific performance
Freehold property

Tenancy of property — Tenants holding option to purchase freehold — Option unregistered — Purported exercise of option — Landlord selling to wholly-owned subsidiary for sole purpose of avoiding obligation to sell to tenants — Sale at undervalue — Whether transaction a sham — Whether

OPTIONS — *continued*
Specific performance — *continued*
Freehold property — *continued*

remedy of specific performance available — Held, transaction genuine — Specific performance available against landlord to direct sale by subsidiary
***Coles (Trustees of the Ward Green Working Men's Club)* v *Samuel Smith Old Brewery (Tadcaster) (an unlimited company) (CA),* [2008]** 2 **159**

OVERRIDING INTEREST — See Mortgage; Title to land

OVERRIDING RIGHTS
Easement
Town and Country Planning Act 1990, s237

Development site — Whether successors in title to local authority entitled to rely upon s237 of 1990 Act where development differing from that proposed at time of local authority acquisition or appropriation — Held, s237 having no application where development not for purpose underlying acquisition or appropriation
***Midtown Ltd* v *City of London Real Property Co Ltd; Joseph* v *City of London Real Property Co Ltd (Ch),* [2005]** 1 **65**

OVERRIDING RIGHTS OF WAY
Statutory powers to override rights
Acquisition of land by statutory authority

Sched 20 to Local Government, Planning and Land Act 1980 — Provisions similar to s237 of Town and Country Planning Act 1990 — Transfer to statutory authority and retransfer — Application to amend pleadings— Whether amendments having chance of success — Whether "acquisition" of land — Whether overriding provisions effective — Whether statutory authority authorised — Whether subsequent owner entitled to rely upon overriding provisions — Whether amendments to pleadings permitted — Held, permission to amend pleadings refused — Amendments having no chance of success — Effective transfer of beneficial ownership and entitlement to register title — Overriding provisions effective — Valid and material resolution — Works by subsequent owner effective to override rights of way
***Ford Camber Ltd* v *Deanminster Ltd (Ch),* [2006]** 3 **81**

PARKING RIGHTS — See Landlord and tenant (Landlord and Tenant Act 1954, Part II)

PARTY WALL
Party Wall etc Act 1996
Award

Appeal procedure — Application under s10(17) to county court challenging award — Whether statutory appeal under CPR 52 — Whether claim under CPR 8 — Whether CPR 8 claim can be ordered under CPR 3.10 to proceed under CPR 52 — Whether indemnity costs — Whether award under 1996 Act enforceable under CPR 70.5 — Held, appeal under s10(17) a statutory appeal subject to CPR 52 — Court should allow amendment to CPR 8 claim or allow claim to proceed as appeal under CPR 52 — Indemnity costs disallowed — Award under 1996 Act not enforceable under CPR 70.5
***Zissis* v *Lukomski (CA),* [2006]** 2 **61**

Legal costs — Commencement of works not authorised by surveyor's award — Adjoining owner preparing legal proceedings for injunction — Proceedings not instituted after undertakings given — Subsequent surveyor's award directing payment of adjoining owner's legal costs of contemplated proceedings — S10(12) and (13) of 1996 Act — Whether surveyors having power to award legal costs — Whether costs arising out of or incidental to party wall dispute — Held, power to award legal costs confined to costs connected with statutory dispute resolution mechanism in dispute arising under 1996 Act — No power to award costs incurred in preparation or prosecution of court proceedings to enforce common law or equitable remedies
***Blake* v *Reeves (CA),* [2009]** 3 **79**

Service — Surveyor's award in party wall dispute sent to parties by post pursuant to s15 of 1996 Act — Appeal against award — Whether 14-day period for actually appealing running from date of posting of award or date or receipt — Whether appeal out of time — Held, service of award by post effected on date when document consigned to post — Time for appeal running from that date — Appeal lodged out of time
***Freetown Ltd* v *Assethold Ltd (QB),* [2012]** 2 **89**

Underpinning works — Validity of appointment of surveyor — Validity of party wall notice — Service under s6 of 1996 Act — Whether defects in notice waived by receiving party — Whether refusal to act effectively by surveyor of one party so as to justify *ex parte* award — Appropriate apportionment of cost of works — Held, appointment of surveyor prior to service of party wall notice valid for purposes of 1996 Act — Notice invalid under s6 where failure to specify site and depth of excavations — Defect waived by failure of other party's surveyor to raise it at earliest opportunity — *Ex parte* award valid — Equal apportionment of cost of works to party wall appropriate where neither party responsible for defect necessitating those works
***Manu* v *Euroview Estates Ltd (CC),* [2008]** 1 **165**

Security

Building owner issuing notice under s6(1) and (2) of 1996 Act of intention to carry out works on own land within three to six metres of adjoining owner's building — Request for security by adjoining owner under s12(1) — Whether security available where no works proposed to adjoining owner's land – Whether s6 works carried out in exercise of rights conferred by Act — Held, security not confined to cases where works carried out to adjoining owner's land — Works on building owner's land under s6(1) and (2) carried out in exercise of rights conferred by Act so as to attract right to request security under s12(1) — Third surveyor's award under s10(12)(c) able to deal with matter of security
***Kaye* v *Lawrence (TCC),* [2011]** 1 **47**

PARTY WALL ETC ACT 1996 — See Party wall

"PAY WHEN PAID" CLAUSE — See Construction (Subcontract)

PERMISSION — See Adverse possession

PERPETUITIES
Option to extend lease
Validity

Perpetuities and Accumulations Act 1964 — Option to extend underlease not becoming exercisable until future date falling outside 21-year perpetuity period — Option assigned to chargee of underlease — Mortgage taking possession and selling underlease — Whether option void as infringing rule against perpetuities — Whether falling within exemption in s9(1) of 1864 Act — Whether option validly assigned to chargee or subsequent purchaser of underlease — Held, lease extension carved out of superior leasehold or freehold interest capable of instituting reversionary interest for purposes of s9(1) — Option to extend underlease falling within s9(1) exemption — Charge of underlease a successor in title of underlessee within s9(1) — Underlease validly assigned to chargee and subsequent purchaser from it — Registration of new title consequent on deed of variation of underlease not affecting that position
***Souglides* v *Tweedie and another (Ch),* [2012]** 2 **95**

PETROLEUM (PRODUCTION) ACT 1934 — See Trespass

PLANNING APPEAL — See Town and country planning (Human rights)

PLANNING AGREEMENT — See also Sale of land
S106 agreement
Implied rights of access

Agreement by estate owners A and B to construct bypass — B holding land required for bypass — Whether normal canons of construction of contracts applying to s106 agreements — Whether rights of access to construct bypass in favour of A necessarily implicit or granted by implied term — Whether proprietary rights — Whether rights capable of protection by notice on registers — Held, normal canons of construction applying to s106 agreements — A having rights of access under s106 agreement as necessarily implicit or by implied term — Rights proprietary and capable of protection under Land Registration Acts
***Stroude* v *Beazer Homes Ltd (Ch),* [2006]** 2 **115**

PLANNING APPLICATION
Decision-making process
Procedural irregularity

Councillors on planning committee attending meeting of objectors although not commenting on application — Two councillors not attending site visit— Councillors abstaining from vote on those grounds — Consents granted — Whether abstention due to erroneous advice from monitoring officer — Whether serious irregularity resulting in unlawfulness of decision reached on vote — Held, evidence disclosing no error in advice — Councillors exercising own judgment in light of advice — No unlawfulness in decision-making process
***R (on the application of Ware)* v *Neath Port Talbot County Borough Council (CA),* [2008]** 1 **55**

PLANNING OBLIGATION – See also Town and country planning
Section 106 agreement
Financial contribution

Money paid to local authority for related highway works — Delay in completion of works — Whether delay can be repudiatory breach — Whether planning obligation extending beyond obligation to pay money

RIGHT TO MANAGE — *continued*
Commonhold and Leasehold Reform Act 2002 — *continued*
Self-contained buildings — continued

RTM company seeking right to manage two blocks of flats within larger development — Whether each block a "self-contained building" — Ss72(1)(a) and 72(2) of 2002 Act — Whether common parts of development compromising "appurtenant property" to which right to manage extending — Held, each block a self-contained building to which right to manage applying since structurally detached — Right to manage extending to all property appurtenant to relevant self-contained building — Such property including parking spaces attaching to each flat in the blocks plus areas over which tenants having incorporeal rights of way and other rights granted by leases — RTM company succeeding to management responsibilities of landlord in respect of such areas
***Gala Unity Ltd* v *Ariadne Road RTM Co Ltd (UT:LC),* [2012] 1 99**

ROOF SPACE — See **Landlord and tenant** (Extent of demise)

SALE OF LAND

Auction
Seller as trustee
Title containing obligation to impose covenants against competition in sales of other properties — Seller not imposing such covenants — Whether breach of seller's duty as trustee for the buyer — Contract — Completion —Whether specific performance — Agent — Whether bidder acting as agent personally liable — Held, seller's duty as trustee not requiring imposition of anti-competition covenants in other sales — Agent personally liable — Specific performance granted on summary judgment
***Englewood Properties Ltd* v *Patel (Ch),* [2005] 2 77**

Charging order
Trustee
Contract for sale and purchase of land — Charging Orders Act 1979 — Contracts exchanged — Purchaser acting as nominee for another — Charging order on property in respect of costs orders made against purchaser — Effect of trust and sub-trust on interest of intermediate trustee — Whether purchaser having interest as trustee upon which order under s2(1)(b)(i) of 1979 Act able to bite — Whether non-party costs order appropriate against purchaser's principal — Held, interest of intermediate trustee not extinguished where trust and sub-trust existing — Vendor holding property on bare trust for purchaser while purchaser holding on bare trust for principal — Purchaser retaining interest as trustee for purposes of charging order — Non-party costs order appropriate where principal supporting litigation
***Nelson* v *Greening & Sykes (Builders) Ltd (CA),* [2008] 1 59**

Completion
Deed or escrow
Legal charge — Mortgage lender advancing funds — Funds remitted to vendors' solicitor and released to vendors — Contracts exchanged — Transfer executed and delivered by vendor — Purchaser granting legal charge in favour of mortgagee — Purchaser failing to pay balance of purchase price — Whether execution and delivery of transfer amounting to effective transfer or transfer made in escrow pending payment of balance — Whether mortgagee having valid charge taking priority over unpaid vendors' lien — Held, where part of purchase price unpaid inference arising that transfer made in escrow pending payment of balance — Inference rebuttable by evidence to contrary — Transfer unconditional where funds released to and used by vendors — Mortgagee's charge valid
***Bank of Scotland plc* v *King (Ch),* [2008] 1 65**

Notice to complete
Compliance — Contract providing for completion by 1pm on completion date failing which purchaser to be treated as having completed on next working day — Whether failure of purchaser to pay purchase moneys by 1pm on last day permissible under notice to complete entitling vendors to rescind — Whether failure to include vendors' legal costs in sum paid having that effect — Held, 1pm cut-off point relevant only to contractual completion date — Breach placing appellant in breach of contract but not giving right to rescind — Payment after 1pm on last day of notice to complete sufficient to comply with notice — Vendor's legal costs not payable as condition of completion such that failure to include them in purchase moneys not giving right to rescind — Specific performance ordered
***Chinnock* v *Hocaoglu (CA),* [2009] 1 61**

Compliance — Contract providing for completion by 1pm on completion date failing which purchaser to be treated as having completed on next working day — Whether completion date in contract to be rectified on ground of common mistake — Whether failure of purchaser to pay purchase moneys by 1pm on last day permissible under completion notices entitling vendors to rescind — Held, purchaser entitled to rectification to amend completion date — Vendor serving valid notice to complete — Tendering of purchase moneys after 1pm on final day sufficient to comply with notice — Absence of vendor's legal costs from moneys tendered amounting to failure to comply with notice such that vendor entitled to rescind
***Chinnock* v *Hocaoglu (Ch),* [2008] 2 77**

Law of Property Act 1925 — Agreement for sale of shares in company owning commercial property — Contractual term that purchaser not obliged to complete in absence of specified documentation – Notice to complete served by vendor — Vendor rescinding contract and forfeiting deposit — Whether vendor ready, willing and able to complete – Whether appropriate to order return of deposit under s49(2) of 1925 Act — Held, vendor obliged to call evidence to support assertion that ready, willing and able to complete where this disputed — Vendor ready, willing and able to complete where able to set up necessary administrative arrangements to enable completion to take place within reasonable time — No return of deposit in case of default by purchaser save in exceptional circumstances — Later sale by vendor to third party at profit not sufficient reason to order return of deposit
***Midill (97PL) Ltd* v *Park Lane Estates Ltd and another (CA),* [2009] 1 65**

Contract
Benefit and burden
Land in ownership of third parties — Vendor contracting to buy from third parties and assigning rights under contracts to purchaser — Purchaser retaining part of sale price pending completion of works under terms of assigned contract — Purchaser subsequently failing to pay balance — Whether third party entitled to recover retention moneys — Whether purchaser bound to observe vendor's obligations under assigned contract as burden appurtenant to benefit received — Held, burden imposed with benefit only where both conferred by same legal transaction and benefit conditional on or reciprocal to burden — Assignment of benefit of contract not made conditional on purchaser undertaking burden — Completion of sale not so conditional — Burden of assigned contract remaining obligation of vendor alone
***Davies* v *Jones (CA),* [2010] 1 82**

Frustration
Sale of block of flats — Purchaser agreeing to sell long lease of porter's flat if no longer needed as such and to pay overage to vendor — Sale of porter's flat prevented by enfranchisement of block under Leasehold Reform, Housing and Urban Development Act 1993 — Whether agreement to be construed so as to require overage payment — Whether agreement frustrated — Held, sale of freehold in porter's flat as part of statutory enfranchisement of entire block falling outside contemplation of parties at date of agreement — Differences between agreed steps under agreement and events leading to enfranchisement so numerous and significant that agreement incapable of applying to such events — Unreasonable to impose absolute liability even where performance of agreement impossible — Agreement discharged by frustration
***Hildron Finance Ltd* v *Sunley Holdings Ltd (Ch),* [2010] 3 1**

Purchase price
Sale of property for development — Price to be fixed sum less costs of obtaining release or variation of restrictive covenants — No planning permission or agreement with covenantees at completion date — Whether deduction to be made from purchase price — Whether implied term that cost of obtaining release to be estimated subject to later correction once agreement reached — Whether court able to reach interpretation different from those advanced by parties — Vendor's proposed term not making commercial sense — No grounds for implying term proposed by purchaser— Declaration in contractual dispute to be made only on basis of submission advanced by a party and tested by adversarial argument
***Anglo-Continental Educational Group (GB) Ltd* v *Capital Homes (Southern) Ltd (Ch),* [2008] 3 96**

Rescission
Land Registration Act 2002 — Contract for sale of registered freehold land on full title basis under general condition 4.3(a) — Special condition requiring purchaser to accept leasehold interest shown in schedule to freehold title previously determined by surrender — Notice and restriction of bankruptcy of lessee shown against leasehold title in land register and bankruptcy petition and order registered in land charges register — S86 of 2002 Act — Whether surrender void by reason of bankruptcy — Whether vendor having notice of bankruptcy — Whether purchaser entitled to rescind on ground that freehold title encumbered by lease — Held, surrender void by reason of bankruptcy proceedings of which landlord having notice by reason of s86 notice on land register — Notice not depending on search of land charges register that landlord under no obligation to make — Purchaser entitled to rescind for failure of vendor to comply with full title guarantee
***Weir* v *Area Estates Ltd (Ch),* [2010] 1 77**

Land Registration Act 2002 — Contract for sale of registered freehold land on full title basis under general condition 4.3(a) — Special condition requiring purchaser to accept leasehold interest shown in schedule to freehold title previously determined by surrender — Surrender in fact ineffective due to pending bankruptcy petition against lessee at relevant time — Whether special condition precluding reliance on continued existence of

TOLERATED TRESPASSER — See **Landlord and tenant** (Housing Act 1980)

TOMLIN ORDER — See **Landlord and tenant (Rent Act 1977)**

TORT

Abuse of process

Judicial review Alleged collateral or improper purpose in seeking judicial review of planning permission for development — Developer bringing claim for damages — Whether tort of abuse of process existing — Whether collateral or improper purpose to bring judicial review claim with aim of obtaining financial advantage compensating for adverse effect of development on neighbouring property — Held, tort of abuse of process existing — Tort established only where predominant purpose in bringing proceedings not to obtain the remedy offered by law but to achieve other object lying outside range of remedies granted by law — Heads of recovery confined to those for malicious prosecution and not extending to general economic loss — Tort of abuse of process unlikely to be established where judicial review proceedings brought with unconditional permission of court

Land Securities plc v ***Fladgate Fielder (a firm) (CA),*** 111

TORTIOUS LIABILITY — See **Negligence**

TOWN AND COUNTRY PLANNING -- See also **Civil practice and procedure** (Abuse of process)

Change of use

Use as dwelling

Enforcement notice alleging change of use of property from dwelling to commercial leisure accommodation — Notice upheld by planning inspector — Whether permitted use as dwelling including use as holiday accommodation — Whether notice expressed too widely — Held, dwelling capable of lawful use for some degree of holiday letting without material change of use occurring — Whether change of use occurring a question of fact and degree depending on particular characteristics of use in question — Change of use occurring where particular use of property for holiday lettings far removed from permitted use as swelling — Terms of enforcement notice upheld where no obvious alternative formulation on material before inspector

Moore v ***Secretary of State for Communities and Local Government (CA),*** **[2012]** 3 **91**

Compensation

Interference with easement

Town and Country Planning Act 1990 — Right of way blocked by development carried out on land appropriated for planning purposes under s122 of 1990 Act — Owner of right of way claiming compensation under s237 — Whether entitled to compensation for "taking" of easement to be assessed by reference to ransom value — Whether compensation confined to diminution in value of dominant tenement by reason of loss of right of way — Held, compensation under s237 to be assessed as it would be under s63 or 68 of Lands Clauses Consolidation Act 1845 or s7 or 10 of Compulsory Purchase Act 1965 in respect of injurious affection — Measure of compensation therefore calculated by reference to diminution in value in retained land — Ransom value not payable

Holliday v ***Breckland District Council (UT:LC),*** **[2012]** 3 **95**

Compulsory purchase

Material considerations

Town and Country Planning Act 1990 — Respondents granting outline planning permission to two rival parties for mixed-use development on same site — Resolution to exercise s226(1)(a) compulsory purchase powers in favour of one — Whether entitled to take into account prospect that development of compulsory purchase order site would development on another land would be cross-subsidise development on another site — Whether cross-subsidy a material consideration under s226(1A) of 1990 Act — Held, s226(1A) requiring consideration of economical, social or environmental "well-being" benefits to wider area than CPO site — Facilitation of desirable development on another site one such benefit — Possibility of cross-subsidising another highly desirable scheme a material consideration

R (on the application of Sainsbury's Supermarkets Ltd) v ***Wolverhampton City Council (CA),*** **[2009]** 3 **94**

Control of advertisements

Deemed consent

Town and Country Planning (Control of Advertisements)(England) Regulations 2007 — London Local Authorities Act 1995 — Advertisements displayed on site for many years — Existing display replaced with digital display showing selection of advertisements for several seconds each — Local authority serving notice under s11 of 1995 Act requiring removal of display — Class 13 in Schedule 3 to 2007 Regulations — Whether display having deemed consent – Whether material alteration in manner of use of site — Whether advertisement comprising sequential display — Whether right to revert to previous permitted display — Held, material alteration in manner of use of site occurring — Sequential display where animated display or structure displaying sequential images — Display "sequential" where one image following another notwithstanding absence of any connection between images — No right to revert to previously permitted display under 2007 Regulations

R (on the application of Clear Channel UK Ltd) v ***Hammersmith and Fulham London Borough Council (CA),*** **[2010]** 1 **89**

Development plan document

Time limit for challenge

Planning and Compulsory Purchase Act 2004 — Application to quash part of development plan document relating to allocation of land — Whether application made within six-week time limit under s113 of 2004 Act — Whether six-week period commencing on date of adoption of plan or following day — Whether local planning authority extending time by terms of adoption statement and notice — Held, six-week period expressly "starting with" relevant date namely date of adoption — Six weeks therefore beginning on day of adoption, not following day — Local planning authority having no power to extend time

Barker v ***Hambleton District Council (CA),*** **[2012]** 2 **101**

Enforcement notice

Building operations

Limitation — Town and Country Planning Act 1990 — Erecting new dwelling without planning permission — Works concealed under straw bales — Bales removed after four years — Enforcement notices — S171Bof 1990 Act — Whether limitation period for enforcement action expiring — Whether erection and removal of straw bales part of relevant building operations — Held, activities not themselves amounting to building operations capable of forming part of intended building operations — Erection and removal of bales concealing building of dwelling in breach of planning control forming part of intended and completed building operations — Operations not substantially completed until bales removed — Enforcement notices issued within relevant four-year time limit

R (on the application of Fidler) v ***Secretary of State for Communities and Local Government (Admin),*** **[2010]** 1 **94**

Non-compliance

Statutory defence — S179 of Town and Country Planning Act 1990 — Notice requiring removal of caravan from land owned by gypsy — Local planning authority laying information alleging non-compliance — Defence of having done all that could be expected to secure compliance with notice — Inability to remove caravan without dismantling it — Lack of available alternative sites — Whether issues of hardship and reasonableness relevant to application of defence — Held, s179(3) concerned only with whether owner of the land capable of complying with notice without assistance of others — Issues of hardship or reasonableness of compliance irrelevant at stage of prosecution for breach of valid enforcement notice

Sevenoaks District Council v ***Harber (DC),*** **[2008]** 2 **91**

Environment

Habitats

Council Directive 92/43/EEC — Conservation (Natural Habitats etc) Regulations 1994 — Application for planning permission to demolish building containing bat roost — Demolition also requiring licence from Natural England — Extent of duty imposed on local planning authority under reg 3(4) of 1994 Regulations — Whether duty to consider at planning permission stage whether requirements of directive met — Whether sufficient to impose condition requiring application for licence — Held, local planning obliged to engage with provisions of directive and consider whether likely to be met — Duty not discharged by making obtaining of licence a condition of planning permission — Insufficient merely to refer in permission to 1994 Regulations and requirement for licence

R (on the application of Woolley) v ***Cheshire East Borough Council (Admin),*** **[2009]** 3 **89**

Human rights

Planning appeal

Art 8 of European Convention on Human Rights — Redevelopment of estate providing accommodation to the elderly — Development requiring residents to leave their homes — Ground 9 in section 8 of Housing Act 1988 — Whether proposed development breaching residents' Art 8 rights — Whether inspector failing to take account of protection available to residents through court eviction process — Held, inspector erring in approach to proportionality — Aim of proposed development proportionate — Means proportionate by reason of consideration of Art 8 rights by courts under ground 9

Atlantic Housing Ltd v ***Secretary of State for Communities and Local Government (Admin),*** **[2008]** 2 **53**

Lawful use certificate

Residential use

Enforcement action — Limitation periods — 1990 Act — Planning permission granted for erection of hay barn for agricultural storage — Building fitted out and used as dwelling — S171B(1) and (2) of 1990 Act — Whether construction of building and its residential use immune to

TOWN AND COUNTRY PLANNING — *continued*
Lawful use certificate — *continued*
Residential use — *continued*

enforcement action after four years — Whether owner entitled to lawful use certificate — Held, residential use of building amounting to change of use from permitted use — No necessity to show actual previous agricultural storage use — Four-year time limit for enforcement applying — Entitlement to lawful use certificate arising once time limit expiring — Erection of building in breach of planning control also immune from enforcement after four years

Welwyn Hatfield Council **v** ***Secretary of State for Communities and Local Government (CA),*** **[2010]** 1 **98**

Planning obligation
Unilateral undertaking

Town and Country Planning Act 1990 — Local planning authority refusing planning permission for residential development but indicating that unilateral undertaking to make financial contributions to infrastructure would overcome objections — Developer providing such undertaking under s106(1) of 1990 Act — Planning permission granted on appeal — Planning inspector expressing view that undertaking unnecessary — Whether undertaking enforceable — Whether unreasonable to enforce— Held, unilateral undertaking enforceable as contractual obligation — Enforceability no longer subject to challenge once undertaking becoming binding according to its own terms — Willingness of inspector to grant permission without undertaking not affecting enforeability of undertaking or depriving it of legitimate planning purpose — Enforcement of undertaking lawful and reasonable

R (on the application of Millgate Developments Ltd) **v** ***Wokingham Borough Council (CA),*** **[2012]** 3 **87**

Town and Country Planning Act 1990 — Local planning authority refusing planning permission for residential development but indicating that unilateral undertaking to make financial contributions to infrastructure would overcome objections — Developer providing undertaking under s106(1) of 1990 Act — Planning inspector granting planning permission on appeal and expressing view that undertaking unnecessary — Whether undertaking enforceable in light of inspector's decision - Whether unreasonable to enforce— Held, undertaking enforceable according to its terms — No condition stipulating that obligations in undertaking to take effect only if inspector holding them necessary — Enforcement not unreasonable where local planning authority validly concluding that obligations continuing to serve useful planning purpose even if not original purpose for which undertaking given

R (on the application of Millgate Developments Ltd) **v** ***Wokingham Borough Council (Admin),*** **[2011]** 1 **81**

Planning permission
Conditions

Condition precedent — Planning permission for erection of flats containing condition requiring commencement of development within five years — Further condition requiring approval of ground levels of flats before works commencing — Preparatory works carried out without obtaining approval of ground levels— Subsequent application for lawful use certificate in respect of development — Whether permission lawfully implemented within five-year period — Whether non-compliance with condition regarding approval of ground levels rendering works unlawful— Held, operations commenced in contravention of condition rendering development unlawful if requirement in question a "condition precedent" going to heart of planning permission — Ground levels of flats a condition precedent by reason of importance and wording of relevant condition — Planning permission not lawfully implemented — No entitlement to lawful use certificate

Greyfort Properties Ltd **v** ***Secretary of State for Communities and Local Government (Admin),*** **[2011]** 1 **87**

Section 106 agreement
Indemnity clause

Developers agreeing to construct roads to form bypass — Developers agreeing to indemnify highway authority in respect of all claims — Whether indemnity clause covering claims for compensation under Part I of Land Compensation Act 1973 — Held, indemnity clause not covering claims for Part I compensation

Wiltshire County Council **v** ***Crest Estates Ltd (CA),*** **[2005]** 3 **19**

Validity

Planning permission granted for office development — Agreement under s106 of Town and Country Planning Act 1990 requiring grant of rights of way over development site to owners of adjoining properties — Adjoining owner seeking declaration as to obligation to grant right of way and proper construction of s106 agreement — Whether grant of declaratory relief appropriate where claimant not party to s106 agreement — Whether s106 agreement invalidated by s2 of Law of Property (Miscellaneous Provisions) Act 1989 — Held, s106 agreement enforceable only by local planning authority against party entering into agreement and those deriving title from it — Inappropriate to grant declaratory relief to third-party claimant in respect of s106 agreement — S2 of 1989 Act not invalidating agreement benefiting third party provided agreement signed by all parties

Milebush Properties Ltd **v** ***Tameside Metropolitan Borough Council (Ch),*** **[2010]** 2 **93**

TOWN AND COUNTRY PLANNING ACT 1990 — See **Overriding rights of way; Planning obligation; Rights to light** (Breach); **Town and country planning** ; **Tree preservation orders**

TOWN AND COUNTRY PLANNING (CONTROL OFADVERTISEMENTS) (ENGLAND) REGULATIONS 2007— See **Town and country planning**

TOWN AND VILLAGE GREENS – See also **Commons and greens**
Class (*c*) green
Registration

Commons Registration Act 1965 — Commons Registration (New Land) Regulations 1969 — Effect of registration of green — Whether relevant inhabitants may use land generally for sports and pastimes on registration — Whether 20-year qualifying period any period of 20 years — Whether any application after 30 January 2001 engaging only amended definition of class (*c*) green — Whether continuity of user in relation to class (*c*) green within amended definition required to date of application or of registration — Whether applications can be amended — Whether less land than applied for capable of registration — Effect of s12 of Inclosure Act 1857 and s29 Commons Act 1876 — Effect of inaccessible areas of land subject to application for registration — Effect of tracks on land capable of or being public rights of way — Held, registration conferring rights to indulge in sports and pastimes — Registration conclusive for purposes of s12 of 1857 Act and s29 of 1876 Act — Application after 30 January 2001 engaging only amended definition — Lawful sports and pastimes must continue to date of application — Applications amendable — Lesser area than that applied for registrable — Guidance on nature of user declined

Oxfordshire County Council **v** ***Oxford City Council (HL),*** **[2006]** 2 **95**

Commons Registration Act 1965 — Commons Registration (New Land) Regulations 1969 — Whether 20-year qualifying period any period of 20 years — Whether any application after 30 January 2001 engaging only amended definition of class (c) green — Whether continuity of user in relation to class (c) green within amended definition required to date of application or registration — Whether applications can be amended — Whether less land than applied for capable of registration — Effect of s12 of Inclosure Act 1857 and s29 of Commons Act 1876 — Effect of inaccessible areas upon land subject to application for registration — Effect of tracks upon land capable of or being public rights of way — Held, registration not conferring status or implying any rights on part of inhabitants — Registration conclusive for purposes of s12 of 1857 Act and s29 of 1876 Act — Application on or after 30 January 2001 engaging only amended definition — Lawful sports and pastimes must continue to date of registration — Applications amendable — Lesser area than that applied for can be registered — Guidance given on many issues concerning registration

Oxfordshire County Council **v** ***Oxford City Council (CA),*** **[2005]** 2 **91**

Registration
Commons Act 2006

User as of right — Application to register part of golf course as green under s15 of 2006 Act — Inspector finding user by local inhabitants not as of right since inhabitants deferring to golf course use — Test of use as of right — Whether deference precluding use being as of right — Held, use of land by local inhabitants for sports and pastimes "as of right" if land enjoyed openly in way a person rightfully entitled would have used it and not by force, stealth or licence of the owner — No additional requirement that it would have appeared to reasonable landowner that users were asserting right to use land — Deference to landowner's use not inconsistent with use as of right — Rights of both landowner and local inhabitants coexisting after registration in such cases — Use as of right shown where deference to golfers a matter of common sense and courtesy

R (on the application of Lewis) **v** ***Redcar and Cleveland Borough Council (SC),*** **[2010]** 1 **13**

User as of right — Application to register part of golf course as town or village green under section 15 of 2006 Act — Inspector finding user by local inhabitants not as of right since deferring to golf course use — Whether "deference" to landowner's use a relevant consideration — Held, user as of right must be such as would give outward appearance to reasonable landowner of being claimed as of right — Deference relevant where competing uses by landowner and local inhabitants — Deference may be such that no apparent assertion of right

R (on the application of Lewis) **v** ***Redcar and Cleveland Borough Council (CA),*** **[2009]** 1 **79**

Commons Registration Act 1965

User as of right — Local authority resolving to register meadow as

TABLE OF STATUTES

References are to year, number and volume,
eg **[2011]** 2 **127** refers to vol 2 of 2011 at p127 and following

TABLE OF STATUTORY INSTRUMENTS

References are to year, number and volume,
eg **[2011]** 2 **127** refers to vol 2 of 2011 at p127 and following

TABLE OF EUROPEAN LEGISLATION

References are to year, number and volume,
eg **[2011]** 2 **127** refers to vol 2 of 2011 at p127 and following